Bioethics, Public Health, and the Social Sciences
for the Medical Professions

Amy E. Caruso Brown
Travis R. Hobart
Cynthia B. Morrow
Editors

Bioethics, Public Health, and the Social Sciences for the Medical Professions

An Integrated, Case-Based Approach

Editors
Amy E. Caruso Brown
Center for Bioethics and Humanities
and Department of Pediatrics
Division of Pediatric Hematology/Oncology
SUNY Upstate Medical University
Syracuse, NY
USA

Travis R. Hobart
Departments of Pediatrics and
Public Health and Preventive Medicine
SUNY Upstate Medical University
Syracuse, NY
USA

Cynthia B. Morrow
Department of Interprofessionalism
Virginia Tech Carilion School of Medicine
Roanoke, VA
USA

ISBN 978-3-030-03543-3   ISBN 978-3-030-03544-0 (eBook)
https://doi.org/10.1007/978-3-030-03544-0

Library of Congress Control Number: 2018965742

© Springer Nature Switzerland AG 2019
This work is subject to copyright. All rights are reserved by the Publisher, whether the whole or part of the material is concerned, specifically the rights of translation, reprinting, reuse of illustrations, recitation, broadcasting, reproduction on microfilms or in any other physical way, and transmission or information storage and retrieval, electronic adaptation, computer software, or by similar or dissimilar methodology now known or hereafter developed.
The use of general descriptive names, registered names, trademarks, service marks, etc. in this publication does not imply, even in the absence of a specific statement, that such names are exempt from the relevant protective laws and regulations and therefore free for general use.
The publisher, the authors, and the editors are safe to assume that the advice and information in this book are believed to be true and accurate at the date of publication. Neither the publisher nor the authors or the editors give a warranty, express or implied, with respect to the material contained herein or for any errors or omissions that may have been made. The publisher remains neutral with regard to jurisdictional claims in published maps and institutional affiliations.

This Springer imprint is published by the registered company Springer Nature Switzerland AG
The registered company address is: Gewerbestrasse 11, 6330 Cham, Switzerland

Preface

The idea that excellence in healthcare requires moving beyond the treatment of disease and of the individual physical body is not a new one. Rudolf Virchow is oft (and variably) quoted for his nineteenth-century observation that medicine is a social science and politics just the practice of medicine on a large scale. What *is* new is the recognition that future healthcare practitioners should be intentionally taught that they have this obligation and systematically trained to develop knowledge, skills, and attitudes in bioethics, population health, and related social sciences. However, there is little consensus regarding the best approaches to teaching medical and health professions students to move beyond the biomedical sciences. Furthermore, standards for assessing quality and outcomes of such efforts have not yet been widely established. Because the emphasis on curricular content beyond biomedicine is a relatively recent development, many faculty may have little experience in their own training upon which to base curricular decisions.

With this textbook, we offer a solution to the problem. We present a case-based approach to understanding how bioethics, population health, and the social sciences impact both patients and populations and provide a structured framework suitable for educators, including course directors and those engaged in curricular design, to use in the classroom. This framework is also adaptable for use directly by students of medicine and the health professions, including those engaged in independent study. The 22 cases in the book address a wide spectrum of patient populations, social and clinical settings, and disease pathologies. They engage learners in rigorous clinical and ethical reasoning, prompting them to make choices based on available information and then providing additional information to challenge assumptions, an approach which simulates clinical decision-making. Each pair of cases shares a core concept in bioethics or public health. All cases integrate six key themes: (1) patient- and family-centered care, (2) cultural humility, (3) evidence-based practice, (4) biases in decision-making, (5) structural competency, and (6) social justice and advocacy.

The book is divided into three major parts. Part I (► Chapters 1 and 2) introduces the book and is intended to optimize its utility and flexibility for a variety of learners and educational settings; Parts II through XII (► Chapters 3, 4, 5, 6, 7, 8, 9, 10, 11, 12, 13, 14, 15, 16, 17, 18, 19, 20, 21, 22, 23, and 24) are the heart of the book, the cases; and Part XIII (► Chapters 25 and 26) addresses challenges and innovative approaches to assessing learning in bioethics, population health, and the social sciences. A glossary of terms broadly important to the entire textbook is included at the end. Each case is divided into 8 or 9 sections: a chapter-specific glossary, learning objectives, case background (in the form of vignette), background questions for self-directed learning, additional case information and questions for discussion, answers to background questions, responses to discussion questions, references, and suggested further reading. The cases were developed iteratively over the past 4 years and, as of publication, have been used by nearly 40 faculty and more than 500 students. The authors of each chapter were selected based upon their expertise in the relevant issues and their ability to make those issues understandable for those just entering the health professions.

It is our belief that learners at all levels will benefit from immersion in the complex ethical, legal, social, and policy issues explored in these cases and will graduate better prepared to provide outstanding care to all patients and communities in our changing world.

Amy E. Caruso Brown
Syracuse, NY, USA

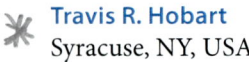
Travis R. Hobart
Syracuse, NY, USA

Cynthia B. Morrow
Roanoke, VA, USA

Contents

I Introduction

1. **Introduction** 3
   Amy E. Caruso Brown

2. **Approaches to Using This Book** 17
   Amy E. Caruso Brown

II Core Principles in Ethics and Public Health

3. **"How Many of These Surgeries Have You Done?"** 39
   Robert S. Olick

4. **"Can't Stop Coughing (But I Need to Get Back to the Shelter by 6)"** 61
   Cynthia B. Morrow

III Community Perspectives

5. **"I Think I'm in Labor"** 79
   Jordana L. Gilman and Sarah Cumbie Reckess

6. **"Why Does My Son Have Lead in His Blood?"** 99
   Travis R. Hobart

IV Healthcare Systems and Comparative Approaches

7. **"Our Baby Is Turning Blue"** 125
   Caitlin M. Nye

8. **"I Have a Touch of Sugar but I Can't Afford My Meds"** 145
   Martha A. Wojtowycz and Ahmed A. Malik

V Decision-Making for Children and Adolescents

9. **"I Don't Want My Child to Get Vaccines"** 165
   Manika Suryadevara and Joseph B. Domachowske

10. **"Our Son's Cancer Is Gone. Why Can't We Stop Treatment?"** 185
    Thomas R. Curran Jr.

VI Trauma, Violence, and Mental Health

11 "He Has a Gun and Wants to Kill Himself" 207
Christopher R. Botash

12 "Bleeding Too Much" (In the Words of a Refugee) 231
Andrea V. Shaw

VII Choices and Care at the End of Life

13 "My Father Would Not Want to Live Like This" 251
Sharon A. Brangman

14 "We're Not Ready to Give Up" 269
Anthony Michael Zepeda

VIII Quality of Life in the Context of Disability

15 "Please Look Beyond My Disability" 289
Jeremy French-Lawyer and Margaret A. Turk

16 "It Runs in the Family" 315
Robert Roger Lebel

IX Medical Mistakes and Patient Safety

17 "I Know Something Is Wrong" 335
Kendra Harris

18 "I'm in Pain!" 349
Theresa Baxter

X Conducting Research

19 "I Don't Want to Be a Guinea Pig" 369
Gregory L. Eastwood

20 "Wait, I'm a Research Subject?" 389
Gregory L. Eastwood

XI Stigma and Marginalization

21 "I Need Blockers So I Don't Turn Into a Girl" 409
Karen L. Teelin

22 "You Don't Understand: He Needs That Bottle" 429
Lauren Hall Mutrie and Janet H. Goode

XII Global Health

23 "They Say My Baby's Head Is Too Small" 453
Amy E. Caruso Brown and Cynthia B. Morrow

24 "I Just Want to Help People and See the World" 477
Andrea V. Shaw

XIII Evaluation and Assessment

25 Evaluating Cases in Context 495
Lauren J. Germain

26 A Practical Framework for Learner Assessment 501
Lauren J. Germain

Supplementary Information
Glossary 508
Index 517

Contributors

Theresa Baxter, MS, FNP-C
Department of Nursing/Anesthesia
SUNY Upstate Medical University
Syracuse, NY, USA
theresabaxter@rocketmail.com

Christopher R. Botash, MD
Department of Psychiatry
SUNY Upstate Medical University
Syracuse, NY, USA
botashc@upstate.edu

Sharon A. Brangman, MD, FACP, AGSF
Center of Excellence for Alzheimer's
Disease and Department of Geriatrics
SUNY Upstate Medical University
Syracuse, NY, USA
brangmas@upstate.edu

Amy E. Caruso Brown, MD, MSc, MSCS
Center for Bioethics and Humanities
and Department of Pediatrics,
Division of Pediatric Hematology/Oncology
SUNY Upstate Medical University
Syracuse, NY, USA
brownamy@upstate.edu

Thomas R. Curran Jr, MD
Center for Bioethics and Humanities
SUNY Upstate Medical University
Syracuse, NY, USA
currant@upstate.edu

Joseph B. Domachowske, MD, FAAP
Departments of Pediatrics, Microbiology and
Immunology, Global Maternal-Child and
Pediatric Health Program at SUNY Upstate
Medical University's Center for Global Health
and Translational Sciences
SUNY Upstate Medical University
Syracuse, NY, USA
domachoj@upstate.edu

Gregory L. Eastwood, MD
Center for Bioethics and Humanities
SUNY Upstate Medical University
Syracuse, NY, USA
eastwood@upstate.edu

Jeremy French-Lawyer, MPH, CHES
Department of Physical Medicine
and Rehabilitation
SUNY Upstate Medical University
Syracuse, NY, USA
frenchlj@upstate.ed

Lauren J. Germain, PhD, MEd
Office of Evaluation, Assessment
and Research and Department of
Public Health and Preventive Medicine
SUNY Upstate Medical University
Syracuse, NY, USA
germainl@upstate.edu

Jordana L. Gilman, MD, MPH
SUNY Upstate Medical University
Syracuse, NY, USA
gilmanj@upstate.edu

Janet H. Goode, JD
Pediatric Advocacy Clinic, The University
of Michigan Law School
Ann Arbor, MI, USA
jhgoode@gmail.com

Kendra Harris, MD, MSc
Department of Radiation Oncology
Tulane Cancer Center
New Orleans, LA, USA
kharris11@tulane.edu

Travis R. Hobart, MD, MPH
Departments of Pediatrics and
Public Health and Preventive Medicine
SUNY Upstate Medical University,
Syracuse, NY, USA
hobartt@upstate.edu

**Robert Roger Lebel,
MS, MA, MD, iv, STM, MD**
Departments of Pediatrics, Obstetrics/
Gynecology, Internal Medicine, and Pathology,
and Center for Bioethics and Humanities
SUNY Upstate Medical University
Syracuse, NY, USA
lebelr@upstate.edu

Contributors

Ahmed A. Malik, MD, MPH
Transitional Year Residency Program
UCF COM/HCA GME Consortium North Florida
Regional Medical Center
Gainesville, FL, USA
ahmedamalik.md@gmail.com

Cynthia B. Morrow, MD, MPH
Department of Interprofessionalism
Virginia Tech Carilion School of Medicine
Roanoke, VA, USA
cbmorrow@vt.edu

Lauren Hall Mutrie, MD, MSc
Department of Pediatrics
Pediatric Hospital Medicine, Northwest
Permanente, Doernbecher Children's Hospital,
and Oregon Health Sciences University
Portland, OR, USA
lehall4@gmail.com

Caitlin M. Nye, MSN, RN-BC
Department of Nursing Recruitment
SUNY Upstate Medical University
Syracuse, NY, USA
nyec@upstate.edu

Robert S. Olick, JD, PhD
Center for Bioethics and Humanities,
SUNY Upstate Medical University
Syracuse, NY, USA
olickr@upstate.edu

Sarah Cumbie Reckess, JD
Center for Court Innovation
Syracuse, NY, USA
sreckess@courtinnovation.org

Andrea V. Shaw, MD
Departments of Internal Medicine
and Pediatrics
SUNY Upstate Medical University
Syracuse, NY, USA
shawan@upstate.edu

Manika Suryadevara, MD
Department of Pediatrics
Division of Pediatric Infectious Diseases,
SUNY Upstate Medical University
Syracuse, NY, USA
suryadem@upstate.edu

Karen L. Teelin, MD, MSEd, FAAP
Department of Pediatrics
SUNY Upstate Medical University
Syracuse, NY, USA
TeelinK@upstate.edu

Margaret A. Turk, MD
Departments of Physical Medicine and
Rehabilitation, Pediatrics, and Public Health
and Preventive Medicine
SUNY Upstate Medical University
Syracuse, NY, USA
turkm@upstate.edu

Martha A. Wojtowycz, PhD
Department of Public Health
and Preventive Medicine
SUNY Upstate Medical University
Syracuse, NY, USA
wojtowym@upstate.edu

Anthony Michael Zepeda, MD, SFHM, FAAFP
Department of Palliative Care
and Family Medicine
University of North Texas Health Science Center,
John Peter Smith Hospital
Fort Worth, TX, USA
mzepeda@gmail.com

Introduction

Contents

Chapter 1   Introduction – 3

Chapter 2   Approaches to Using This Book – 17

Introduction

Amy E. Caruso Brown

1.1 Introduction – 4

1.2 **Medical Education "Beyond" Biomedical Science – 4**
1.2.1 Why Integrate Bioethics, Population Health, and the Social Sciences? – 5
1.2.2 Why Employ a Case-Based Learning Approach? – 5
1.2.3 Filling the Gap – 6

1.3 **The Six Key Themes – 6**
1.3.1 Patient- and Family-Centered Care – 6
1.3.2 Cultural Humility – 7
1.3.3 Evidence-Based Practice – 8
1.3.4 Cognitive Biases in Decision-Making – 9
1.3.5 Structural Competency – 10
1.3.6 Social Justice and Advocacy – 12

1.4 **Conclusion – 13**

References – 14

1.1 Introduction

» Medicine is not a science. It uses science in the service of disease prevention, health promotion, and healing of those with disease. Western medicine is strongly rooted in scientific inquiry, testing, and application, but excellence in the practice of medicine and other health professions requires diverse tools in addition to biomedical science, including knowledge and skills in ethics, law, economics, public policy, social medicine, and population health. It is fundamentally based on a commitment to a broad notion of health care which encompasses not only the treatment of disease but the prevention of illness and a commitment to care even when disease cannot be effectively treated.[1]

Despite growing recognition of the validity of these assertions, there is little consensus regarding the best approaches to teaching medical and health professions students to move beyond the biomedical sciences. Furthermore, standards for assessing quality and outcomes of such efforts have not yet been widely accepted [1–6]. Accreditation bodies, such as the Liaison Committee on Medical Education (LCME), and professional organizations, such as the Association of American Medical Colleges (AAMC), have traditionally provided only broad recommendations, leaving the details of implementation—content, objectives, and pedagogical approaches—to the discretion of individual institutions [7, 8]. Because the emphasis on curricular content beyond biomedicine is a relatively recent development, many faculty, including course directors and education administrators, may have little experience in their own training upon which to base curricular decisions.

This book offers one solution to this problem. It presents a case-based approach to understanding how bioethics, population health, and the social sciences impact both individual patients and populations and provide a structured framework suitable for educators, including course directors and those engaged in curricular design,

to use in the classroom. This framework is also designed for use directly by students of medicine and the health professions, including those engaged in independent study. The 22 cases in the book address a wide spectrum of patient populations, clinical settings, and disease pathologies. They engage learners in rigorous clinical and ethical reasoning, prompting learners to make choices based on available information and then providing additional information to challenge assumptions, simulating clinical decision-making. Each pair of cases shares a core concept in bioethics or public health. All cases integrate six key themes: (1) patient- and family-centered care, (2) cultural humility, (3) evidence-based practice, (4) biases in decision-making, (5) structural competency, and (6) social justice and advocacy.

1.2 Medical Education "Beyond" Biomedical Science

The idea that excellence in health care requires moving beyond the treatment of disease and of the individual physical body is not a new one. Rudolf Virchow is oft (and variably) quoted for his nineteenth-century observation that medicine is a social science and politics is akin to the practice of medicine on a large scale [9, 10]. What is new is the recognition that future healthcare practitioners should be intentionally taught that they have this obligation and systematically trained to develop knowledge, skills, and attitudes in bioethics, population health, and related social sciences [11].

In order to develop a systematic approach to training healthcare practitioners in these domains, we identified the six aforementioned themes as being foundational to excellence in the practice of medicine, and we ensured that each is incorporated into every case. While each theme arises from a distinct intellectual history and philosophy, they necessarily intersect. In isolation, they cannot be fully understood nor effectively applied in practice. Furthermore, the four core principles of bioethics—respect for autonomy, beneficence, non-maleficence, and justice—each have bearing on the themes, as do the socio-ecological model of health and other core principles in public health. ◘ Table 1.1 illustrates the parallels between the six themes, the "four principles" approach to biomedical ethics [12, 13], and the

---

1 This quote has appeared at the start of every course syllabus since 2015 and is attributed to Kathy Faber-Langendoen, Chair of the Center for Bioethics and Humanities at SUNY Upstate Medical University.

Table 1.1 Tying it all together: parallels between the Charter on Medical Professionalism [14], the "four principles" approach [12, 13], and the six themes in this book

Principles in the "Charter on Medical Professionalism"	"Four principles" approach to biomedical ethics	Six themes in this book
Principle of patient autonomy	Autonomy or respect for persons	Patient-centered care Cultural humility
Principle of primacy of patient welfare	Beneficence Non-maleficence	Evidence-based practice Cognitive biases
Principle of social justice	Justice	Structural competence Social justice and advocacy

three principles enshrined in the 2002 "Charter on Medical Professionalism," a document endorsed by more than 108 professional organizations [14].

other healthcare professionals face and aligning the objectives of the entire set of cases with institutional graduation competencies [15].

1.2.1 Why Integrate Bioethics, Population Health, and the Social Sciences?

Prior to 2015, Upstate Medical University had, for many years, required first year medical students to take three consecutive courses oriented toward the practice of medicine: one in clinical skills, one in ethics and law in medicine, and one in population health and preventive medicine. In 2015, the decision was made to integrate the ethics and population health courses, with separate courses dedicated to clinical skills and to clinical reasoning. There was a powerful rationale for uniting these disciplines in a single course. The overlap in content was evident—both courses, for example, included sessions on healthcare finance, cultural humility, and physician advocacy—but each discipline brought vital perspective to the issues discussed. When the material was presented only from a bioethical perspective, the focus of discussion tended to be on the individual and critical contextual information about the community and population was lost; when it was presented only from a public health perspective, ethical concerns were too easily overlooked. Questions of law, policy, and justice united the two. By integrating bioethics and population health, we were able to cover more materials and provide more opportunities for in-depth discussions while still grounding the content in the real issues that physicians and

1.2.2 Why Employ a Case-Based Learning Approach?

Case-based learning (CBL) and its predecessor problem-based learning (PBL) are student-centered pedagogic methods. McMaster University, where PBL was pioneered in the late 1960s, defines CBL as:

> ...a form of active learning that involves solving and examining real-world problems in small groups with guided instruction. The case or scenario can be acted out through role playing to deepen learning and gain additional understanding of the problem. CBL is a more structured approach to learning than problem based learning (PBL) in that it provides a specific real-world problem to be solved with guided inquiry, compared to PBL, which provides...a more general problem with more open inquiry [16].

CBL-based courses and curricula promote the development of skills essential for physicians and healthcare practitioners, including critical thinking, problem-solving, and communication in collaborative environments. In a recent narrative review, Schmidt, Rotgans, and Yew argued that "…there is considerable support for the idea that PBL and CBL works because it encourages the activation of prior knowledge in the small-group setting and provides opportunities for elaboration on that knowledge" [17]. Well-designed cases allow learners to experience intentional complexity, better preparing them for encounters and challenges that they will face in their careers. Cases can also simulate the decision-making process such that learners have a better understanding of how information is collected and synthesized to support a decision, as well as the omnipresence of uncertainty and the potential consequences of various options.

Although existing studies have not clearly demonstrated superiority of PBL and CBL over traditional methods of instruction in the biomedical sciences, it is notable that, by the most ubiquitous outcome measure (standardized test performance), PBL and CBL are at least as effective as other approaches [18, 19]. While there are significantly fewer studies focused on ethics or social determinants of health, this finding does seem to be applicable beyond the biomedical sciences [20]. A meta-analysis of 13 studies compared the physician core competencies achieved by traditional versus case-based learning curricula and found that students educated via CBL models were stronger in four areas: teamwork skills, appreciation of social and emotional aspects of health care, appreciation of legal and ethical aspects of health care, and appropriate attitudes toward personal health and well-being [21].

1.2.3 Filling the Gap

Despite the increasing adoption of problem- and case-based learning in medical and health professions education, case materials are still largely developed, implemented, and assessed within single institutions, if they are assessed at all. While biomedical cases are more readily available in other textbooks and through online clearinghouses, such as MedEdPortal, cases focused on bioethics, population health, and social sciences are rare. In addition, the few case studies that have been published are not suitable for novice learners. With these 22 case studies, and accompanying guidance on how to implement and evaluate them in a medical or health professions curriculum, we offer readers new tools to achieve excellence in teaching the practice of medicine.

1.3 The Six Key Themes

Each theme is present in every case, sometimes explicitly and sometimes implicitly.

1.3.1 Patient- and Family-Centered Care

Patient- and family-centered care (and the related term person-centered care) is the predominant model of contemporary health care, though the quality of its implementation may vary from institution to institution. It is defined by the Institute of Medicine as "care that is respectful of, and responsive to, individual patient preferences, needs and values, and ensuring that patient values guide all clinical decisions" [22]. It is typically contrasted with older models that are now perceived as physician-centric and paternalistic. Eight principles of patient-centered care, first published in 1993, are still widely referenced today [23]. These include:
1. Respect for patients' values, preferences, and expressed needs
2. Coordination and integration of care
3. Information and education
4. Physical comfort, including environment and pain management
5. Emotional support, including alleviation of fears
6. Involvement of family and friends
7. Continuity and transition
8. Access to care

> **Box 1.1 Patient- and Family-Centered Care**
> Patient- and family-centered care is the predominant model of contemporary health care.

1.3.1.1 What Does Patient-Centered Care Look Like in the Context of These Case Studies?

First, every case starts with an encounter between a patient or family and a physician or other healthcare practitioner (discussed in ▶ Chapter 2). Although all the patient's or family's needs, values, and preferences will not be immediately apparent, at least one will be obvious up-front, setting up the question or problem at the heart of each case.

As teachers and learners move through the case, they will have to answer recurring questions related to patient-centered care, such as:
– "What would you tell this patient or family?"
– "What are your obligations to this patient or family?"

- "How would you proceed while respecting the patient's autonomy?"
- "How would you support this patient or family?"

They also will have opportunities to learn more about the patient's backstory, which might include information about the patient's family, education, occupation, socioeconomic status, culture, or religious beliefs. Some cases include events in which care that is *not* patient-centered might also occur, allowing learners to contrast their ideals with realities and identify opportunities for improvement.

1.3.1.2 What Knowledge, Skills, and Attitudes Are Learners Expected to Develop Related to This Theme?

In every case, learners are encouraged to demonstrate respect for individual autonomy (while sometimes considering situations in which respecting the autonomy of one person might harm another), to recognize the primacy of the patient's welfare, to demonstrate responsiveness to patient needs that supersedes self-interest, and to appreciate the importance of honesty, integrity, and compassion in all interactions with patients' families, colleagues, and others with whom physicians and healthcare professionals must interact in their professional lives.

1.3.2 Cultural Humility

Cultural humility was first described as a concept in medicine by Melanie Tervalon and Jan Murray-Garcia who sought to address the problems and inadequacies of "cultural competence" [24]. Cultural competence was focused on the acquisition of knowledge and development of specific skills to be deployed in interactions with patients of a given culture. Consequently, models of cultural competence tended to falsely suggest that healthcare practitioners could become "competent" in cultures other than their own while also perpetuating misconceptions of cultures as monolithic and experienced or interpreted identically by all members.

Cultural humility focuses instead on ideals of self-reflection and lifelong learning, encouraging healthcare practitioners to recognize and appreciate differences of cultural identity among themselves, their patients, and their colleagues. In the construction of this textbook, we have particularly emphasized the recognition of the culture of medicine. Every clinical encounter involves at least three cultures: the culture of the patient, the culture of the healthcare practitioner, and the culture of medicine. Often, other cultures are involved too—patients and healthcare practitioners may have multicultural backgrounds themselves, and healthcare institutions have distinct institutional cultures of their own.

> **Box 1.2 Cultural Humility**
> Cultural humility is essential to patient- and family-centered care.

1.3.2.1 What Does Culture Look Like in the Context of These Case Studies?

Every patient and family in these cases has a distinct cultural background; however, depending on the perspective of the teacher and learner, not every case may have an obvious "other" (a person apparently from a background that differs from the learner), and the patient's culture, though important, is frequently not central to the conflict, just as in real life. In these cases, learners may need guidance to recognize that cultural humility is still important.

As teachers and learners move through the case, they will have to answer recurring questions related to culture, such as:
- "Is it important to ask about faith or culture in this situation?"
- "How might understanding the patient's culture help or change what you do?"
- "Do your beliefs influence your feelings about this case or what you think should be done? How so?"

They will also have frequent opportunities to reflect on ways in which medicine and medical practice are grounded in culture and tradition as much as in science.

1.3.2.2 What Knowledge, Skills, and Attitudes Are Learners Expected to Develop Related to This Theme?

In every case, learners are encouraged to appreciate the three cultures involved in any clinical interaction, to recognize when the culture of medicine is contributing to a problem, to describe their own values and explain how those values affect their responses to certain situations, and to avoid imposing values on patients.

1.3.3 Evidence-Based Practice

The concept of evidence-based practice (EBP) arose from evidence-based medicine (EBM) [25]. In clinical models of evidence-based practice, physicians and healthcare practitioners synthesize three types of information: (1) the best available evidence, drawn from scholarly research, to support a decision, approach, test, or intervention; (2) clinical judgment and experience, supporting the practitioner's assessment of the patient's or family's situation and needs; and (3) the patient's or family's preferences and values [26]. However, principles of evidence-based practice can readily be applied to population health and healthcare policy [27, 28]. In those settings, policymakers and others seek the same three types of information: (1) the best available evidence, drawn from scholarly research, to support a decision, program, policy, or law; (2) professional judgment and experience, supporting the assessment of the community's or society's situation and needs; and (3) the community's or society's preferences and values.

At the heart of evidence-based practice are the core ethical principles of beneficence and non-maleficence: the Sicily statement on evidence-based practice notes that "Health care delivered in ignorance of available research evidence, misses important opportunities to benefit patients and may cause significant harm" [26].

> **Box 1.3 Evidence-Based Practice**
> Effective evidence-based practice requires the skills of patient- and family-centered care and cultural humility.

1.3.3.1 What Does Evidence-Based Practice Look Like in the Context of These Case Studies?

Evidence-based practice is a process [25, 26]. The following steps have been adapted to apply specifically to the cases in this book:

1. **Assess** the patient, family, and/or community and ask the question(s): What is the problem or dilemma? What is known and what is not known? What additional information would be helpful?
2. **Acquire** the evidence: What research has been done to address the question(s)?
3. **Appraise** the evidence: What is the strength of the available research? What are the limitations of the available research?
4. **Apply** the evidence: What factors, unique to this patient, family, or community, must be considered? What decision(s) are supported after considering these factors?
5. **Evaluate** the outcome: Was the desired outcome achieved? For all stakeholders? What are the next steps?

As teachers and learners move through the case, they will have to answer recurring questions related to evidence-based practice that draw on the aforementioned steps. They will also be prompted to think about how they might obtain better evidence, whether more research would help to address the underlying problems, and why obtaining evidence might be difficult. Examples of such questions include:

- "How would you design a study to answer this question?"
- "Why do you think more studies have not been done to address this question?"
- "What are some of the challenges of evaluating the impact of policy change?"
- "What would you want to know if you were designing a law or policy to address this issue?"

1.3.3.2 What Knowledge, Skills, and Attitudes Are Learners Expected to Develop Related to This Theme?

In every case, learners are encouraged to recognize the limitations and imperfections of medicine and to appreciate the importance of high-quality evidence in order to support optimal medical, ethical, and policy decisions. Depending on the design of the course (see ► Chapter 2), the structure of the cases allows learners to engage in self-directed learning and independent inquiry, developing information-seeking and critical appraisal skills.

1.3.4 Cognitive Biases in Decision-Making

Cognitive bias is a systematic pattern of deviation from rationality in judgment [24]. Such biases are adaptive in some contexts (e.g., as shortcuts, increasing efficiency and reducing cognitive workload), may affect memory and perception, and can occur at both the group and individual levels. They can be broadly grouped into four categories: biases related to limitations of memory, those related to the need to act quickly, those related to insufficient available information, and those related to the presence of too much information (◘ Fig. 1.1) [29, 30].

Several types of cognitive bias have been shown to affect decision-making by both patients and healthcare practitioners, though more attention is paid to healthcare practitioners [31]. These include [31, 32]:

- **Anchoring:** The tendency to lock onto salient features in the patient's initial presentation and then fail to adjust initial impression in the light of later information
- **Availability bias:** The tendency to judge things as being more likely or common if they readily come to mind, affected by recent or significant cases
- **Information bias:** The tendency to seek more information even when it will not affect action

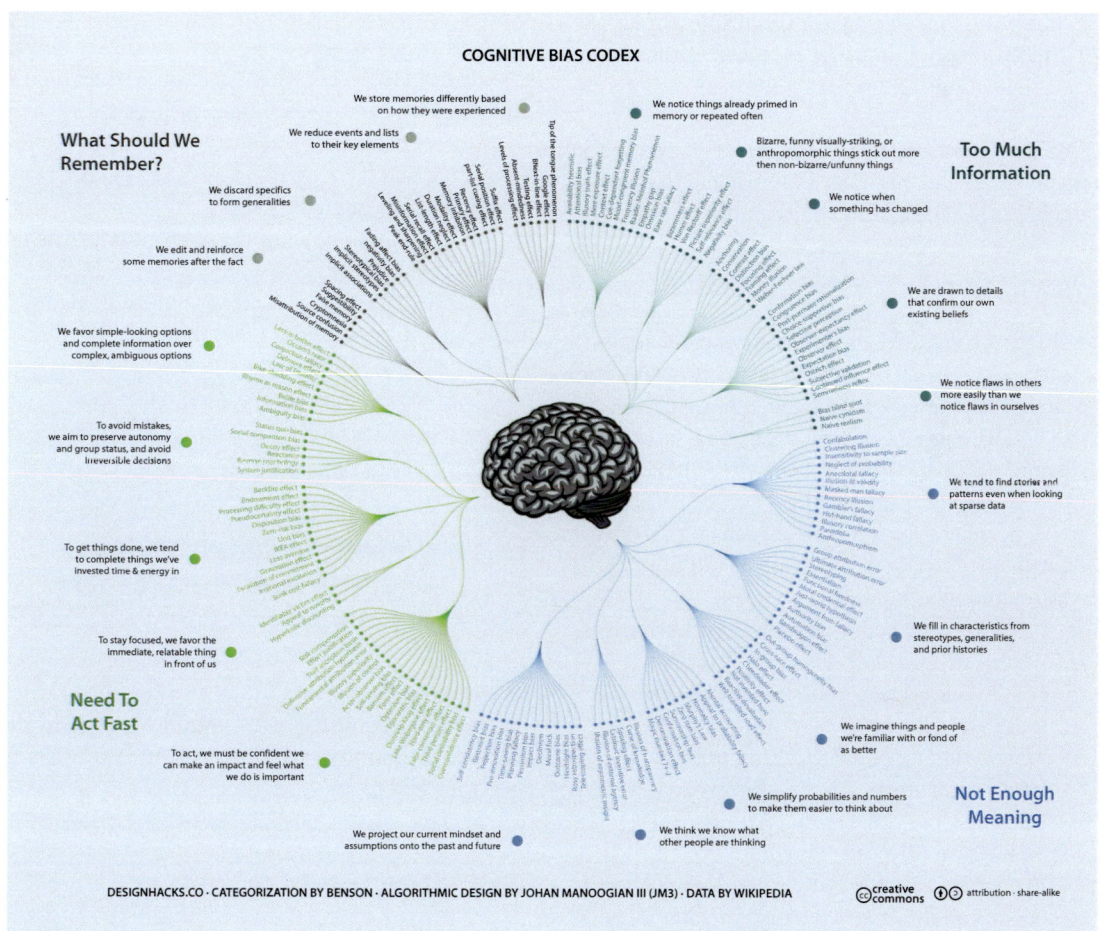

◘ Fig. 1.1 Complete (as of 2016) list of cognitive biases, arranged and designed by John Manoogian III (jm3). (Categories and descriptions originally by Buster Benson. ► https://commons.wikimedia.org/wiki/File:The_Cognitive_Bias_Codex_-_180%2B_biases,_designed_by_John_Manoogian_III_(jm3).png Reprinted under terms of ► https://creativecommons.org/licenses/by-sa/4.0/)

- **Confirmation bias**: The tendency to look for confirmatory evidence to support a diagnosis, rather than looking for evidence to disprove it
- **Premature closure**: The tendency to accept a diagnosis as final before it has been fully verified
- **Framing effect**: The tendency to have one's assessment influenced by the way a situation is initially framed (by patient, nurse, previous healthcare practitioner, etc.)
- **Fundamental attribution error**: The tendency to judge and blame patients for sickness (dispositional causes) rather than examine the circumstances (situational factors) that might have been responsible
- **Representativeness**: The tendency to quickly identify prototypical manifestations of disease (beginning early in health professions training in the context of multiple-choice questions), often using race and ethnicity associations and missing atypical variants

Implicit, or unconscious, bias is also a type of cognitive bias. Unconscious bias refers to relatively unconscious and automatic features of prejudiced judgment and social behavior; it is usually contrasted with explicitly endorsed beliefs [33]. Implicit biases can interact with other cognitive biases in a number of ways to produce decisional errors in health care [34–36]. There is growing recognition that implicit biases lead to explicit harms, including the undertreatment of pain among racial and ethnic minority patients and disparities in application of screening recommendations [37–39].

A number of cognitive "de-biasing" strategies have been described [31]. Suggested approaches include the following:
- Practice forced consideration of alternatives.
- Routinely ask, "What else might this be?"
- Provide adequate time for decision-making.
- Reflect on one's own thinking process (metacognition).
- Elicit and provide feedback on decision-making rapidly and reliably.
- Recognize situations in which bias is likely to occur and develop strategies in advance.
- Decrease reliance on memory.
- Use decision aids when available.

> **Box 1.4 Cognitive Bias**
> Recognizing and ameliorating cognitive biases rests on a strong foundation in evidence-based practice, as well as the same skills of reflective practice that are fundamental to patient- and family-centered care and to cultural humility.

1.3.4.1 What Does Bias in Decision-Making Look Like in the Context of These Case Studies?

Within the cases, bias is present at a number of levels. First, the events of each case will demonstrate that healthcare practitioners are not perfect: decisional errors will sometimes be made as a consequence of bias. Second, learners will sometimes give responses or suggest approaches that reveal their own biases. Some cases have been intentionally designed to elicit this and promote reflection on it. In others, teachers will have to decide when and how to intervene to encourage learners to recognize their biases.

As teachers and learners move through the case, they will have to answer recurring questions related to bias, such as:
- "When you first heard this patient's story, what did you assume about the patient? Why?"
- "Some groups of patients are more likely to be given a particular label than other groups of patients. Why might this be?"
- "Do you think that the outcome of this patient's case could have been prevented or changed? Why or why not?"

1.3.4.2 What Knowledge, Skills, and Attitudes Are Learners Expected to Develop Related to This Theme?

In every case, learners are encouraged to recognize when common biases might be at play in the context of a clinical scenario and to take steps to ameliorate biases.

1.3.5 Structural Competency

Structural violence, a term typically ascribed to Johan Galtung, describes phenomena in which an institution, policy, or other social structure causes harm by preventing people from meeting basic needs [40]. Structural or institutionalized racism is a specific form of structural violence, described by Camara Jones, as follows:

> Institutionalized racism manifests itself both in material conditions and in access to power. With regard to material conditions, examples include differential access to quality education, sound housing, gainful employment, appropriate medical facilities, and a clean environment. With regard to access to power, examples include differential access to information (including one's own history), resources (including wealth and organizational infrastructure), and voice (including voting rights, representation in government, and control of the media). It is important to note that the association between socioeconomic status and race in the United States has its origins in discrete historical events but persists because of contemporary structural factors that perpetuate those historical injustices. In other words, it is because of institutionalized racism that there is an association between socioeconomic status and race in this country [41].

Structural competency builds upon these concepts to address stigma and inequity in health care by directing attention to the forces that influence health, healthcare interactions, and health outcomes beyond the level of the individual patient or healthcare practitioner [42–44]. It provides a framework for how healthcare practitioners' understanding of structural violence, racism, and vulnerability can guide practice and improve individual and population health. The structurally competent healthcare practitioner has:

> ...the trained ability to discern how a host of issues defined clinically as symptoms, attitudes, or diseases (e.g., depression, hypertension, obesity, smoking, medication "non-compliance," trauma, psychosis) also represent the downstream implications of a number of upstream decisions about such matters as health care and food delivery systems, zoning laws, urban and rural infrastructures, medicalization, or even about the very definitions of illness and health [43].

> **Box 1.5   Structural Competency**
> Structural competency builds on the previous themes. It moves beyond cultural humility and cognitive bias, which tend to focus on individuals, but integrates principles from each. Evidence-based practice can be directed toward structural interventions.

1.3.5.1 What Does Structural Competency Look Like in the Context of These Case Studies?

Every case begins at the level of an encounter between an individual patient or family and healthcare practitioner. These encounters, however, do not occur in a vacuum—these hypothetical people exist within communities, institutions, and structures—and, as in the real world, their experiences and choices are constrained by structures. Some cases explicitly address racism, as well as xenophobia, homophobia, sexism, and ableism. The role of adverse childhood experiences, chronic stress, and poverty on health is also woven throughout the book [45]. Background questions present learners with the opportunity to research aspects of history and law that are relevant to structural inequalities today. As teachers and learners move through a case, they will acquire new information regarding structural factors affecting the patient's health and well-being and be encouraged to revise their impressions in light of this broader perspective. They will also be asked to consider how they might address the patient's structural vulnerability.

As teachers and learners move through the case, they will have to answer recurring questions related to structural competency, such as:

- "How might factors outside the clinic affect this patient's health and well-being?"
- "What barriers might this patient face in accessing health care?"
- "Given what you know, what choices do you think this patient has?"

- "Do you think the usual medical recommendations for this situation will be feasible for this patient?"
- "What resources in the community might be helpful to address this issue?"

1.3.5.2 What Knowledge, Skills, and Attitudes Are Learners Expected to Develop Related to This Theme?

In every case, learners are encouraged to analyze the case in context of the relevant social determinants of health, to consider the role of structures and institutions in affecting health, and to recognize and address the needs of the healthcare population that one serves.

1.3.6 Social Justice and Advocacy

Although the concept of justice is one of the core principles of bioethics, it cannot be realized through individual medical practice alone or even through improvements in care at a single healthcare institution [46]. Substantive progress to address and eliminate inequities in health and health care requires "interventions outside the clinic" [47]. Social justice refers to conceptualizations of fairness between individuals and societies. In the context of health care, social justice means that "no one is denied the possibility to be healthy for belonging to a group that has historically been economically/socially disadvantaged" [48].

According to the World Health Organization (WHO):

» Health has special meaning to individuals and communities at large. Good health is necessary for human well-being, providing intrinsic value for comfort, contentment and pursuit of the joys of life. But good health does more than that. It is important in allowing individuals to exercise a range of human rights—both civil and political (e.g. physical integrity, personal security, political participation), social and economic (e.g. employment, education and family life). Just as important, health is necessary for well functioning societies. If a population does not have a decent level of health, it is very difficult to ensure economic prosperity, political participation, collective security and so forth [49].

Two closely related concepts are health equity and health disparities. Health equity has been defined as "[health] differences which are unnecessary and avoidable but, in addition, are considered unfair and unjust," more recently as "systematic disparities in health (or in the major social determinants of health) between groups with different levels of underlying social advantage/disadvantage—that is, wealth, power, or prestige," and perhaps most simply as "social justice in health" [48, 50, 51]. Health disparities are measurable entities that can be used to assess progress toward health equity or the extent to which health inequity persists [48].

These ideas are increasingly recognized throughout medical and health professions education [52, 53]. The Charter on Medical Professionalism, endorsed by more than 90 medical professional societies since its publication in 2002, identifies three central principles that then define the obligations of physicians: (1) the principle of primacy of patient welfare, (2) the principle of patient autonomy, and (3) the principle of social justice [54, 55].

Although the obligation of physicians and other healthcare practitioners to engage in advocacy to promote health in social justice is central to this book, advocacy skills themselves are beyond our scope.

> **Box 1.6 Social Justice and Advocacy**
> Social justice and advocacy, for patients and populations, by healthcare practitioners, require the synthesis of the first five themes.

1.3.6.1 What Does Social Justice Look Like in the Context of These Case Studies?

An understanding of the centrality of social justice and advocacy to the mission of medicine and the health professions is found near the conclusion of each case. In nearly every scenario, learners have opportunities to consider whether systemic change is necessary to address the injustice or inequity described in the case. As teachers and learners move through the case, they will have to answer recurring questions related to social justice and advocacy, such as:

- "Do healthcare practitioners have an obligation to advocate for change in situations like these?"
- "What might you do if you wanted to change the law or policy as it relates to this situation?"

1.3.6.2 What Knowledge, Skills, and Attitudes Are Learners Expected to Develop Related to This Theme?

In every case, learners are encouraged to recognize that physicians are accountable not only to patients but also to the community, to the society, and to the profession, to appreciate why physicians and other healthcare professionals have an obligation to address injustice, to acquire the skills to be able and willing to speak up in the face of injustice, and to learn where to begin in order to advocate for change when one witnesses a problem in a community or healthcare system that one wishes to address.

1.4 Conclusion

This chapter introduced the purpose of this book: the gaps within medical education that it seeks to address, the rationale for integration of bioethics and population health and for the use of case-based learning, the origins of the cases, and the six key themes that appear in every case. This chapter, along with ▶ Chapter 2, Approaches to Using This Book; ▶ Chapter 25, Evaluating Cases in Context; and ▶ Chapter 26, A Practical Framework for Learner Assessment, provides readers with the tools to optimize the use of the cases in medical and health professions education.

- **Editors' Note**

On the Origins and Development of the Cases

The cases in this textbook were developed iteratively over the past 4 years and, as of this book's publication, have been used by nearly 40 faculty and more than 500 students. Historically, both the ethics and public health courses at our institution were taught through a combination of traditional lectures and small group discussions; thus, both had some pre-existing cases designed for use with small groups of students. During the first year of the integrated course, some of these were revised and integrated in a simple format: a brief vignette, followed by four to five questions. Students were expected to have prepared for the discussion by attending the relevant lecture and completing several pages of assigned readings on related topics, and some small group discussions involved questions based on readings alone, without an attendant case.

When the course transitioned to a format that primarily relied on case-based and self-directed learning, the editors (two of whom were serving as the course directors and all of whom were leading small groups) extensively rewrote the existing course materials, with input from many other faculty members. Cases were expanded beyond the initial vignette into rich narratives and detailed case studies. Students were given background questions (explained in detail in ▶ Chapter 2) to research prior to small group discussions. Guides for facilitators, containing notes for how to approach specific topics, were developed at the request of faculty who did not always feel that they were "expert" in the complex array of issues covered. Systematic evaluation, the subject of ▶ Chapter 25, led to further revision and refinement of case content.

It is these case studies that provide the backbone for the chapters and cases here. The authors of each chapter were selected based upon their expertise in the relevant issues and their ability to make those issues understandable for those just entering the health professions. In some cases, they were co-authors of the earlier cases (▶ Chapters 3, 4, 6, 8, 16, 19, and 20) or directly involved in the real case upon which the chapter was based (▶ Chapters 10, 12, and 21); in others, we invited clinician-educators with relevant expertise to adapt the existing cases in the direction they thought most appropriate (▶ Chapters 7, 9, 11, 13, 14, 15, 17, 22, 23, and 24); and in still others, we asked experts to develop new cases on topics that had previously been neglected (▶ Chapters 5 and 18).

The writing and editing process was therefore highly collaborative, and the editors had prominent roles in the evolution of each case, from inception to implementation in the classroom to publication in this book. Where the cases drew inspiration from older versions developed by other faculty, from real events in public health and medical research, or from published articles, such influences are listed in the acknowledgments for the chapter, along with other colleagues who assisted with revisions or research. The cases are stronger for the many faculty and scholars who have had input into their development and have also been informed by feedback from the small group facilitators and medical students who participated in discussions of them over the past 4 years.

References

1. Kasper J, Greene JA, Farmer PE, Jones DS. All health is global health, all medicine is social medicine: integrating the social sciences into the preclinical curriculum. Acad Med. 2016;91(5):628–32.
2. Dao DK, Goss AL, Hoekzema AS, Kelly LA, Logan AA, Mehta SD, Sandesara UN, Munyikwa MR, DeLisser HM. Integrating theory, content, and method to foster critical consciousness in medical students: a comprehensive model for cultural competence training. Acad Med. 2017;92(3):335–44.
3. Hansen H, Metzl JM. New medicine for the US health care system: training physicians for structural interventions. Acad Med. 2017;92(3):279–81.
4. Avci E. Drawing on other disciplines to define quality in bioethics education. Qual High Educ. 2017;23(3):201–12.
5. Irby DM, Hamstra SJ. Parting the clouds: three professionalism frameworks in medical education. Acad Med. 2016;91(12):1606–11.
6. Carrese JA, Malek J, Watson K, et al. The essential role of medical ethics education in achieving professionalism: the Romanell Report. Acad Med. 2015;90(6):744–52.
7. Liaison Committee on Medical Education. Functions and structure of a medical school: standards for accreditation of medical education programs leading to the M.D. degree. LCME; 2016 [cited 2017 Sept 1]. Available from: http://lcme.org/publications/#Standards.
8. Association of American Medical Colleges. Core entrustable professional activities for entering residency: curriculum developers' guide. Washington, D.C.: AAMC; 2014 [cited 2018 July 26]. Available from: https://members.aamc.org/eweb/upload/Core%20EPA%20Curriculum%20Dev%20Guide.pdf.
9. Meili R, Hewett N. Turning Virchow upside down: medicine is politics on a smaller scale. J R Soc Med. 2016;109:256–8.
10. Mackenbach JP. Politics is nothing but medicine at a larger scale: reflections on public health's biggest idea. J Epidemiol Community Health. 2008;63(3):181–4.
11. Westerhaus M, Finnegan A, Haidar M, Kleinman A, Mukherjee J, Farmer P. The necessity of social medicine in medical education. Acad Med. 2015;90(5):565–8.
12. Beauchamp TL, Childress JF. Principles of biomedical ethics. Oxford, MA: Oxford University Press; 2001.
13. Beauchamp TL. The 'four principles' approach to health care ethics. In: Ashcroft RE, Dawson A, Draper H, JR MM, editors. Principles of health care ethics. Chichester: John Wiley & Sons, Ltd; 2007. p. 3–10.
14. ABIM Foundation. American Board of Internal Medicine. Medical professionalism in the new millennium: a physician charter. Ann Intern Med. 2002;136(3):243.
15. Englander R, Cameron T, Ballard AJ, Dodge J, Bull J, Aschenbrener CA. Toward a common taxonomy of competency domains for the health professions and competencies for physicians. Acad Med. 2013;88(8):1088–94.
16. McMaster University. Case Based Learning. [Internet]. Hamilton (ON): McMaster University. [cited 2018 May 24]. Available from: http://cll.mcmaster.ca/resources/C/Case_Based_Learning.html.
17. Schmidt HG, Rotgans JI, Yew EH. The process of problem-based learning: what works and why. Med Educ. 2011;45(8):792–806.
18. Walker A, Leary H, Hmelo-Silver C, Ertmer PA, editors. Essential readings in problem-based learning: exploring and extending the legacy of Howard S. Barrows. West Lafayette (IN): Purdue University Press; 2015.
19. McLean SF. Case-based learning and its application in medical and health-care fields: a review of worldwide literature. J Med Educ Curric Dev. 2016;3:JMECD-S20377.
20. Harasym PH, Tsai TC, Munshi FM. Is problem-based learning an ideal format for developing ethical decision skills? Kaohsiung J Med Sci. 2013;29(10):523–9.
21. Koh GC, Khoo HE, Wong ML, Koh D. The effects of problem-based learning during medical school on physician competency: a systematic review. CMAJ. 2008;178:34e41.
22. Hurtado MP, Swift EK, Corrigan JM. Crossing the quality chasm: a new health system for the 21st century. Washington, D.C.: Institute of Medicine, Committee on the National Quality Report on Health Care Delivery; 2001.
23. Gerteis M. Through the patient's eyes: understanding and promoting patient-centered care. Camden (MA): Picker Institute; 1993.
24. Tervalon M, Murray-Garcia J. Cultural humility versus cultural competence: a critical distinction in defining physician training outcomes in multicultural education. J Health Care Poor Underserved. 1998;9(2):117–25.
25. Guyatt G, Cairns J, Churchill D, Cook D, Haynes B, Hirsh J, et al. Evidence-based medicine: a new approach to teaching the practice of medicine. J Am Med Assoc. 1992;268(17):2420–5.
26. Dawes M, Summerskill W, Glasziou P, Cartabellotta A, Martin J, Hopayian K, et al. Sicily statement on evidence-based practice. BMC Med Educ. 2005;5(1):1.
27. Green LW. Public health asks of systems science: to advance our evidence-based practice, can you help us get more practice-based evidence? Am J Public Health. 2006;96(3):406–9.
28. Brownson RC, Fielding JE, Maylahn CM. Evidence-based public health: a fundamental concept for public health practice. Ann Rev Public Health. 2009;30:175–201.
29. Kahneman D, Tversky A. Subjective probability: a judgment of representativeness. Cogn Psychol. 1972;3(3):430–54.
30. Benson B, Manoogian J. The Cognitive Bias Codex - 180+ biases [image on the internet]. 2016. Available from: https://commons.wikimedia.org/wiki/File:The_Cognitive_Bias_Codex_-_180%2B_biases,_designed_by_John_Manoogian_III_(jm3).png.

31. Croskerry P. The importance of cognitive errors in diagnosis and strategies to minimize them. Acad Med. 2003;78(8):775–80.
32. Blumenthal-Barby JS, Krieger H. Cognitive biases and heuristics in medical decision making: a critical review using a systematic search strategy. Med Decision Making. 2015;35(4):539–57.
33. Greenwald AG, Banaji MR. Implicit social cognition: attitudes, self-esteem, and stereotypes. Psychol Rev. 1995;102:4–27.
34. Godsil RD, Tropp LR, Goff PA, Powell JA. The science of equality, vol. 1: addressing implicit bias, racial anxiety, and stereotype threat in education and healthcare. Perception Institute; 2014 [cited 2016 August 11]. Available from: http://perception.org/app/uploads/2014/11/Science-of-Equality-111214_web.pdf.
35. Staats C, Capatosto K, Wright RA, Contractor D. State of the science: implicit bias review. Columbus (OH): Kirwan Institute for the Study of Race and Ethnicity; 2014.
36. Zestcott CA, Blair IV, Stone J. Examining the presence, consequences, and reduction of implicit bias in health care: a narrative review. Group Process Intergroup Relat. 2016;19(4):528–42.
37. Hoffman KM, Trawalter S, Axt JR, Oliver MN. Racial bias in pain assessment and treatment recommendations, and false beliefs about biological differences between blacks and whites. Proc Natl Acad Sci. 2016;113(16):4296–301.
38. Horner-Johnson W, Dobbertin K, Andresen EM, Iezzoni LI. Breast and cervical cancer screening disparities associated with disability severity. Womens Health Issues. 2014;24(1):e147–53.
39. Holden CD, Chen J, Dagher RK. Preventive care utilization among the uninsured by race/ethnicity and income. Am J Preventive Med. 2015;48(1):13–21.
40. Galtung J. Violence, peace, and peace research. J Peace Res. 1969;6(3):167–91.
41. Jones CP. Levels of racism: a theoretic framework and a gardener's tale. Am J Public Health. 2000;90(8):1212.
42. Metzl JM. Structural competency. Am Q. 2012;64(2):213–8.
43. Metzl JM, Hansen H. Structural competency: theorizing a new medical engagement with stigma and inequality. Soc Sci Med. 2014;103:126–33.
44. Metzl JM, Petty J. Integrating and assessing structural competency in an innovative prehealth curriculum at Vanderbilt University. Acad Med. 2017;92(3):354.
45. Dong M, Anda RF, Felitti VJ, Williamson DF, Dube SR, Brown DW, Giles WH. Childhood residential mobility and multiple health risks during adolescence and adulthood: the hidden role of adverse childhood experiences. Arch Pediatr Adolesc Med. 2005;159(12):1104–10.
46. Beauchamp TL, Childress JF. Justice. In: Principles of biomedical ethics. 7th ed. Oxford, England: Oxford University Press; 2013. p. 292–301.
47. Woolf SH. Progress in achieving health equity requires attention to root causes. Health Aff. 2017;36(6):984–91.
48. Braveman P, Gottlieb L. The social determinants of health: it's time to consider the causes of the causes. Public Health Rep. 2014;129(Suppl 2):19–31.
49. Gostin LO. Health and social justice. Bull World Health Organ. 2011;89:78.
50. Whitehead M. The concepts and principles of equity and health. Health Promot Int. 1991;6(3):217–28.
51. Braveman P, Gruskin S. Defining equity in health. J Epidemiol Community Health. 2003;57(4):254–8.
52. Woollard B, Boelen C. Seeking impact of medical schools on health: meeting the challenges of social accountability. Med Educ. 2012;46(1):21–7.
53. DasGupta S, Fornari A, Geer K, et al. Medical education for social justice: Paulo Freire revisited. J Med Humanit. 2006;27(4):245–51.
54. ABIM Foundation. ACP—ASIM Foundation, and European Federation of Internal Medicine. Medical professionalism in the new millennium: a physician charter. Ann Intern Med. 2002;136(3):243–6.
55. Blank L, Kimball H, McDonald W, Merino J. Medical professionalism in the new millennium: a physician charter 15 months later. Ann Int Med. 2003;138(10):839–41.

Approaches to Using This Book

Amy E. Caruso Brown

2.1 **Overview – 18**

2.2 **Overall Structure of the Book – 18**

2.3 **Structure of the Cases and Chapters – 18**
2.3.1 Glossary – 18
2.3.2 Learning Objectives – 18
2.3.3 Case Background – 18
2.3.4 Background Questions – 24
2.3.5 Additional Case Information and Questions for Discussion – 24
2.3.6 Answers to Background Questions – 25
2.3.7 Responses to Discussion Questions – 25
2.3.8 References – 25
2.3.9 Further Reading on This Topic – 25

2.4 **How to Use a Case – 25**
2.4.1 A Note About Flexibility – 25
2.4.2 For Course Directors and Others Responsible for Curricular Design – 26
2.4.3 For Discussion Facilitators – 29

2.5 **Trigger Warnings and Brave Spaces – 34**

References – 35

© Springer Nature Switzerland AG 2019
A. E. Caruso Brown et al. (eds.), *Bioethics, Public Health, and the Social Sciences for the Medical Professions*, https://doi.org/10.1007/978-3-030-03544-0_2

2.1 Overview

This book was designed to optimize flexibility for educators to adapt the cases to their learners' needs while sharing our experience in teaching this material. Here, we offer a guide to possible approaches. This chapter provides an overview of the structure of the book and of the individual cases that comprise the chapters in Part II; guidance for course directors and discussion facilitators on how to use individual cases, in a question-and-answer format; and a brief discussion of the sensitive nature of the content, trigger warnings, and safe spaces.

2.2 Overall Structure of the Book

The book is divided into three major sections. Part I (▶ Chapters 1 and 2) introduces the book and is intended to optimize its utility; Parts II through XII (▶ Chapters 3, 4, 5, 6, 7, 8, 9, 10, 11, 12, 13, 14, 15, 16, 17, 18, 19, 20, 21, 22, 23, and 24) are the heart of the book—the cases as discussed below—and Part XIII (▶ Chapters 25 and 26) addresses challenges and innovative approaches to assessing learning in bioethics, public health, and the social sciences. A glossary of terms broadly important to the entire textbook is included at the end of the book.

Parts II through XII each include two chapters that share a related subject in bioethics, public health, and the social sciences (e.g., community perspectives or healthcare systems). Although these topics may make peripheral appearances in other chapters, they are central to the learning objectives of the cases within their respective parts. Each chapter contains a single case, organized into eight or nine subsections as discussed in more detail below. The following tables are available to help users of this book locate specific cases and recognize topical connections between cases. Each table organizes and classifies cases according to a different group of qualities.

- ▫ Table 2.1a, b Core concepts by chapter
- ▫ Table 2.2 Cases by diagnosis and healthcare features

2.3 Structure of the Cases and Chapters

Each chapter in Part II is divided into the following sections, in the same order.

2.3.1 Glossary

This subsection is not present in every chapter. Chapter-specific glossaries contain definitions that are critical to the content of the specific case, including medical or public health terminology. Terms that have different uses or where misunderstandings are prevalent outside of health care are included here. In addition, there is an overall glossary for the book that includes definitions for key terms that appear in multiple cases.

2.3.2 Learning Objectives

The next subsection contains three to five learning objectives, describing what learners can expect to accomplish by the conclusion of the chapter. These objectives presume that learners are not simply reading the chapter but are engaged in active learning through self-directed study and guided case analysis with faculty and peers. Instructors assessing students should find that most, but not all, of the learning objectives are straightforward to evaluate and measure. Part XIII addresses evaluation and assessment in more detail.

2.3.3 Case Background

The third subsection contains a brief introduction to the case, structured as a clinical vignette. Each case begins at the level of the encounter between an individual patient and a physician or other healthcare practitioner and provides enough information that learners should be able to identify at least one potential issue or dilemma facing the patient and provider. The "you" in each case background is always the healthcare practitioner (HCP); learners are asked to imagine themselves in this role during the discussion of the case, although they are also asked to practice taking the perspective of other stakeholders in the case.

Information in the case background will include the level of training, specialty, and practice situation of the healthcare practitioner, if relevant. There is considerable variation in these characteristics: the protagonists range from medical students to practicing physicians, from primary care physicians to surgical subspecialists, and from urban academic practitioners to rural private practitioners. In all chapters, the case can be readily adapted to involve

Approaches to Using This Book

Table 2.1a Core concepts by chapter. The six key themes—patient- and family-centered care, cultural humility, evidence-based practice, cognitive biases, structural competency, and social justice and advocacy—are covered in every chapter, as are the basic ethical principles of autonomy, beneficence, non-maleficence, and justice. The socioecological model can be applied to every case but is done so explicitly where noted below

Core concept	Covered in chapters
Access to care/lack of insurance	▶ 3, 5, 8, 10, 14, 15, 16, 18, 22, 23
Confidentiality	▶ 4, 5, 11, 12, 16
Conflict of interest	▶ 19, 20
Conscientious objection/duty to care	▶ 4, 9, 14, 16, 23
Decisional capacity	▶ 7, 10, 11, 13
Disagreement between family members	▶ 13, 16, 21
Disclosure/breaking bad news	▶ 3, 14, 17, 23
Disease surveillance	▶ 4, 6, 9, 11
End-of-life care	▶ 13, 14
Environmental factors	▶ 6, 15, 22
Health disparities	▶ 3, 5, 6, 8, 12, 15, 18, 22, 23, 24
Healthcare finance	▶ 8, 14, 17, 18, 24
Health literacy	▶ 3
Homelessness/housing insecurity	▶ 4, 6, 22
Informed consent	▶ 3, 16, 19, 20, 21
Medical error/malpractice/patient safety	▶ 3, 7, 15, 17, 20
Medical futility	▶ 13, 14
Pediatric decision-making standards (best interest standard/harm principle/right to an open future)	▶ 7, 9, 10, 16, 19, 21
Poverty	▶ 5, 6, 8, 12, 22, 24
Public health system (public health agencies, laws, and regulations to protect health including ACA)	▶ 3, 4, 6, 7, 8, 9, 11, 14, 19
Quality of life	▶ 10, 13, 14, 15, 19
Racism (structural/institutionalized/interpersonal)	▶ 3, 5, 6, 8, 18
Refugees and migrants	▶ 12, 22
Refusal of treatment	▶ 9, 10, 11
Religious beliefs	▶ 9, 10, 13, 24
Reproductive rights	▶ 5, 7, 12, 16, 23
Research ethics	▶ 6, 9, 19, 20, 21, 23
Resource allocation	▶ 12, 14, 18, 19, 23, 24
Scope of practice	▶ 3, 7, 15, 16, 24
Screening/prevention	▶ 4, 6, 7, 8, 9, 11, 15, 18, 22, 24

(continued)

Table 2.1a (continued)

Core concept	Covered in chapters
Socioecological model	▶ 4, 6, 11, 12, 14, 15, 18, 22
Study design	▶ 6, 7, 11, 20, 21, 23
Surrogate decision-making	▶ 13, 14
Trauma-informed care	▶ 12
Trust/mistrust	▶ 3, 5, 9, 10, 12, 13, 17, 18, 19

Table 2.1b Core concepts by chapter. The six key themes—patient- and family-centered care, cultural humility, evidence-based practice, cognitive biases, structural competency, and social justice and advocacy—are covered in every chapter, as are the basic ethical principles of autonomy, beneficence, non-maleficence, and justice. The socioecological model can be applied to every case but is done so explicitly where noted below

Chapter	Section	Key concepts covered
▶ 3	Core principles	Access to care, disclosure, health disparities, health literacy, informed consent, medical error/malpractice/patient safety, scope of practice, public health system, racism, trust/mistrust
▶ 4		Confidentiality, disease surveillance, duty to care, homelessness/housing insecurity, public health system, screening/prevention, socioecological model
▶ 5	Community health	Access to care, confidentiality, health disparities, poverty, racism, reproductive rights, trust/mistrust
▶ 6		Disease surveillance, environmental factors, health disparities, homelessness/housing insecurity, poverty, public health system, racism, research ethics, screening/prevention, socioecological model, study design
▶ 7	Healthcare systems	Medical error/malpractice/patient safety, pediatric decision-making standards, public health system, reproductive rights, scope of practice, screening/prevention, study design
▶ 8		Access to care, health disparities, healthcare finance, poverty, public health system, racism, screening/prevention
▶ 9	Pediatric issues	Duty to care, disease surveillance, pediatric decision-making standards, public health system, refusal of treatment, research ethics, religious beliefs, screening/prevention, trust/mistrust
▶ 10		Access to care/lack of insurance, decisional capacity, pediatric decision-making standards, quality of life, refusal of treatment, religious beliefs, trust/mistrust

Table 2.1a (continued)

Chapter	Section	Key concepts covered
▶ 11	Trauma, violence, and mental health	Confidentiality, decisional capacity, disease surveillance, public health system, refusal of treatment, screening/prevention, socioecological model, study design
▶ 12		Confidentiality, health disparities, poverty, refugees and migrants, reproductive rights, resource allocation, socioecological model, trauma-informed care, trust/mistrust
▶ 13	The end of life	Decisional capacity, disagreement between family members, end-of-life care, medical futility, quality of life, religious beliefs, surrogate decision-making, trust/mistrust
▶ 14		Access to care/lack of insurance, conscientious objection/duty to care, disclosure/breaking bad news, end-of-life care, healthcare finance, medical futility, public health system, quality of life, resource allocation, socioecological model
▶ 15	Disability and quality of life	Access to care/lack of insurance, environmental factors, health disparities, medical error/malpractice/patient safety, quality of life, scope of practice, screening/prevention, socioecological model
▶ 16		Access to care/lack of insurance, confidentiality, conscientious objection/duty to care, disagreement between family members, informed consent, pediatric decision-making standards, reproductive rights, scope of practice
▶ 17	Patient safety and medical mistakes	Disclosure/breaking bad news, healthcare finance, medical error/malpractice/patient safety, trust/mistrust
▶ 18		Access to care/lack of insurance, health disparities, healthcare finance, racism, resource allocation, screening/prevention, socioecological model, trust/mistrust
▶ 19	Research ethics	Conflict of interest, informed consent, pediatric decision-making standards, public health system, quality of life, research ethics, resource allocation, trust/mistrust
▶ 20		Conflict of interest, informed consent, medical error/malpractice/patient safety, research ethics, study design
▶ 21	Stigma and marginalization	Disagreement between family members, informed consent, pediatric decision-making standards, research ethics, study design
▶ 22		Access to care/lack of insurance, environmental factors, health disparities, homelessness/housing insecurity, poverty, refugees and migrants, screening/prevention, socioecological model
▶ 23	Global health	Access to care/lack of insurance, conscientious objection/duty to care, disclosure/breaking bad news, health disparities, reproductive rights, research ethics, resource allocation, study design
▶ 24		Health disparities, healthcare finance, poverty, religious beliefs, resource allocation, screening/prevention, scope of practice

Table 2.2 Cases by diagnosis and healthcare features

Chapter	Primary diagnosis	Specialty*	Practice setting	Trainees present	Interprofessional elements
3	Abdominal aortic aneurysm	Surgery	Hospital	Resident[a]	
4	Tuberculosis	Emergency medicine (Infectious diseases)	Hospital/emergency department (ED)	Medical student[a]	County health department Infection control office
5	Preterm labor	Obstetrics and gynecology	Private practice affiliated with a hospital, jail		Physician assistant Nurse Corrections officer
6	Lead poisoning	General pediatrics	Outpatient clinic in a poor urban area		Early intervention program County health department
7	Congenital heart disease	Emergency medicine	Rural hospital ED		Direct-entry midwife
8	Type 2 diabetes	Emergency medicine	Hospital ED		
9	Well-child visit	General pediatrics	Private practice		
10	Curable brain tumor (pineal germinoma)	Pediatric oncology	Academic medical center		Naturopathic practitioner Ethics consultant
11	Depression, suicidality	Family medicine (Psychiatry)	Outpatient private practice		Psychologist Emergency medical services (EMS) Police
12	Dysmenorrhea	Family medicine (Obstetrics and gynecology)	Outpatient private practice		Interpreter services Refugee case manager Psychologist
13	Alzheimer disease	Critical care medicine	Intensive care unit		Ethics consultant Social worker
14	Metastatic breast cancer	Medical oncology (Palliative care) (Radiation oncology)	Outpatient oncology clinic, intensive care unit		
15	Multiple sclerosis	Internal medicine (Physical medicine/rehabilitation) (Neurology)	Outpatient setting		Physical therapy Occupational therapy

Approaches to Using This Book

#	Topic	Specialty*	Setting	Trainee[a]	Other
16	Family history of Huntington disease	Medical genetics (Obstetrics/gynecology)	Outpatient setting		Genetic counselor
17	Sepsis, leukemia	Internal medicine (Hematology/oncology)	Inpatient unit	Intern[a]	Nursing Family advisory committee Rapid response team
18	Opioid use disorder	Pain medicine/opioid care (Hematology) (Cardiothoracic surgery)	Inpatient units in a tertiary academic medical center		
19	Duchenne muscular dystrophy	Neurology	Outpatient setting		Pharmaceutical company
20	Not applicable	Radiology (Internal medicine) (Surgery)	Residency program	Intern[a]	Institutional review board
21	Gender dysphoria	Adolescent medicine	Outpatient setting	Resident[a] Medical student	Family therapist Ethics consultant Medicolegal partnership Case manager Private insurance company
22	Obesity	Family medicine	Urban safety-net clinic	Medical student	Social worker Medicolegal partnership
23	Zika in pregnancy Congenital Zika syndrome	Obstetrics/gynecology (Pediatric critical care)	Public clinic in Brazil		Nursing assistant Physical therapy Family advocacy group Public health authorities
24	Not applicable	Not applicable	Temporary clinic in lower-middle-income country	Medical student[a]	Other medical mission volunteers

*The specialties noted in parentheses under "Specialty" represent specialties that are present in and relevant to the case but not the protagonist (the "you")
[a]Trainee is the protagonist (the "you") in these cases. In all other cases, learners are directed to envision themselves as fully trained and licensed healthcare practitioners

a physician assistant (PA), nurse practitioner (NP), or other advanced practice provider (APP) in lieu of a physician; most cases also provide opportunities to discuss issues faced by those in other nursing or allied health roles. However, since learners should be taking on the role of the healthcare practitioner, the practitioner's own cultural, religious, and ethnic background is never provided, and learners are encouraged to reflect on the values they bring to the health professions.

2.3.4 Background Questions

The fourth subsection contains seven to eight background questions. These questions are designed for self-directed learning and class preparation (discussed further in "How to Use a Case: For Course Directors" below). They cover core concepts necessary for learners to effectively engage in a thoughtful exploration of the situation presented in the case background. Background questions are intended to be objective and factual in nature, often relating to health policy or law, public health systems and practices, or epidemiology, biostatistics, and study design. Occasionally, they relate to treatment and diagnosis, although in general, these cases have been designed to be manageable for learners with minimal medical knowledge. The questions are designed to ensure that learners can reasonably find answers through scholarly research and use of reliable sources of information, such as peer-reviewed academic journals and relevant government websites (e.g., Centers for Disease Control and Prevention).

In settings where self-directed learning to prepare for the case is not feasible, the background questions may be useful as an outline for the content of one or more lectures occurring prior to the discussion of the case.

2.3.5 Additional Case Information and Questions for Discussion

The fifth subsection is primarily intended as a guide for instructors facilitating a discussion of the case, typically in a small group or team-based learning setting. This section includes four types of materials:
- *Case Events:* Case events describe "what happens next" in the evolution of the situation introduced in the case background. Although they are always in chronological order, the time frame varies from case to case; some cases occur entirely within the context of a single visit, while others span years. Case events often describe the consequences of decisions made in response to earlier events and information. They may sometimes introduce new stakeholders or other settings; for example, in some chapters, a family member who has differing views on the medical decision-making is introduced later in the case. In others, a case event will involve the provider seeing an entirely different patient, whose situation prompts reflection on the original patient's care.
- *Discussion Questions:* Discussion questions follow each case event and are designated with Arabic numerals. They are broad, open-ended questions. Some discussion questions prompt learners to make concrete decisions regarding the next steps they would take in the role of the healthcare provider, while others encourage exploration of the ethical or public health principles at play in the case. They support the learning objectives. In contrast to background questions, which require factual knowledge to answer, responses to discussion questions can be reached through guided inquiry, reasoning, and critical thinking.
- *Follow-Up Questions:* Follow-up questions follow many, but not all, discussion questions and are designated with lowercase letters. These are narrower in scope and are often written to elicit more nuanced responses to discussion questions. Some instructors and learners will find that the conversation resulting from the discussion questions organically encompasses the follow-up questions, without those questions being explicitly asked. However, they may be useful with quieter groups or when all learners seem to agree. Some follow-up questions relate content from the background questions to the current case.
- *Self-Reflection Questions:* Every case has at least one self-reflection question (labeled as such), which is intended to encourage learners to reflect on their own values, beliefs, and experiences, as well as to elicit reflection on personal biases and judgments. There are no correct answers to self-reflection questions; the best answer is an honest one that leads

learners to a deeper understanding of the case and of their future or present role as healthcare practitioners. Self-reflection questions may be sensitive, and some of them openly ask learners to acknowledge emotional reactions to the case. These questions are rooted in the belief that awareness of bias and emotion in medicine is an important step to providing optimal, unbiased, and compassionate care.

- *Creative Problem-Solving Questions:* Most cases include one creative problem-solving question (labeled as such). Like self-reflection questions, these questions do not have "right" answers; instead, they ask learners to consider some of the more intractable problems facing health care today—typically problems that have not been solved despite considerable attention. They present opportunities for learners to brainstorm new ways of thinking about and approaching these issues. Learners may find them frustrating, or they may respond with proposals that have already been tried and failed. The goal of such questions is to challenge learners to explore systemic solutions.

2.3.6 Answers to Background Questions

The sixth subsection contains in-depth answers to each of the background questions. The answers are intended to be provided to instructors as a means of checking the effectiveness of learners' self-directed learning, but instructors may also consider providing them to learners as a form of feedback. Answers are particularly useful for instructors with limited expertise in the specific topic of an individual case.

2.3.7 Responses to Discussion Questions

Because each case is complex and multilayered, a comprehensive analysis of all the potential issues is beyond the scope of this textbook; however, the seventh subsection contains thorough responses to the discussion and follow-up questions. References are provided to support factual arguments. Where multiple points of view exist within medicine and bioethics, counterarguments are provided.

This section also includes three types of boxes:

- *Teaching Tips:* Teaching tips are targeted at instructors only (not learners). They provide insights and suggestions for how best to facilitate a discussion of the material, including advice for how to manage group dynamics and encourage participation from quieter learners.
- *Personal Perspectives:* Personal perspectives are brief excerpts from previously published firsthand narratives that relate to the case. They are written by patients, family members, physicians, or other healthcare practitioners and may be drawn from memoirs and anthologies, newspapers and newsmagazines, or narrative sections of academic journals. They can be read aloud or otherwise shared with learners during the discussion, or instructors may wish to obtain and assign the full narrative to learners before or after the discussion. The excerpts are intended to provide deeper insight into the lived experiences of those who have faced similar issues.
- *Case Conclusions:* Some cases, particularly those based upon real events, conclude with additional information about "what really happened" or about events occurring after the timeline of the case ends. In such chapters, the decision to withhold this information until the end is deliberate and intended to avoid prematurely biasing the discussion toward a specific conclusion.

2.3.8 References

All references cited in the chapter are listed in the eighth subsection.

2.3.9 Further Reading on This Topic

Suggested further reading is included in the ninth and final subsection of each chapter.

2.4 How to Use a Case

2.4.1 A Note About Flexibility

The cases in this book have been designed to allow for maximal flexibility. They can be used across many educational settings, by faculty from diverse professional backgrounds, for learners at any level and in any health professions program, and with a variety of pedagogic strategies. They can provide a framework to develop and support a comprehensive course in bioethics and population health for students of medicine and the health professions, as has been done at our medical school. Alternatively, individual cases can supplement other courses, clerkships, or clinical rotations. (The tables in this chapter can be used to identify cases involving specific specialties, diagnoses, or patient characteristics.) Although originally designed for small group discussion and active learning, they can be readily adapted for online instruction, for reflective or analytical writing, or for independent study, as well as for formative or summative assessment (as a written or oral examination or as a script for standardized patient encounter).

Before adopting these cases, course directors and others responsible for curricular design will want to consider the bigger picture of the curriculum. The following questions do not have answers in the medical or education literature to support best practices; however, they are necessary considerations when designing a course in bioethics, population health, and the social sciences.

- **What is the goal of such courses?** Is the purpose to merely introduce learners to the core principles, in the hope that they will recognize their relevance later in training, to provide a foundation upon which to build clinical skills, to influence or guide the type of clinician that learners become, or to support some other, more concrete objective?
- **What should be taught?** Should students be able to choose from a selection of courses in disciplines such as epidemiology, medical law, or health humanities, fitting them together to create a personalized education? Or should they be required to have a single uniform curricular experience? Should courses provide a broad survey or a deep dive into "selected topics"?
- **What pedagogical approach is best?** Should the primary mode of instruction be passive (e.g., lectures, readings), active (e.g., small groups), or a combination? How much, if any, self-directed learning should be included? Is there a role for experiential learning?
- **How should content be organized?** Should it be presented by discipline (bioethics, law, research methods, etc.), organ system, patient population (elderly, LGBTQ, refugees, poor, etc.), specialty, or other approach?
- **Who should teach?** Are such courses best led by physicians, clinicians providing interprofessional perspectives (e.g., nurses or social workers teaching medical students), faculty with expertise in specific disciplines (e.g., bioethics or public health), or peers or near-peers of learners?
- **When should the content be taught?** Should the content begin on the first day of a health professions program or concurrent with the start of clinical learning experiences? Should some exposure to such material be required for admissions? Should it be taught concurrently with other learning experiences or during dedicated blocks, such as intersessions?

Once these questions have been addressed, course directors will be able to move forward with the adaptation of the cases to best meet the needs with their learners. The next sections address practical questions regarding implementation of cases for course directors and for facilitators of case discussions.

2.4.2 For Course Directors and Others Responsible for Curricular Design

1. *What qualifications are necessary for an instructor to facilitate a discussion of a case?* In general, we have found that facilitators do best if they meet at least two of the following three criteria *and*—in most cases—if they are able and willing to pursue continuing education and professional development in the missing criterion:
   (1) *Prior experience teaching in small group settings and using a case-based learning approach.*
   We strongly recommend that facilitators without this experience either be paired with an experienced facilitator who can provide mentoring and regular feedback or participate in training in small group facilitation and case-based learning prior to teaching.

(2) *Expert knowledge in at least one of the disciplines represented here* (e.g., bioethics, public health, health humanities, or related social sciences, such as medical anthropology or medical sociology).

In our experience, facilitators do not need expertise in all disciplines, but it is important that they have depth of knowledge in at least one discipline and experience beyond the classroom. For example, a Master of Public Health degree is not necessarily sufficient if the facilitator does not have related experience beyond completion of the degree program. It has been our observation that faculty who are truly expert in one discipline are more comfortable recognizing and acknowledging their limitations in other disciplines and taking steps to learn more (e.g., assigning a learner to research a question that the facilitator was unable to answer).

(3) *Practicing clinicians (e.g., physicians, nurse practitioners, etc.).*

If the design of the course allows for facilitators to teach in pairs, it can be beneficial to pair a skilled educator who is expert in bioethics, public health, or a related discipline with a clinician. Students entering the health professions, particularly early in their training, before they have acquired significant clinical exposure, often have a strong desire to hear "what it is really like" from clinicians. On the other hand, facilitators in these circumstances need to share their own experiences judiciously and avoid permitting small group sessions to turn into question-and-answer sessions or career advising workshops.

Clinicians do not need to have the same professional background as the learners. In our course, we have had several registered nurses, nurse practitioners, and nurse-midwives effectively teach medical students, offering interprofessional exposure that is otherwise rare. We also had many hospital social workers teach. All facilitators were most successful if they also met the above criteria.

Finally, commitment to teaching is extremely important. Small group facilitation and case-based teaching are acquired skills. Comfort and confidence facilitating these cases is built incrementally. Learners report far more satisfaction with facilitators who are able to "put away the script," but this requires practice and familiarity. Whenever possible, we encourage course directors to select facilitators who are able to commit to teaching several or all course sessions and to returning to teach the same course year after year.

2. *What educational background is necessary for learners to approach these cases?*

The cases were initially developed for first-year medical students. They have also been used with second-, third-, and fourth-year medical students and with medical residents. Even visiting high school and college students have been able to actively participate in case discussions. Cases are not limited for use with future physicians; they can be readily adapted for other health-profession students engaged in direct patient care, as well as for students of public health and healthcare policy. The answers and responses are written to be comprehensible by a learner with little to no medical knowledge, though they expect that the learner is cognitively and emotionally ready to acquire such knowledge *and* has a sincere interest in doing so. Furthermore, even many healthcare practitioners in practice will find something of value in a case discussion.

3. *How many learners are optimal for a small group discussion?*

Instructors have successfully facilitated these case discussions in groups with as few as 6–8 learners and as many as 16. As group sizes exceed 18–20 learners, it can be difficult to ensure that all learners have an opportunity to participate meaningfully. Active participation—responding to discussion questions, asking questions of peers, and responding to questions from peers—is a critical part of the learning experience. Learner satisfaction tends to be higher in smaller groups. Some research in group dynamics and learning has suggested that groups of 5–10 learners are ideal for generating and criticizing ideas. Such groups are small enough to avoid splintering of the group but large enough that strong learners can "draw in" weaker learners. Maintaining the group's focus may become more difficult in groups of more than 10 learners; it is also more difficult to sustain a sense of support and safety within the group. Furthermore, reticent learners are more easily able to "hide" (i.e., not participate meaningfully), and more outspoken learners may dominate the discussion, leading to an artificial sense of group consensus or "groupthink" [1].

In our experience, groups of medical students with fewer than eight learners had insufficient creative resources to independently analyze the case and required more input from small group facilitators; however, this was not necessarily true with more advanced learners (e.g., resident physicians). If larger groups are necessary (12 and above), we suggest taking opportunities to separate learners into pairs or threes for brief discussions before reporting back to the whole group. Alternatively, learners can be asked to write and submit responses to the self-reflection and creative problem-solving questions (briefly, during the session itself or in more depth as a take-home assignment), allowing everyone to engage with these questions.

4. *How should the background questions be approached?*
As noted in the previous section, the background questions are designed for independent study, preparation, and research but can be adapted in various ways.

If using the background questions for self-directed learning, any combination of the following approaches can be considered:
- Learners can be assigned one or more questions (up to all eight).
- Learners can work individually, in pairs, or in small groups to answer the questions.
  - We suggest individual or pair approaches if each learner only has one or two questions to research.
- Learners can present their findings to their peers and facilitators.
  - Presentations can be oral (taking place within the small group itself, just prior to the case discussion) or written (submitted to facilitators or shared with peers via an online bulletin board discussion).
- If findings are not shared among all learners or not assessed for accuracy by faculty, we suggest that facilitators share the answers to the background questions of the chapter with the learners prior to the case discussion.

We have had success in assigning medical students to work together in pairs to research the answer to one question per case. They are asked to bring notes and scholarly current sources to class; they cannot use slides or other visual aids and are asked not to read directly from their notes. They then have 2 minutes to succinctly present key findings to their peers. At the start of the course, we explain that we fully expect they will have discovered more material relevant to the question than can be covered in 2 minutes. We ask them to learn how to prioritize information and speak concisely—skills that will become acutely relevant when they move on to clinical clerkships in their third year. However, we also tell them that the "excess" information they uncover is very likely to be relevant during the case discussion. Facilitators are prepared to call on learners whose background question relates to the current discussion question. This strategy is particularly effective in promoting active participation by all learners because it means that one or two learners typically have access to relevant information that no one else has. For learners who are reluctant to speak if they are unsure that they are adding meaningfully to the discussion, this can be crucial.

Alternatively, the background questions can be used:
- As an outline to guide preparation of a lecture given to all learners prior to the case discussion
- As a source of references for readings assigned to learners prior to the case discussion
- As a required reading prior to the case discussion

5. *How much time should be allotted for each case?*
As discussed previously, cases are designed to be flexible. A skillfully facilitated, thorough exploration of the content in the additional case information and questions for discussion section can be undertaken in 2 hours;

however, course directors can also select key aspects of the case to emphasize (or permit small group facilitators to make such choices), shortening the discussion to about 1–1.5 hours. Most facilitators will notice variation between cases and groups of learners; some will naturally engage more enthusiastically with certain topics, and the presence of a learner with personal or professional experience with a topic may lead to a deeper (and longer) discussion.

Some course directors may choose to discuss both paired cases in a 3-hour session or to divide the discussion of a single case over two or more sessions, allowing learners to write reflectively or pursue other self-directed learning activities between the two sessions.

It should be noted that the amount of time required depends in part on the group size; larger groups require more time to give all learners a chance to speak, but too much time, coupled with too many learners, can result in learners repeating what has already been said. Our students have reported frustration and boredom when this occurs, as well as misperceptions that the sessions are overly long—when, in fact, extra time is necessary to allow all learners to articulate their reasoning.

6. *Should learners be given access to the answers to background questions and responses to case discussions?*
This is at the discretion of the course directors. In addition to the above, cases can also be adapted for other pedagogical formats, including essay questions, oral examinations, and simulated patient encounters. Further information about assessment is provided in ▶ Chapters 25 and 26.

2.4.3 For Discussion Facilitators

1. *What do facilitators need to know to be effective?*
The most important skills to successfully implement these cases are those of small group facilitation. With adequate time to become familiar with the case, even facilitators who have little knowledge regarding the subject matter can be very effective teachers. We encourage facilitators to seek out opportunities for professional development in this area, particularly those which include formal observation of the facilitator and feedback on performance.

Facilitators need to allow themselves sufficient time to become truly comfortable with the case—this is discussed in the next question (see "How Should I Prepare?" below)—but they do not need to be experts in every aspect of every case; in fact, such breadth and depth of knowledge is highly improbable. Facilitators do need to be willing to recognize gaps in their knowledge and be willing to acknowledge such gaps. They also need patience and willingness to allow for occasional awkward silence.

Finally, facilitators need to be willing to practice the same skills of self-reflection and lifelong learning that we ask of our students, trainees, or other learners. They need to be self-critical and humble in order to identify gaps in their own knowledge, as well as curious and willing to seek out new information to fill those gaps. They need to be willing to learn from their learners and—if facilitating with a partner—to respect their partner's unique expertise and contribution to the group.

2. *How should facilitators prepare for a given case discussion?*
Preparation is key to a successful discussion. Depending on level of experience with this style of teaching and familiarity with the issues raised by the case, this process may take as little as 1–2 hours or as long as 6–8 hours. We recommend the following approach.

If a facilitator will be teaching with a partner (another faculty member or a learner who will be serving as discussion leader), we recommend that both facilitators meet prior to the session to review the case and decide on roles. Often, one facilitator will lead the discussion, while the other observes and assesses learner participation. Alternatively, facilitators may want to divide the case by areas of expertise. For instance, a facilitator who is a clinician might lead sections of the case that involve the patient-healthcare practitioner interaction, while the second facilitator leads sections involving research or

> **Box 2.1 Preparing to Facilitate a Case**
>
> 1. Start by reading through the learning objectives, case background, background questions, and discussion questions. Take time to think about how you might answer these questions yourself, including the self-reflection questions. Consider your areas of expertise as well as any gaps in your knowledge. Think about the relationship between the learning objectives and the case. Does the case resonate with your previous experience in health care? Which parts of the case seem most interesting or provocative?
> 2. Then read the answers and responses sections carefully. Were you surprised by anything? Do you disagree with any of the arguments presented? Did you become aware of any of your own biases? Was anything confusing or unclear? Now is the time—well before facilitating a discussion—to think about and answer these questions, to the best of your ability. Seek out additional resources, including discussions with other faculty, to better understand concepts that are new to you.
> 3. Read the case again, at least once, paying attention to the teaching tips and the different types of questions. Return to this chapter for reference if you have forgotten the purposes of each question type.
> 4. If your course director requires learners to complete any additional readings or other preparatory work prior to the session, you should do the same. Similarly, if the learners attend or otherwise view lectures prior to the session, it is helpful if you also attend the lecture or familiarize yourself with its content.
> 5. Finally, read through the entire case and annotate it. Highlight questions that you think are particularly important for your learners and want to be certain you use. You may find that you think a question might better serve your learners if asked earlier or later—you can make notes to remind yourself to ask the questions in a different order. If the case includes a major concept that you have not taught before, we suggest practicing. Solicit feedback from colleagues who are more familiar with it. Try to anticipate areas where learners might struggle to understand a concept or where discussion might be harder to stimulate.

public policy. Facilitators who know each other well and have taught together before might choose a more free-form approach.

3. *What is the facilitator's role during the case discussion?*
   The facilitator's role is primarily to keep the discussion on track by sharing case events and asking questions at appropriate times. The facilitator may also need to actively manage interpersonal dynamics at times, depending on the learners and their relationships to one another. Facilitators, for instance, should directly seek responses from learners who seldom participate, while encouraging more vocal learners to make space for others by listening or asking questions of their peers. Occasionally, facilitators will need to intervene privately, outside of the classroom, when it becomes clear that two or more learners are not able to interact productively in the small group. At other times, it may be helpful to intervene more publicly when the problem lies with group dynamics (e.g., when one subgroup of students consistently interrupts and speaks over another subgroup of students) or when one or more students engage in behavior that is unprofessional, disrespectful, or hostile (e.g., comments that are clearly biased against one or more groups). In the latter case, we believe that it is important to help facilitators to feel comfortable directly addressing such behavior when it arises; it is important to send a clear message to all students that such behavior is unacceptable and will not be tolerated.

   It is appropriate and sometimes necessary to correct misinformation, but it is beneficial to give learners a chance to correct one another first: they often will. Facilitators should avoid lecturing on any topic and should generally not speak for more than a few minutes at a time. If a learner appears to be struggling with an issue that is tangential to the case's learning objectives, we recommend having the facilitator ask the learner to meet with the facilitator or course director after class for further discussion, to avoid derailing the entire session. Alternatively, if many learners appear to have the same source of confusion (e.g., they are having difficulty understanding a concept in biostatistics that is only a small part of the case), the facilitator might consider providing them with supplemental materials (e.g., online modules, readings, problem sets) or holding a separate review session to teach that concept. If all learners seem to agree upon a point for which there is, in fact, substantive debate, the facilitator may want to share the counterargument and ask learners to respond to it. This approach may also allow learners with viewpoints in the minority to feel comfortable speaking up.

Finally, facilitators should use caution when interjecting their personal opinions or experiences. Such interjections seem to fall into three categories: (1) facilitators who are clinicians have professional experience with situations similar to those in the case; (2) facilitators have personal experience (themselves or their close friends or family) with situations similar to those in the case; and (3) facilitators have strongly held values, beliefs, or opinions regarding situations similar to those in the case. Each category has potential risks and benefits; thus, we do not suggest completely avoiding such interjections but approaching them warily.

For instance, in the first category, learners often respond enthusiastically to stories of clinical experience. Facilitators sharing such stories have opportunities to share things done well and lessons learned from things done poorly and to model reflective practice for learners. On the other hand, learners may follow up with rapid-fire questions about "what it's like in the real world" that can lead the facilitator into storytelling mode and away from the learning objectives for the case. In the second category, facilitators who are willing to share a personal experience are modeling openness and demonstrating trust in their learners, which can engender the same behavior from learners; however, it is important to avoid creating an environment in which learners simply take turns sharing personal stories, without engaging in the critical thinking and reasoning that is central to this book. In the third category, learners may benefit from hearing facilitators explain how they accept differences between patients' values and their own or from hearing facilitators share the reasoning behind their position on a difficult topic (e.g., aid in dying). Again, caution is required, because facilitators want to encourage learners to feel that they can disagree and that the facilitators' conclusion (in such cases) is not the only "right" one.

Facilitators should model the dialogic skills they expect of the learners, as illustrated in the next question ("Should I set ground rules?"). In most discussions, learners will ask at least one question that is outside the facilitators' scope of knowledge and expertise. Facilitators who are willing to admit when they do not know the answer can contribute to the creation of an active learning environment. Depending on the question and context, the facilitator might suggest that the learner look up the answer on a laptop computer or tablet (if available and if the question asks for readily available facts). If that is not appropriate, facilitators can decide whether it is something the learner should research and report back regarding or whether the facilitators should investigate and update the class later.

4. *Should I set ground rules for the discussion? Which ones?*

Yes—we strongly recommend this. If facilitators have extended periods of time (e.g., semester, year, etc.) with the same learners, we suggest asking the learners to propose and agree upon ground rules. This approach lets them take ownership of the group and their learning environment. Otherwise, in the first few minutes of the session, facilitators can quickly review preset ground rules.

The following is one possible set of ground rules, used by faculty in our course:
- Listen carefully, with an open mind to the contributions of others.
- Ask for clarification when you do not understand.
- Use evidence and logic when challenging others' assertions and ideas.
  - Be willing to change your mind when others demonstrate errors in your evidence or logic.
- Critique ideas and positions, not people.
- Explain the relevance of issues when it is not obvious to your peers.
- Be concise and give others time to speak.
  - Only repeat a point when you have something to add.

The content of this book can be sensitive. Both facilitators and learners may have intimate personal experiences with some of the issues discussed, and in some settings, it may be appropriate to include confidentiality among the ground rules. The next section,

"Trigger Warnings and Brave Spaces" addresses these concerns in more detail.

5. *What are some common challenges that facilitators face? How can facilitators overcome them?*

Facilitators commonly face some challenges when beginning to facilitate discussion of these cases. Barriers and strategies for overcoming them are presented in ▶ Box 2.2.

> **Box 2.2 Overcoming Barriers to Successful Facilitation**
>
> 1. *Being insufficiently prepared for the session*
>   - Reread Question 2 above.
>   - Begin preparation at least 1–2 weeks prior to the session, and allot 1–2 hours over 3–4 days for the preparation process so that you have time to ask questions to the course director if necessary.
>   - Review the material again the day before the session.
>   - Do not assume that preparation will be quick or easy simply because you have prior experience in the topic of the case.
> 2. *Failing to recognize and acknowledge personal biases*
>   - Inquire about your institution's opportunities for professional development related to recognition of biases. Consider taking an implicit association test and discussing the results with colleagues who have experience in this area.
>   - Take time during the preparation process to consider aspects of the case that resonate with prior experiences and consider whether you might be biased because of those experiences.
>   - Pay special attention to cases in which your values and beliefs may not align with the hypothetical patient or family, and consider discussing with the learners how you reconcile this and avoid values imposition.
>   - Personal beliefs and professional obligations are not always synchronous, but learners—as future healthcare practitioners—and facilitators, especially current healthcare practitioners, need to recognize the limits of conscientious objection (e.g., obligation to refer a patient for an abortion to another provider if the healthcare practitioner does not perform abortions, regardless of personal beliefs). When preparing for a session, observe where controversy or multiple viewpoints are noted and where there is ethical and/or legal consensus, and avoid spending too much time discussing areas simply because your personal beliefs are not congruent.
>   - Be aware of ways in which medical educators inadvertently perpetuate shame, stigma, bias, or stereotyping in health care. Review ◘ Table 2.3 for examples.
>   - When learners bring issues of this nature to your attention, listen, learn, and, when appropriate, apologize.
> 3. *Feeling compelled to cover every aspect of the case in a session that is too short to allow for this*
>   - Reread ▶ Chapters 1 and 2 of this book! It is not necessary to cover every aspect of the case.
>   - With good preparation, you should be able to identify aspects of the case that you will skip if limited on time, considering (1) the learning objectives, (2) what your learners have discussed in past sessions, and (3) where your learners are in their educational journeys.
>   - Learners will often spontaneously bring up questions, issues, or arguments that appear later in the chapter. Discussions are rarely linear and there are multiple paths to elucidating or resolving the issues in each case. With adequate preparation and attentive listening, you will recognize this when it is happening. In some cases, it is appropriate to ask a learner to postpone the line of inquiry until later in the case, for example, when a crucial piece of information in the case events has yet to be revealed. In most cases, however, it is best to allow the discussion to flow naturally, simply noting which of the later questions have already been answered.
> 4. *Reading aloud from the responses to discussion questions*
>   - Again, preparation is key, and practice may be helpful to feel comfortable explaining less-familiar concepts. Try to anticipate questions and write succinct explanations in your own words.
>   - If asked a question that you are not comfortable explaining without reading from the case, give other learners an opportunity to answer first, and then consider (1) suggesting that learners research it or (2) referring them to the course directors for further explanation and discussion. Acknowledge that the question is out of your scope of expertise.
> 5. *Allowing some learners to monopolize the discussion*
>   - As noted in Question 1, small group facilitation is an

Approaches to Using This Book

> art. Deliberate practice with reflection and feedback is necessary to develop group management skills.
> - Memorize your learners' names quickly so that you can call on them individually.
> - Prevent the problem by including hand-raising in the ground rules or using other nonverbal gestures. Some groups use an object that is passed from speaker to speaker, while others have adopted hand signals based on whether the next speaker wants to address a point currently being made or speak to a new point.
> - Pay attention to the layout of the room and assure that all learners are sitting in a circle. Restructure the room and furniture as much as possible, and rearrange seating if dominant learners tend to take optimal seats, while pushing more reticent learners to the periphery. It may help if the facilitator moves around the room, sitting next to a different learner during each session. If a circular arrangement is truly not possible, ask learners to rotate seats so that dominant learners are not always sitting in the front.
> - If a learner resists intervention within the group setting, meet privately with the learner to discuss the issue. In our course, students' participation assessment includes items regarding their listening skills and willingness to give space to other students, and students are informed of this at the start of the course (further discussed in ▶ Chapter 25).

Table 2.3 Examples in which medical educators inadvertently perpetuate shame, stigma, bias, or stereotyping in health care

Indicator	Examples that may promote stereotypes, bias, shame, and stigma
Race	Referencing associations between race and disease incidence without context Assuming that racial-minority patients are poor or uneducated
Ethnicity	Assuming Latinx patients are undocumented immigrants/migrant workers Stating or implying that all patients from a particular culture participate in certain practices or reject certain medical interventions (e.g., "Muslim women are not permitted to be examined by male physicians")
Gender	Assuming that families are nuclear, with heterosexual, married parents and biological offspring Including maternal age as a risk factor for diseases/conditions while failing to list other risk factors that are epidemiologically more important Assuming that gender is binary Failing to use preferred pronouns for gender-nonconforming patients, including those in clinical vignettes
Sexuality	Using stigmatizing language in clinical vignettes or discussions of history-taking such as "The patient admitted to having sex" Taking a sexual history that does not account for the full spectrum of sexual identities and encourages categorization Labeling sexual identities and behaviors as "high risk" Conflating gender identity with sexual orientation
Disability	Failing to recognize that most people with disabilities regard their quality of life as comparable to those without disabilities Using diagnosis-first language (e.g., "autistic male" vs. "male with autism") Assuming that preventive health is not as important to patients with disabilities
Mental health	Assuming that all mentally ill patients are potentially violent Failing to recognize how some might value neurodiversity while still desiring treatment for symptoms that cause suffering

(continued)

Table 2.3 (continued)

Indicator	Examples that may promote stereotypes, bias, shame, and stigma
Age	Focusing only on declining health/quality of life and need for advance directives/limitations of care Ignoring positive portrayals of aging and geriatric care Neglecting consideration of sexual health at all ages
Substance use	Using language of personal responsibility/self-control to discuss addiction, rather than treating it as a disease Referring to a patient as an "addict"
Weight	Emphasizing personal responsibility in discussions of obesity at the expense of important genetic/epigenetic, social, and structural risk factors Assuming that all overweight and obese are unhealthy, when it is more complicated biologically
Immigration status	Focusing only on language barriers in clinical encounters between physicians and patients who are immigrants
Poverty	Presenting race as a risk factor for disease occurrence or outcome without explaining role of poverty, access to health care, etc. Presenting poor people as lazy or lacking in character

2.5 Trigger Warnings and Brave Spaces

Trigger warnings are statements presented at the start of a piece of content alerting the user to the fact that it contains potentially distressing material. They have been used, with some debate, throughout higher education, largely in classes that contain graphic depictions (e.g., in literature or film) or discussions of material that it is likely to cause severe distress for people who have experienced previous traumas, such as child abuse, sexual violence, or military combat [2].

Many of the cases in this book involve sensitive and potentially triggering content. Although we have chosen not to place individual trigger warnings on each chapter, the nature of the content is evident from the chapter titles and the tables. We encourage faculty to carefully consider whether they wish to use case-specific warnings. Our decision was rooted in the most fundamental aspects of medicine and health care—particularly, the reality that someone entering medicine and the health professions will very soon be asked to take care of any patient who walks through the door and to put that patient's welfare above all others, including, for the most part, the healthcare practitioner's own. That patient might be in the emergency department after experiencing intimate partner violence; on the psychiatric unit after a suicide attempt; in the labor and delivery unit, giving birth for the first time with a history of female genital cutting; or in any number of other healthcare settings. No specialty is immune from encounters with trauma and violence. It is our belief that students need to understand and accept this commitment to the profession. In our course, we do make allowances for the rare student who is actively in treatment with a psychiatrist, psychologist, or other therapist, related to a history of trauma that appears in one of the cases, permitting such students to defer the case until later in their training.

The ideal setting in which to facilitate discussions of these cases is a "brave space"—distinguished from a "safe space" by the acknowledgment that there are risks inherent in these discussions and that optimal learning requires successful navigation, not avoidance, of such risks. In a brave space, "those who enter…have the courage to… take risks because they know they will be taken care of—that painful or difficult experiences will be acknowledged and supported, not avoided or eliminated" [3].

Conclusion

The approaches described are not comprehensive, but they are intended to both provide a solid foundation for educators to incorporate the cases into their bioethics, public health, and social science curricula and to illustrate ways in which the cases can be adapted to meet the needs of institutions and the learners they serve.

References

1. Group dynamics – How group size affects function [Internet]. London (UK): London Deanery; 2012 [cited 2018 Feb 28]. Available from: http://www.faculty.londondeanery.ac.uk/e-learning/small-group-teaching/group-dynamics-how-group-size-affects-function.
2. Manne K. Why I use trigger warnings. The New York Times [Internet]. 2015 Sept 19 [cited 2018 Feb 18]. Available from: https://www.nytimes.com/2015/09/20/opinion/sunday/why-i-use-trigger-warnings.html.
3. Cook-Sather A. Creating brave spaces within and through student-faculty pedagogical partnerships. Teach Learn Together High Educ. 2016;1(18):1.

Core Principles in Ethics and Public Health

Contents

Chapter 3   "How Many of These Surgeries Have You Done?" – 39

Chapter 4   "Can't Stop Coughing (But I Need to Get Back to the Shelter by 6)" – 61

"How Many of These Surgeries Have You Done?"

Robert S. Olick

3.1 Background Questions – 40

3.2 Additional Case Information and Questions for Discussion – 40

3.3 Answers to Background Questions – 42

3.4 Responses to Discussion Questions – 51

References – 59

Learning Objectives

1. Describe the ethical and legal models of informed consent and begin to apply these models at the bedside.
2. Analyze the nature and extent of the duty to obtain informed consent.
3. Begin to appreciate the complex relationship between ethics, law, and medicine.
4. Begin to demonstrate the ability to take the perspective of the patient or family in the healthcare practitioner-patient-family interaction.

> **Case Background**
>
> *You are a first-year surgical resident. You have been instructed by the attending surgeon to get consent today from a patient who was admitted overnight. The patient, Patrice West, is a 73-year-old woman who was admitted after an abdominal aortic aneurysm (AAA) was identified. She is currently stable but urgently requires a procedure to repair the AAA. You have never seen an AAA repair; however, you think you understand enough of the technique to give Ms. West a simple explanation. You feel energized to be able to assist in a surgery that is new to you.*

3.1 Background Questions

1. What types of information are physicians and other healthcare practitioners legally required to tell patients in the informed consent process? Find a court case from the last 20 years that addresses informed consent in your state. What was the decision in that case? What is the standard of disclosure applied in the case?

2. Are physicians legally required to provide information about their professional experience; for example, how many times the surgeon has done the recommended procedure? Are they legally required to disclose HIV (human immunodeficiency virus) status, history of drug or alcohol abuse, disability, financial conflicts of interest, or other personal characteristics?

3. Under the Affordable Care Act, the Centers for Medicaid and Medicare Services developed the "Physician Compare" Website. What was its intent? Quality measures were included in the available data. What are the limitations of these data? Can you find information about local surgeons on this site?

4. What is the definition of a surgical complication? Identify the known rates of complications for an abdominal aortic aneurysm (AAA) repair.

5. What is the ethical model of informed consent? How much is it reasonable to expect patients to understand? What is health literacy?

6. How does the cultural background of the patient affect the informed consent process? Identify two examples. Drawing from these two examples, how should the informed consent process be modified to accommodate patients cultural backgrounds?

7. What are the four recognized exceptions to informed consent, that is, situations in which informed consent is not required? Describe each of these exceptions.

3.2 Additional Case Information and Questions for Discussion

When you enter Ms. West's room, her mother, her brother, and three adult children are all crowded into the hospital room. You hand Ms. West the standard procedural consent form for an open abdominal surgery and start the conversation. Before you can say much, Ms. West's family starts to interrupt with questions. Her daughter wants to know how many of these procedures your attending surgeon has performed. Her brother wants to know whether you will be in the operating room (OR) and if so, what your role will be. Her older son says while in the waiting room, he did an Internet search of AAA repairs and wants to know why an endoscopic repair is not being recommended. Her younger son asks simply, "Could my mom die?"

1. How would you approach this situation? What would you say or do first in response to the family's questions?

2. Should you have told the attending surgeon that you had never seen the procedure performed and therefore should not be the person to obtain consent?
   (a) Self-reflection: What do you think you *would* do in real life? Do you think you would speak up?

Contemplating the family's first question, you are unsure how many similar procedures the surgeon has performed before, but you know that she completed fellowship training only 6 months prior.

3. Does this change your view about obtaining informed consent?

4. Do surgeons, or healthcare practitioners in general, have a duty to inform patients of their prior experience performing the procedure?
   (a) Do you think that reasonable patients would want to know how often their surgeon has done the recommended procedure?
   (b) Should patients always choose the most experienced provider? Why or why not?
   (c) You were asked a direct question about surgical experience. What would you say?
   (d) What are the ethical and legal consequences of lying?

5. How would medical education be affected if patients refused to see healthcare practitioners in training and junior healthcare practitioners because they lack experience?

6. How should you respond to the question, "Could my mom die?"

Public reporting of quality measures in patient outcomes has become more common. As an example, since 1994, the New York State Department of Health has collected and made available patient mortality data for interventional cardiologists [1]. Some health systems have chosen to generate public "report cards" for their organizations.

7. Do you think that there is a duty to share surgeon-specific morbidity and mortality rates for procedures, allowing patients to compare surgeons?
   (a) If so, to whom would you compare?

8. Patients may want to know other "personal characteristics" such as drug or alcohol abuse, physical or mental impairment, or HIV status. Is there an affirmative duty to share this information in the informed consent process?
   (a) If the patient asks, are you obligated to provide an honest answer?
   (b) Creative Problem-Solving: Are there better alternatives that balance patient welfare and healthcare practitioner privacy?

9. Self-reflection: Of the above (provider experience, provider or institutional outcomes for a procedure, provider personal characteristics), what, if anything, would *you* want to know, for yourself or your family member?
   (a) Does it, or should it, matter if some personal characteristics (e.g., a prior history of alcohol abuse or HIV status) might trigger patients' biases, rather than lead to more informed decisions that produce better outcomes?

10. If your answers to any of the above questions were affirmative, do you think the law should be changed to reflect this?

About 3 months earlier, your co-resident admitted a patient with the same condition as Ms. West. You were not involved with that patient's care but you heard the story from your co-resident. You heard that the patient's daughter was a prominent physician in New York City and that the patient was privately transferred to a larger hospital to consult with a nationally known vascular surgeon. You do not know very much about Ms. West's family—just that she has Medicaid, not private insurance.

❓ 11. **Is there an obligation to transfer Ms. West to a better-equipped hospital with a more experienced surgical staff?**
   (a) Should a decision to discuss transfer take into account whether the family has the resources and ability to travel to a more distant center to see a specialist?
   Is it important that Ms. West has Medicaid, rather than private insurance?

As you attempt to tactfully navigate the family's questions, Ms. West's mother looks increasingly distressed. She finally says that she is 94 years old and "in her day" it was not uncommon for residents to be allowed to operate unsupervised on poor black patients like her family.

❓ 12. **How should you respond?**

❓ 13. **What can you, as a healthcare provider, do to address medical mistrust?**

Residents are much more closely supervised now and institutionally sanctioned discrimination no longer openly occurs. However, many residents still seek out academic programs and teaching hospitals in underserved communities because they feel they are allowed to "do more" than in private hospitals serving more affluent and educated populations.

❓ 14. **Is this a just system? Why or why not?**

❓ 15. **Self-reflection: If you were having surgery, would you allow a resident to assist? What if the patient was your parent? Your child?**

3.3 Answers to Background Questions

✅ 1. **What types of information are physicians and other healthcare practitioners legally required to tell patients in the informed consent process? Find a court case from the last 20 years that addresses informed consent in your state. What was the decision in that case? What is the standard of disclosure applied in the case?**

The doctrine of informed consent can be understood as the union of two concepts: (1) disclosure of information so the patient possesses relevant facts and (2) obtaining the patient's voluntary authorization to proceed with diagnosis or treatment—in other words the patient's consent. This question concerns the physician's or other healthcare practitioner's fundamental duty to disclose information to patients. Additional features of informed consent are addressed in question 2.

It is well-established in law and ethics that in the physician-patient relationship, the physician must make certain "core" disclosures to patients as part of the process of enabling patients to make informed decisions to consent to or refuse treatment. Physicians must provide patients with relevant information about (1) the nature of their medical condition, including diagnosis and prognosis; (2) the risks, benefits, and burdens of the recommended treatment or procedure; and (3) the risks, benefits, and burdens of reasonable alternatives, including the option of no treatment. In turn, informed decision-making means that patients understand and reason about this information in order to reach a decision to consent to (to authorize) or refuse the treatment or intervention. Further, reasoning to a decision requires that the patient have a set of values that shape their goals for treatment. Disclosure is an affirmative duty that recognizes the knowledge and power divide between physician and patient. It would run counter to patient autonomy and well-being to expect patients to present their physicians with a long list of questions or to limit information sharing to the physician's responses only to specific patient queries [2].

The information to be disclosed is what is reasonable and relevant under the circumstances. Generally, physicians are not required to disclose very remote risks that are highly unlikely to occur. However, the law in some states does require telling patients about very serious risks of death or severe disability, even if these risks are highly unlikely. The parameters of what is reasonable and relevant mean that it is not necessary to develop and present to patients a laundry list of all conceivable risks, though some surgical consent forms give that appearance.

The phrase "standard of disclosure" refers to the nature and scope of information physicians must by law tell their patients. Most of informed consent law initially emerged from case law. In these cases patients suffered a bad outcome from a treatment or procedure and, looking back, claimed that their injury resulted from an undisclosed risk of the procedure, and further that had they been informed of the risk they would have chosen a different approach to treatment or refused the procedure and treatment altogether. In order to determine whether the physician had a duty to disclose the risk information in question, courts had to decide on a standard of disclosure that defines the scope of the duty. Two legal standards have emerged nationally. The first and older standard is the professional, physician-oriented standard that requires physicians to disclose information that is reasonable and customary for physicians in the same or similar circumstances (e.g., the same specialty, same type of patient) to tell patients. The more recent standard is the objective reasonable patient-oriented standard that requires physicians to disclose information that a reasonable patient would want to know under the circumstances (e.g., the same or similar medical condition and treatment options). As one can imagine, in a contested case, the professional standard tends to favor physicians, as it would be other physician-experts who are called upon to attest what is customary and standard practice. By contrast, the reasonable patient-oriented standard is more favorable toward patients, as it would be lay people imagining themselves in the patient's place who are called upon to attest or determine (perhaps as members of the jury) what a reasonable patient would want to know under the circumstances. At present, states across the country are evenly divided: about half of states follow the professional standard, and about half of states follow the reasonable patient standard [3]. However, in practice physicians more often adopt the reasonable patient standard of disclosure because it is more closely aligned with the ethical model of informed consent, discussed below.

2. **Are physicians and other healthcare practitioners legally required to provide information about their professional experience; for example, how many times the surgeon has done the recommended procedure? Are they legally required to disclose HIV (human immunodeficiency virus) status, history of drug or alcohol abuse, disability, financial conflicts of interest, or other personal characteristics?**

The law of battery holds that any unauthorized touching of the body is an offense and a violation of law. In health care this means that laying hands on the patient to do a physical exam or providing a treatment intervention without consent is a violation of the patient's rights of consent and refusal and a battery. The common practice to obtain written consent to surgery on the surgical consent form developed in large measure to protect surgeons from claims of battery. The typical consent form names the surgeon selected by the patient. But whether and how to ensure that this is an informed choice has received scant attention.

While there is a strong autonomy-based argument that the typical patient should be told about surgical experience, the question here is whether the law of informed consent includes (or should include) an affirmative duty to tell patients about surgical experience.

Overall, a small number of courts support the principle that informed consent should include disclosure of surgical experience. These same cases offer some support for a legal duty to disclose surgical outcomes and to refer patients elsewhere if another hospital is significantly better equipped for this type of surgery. The leading case is *Johnson v. Kokemoor* (discussed in the box at the end of this chapter). However, more often courts have confined informed consent to disclosure of clinical information; in other words, they have maintained the traditional parameters of informed consent described above. Another lesson from this body of cases is that if the patient asks about experience or batting average, there is a duty to answer truthfully [4].

There are other "personal characteristics" that patients may want to know about, such as the surgeon's HIV status, history of alcohol or drug use, disability that impairs surgical performance (consider the neurosurgeon with even a slight loss of fine motor skills), or financial conflicts of interest (is the physician being paid to prescribe this company's drug or use its medical device?) Each of these conditions (and there may be others) poses a potential risk to patients. As with surgical experience, from the standpoint of respect for patient autonomy, there is a strong argument that patients should be told about these risks and be given the opportunity to consider them in their choice of surgeon. However, there is no ethical or legal consensus regarding these questions. Only a handful of cases have presented this view in court.

In one case early in the HIV epidemic, the court ruled that a surgeon had an obligation to inform his patients that he had AIDS (acquired immunodeficiency syndrome) [5]. Around the same time, the U.S. Centers for Disease Control and Prevention (CDC) mandated universal precautions in the care of patients to prevent transmission of HIV and issued widely adopted policies further designed to protect both patients and healthcare providers. Current CDC guidelines call on hospitals to implement infection-control programs and to establish expert review panels to evaluate whether an HIV-positive physician should be permitted to perform exposure-prone procedures. The guidelines state that "HCWs [health care workers] who are infected with HIV or HBV [hepatitis B virus] (and are HBsAg positive) should not perform exposure-prone procedures unless they have sought counsel from an expert review panel and been advised under what circumstances, if any, they may continue to perform these procedures. Such circumstances would include notifying prospective patients of the HCW's seropositivity before they undergo exposure-prone invasive procedures" [6].

Courts generally do not require physicians to tell patients about drug or alcohol abuse, unless it is a current problem that increases the risks of physician impairment and incompetence [7]. Some law supports the view that potential financial conflicts of interest must be shared with patients, in both the clinical and research context [8]. Overall though, the law has maintained its traditional focus on disclosure of medical information and has not done much to expand informed consent to impose an affirmative duty to tell patients about so-called personal characteristics of the physician.

3. **Under the Affordable Care Act, the Centers for Medicaid and Medicare Services (CMS) developed the "Physician Compare" Website. What was its intent? Quality measures were included in the available data. What are the limitations of these data? Can you find information about local surgeons on this site?**
The Affordable Care Act included a requirement for CMS to create and make accessible the "Physician Compare" database. The intent of the database is "to help consumers make informed decisions about their health care and to create clear incentives for physicians to perform well." The "Physician Compare" Website has general information about physicians who participate in Medicare including their clinical training, specialties, board certification, and practice information (e.g., site information). In addition, the Website includes information about quality measures and participation in certain programs such as "Maintenance of Certification Program" [9].

More than 50 quality measures are available for both group practices and for individuals. Examples of quality measures available for both group practices and individual clinicians include rate of influenza vaccination, rate of giving statins to patients at risk for heart disease, and assessment of spirometry evaluations in patients with chronic obstructive pulmonary disease.

The data for the quality measures were obtained through multiple sources including but not limited to claims data and data from required registries. The limitations on the data include that they represent Medicare patients only; that claims data represent claims rather than actual delivery of care and that there is necessarily a significant delay in reporting of those data; and that some of the performance data are self-reported [10].

4. **What is the definition of a surgical complication? Identify the known rates of complications for an AAA repair.**
The key elements to consider for a complication of a surgical or medical procedure include (1) that there was an unexpected outcome that impacted the patient and (2) this outcome was the result of the procedure being done. Dindo and his colleagues defined complications "as any deviation from the normal postoperative course. This definition also takes into account asymptomatic complications such as arrhythmia and atelectasis. A sequela is an 'after-effect' of surgery that is inherent to the procedure (e.g., inability to walk after an amputation of the leg)" [11].

If surgery is indicated for a person with an infrarenal abdominal aortic aneurysm, there are two main surgical approaches: an open repair through a large abdominal incision and an endovascular repair typically through a femoral catheter. The latter is considered a less risky procedure for patients who qualify. Both procedures can have complications associated with bleeding, wound or graft infections, kidney damage, blood clots with potential loss of blood flow to the legs, and spinal cord injuries. Other complications associated with open repairs include cardiac complications (heart attacks and arrhythmias), bowel injury, and pulmonary complications. Other complications associated with endovascular repair include blood vessel damage or damage to other organs as a result of the catheter and endoleaks (ongoing leaking out of the graft) [12, 13].

The distinctions and relationships between complications, medical errors, and negligence are explored in ▶ Chapter 17.

5. **What is the ethical model of informed consent? How much is it reasonable to expect patients to understand? What is health literacy?**
The ethical model of informed consent is grounded in the physician's duty to promote patient autonomy in the physician-patient interaction. Respect for patient autonomy also promotes patient well-being, as it fosters healthcare decisions based on the patient's own values and personal goals. Further, respect for patients fosters trust; this in turn encourages patients to be forthcoming in describing their health concerns and goals for care and treatment. The ethical model emphasizes that the physician's role is not merely to provide information, but also to facilitate and assist the patient to understand the nature of their medical condition and the risks, benefits, and burdens of treatment alternatives, in order to make voluntary autonomous informed decisions. The requirement of voluntariness recognizes that patients do not make their own decisions if they are coerced or unduly manipulated by others (perhaps a controlling family member). Most patients act voluntarily, but they cannot exercise their autonomy if they do not understand essential information about their medical condition and diagnostic or treatment options. A rigorous commitment to respect for patient autonomy means engaging the patient to take a thorough history and recognizing and responding to the patient's personal, subjective concerns, and

informational needs. Another element of the ethical model of informed consent is the call to explain things in language the patient can understand. The ethical model emphasizes the quality of consent through the promotion of patient autonomy in the physician-patient interaction [14]. By contrast, the rationale for the legal model also rests on respect for autonomy, voluntary choice, and self-determination, but its near-exclusive focus on information disclosure does little to promote the quality of the consent process.

The gap between the theory and practice of informed consent has been much discussed [15]. Concerns about the quality of the consent process often focus on how much patients should understand to make informed decisions. Patients should have at least adequate understanding for the decision to be made and should know that they have the right to consent to or refuse the physician's recommendation. But there is no minimum standard of comprehension that is widely accepted. To expect full understanding, akin to that of the physician, would be too demanding and would mean that few patients would be able to make truly informed decisions. Faden and Beauchamp, leading experts on informed consent, propose a benchmark of "substantial understanding," meaning that patients' understanding of the relevant information and of their rights is substantially accurate [14]. But this threshold and expectation may be too high, as patients often do not have a highly accurate and thorough understanding (let alone a full understanding), especially when the relevant clinical information is complex.

Whether patients possess sufficient health literacy to give meaningful informed consent is a complex matter. The 2004 report on health literacy from the Institute of Medicine found that "[n]early half of all American adults—90 million people—have difficulty understanding and acting upon health information" [16]. Health literacy can be described as the extent to which patients can access, process, and reason about health information. Health literacy compasses a range of skills and abilities: reading, writing, communication, and comprehension of medical information and of numbers and statistics (numeracy). Facility with computers and other computer devices is an increasingly important element of literacy. Limited health literacy has been found to be more common in the African American and Hispanic populations and among the elderly. It has also been associated with decreased attention to preventive care, lower immunization rates, and higher rates of non-compliance such as taking medications as prescribed [17].

There is a growing number of proposed tools and approaches intended to improve patient understanding and the informed consent process [18]. Consent forms and patient education materials have been revised, produced, and evaluated at a more appropriate reading level for the average patient (the eighth grade level is an accepted target). One oft-used measure of understanding and literacy is the patient's ability to teach back (repeat back) what has just been discussed with the physician [19]. What is reasonable to expect of patients will vary with the nature and complexity of the patient's condition and treatment options, and the physician will need to assess in each case whether the patient has sufficient understanding to make an informed decision. Here again, good communication will reveal clues as to the patient's level of health literacy. Another factor that plays an important role for many patients with respect to health literacy and informed consent is the patient's cultural background.

6. **How does the cultural background of the patient affect the informed consent process? Identify two examples. Drawing from these two examples, how should the informed consent process be modified to accommodate patients' cultural backgrounds?**

Broadly defined, "[c]ulture refers to integrated patterns of human behavior that include the language, thoughts, actions, customs, beliefs, and institutions of racial, ethnic, social or religious groups" [20]. Cultural differences have been shown to be related to racial and ethnic disparities in health and health outcomes [21]. Here we briefly mention three common sorts of issues that can arise in the informed consent process when the patient (or the patient's family) is of a significantly different culture from the physician's.

First, and directly related to health literacy, when the patient is not fluent in the English language, use of a translator may be warranted. Patients have a legal right to a translator, and hospitals are required to make translator services available. Use of family members as translators is generally discouraged, as there is potential for families to share information in accordance with their own views of their loved one's best interests. As with health literacy in general, physicians should strive to avoid medical jargon. In some cases use of translated written materials may be an effective tool.

Second, some cultures place less importance on patient autonomy. For some patients it is family members who are the primary decision-makers, and this may extend to insistence that a grave prognosis not be shared with the patient [22]. In Amish culture, where some aspects of Western medicine are not accepted, it is men and elders of the community who are typically recognized as deciders. In some cases, decision-making may not be fully patient-centered, as community values and resources can play a significant role where the costs of care are very high.

Third, patients sometimes make decisions based on their personal religious or ethnic values that seem contrary to the patient's own best interests. Perhaps the best known example is refusal of lifesaving blood transfusions by members of the Jehovah's Witness faith [23]. In end-of-life care, religious faith is the reason some patients insist that "everything be done," even when facing extremely poor quality of life. And some patients reject their physician's recommendation in favor of culturally embedded alternative remedies, while others may harbor deep distrust of the healthcare system or broadly reject Western medicine.

Education and training in cultural awareness and humility for medical students and residents with the goal of promoting a process of culturally sensitive patient-centered informed consent has received considerable attention in recent years. Some emphasize the physician's duty to become familiar with different cultures in their patient population. Other initiatives focus on how to frame the physician-patient interaction to elicit how social, economic, religious, and cultural beliefs influence the patient's understanding of their illness, their health values, goals, and behaviors, as well as the role of family in care decisions [24]. A team approach that includes religious or community leaders, spiritual care, and social work may be considered. The healthcare practitioner should remember that patients from particular ethnic or religious groups are always individuals with their own views and interpretations of the place of culture in their health care. For example, even a self-identified devout Catholic may not adhere to all of the official tenets and teachings of the Catholic faith. This subject is also explored in ▶ Chapters 10 and 13.

7. **What are the four recognized exceptions to informed consent, that is, situations where informed consent is not required? Describe each of these exceptions.**
   There are four generally recognized exceptions to the requirement of informed consent. They are (1) decisional incapacity, (2) waiver, (3) therapeutic privilege, and (4) emergencies [25].
   - **Decisional (in)capacity**: In order to make informed decisions, patients must have the capacity to put autonomy into action. Decisional capacity means that the patient possesses the ability to understand and appreciate (1) the nature of his or her medical condition, including diagnosis and prognosis; (2) the benefits, risks, and

burdens of the recommended treatment and reasonable alternatives, including the option of no treatment; and (3) to reach and express an informed decision. Patients are legally presumed to have decisional capacity if they are 18 years of age or older. Under law, adolescents have the right to make certain healthcare decisions, such as those concerning sexual activity and reproduction, when under the age of 18. (Decision-making in pediatrics is further discussed in ▶ Chapters 7, 9, and 19. Decision-making for and by adolescents is discussed in ▶ Chapters 5, 10, and 21).

Decisional incapacity is the most important and most frequently encountered exception to the rule that informed consent or refusal must come from the patient herself. When the patient appears to lack capacity, it is the responsibility of the attending physician, both ethically and legally, to undertake a closer examination to assess capacity. If the patient is deemed to lack capacity, the locus of decisional authority shifts away from the patient to an appropriate surrogate decision-maker, such as a healthcare proxy, spouse, or adult child. Often, the attending physician will request a consultation from a psychiatrist to further evaluate the patient's capacity ("call psych"), but in most cases it is up to the attending physician to make the final determination of capacity. Sometimes the law requires consultation with psychiatry or an appropriately trained professional if the patient suffers from intellectual disability or has a history of developmental disability.

Several important points about capacity should be kept in mind. Capacity is decision-specific; it is about the patient's abilities to make a particular decision and may vary depending on the nature and complexity of the decision. A patient may have capacity to make one decision (e.g., "my son should make decisions for me if I become unable to do so") but not other more complex decisions (e.g., "should I have surgery to repair my aneurysm?"). Capacity is about the patient's ability to make decisions at a particular time. Some patients have fluctuating capacity; they are unable to make a treatment decision today, but tomorrow when antibiotics have taken effect or pain medications have been reduced, decisional capacity may be restored. Respect for patient autonomy includes treating cognitive impairment to enhance opportunities to make informed decisions. Lastly, it is often said that capacity is not a status-based judgment. Patients are not to be deemed to lack capacity simply because they are depressed, disabled, elderly, or have an intellectual disability. Rather, these conditions should be taken as indications that call for a closer look at the patient's capacity.

Physicians tend to accept that patients have capacity and are making informed decisions when they agree with the recommended course of treatment and are more inclined to question capacity when patients disagree. Physicians may believe that the patient who refuses the recommended treatment is choosing contrary to their own best interests and sometimes deem the patient to lack capacity for this reason. However, patients do not lack capacity simply because they disagree with the physician's recommendation. This may be an indication that capacity should be carefully assessed, but respect for autonomy means that patients have the right to make the "wrong" decision, based on their own values and goals.

Finally, capacity should be distinguished from the legal term "competence," though they are often used interchangeably. When courts determine that a person is incompetent, this typically means the person is unable to manage their affairs and to make a range of types of decisions for himself, and someone else needs to be appointed to act on the person's behalf. By contrast, physicians make determinations about patient capacity every day without any involvement of the court system [26].

- **Waiver**: Suppose the patient says, "Wait, wait, don't tell me, Doctor. Tell me what to do, I trust you." Perhaps, the patient frames it as, "Doctor, what would you do? That's good enough for me." Does the physician still have a duty to provide all relevant information, or has this been waived by the patient? Can the physician proceed to obtain the patient's signature on the consent form or otherwise document consent in the progress notes? Taking respect for patient autonomy and the right to decide seriously means the physician should continue to engage the patient in a dialogue that presents relevant information and clarifies the risks, benefits, and burdens of the recommended intervention. At the same time, to honor the patient's waiver is also to respect an autonomous choice not to know.

Reliance on the patient's waiver of information is disfavored in both law and ethics. However, there may be rare cases where accepting a waiver of information is appropriate. To rely on a waiver, patients must voluntarily and intentionally give up the right to more information, knowing that they are entitled to receive the information and that the decision to accept or refuse treatment belongs to them and not to the physician. It must be a knowing waiver. Thus, persistent refusal of information by itself does not constitute waiver. Nor is the patient who simply avoids the conversation waiving the right to know. This is not to say that physicians have an affirmative duty to recite these rights for the patient, but rather when the patient indicates a desire not to know, then it should be a prompt to establish that the patient knows their rights of informed consent. One approach would be to refer to and discuss the patients' bill of rights that is typically included in patient intake materials and posted in the halls of hospitals. To go one step further, to rely on a waiver physicians should offer accurate information (not downplay the risks) and assess whether the patient is refusing specific facts, such as prognosis, the nature of the procedure, the risks and likelihood of a good or bad outcome, or the side effects. The informed consent conversation including the patient's waiver and the signing of any consent forms should be clearly documented in the patient's record.

Other important questions about waiver may arise when patients choose to look to family members to make treatment decisions for them ("Doc, ask my daughter"). In some cultures it is common for families, not patients, to be the deciders. Families are trusted to decide in the patient's best interests and in accordance with the patient's wishes. Sometimes patients want to know but not decide; other times they may direct that family be informed and also give consent [27]. When the patient is from a different ethnic or cultural background that values family authority over patient autonomy, the physician should clearly establish this preference before looking to family for consent. Often it will be ethically appropriate to respect patient and family cultural norms and accept this choice as a form of waiver. However, in some cases, such as where the family is making a decision contrary to the patient's wishes or their best interests, the physician must act as a patient advocate and explore more rigorously the family dynamics and the basis for the family's treatment decisions.

- **Therapeutic Privilege**: If healthcare practitioners feel that sharing information about diagnosis or prognosis will be too upsetting to the patient, they may consider not disclosing the seriousness of the patient's condition or the risks of surgery in order to avoid doing more harm than good. A variation of this concern might be understating the severity of the risks of treatment. The concept that arguably supports this approach is therapeutic privilege, the idea that healthcare practitioners have the privilege to not disclose certain information if, in their judgment, doing so would be harmful to the patient or that disclosure would not be in the patient's best interests. Therapeutic privilege held important currency in the past; however, today the doctrine is disfavored in both ethics and law. Withholding important information would be a violation of patient autonomy and undermines the right of informed consent. Further, respect for autonomy also means giving patients the opportunity to judge for themselves whether some information would be too upsetting. There is additional concern that recognition of therapeutic privilege would too often become an excuse to avoid difficult or time-consuming interactions with patients. Nor should the privilege be invoked paternalistically because the patient disagrees with the physician's recommendation, but might be persuaded to agree based on incomplete information. Note that the doctrine of therapeutic privilege would give the physician permission to limit disclosure by withholding certain information. By contrast, under the waiver exception, information is offered and the patient chooses not to know more.

  Sometimes the issue of therapeutic privilege arises from the family's request that certain information be withheld from the patient ("Don't tell dad he has a terminal illness, he'll be devastated."). As noted in the discussion of waiver, the request may sometimes be grounded in the family's ethnic or cultural norms that family decision-making takes priority over the personal autonomy of the patient. In some Asian cultures, bad news is withheld from loved ones to preserve the benefits of hope and hopefulness. In such instances the physician must balance respect for patient autonomy with both patient welfare and appropriate respect for the family who know the patient best. Putting aside the family's concerns in favor of patient autonomy can be a difficult and sensitive choice, especially where the patient is likely to lose decisional capacity with progression of their disease and family will soon assume responsibility for making treatment decisions. Still, the general rule is that patients should be told and the right to informed consent should be honored. One approach here would be to ask the patient if they want to know everything. If not, perhaps they will defer to family. The question of waiver, rather than therapeutic privilege, would then become the better way to understand the consent process. In rare cases a grave prognosis may truly be so traumatic as to impair the patient's capacity to make treatment decisions or may put the patient at imminent risk, for example, the patient who has manifested suicidal behavior related to their medical condition. But invocation of therapeutic privilege to withhold information should be narrowly and strictly construed.

- **Emergency treatment**: In both ethics and law, the emergency treatment exception to informed consent is well-established. When patients suffer a sudden, serious injury or event and are in imminent need of medical attention, preventing death or serious harm takes precedence over obtaining informed consent, or even uninformed consent, from the patient. This principle also rests on the practical reality that often there is no time to engage the patient in conversation; treatment must begin right away. A further rationale for the emergency exception rests on the idea that consent is implied by the circumstances, meaning that we have made a societal judgment that under the circumstances emergency medical technicians, emergency physicians, and other personnel have permission to treat without stopping to seek consent from the patient.

This is not to say, however, that all "emergencies" are the same. Many ambulatory visits to the emergency room are for nonurgent conditions, for example, a broken arm or a laceration in need of stitching. In these instances there is typically time and opportunity to explain the nature of the medical condition and the recommended treatment and to obtain informed consent. Under these and other circumstances where the patient appears capable of giving informed consent, emergency physicians should be especially attentive to the patient's decisional capacity. Compared to other specialties, emergency physicians tend to see a higher proportion of patients whose mental state may be compromised by drugs, alcohol, pain, or anxiety. Also important, patients' interactions with emergency physicians are typically first-time encounters between strangers, with no prior relationship characteristic of primary care practice or patients with chronic illness. Hence, the emergency physician often must rely solely on the patient's understanding and his or her expressions of values and goals in the immediate time-limited circumstances [28].

3.4 Responses to Discussion Questions

When you enter Ms. West's room, her mother, her brother, and three adult children are all crowded into the hospital room. You hand Ms. West the standard procedural consent form for an open abdominal surgery and start the conversation.

Before you can say much, Ms. West's family starts to interrupt with questions. Her daughter wants to know how many of these procedures your attending surgeon has performed. Her brother wants to know whether you will be in the operating room (OR) and if so, what your role will be. Her older son says while in the waiting room, he did an Internet search of AAA repairs and wants to know why an endoscopic repair is not being recommended. Her younger son asks simply, "Could my mom die?"

1. **How would you approach this situation? What would you say or do first in response to the family's questions?**

> **Box 3.1 Teaching Tip**
> Encourage learners to think about what they would want to be told if they, or a loved one, were undergoing surgery. Ask them to consider all aspects of informed consent and whether they think a first-year surgical resident is able to address every aspect.

The resident in this situation should take a step back and consider several things. Has she clearly introduced herself and explained her role? Has she asked each family member to introduce himself or herself? Has she given Ms. West the opportunity to speak with her privately or with only one or two family members present? Though the resident may feel less pressure one-on-one, this is ultimately the patient's choice, and the resident should not insist that anyone leave. Furthermore, it is generally recommended that patients have at least one person present to help them take notes and remember all aspects of the discussion. Does everyone—including the resident—have a place to sit comfortably during the discussion? Once these steps have been taken, then she can proceed.

There is no single "right" way to start. One good approach would be to reassure Ms. West's children that the resident will answer all their questions, or find and return with answers to any she cannot answer immediately, including the last question (about whether the patient might die), but that the resident wants to start by asking Ms. West what she already knows about her condition and the procedure.

Starting with this question will allow the resident to (1) avoid repeating information that Ms. West already understands; (2) correct misconceptions upfront, before they can persist; and (3) appreciate something of Ms. West's educational background.

When the resident finally begins to explain the procedure, she should use lay language that is easy to understand and pause frequently to make space for questions [29]. She should also allow time for "teaching back," asking Ms. West to explain in her own words what she has just been told. She needs to be able to explain the risks and benefits of the proposed surgical intervention, *as well as* any alternatives and the risks and benefits of such alternatives.

> **Box 3.2   Personal Perspective**
>
> In the essay "Treating Patients as Partners, by Way of Informed Consent," the surgeon Pauline Chen describes an encounter in which she fails to do this [29]. She writes:
>
> "Have you ever had a paracentesis?" I asked, pulling out a consent form for Pete to sign.
>
> "No," he answered between short labored breaths. "Does it hurt?"
>
> I tried to reassure him by explaining how I would numb him first. But as I began describing the anesthetic, the bee sting prick of the needle and the pressured sensation of medication infiltrating the flesh, I felt myself slipping into a familiar spiel, the same one I had delivered to all the other patients with intractable ascites. I pointed to the quadrant on his belly where I would work, estimated the amount of fluid I would pull out, and then reeled off the standard catalog of complications for this procedure.
>
> Pete looked away from me and stared at the consent form. Yet even as I watched his brows knit together, his eyes widen then wince, I kept on talking. I had gone into my informed consent mode—a tsunami of assorted descriptions and facts delivered within a few minutes. If Pete had wanted me to pause and linger over something, I never knew. He couldn't get a word in edgewise.
>
> "So," I finally asked him at the end of my monologue, "do you have any questions?" Even as that sentence came out of my mouth, I knew what his answer would likely be.

✓ 2. **Should you have told the attending surgeon that you had not seen the procedure performed previously and therefore should not obtain consent?**
Yes. Never having witnessed such a surgical procedure necessarily limits the resident's ability to fully engage the patient and her family, as the resident is likely to be unable to adequately respond to certain questions that the family may have during the consent process such as, "What have your experiences been with this procedure?" If the resident is not prepared to engage in the informed consent interaction with the patient and family, then he cannot do so ethically. The first obligation is to the patient: to her role and authority in the decision-making process and to promoting her well-being and best interests. Residents may feel embarrassed to tell attending physicians that they lack the necessary knowledge and experience to obtain consent, but this obstacle must be overcome.

(a) **Self-reflection:** What do you think you *would* do in real life? Do you think you would speak up?

> **Box 3.3   Teaching Tip**
> "Speaking up" is explored in-depth in ▶ Chapter 20

Contemplating the family's first question, you are unsure how many similar procedures the surgeon has performed before, but you know that she completed fellowship training only 6 months prior.

✓ 3. **Does this change your view about obtaining informed consent?**
This indicates that the resident has some more information to share with the patient and her family. The resident can tell them about the attending physician's training and may be able to make some inferences about the surgeon's experience with particular procedures. It does not answer the specific question of how often the surgeon has performed this procedure. Providing this information opens

up the possibility that the patient and her family may be happy that they have a younger, more recently trained surgeon, who is likely familiar with the most current surgical techniques. It also opens the resident up to more questions about the attending surgeon's experience, given that she has only recently completed fellowship training. While the resident should be prepared to answer these types of questions, she still needs to do more research before the family's question can be fully answered.

4. **Do surgeons, or healthcare practitioners in general, have a duty to inform patients of their prior experience performing the procedure?**
To some extent, there are strong ethical grounds for surgeons and physicians to volunteer how often they have done the procedure, but it is more firmly established that one's knowledge and training should be explained. Indeed it is hard to imagine explaining the surgery to the patient without demonstrating one's knowledge of the procedure, and this is a natural place for the surgeon to comment on his or her training and experience with the procedure. In this case, the question has already been asked, so it is incumbent on the surgeon to give a forthright answer.

   (a) **Do you think that reasonable patients would want to know how often their surgeon has done the recommended procedure?**
   Increasingly, patients seek information about a surgeon's credentials or the hospital's surgical program online. There is substantial evidence that in addition, most patients would want to know how often the surgeon has performed their procedure and would take this into account in their choice of surgeon. In two studies of emergency room patients at Vanderbilt University, approximately two-thirds of patients wanted to know about the resident's experience with the procedure [30] and what the role of medical students would be [31]. Another study of cancer patients at Johns Hopkins University found that most patients rated as "very important" knowing how many times the surgeon had done their procedure and how many years the surgeon had been in practice [32]. Other studies also reveal a strong interest in knowing whether this would be a first-time procedure for the resident [33].
   Regardless of an individual healthcare provider's view of what a reasonable patient would want to know, or the legal standard of disclosure in a particular state, what matters to the patient should not be ignored. The patient's personal, and perhaps unusual or idiosyncratic, concerns can play a pivotal role in the decision to accept or refuse the recommended treatment plan. Consider, for example, the rock guitarist who is especially sensitive to even the slightest loss of dexterity following shoulder surgery or the master chef who cannot afford even the slightest loss of taste as a side effect of medication. Perhaps in this case the family knows firsthand of a surgery that went bad with an inexperienced surgeon. The physician may and should become aware of such concerns through the taking of a good social history and the process of communication and interaction with the patient.

   (b) **Should patients always choose the most experienced provider? Why or why not?**
   This argument has intuitive appeal. After all, who would not choose the more experienced surgeon over the first-year resident? Who would not recommend that a family member go to see the surgeon with the best track record of good outcomes for the procedure they are facing? However, the assertion that patients would routinely flock to more experienced surgeons seems

to be based more on speculation than factual evidence. One reason is that focus on surgical experience alone oversimplifies the way in which people choose physicians. Patients typically look to primary care physicians, family, friends, or co-workers for recommendations. Some patients consider the surgeon's medical school, certification, or other professional qualifications. For some, hospital reputation is important. For others, gender and ethnicity are important considerations. Further, once relationships between patients and providers are established, trust, familiarity, and rapport may lead patients to choose to continue care, even if given the option to transfer to a more experienced provider.

Cost and location can also be major considerations. Insured patients have strong financial incentives to choose from the panel of surgeons that participate in their health plan and thus face restricted choice. Further, staying in network and the higher costs of going out of network also limit patients' choice of hospitals. In sum, many patients want to know about surgical experience, but this may be only one of a number of factors taken into account when choosing a surgeon.

Finally, experience is not the only factor in determining physician competence and driving outcomes. Some senior physicians will have little recent, direct experience performing procedures, if they have primarily supervised residents and fellows; others, particularly outside of academic institutions, may not be as up-to-date on changing practice standards. And training with new and innovative equipment and techniques, much of which occurs post-residency, can be even more uneven. Pointing to these variations in surgical practice, some argue that experience and outcome statistics are misleading and should not be routinely shared with patients.

(c) **You were asked a direct question about surgical experience. What would you say?**
The resident could simply say that she does not know, but a better response is to also offer to ask the attending physician to come talk to the patient personally. Alternatively, the resident could offer to find out and report back to the patient, proceed to obtain the data to the best of his ability, and be prepared to explain it. If the resident does seek out the answer, both ethically and legally, she has an obligation to tell the truth. Truth-telling is a core ethical value in medicine. There are resources available to the resident, including the library, patient educator, the Internet, and of course the attending surgeon.

(d) **What are the ethical and legal consequences of lying?**
Lack of honesty in the physician-patient-family relationship can mean loss of trust and rapport in that relationship. In response, patients may be less forthcoming about their condition, their concerns, and goals for treatment, and the quality of care can be compromised. In some cases, when trust is broken, patients may choose to find another physician. Fraud, misrepresentation, and manipulation in the informed consent process can lead to professional disciplinary actions and even lawsuits from patients for violation of the right of informed consent or fraud.

5. **How would medical education be affected if patients refused to see healthcare practitioners in training and junior healthcare practitioners?**
Medical education, and the next generation of well-trained qualified healthcare practitioners, requires that most patients be willing to allow trainees to participate in their care. Most patients, most of the time, do allow trainees to participate and

to learn in their care and treatment with the understanding that the trainees have adequate supervision. (Supervision of trainees is discussed further in ▶ Chapter 20.) Most identify with the fact that to acquire the skills and abilities of any medical profession takes practice. Experience matters on the learning curve. Another perspective, perhaps implicit in patients' willingness to have residents and students involved in their care, is that doing so is part of the social contract and that the social contract entails obligations to future generations to assure that they will benefit from a high-quality healthcare system. If done appropriately and honestly, most patients will allow students and trainees to assist in their care.

> **Box 3.4 Teaching Tip**
> It may be appropriate to explore the limitations of this idea now or later in the discussion. For instance, is a physician-parent justified in refusing to allow trainees to participate in his child's surgery? Should a physician who is a patient, say in labor, decline to have trainees present for the birth? All trainees, or just those known to her or whom she might later supervise? What about a child who has been sexually abused and whose parents would like to limit the number of people examining her?

6. **How should you respond to the question, "Could my mom die?"**
   If there is a meaningful risk the patient could die—whether in the operating room, in recovery, or after discharge (what is the expected survival post-discharge?)—then these risks need to be presented to the patient with opportunity for the patient to ask questions. The patient cannot make an informed decision to have or decline the surgery without considering this possibly significant risk. For Ms. West, who needs her AAA repaired, the risk of death is significant and must be discussed. Many states require by law that the risk of death be disclosed even if that risk is considered remote.

As with any procedure, the physician who is obtaining consent from the patient needs to tell her of the risks inherent with an AAA as well as the benefits and risks of the indicated surgical procedure. Prior risk factors pertaining to a patient's history and habits seem to dramatically affect morbidity and mortality associated with AAA repair. If surgical repair has been recommended, it is reasonable to assume that Ms. West is at risk for rupture of her AAA. Without surgical intervention, rupture of an AAA will result in death. Given the gravity of the underlying medical condition, it is imperative that the risks of the AAA repair are shared in the context of the life-threatening condition. Information about complications should be shared with the family only with Ms. West's consent.

Public reporting of quality measures in patient outcomes has become more common. As an example, since 1994, the New York State Department of Health has collected and made available patient mortality data for interventional cardiologists [1]. Some health systems have chosen to generate public "report cards" for their organizations.

7. **Do you think that there is a duty to share surgeon-specific morbidity and mortality rates for procedures, allowing patients to compare surgeons?**
   Although some commentators have argued that surgeons should have a legal duty to share their morbidity and mortality rates, for the most part, the law imposes no such obligation. Ethically, there are several different arguments that apply here. Those in favor argue that this information is important to patients because they are not just consenting to a specific procedure, but to a specific procedure done by a specific provider, and this information is part of the risks, benefits, and consequences of the procedure. It is therefore relevant to making an informed decision. On the other hand, those against such a duty emphasize that understanding these statistics is a complex matter and may confuse more than help patients to make informed decisions.

They emphasize, for example, variables of patient mix, different patients' overall condition and comorbidities, and the quality of surgical centers. Further questions to consider are whether the resident can in fact obtain comparative statistics for other surgeons who have done this procedure and whether that information, if available, will be current and up-to-date.

(a) **If so, to whom would you compare?** Comparison might take place between physicians within the same institution, between physicians or institutions in the same geographic area, or between physicians or institutions in the USA (or even globally). All of these groups may be of interest to some patients, but many patients may not have the time or resources to get second opinions from physicians further afield. Furthermore, they may themselves differ in clinically significant ways from the patient population in the geographic area or nationally—an important factor that can limit the usefulness of comparisons to other surgeons.

8. **Patients may want to know other "personal characteristics" such as drug or alcohol abuse, physical or mental impairment, or HIV status. Is there an affirmative duty to share this information in the informed consent process?** The law generally does not require physicians to volunteer to tell patients about drug or alcohol abuse or disability. Rather, the traditional boundaries of informed consent have been maintained. Concerns for the privacy of physicians and the sometimes fuzzy line between one's professional and personal life also counsel against requiring disclosure. Still, there may be an obligation to do so if there is a current problem that increases the risks of physician impairment and incompetence.

(a) **If the patient asks, are you obligated to provide an honest answer?** Some argue that, if asked, the healthcare practitioner should always answer truthfully. After all, truth-telling is a core value in medicine. However, it is less clear that these sorts of personal characteristics represent current risks for the patient that the patient is entitled to know, particularly when compared to surgical inexperience. Questions about drugs, drinking, or disability can conflict with healthcare practitioners' interests in protecting their privacy. For these reasons, others argue that healthcare practitioners may properly decline to answer such questions or answer only when the healthcare practitioner's current condition compromises the ability to provide competent and appropriate care. It is important to note that state regulatory agencies (e.g., medical boards) have policies in place to protect patients from impaired healthcare practitioners. There is significant variation from state to state. For example, some states require that healthcare practitioners report colleagues who are suspected of performing procedures while impaired, but others only encourage such reporting. Healthcare practitioners should familiarize themselves about their state's policies.

(b) **Creative Problem-Solving: Are there better alternatives that balance patient welfare and physician privacy?**

> **Box 3.5   Teaching Tip**
> Consider asking the learner(s) who researched Background Question 4 to comment on how easy or difficult it was to find information about the procedure and about local surgeons. How feasible would finding such information be for the typical patient or family? Thinking about it from this perspective may affect the response.

9. **Self-reflection: Of the above (provider experience, provider or institutional outcomes for a procedure, provider**

personal characteristics), what if anything would *you* want to know, for yourself or your family member?
(a) How do you think the reasonable patient would answer these questions?

✓ 10. **If your answers to any of the above questions were affirmative, should the law be changed to reflect this?**
If reasonable patients would want to know, then there is a strong ethical argument for a duty to disclose and hence a strong argument for making disclosure a legal obligation. But there are also reasons not to turn disclosure into an affirmative legal obligation. A leading objection to imposing an affirmative duty to tell patients about surgical experience and outcomes ("batting average") is that patients would regularly choose more experienced surgeons or those with a stronger track record. In turn, medical education and training would suffer, the next generation of surgeons would be less qualified, and the next generation of patients would be at risk. Much the same argument has been made in even stronger terms in opposition to mandated disclosure of surgical batting averages. It is far from clear that this same argument applies to physicians impaired by personal challenges of alcohol, drugs, or disability, as these conditions apply in particular cases rather than across a wide group of surgeons in training. However, as just noted, there are other concerns about physician privacy to be considered.

About 3 months earlier, your co-resident admitted a patient with the same condition as Ms. West. You weren't involved with that patient's care but you heard the story from your co-resident. You heard that the patient's daughter was a prominent physician in New York City and that the patient was privately transferred to a larger hospital to consult with a nationally known vascular surgeon. You don't know very much about Ms. West's family—just that she has Medicaid, not private insurance.

✓ 11. **Is there an obligation to transfer Ms. West to a better-equipped hospital with a more experienced surgical staff?**
As long as the center meets the standard of care, there is no duty to refer elsewhere. Certainly, if a patient wants a second opinion, physicians have a duty to allow the patient adequate time and even to assist in finding a consulting surgeon, as long as the risks and benefits of delaying care are understood. For example, will delaying surgery for a second opinion increase the patient's risk of a bad outcome?
 If the level of experience, skills or equipment, or track record of successful outcomes is substandard, the facility should not offer that surgery. There should be outcomes data, and accreditation standards that are met, to ensure the hospital is safe and the procedures are of high quality and successful.
(a) **Should a decision to discuss transfer take into account whether the family has the resources and ability to travel to a more distant center to see a specialist? Is it important that Ms. West has Medicaid, rather than private insurance?**
If the hospital meets the criteria established above, there is no obligation to have the discussion. If, however, a transfer is either medically indicated or requested by the patient, the patient's insurance status should not influence the decision to have the discussion. In the former situation, the transfer should be a covered service (e.g., if the rationale is transfer to a higher level of care); in the latter situation, if the transfer is not eligible for coverage, the patient would have to be able and willing to cover the costs personally. That is not for the physician to decide without the patient's input. The physician may not have full information regarding the family's values, beliefs, interests, and resources *until* all options have been explained.
 Patients are entitled to decide for themselves about the burdens they

are willing to assume to pursue care elsewhere, including the personal, social, and financial burdens. The patient's social situation, including financial resources and family circumstances, certainly matters. They should be considered but only to the extent that such considerations help patients to access the best available care and achieve desired health outcomes. Circumstances should not be "considered" in order to withhold options.

As you attempt to tactfully navigate the family's questions, Ms. West's mother looks increasingly distressed. She finally says that she is 94 years old and "in her day" it was not uncommon for residents to be allowed to operate unsupervised on poor black patients like her family.

✓ 12. **How should you respond?**
The resident should explore Ms. West's mother's fears and the implied message that the family does not trust the care she is receiving and feels she is being treated differently. It would be beneficial to address this with the attending physician, preferably before giving the impression that they are being disrespected by not having the attending surgeon perform the informed consent. Currently, all procedures performed in the OR or with significant risk must have an attending present, as per the American College of Surgeons (ACS) Statement on Principles [34].

The primary attending surgeon is personally responsible for the patient's welfare throughout the operation. In general, the patient's primary attending surgeon should be in the operating suite or should be immediately available for the entire surgical procedure. There are instances consistent with good patient care that are valid exceptions. However, when the primary attending surgeon is not present or immediately available, another attending surgeon should be assigned to be "immediately available."

✓ 13. **What can you, as a healthcare provider, do to address medical mistrust?**
Providers should be professional, honest, and trustworthy and should treat every patient as if the patient is one of their own family members. They also must be willing to speak up if they see ethical problems or lapses in standard of care, even as medical students.

Residents are much more closely supervised now and institutionally sanctioned discrimination no longer openly occurs. However, many residents still seek out academic programs and teaching hospitals in underserved communities because they feel they are allowed to "do more" than in private hospitals serving more affluent and educated populations.

✓ 14. **Is this a just system? Why or why not?**
Many believe this is not a just system because it disproportionately burdens disadvantaged patients. While many patients may in fact benefit from a student's or resident's interest in their care, patients may experience longer visits (particularly if they have to see a student, then a resident, then an attending). It has often been shown that patients on Medicaid have worse health outcomes than those on private insurance, though there are likely many factors contributing to this. It is important for these health systems to make every effort to provide the most beneficial treatment to every patient.

On the other hand, academic medical centers offer certain benefits. Residents can be very rich in current knowledge and experience, and studies often show that physicians at academic centers are more likely to follow the most up-to-date treatment recommendations than those in private practice.

✓ 15. **Self-reflection: If you were having surgery, would you allow a resident to assist? What if the patient was your parent? Your child?** [35]

Box 3.6 Case Conclusion

The case of Ms. West that appears above is based on the Wisconsin case of Johnson v. Kokemoor, the leading court case ruling that there is an affirmative duty to disclose surgical experience and outcomes [35]. There the patient underwent surgical repair of a basilar bifurcation aneurysm of the brain. Although the clipping of the aneurysm was technically successful, postoperatively the patient lost bowel and bladder control, was unable to walk, was "rendered an incomplete quadriplegic," and suffered other impairments.

In her lawsuit the patient claimed that the surgeon had breached three related duties: (1) to tell her about his experience performing this procedure, (2) to tell her about his morbidity and mortality statistics in comparison to those of other surgeons doing this procedure (sometimes called comparative provider statistics), and (3) to refer her to another hospital with more experienced surgeons.

In the first ruling of its kind, the Wisconsin court agreed on all three points. The core idea in the court's ruling was that this information would be relevant (the law uses the term "material") to a patient's choice of surgeon to repair a basilar bifurcation aneurysm of the brain. Failure to provide this information constituted a failure to respect patient autonomy and violated the patient's right to make an informed choice of surgeon. It was also important in this case that the surgeon had misrepresented his experience, claiming he had done the procedure "dozens" of times, when in fact he had only performed two such aneurysm repairs. This gross and intentional misrepresentation constituted fraud.

The following further summarizes the key ruling in the case:

'[A] patient cannot make an informed, intelligent decision to consent to a physician's suggested treatment unless the physician discloses what is material to the patient's decision, i.e., all of the viable alternatives and risks of the treatment proposed.' In this case the information regarding a physician's experience in performing a particular procedure, a physician's risk statistics as compared with those of other physicians who perform that procedure, and the availability of other centers and physicians better able to perform that procedure would have facilitated the plaintiff's awareness of 'all of the viable alternatives' available to her and thereby aided her exercise of informed consent. [35]

To more fully bring disclosure of surgical batting averages within the terms of the statute, the court stressed that "When different physicians have substantially different success rates, whether surgery is performed by one rather than another represents a choice between 'alternate, viable medical modes of treatment' under the Wisconsin statute. Surgical experience and outcomes are part of the risk assessment essential to the choice of medical treatment" [35].

References

1. Narins CR, Dozier AM, Ling FS, Zareba W. The influence of public reporting of outcome data on medical decision making by physicians. Arch Intern Med. 2005;165(1):83–7.
2. Beauchamp T. Informed consent: its history, meaning, and present challenges. Camb Q Healthc Ethics. 2011;20(4):515–23.
3. King JS, Moulton B. Rethinking informed consent: the case for shared medical decision-making. Am J Law Med. 2006;32(4):429–501.
4. Sawicki N. Modernizing informed consent: expanding the boundaries of materiality. Univ of Illinois L Rev. 2016;821–71.
5. Estate of William Behringer v The Medical Center at Princeton [1991] 592 A.2d 1251 (NJ Super).
6. Centers for Disease Control and Prevention. Recommendations for preventing transmission of Human Immunodeficiency Virus and Hepatitis B Virus during exposure-prone invasive procedures [Internet]. Atlanta (GA): CDC; 1991 July 12 [cited 2017 Dec 14]. Available from: www.cdc.gov/mmwr/preview/mmwrhtml/00014845.htm.
7. Albany Urology Clinic, PC v. Cleveland [2000] 528 S.E.2d 777 (Ga.).
8. Moore v. Regents of Univ. of California [1990] 793 P.2d 479 (Cal.).
9. Centers for Medicare & Medicaid Services. Physician compare initiative [Internet]. Baltimore (MD): CMS. [cited 2018 Jan 15]. Available from: https://www.cms.gov/Medicare/Quality-Initiatives-Patient-Assessment-Instruments/physician-compare-initiative/index.html.
10. Reese S. How will the CMS Physician compare website affect your practice? [Internet]. Medscape; 2013 Nov 15 [cited 2018 Jan 15]. Available from: https://www.medscape.com/viewarticle/813124.
11. Dindo D, Demartines N, Clavien P. Classification of surgical complications a new proposal with evaluation in a cohort of 6336 patients and results of a survey. Ann Surg. 2004;240(2):205–13.
12. Maleux G, Koolen M, Heye S. Complications after endovascular aneurysm repair. Semin Interv Radiol. 2009;26(1):3–9.
13. Johns Hopkins Medicine. Abdominal aortic aneurysm repair [Internet]. Baltimore (MD); 1991 July 12 [cited 2018 Jan 17]. Available from: https://www.hopkinsmedicine.org/healthlibrary/test_procedures/cardiovascular/abdominal_aortic_aneurysm_repair_92,P08291.

14. Faden R, Beauchamp TL. A history and theory of informed consent. New York: Oxford University Press; 1986.
15. Grady C. Enduring and emerging challenges of informed consent. NEJM. 2015;372(9):855–62.
16. National Institutes of Medicine. Report brief health literacy: a prescription to end confusion. Washington, D.C.: Institute of Medicine; 2004.
17. Hersh L, Salzman B, Snyderman D. Health literacy in primary care practice. Am Fam Physician. 2015;92(2):118–24.
18. Siegal G, Bonnie R, Appelbaum P. Personalized disclosure by information-on-demand: attending to patients' needs in the informed consent process. J Law Med Ethics. 2012;40(2):359–67.
19. Ubel P, Scherr K, Fagerlin A. Empowerment failure: how shortcomings in physician communication unwittingly undermine patient autonomy. Am J Bioeth. 2017;17(11):31–9.
20. Association of American Medical Colleges. Cultural competence education. Washington, D.C.: AAMC; 2005. Available from: www.aamc.org/download/54338/data/
21. Institute of Medicine. Unequal treatment: what health care system administrators need to know about racial and ethnic disparities in healthcare. Washington, D.C.: Institute of Medicine; 2002.
22. Partain D, Ingram C, Strand J. Providing appropriate end-of-life care to religious and ethnic minorities. Mayo Clin Proc. 2017;92(1):147–52.
23. West JM. Ethical issues in the care of Jehovah's Witnesses. Curr Opin Anaesthesiol. 2014;27(2):170–6.
24. Betancourt JR. Cultural competence and medical education: many names, many perspectives, one goal. Acad Med. 2006;81(6):499–501.
25. Berg JW, Appelbaum PS, Lidz CW, Parker LS. Informed consent: legal theory and clinical practice. 2nd ed. New York: Oxford University Press; 2001. chapter 4.
26. Ganzini L, Volicer L, Nelson WA, Fox E, Derse AR. Ten myths about decision-making capacity. J Am Med Dir Assoc. 2005;6(3 Suppl):S100–4.
27. Siddiqui S. Family loyalty as a cultural obstacle to good care: the case of Mrs. Indira. J Clin Ethics. Spring 2017;28(1):67–9.
28. Moskop JC. Informed consent in the emergency department. Emerg Med Clin North Am. 1999;17(2):327–40.
29. Chen P. Treating patients as partners, by way of informed consent. The New York Times [Internet]. 2009 July 30 [cited 2018 Feb 4]. Available from: http://www.nytimes.com/2009/07/30/health/30chen.html?pagewanted=all.
30. Santen SA, Hemphill RR, McDonald MF, Jo CO. Patients' willingness to allow residents to learn to practice medical procedure. Acad Med. 2004;79:144–7.
31. Santen SA, Hemphill RR, Spanier CM, Fletcher ND. "Sorry, it's my first time!" Will patients consent to medical students learning procedures. Med Educ. 2005;39:365–9.
32. Ejaz A, Spolverato G, Bridges JF, Amini N, Kim Y, Pawlik TM. Choosing a cancer surgeon: analyzing factors in patient decision making using a best-worst scaling methodology. Ann Surg Oncol. 2014;21: 3732–8.
33. Lee Char SJ, Hills NK, Lo B, Kirkwood KS. Informed consent for innovative surgery: a survey of patients and surgeons. Surgery. 2013;153(4):473–80.
34. American College of Surgeons. Statement of principles [Internet]. Chicago (IL): ACoS; 2016 Apr 12 [cited 2018 Feb 4]. Available from: https://www.facs.org/about-acs/statements/stonprin.
35. *Johnson v. Kokemoor* [1996] 199 Wis.2d 615, 545 N.W.2d 495 (Wis.).

Further Reading on this Topic

Faden RR, Beauchamp TL. A history and theory of informed consent. New York: Oxford University Press; 1986.

Feinberg RS. The impaired physician: medical, legal and ethical analysis with a policy recommendation. Nova Law Rev. 2010;34(3):595–628.

Furrow BR. Must physicians reveal their wounds? Camb Q Healthc Ethics. 1996;5:204–13.

Grisso T, Appelbaum PS. Assessing competence to consent to treatment: a guide for physicians and other health professionals. New York: Oxford University Press; 1998.

Katz J. The silent world of doctor and patient. New York: The Free Press; 1984.

Kutner MA. The health literacy of America's adults: results from the 2003 National Assessment of Adult Literacy. Washington, D.C.: Department of Education, National Center for Education Statistics; 2006.

Mason NC, O'Neill O. Rethinking informed consent in bioethics. Cambridge: Cambridge University Press; 2007.

Schenker Y, Meisel A. Informed consent in clinical care: practical considerations in the effort to achieve ethical goals. JAMA. 2011;305(11):1130–1.

"Can't Stop Coughing (But I Need to Get Back to the Shelter by 6)"

Cynthia B. Morrow

4.1　Background Questions – 62

4.2　Additional Case Information and Questions for Discussion – 62

4.3　Answers to Background Questions – 64

4.4　Responses to Discussion Questions – 67

　　　References – 75

■ **Glossary**

Latent tuberculosis (TB) infection - Refers to a state in which a person has been exposed to *Mycobacterium tuberculosis*, as demonstrated by an immune response, but has no evidence of disease. People with latent tuberculosis (TB) infection cannot transmit TB.

TB disease (also referred to as "active TB") - Refers to a state in which a person has clinical evidence of TB. While TB disease most often affects the lungs, TB can also affect other parts of the body, such as the brain, spine, and kidneys.

Infectious TB disease - Refers to a state in which a person has clinical evidence of TB that is considered communicable. Pulmonary and laryngeal TB are communicable (infectious TB disease), whereas TB disease of the brain, spine, and kidneys is generally not considered to be communicable.

Learning Objectives
1. Apply the socioecological model to transmission of tuberculosis.
2. Analyze an infectious disease outbreak.
3. Describe the balance between autonomy (including confidentiality) and justice with respect to disease reporting, communicable disease investigations, and mandated treatment for communicable diseases.

> **Case Background**
>
> *You are a medical student working in the emergency department when a 45-year-old man, James Butler, presents with a cough and fever. Following protocol, the staff placed him in a room by himself. On review of symptoms, Mr. Butler states that he has had night sweats for the past month. He also reports some weight loss, but he attributes this to being homeless and not always having food available to eat.*

4.1 Background Questions

1. Define an epidemic and describe essential elements of an outbreak. What is a contact case investigation?
2. What is "directly observed therapy" for tuberculosis?
3. What are known risk factors for tuberculosis (TB) infection and active disease?
4. What are risk factors for becoming homeless?
5. What policies/procedures should hospitals have in place to prevent TB transmission in the hospital?
6. What legal authority, if any, is there to address non-compliance if a person with active pulmonary TB refuses to take medication?
7. Are there public health exemptions to the Health Insurance Portability and Accountability Act (HIPAA) for contagious diseases? Why?

4.2 Additional Case Information and Questions for Discussion

Mr. Butler's symptoms and history lead you to consider TB in your differential diagnosis. In order to complete a thorough history, you apply the socioecological model to review known risk factors for TB.

1. **Intrapersonal:** Considering what you should ask Mr. Butler, what individual factors would you take into account for this case?
2. **Interpersonal:** What role, if any, do family, friends, and social networks play in the transmission of tuberculosis? How would you assess this with Mr. Butler?
3. **Organizational:** How do organizational factors impact the risk of tuberculosis within a community?

You learn that Mr. Butler was born in the United States and has not travelled outside of the country. He is a smoker and states that in the past he was a heavy drinker and occasionally used intravenous drugs. He recalls being told that he had "a positive TB test" at a shelter in a different city but he did not take any medications. He denies any other pertinent past medical history.

Mr. Butler does not have any contact with family members. He "hangs out" with a few guys he knows from the local shelters. He usually, but not always, spends nights at one of two local homeless shelters. He shares that he has not been there for the past few weeks. He works day jobs when he can find them but that is not often.

Based on his clinical presentation and his history, you are concerned that Mr. Butler may have tuberculosis. You recommend a chest X-ray (CXR) and that he be placed in isolation. Your attending physician agrees. The CXR shows right upper lobe consolidation with mediastinal and hilar lymphadenopathy, suggestive of TB.

The patient is admitted to an airborne infection isolation room in the hospital. Sputum samples are sent to the lab. Your attending physician suggests that your fellow medical student also examine the patient, as a learning opportunity. However, when the student hears the complaint and suspicion of TB, he anxiously states, "I have young children at home, and my dad is getting chemotherapy. I don't want to go in there."

4. How would you respond to your colleague's concerns?

5. Consider a more likely scenario in which a patient with acquired immunodeficiency syndrome (AIDS) is admitted for treatment of disseminated cytomegalovirus (CMV). Standard precautions are typically used for patients with CMV who are hospitalized. However, CMV infection in pregnancy can cause devastating birth defects; therefore, it is not unusual for pregnant healthcare practitioners to avoid providing direct patient care, if other providers are available.
    (a) Self-reflection: In this scenario, how would you feel about the examination if you or your colleague were pregnant?
    (b) Do you have the right as a medical or health professions student or trainees to refuse to examine, or be present during the examination of, a patient?
    (c) How does the prevalence of disease affect our risk perception and risk tolerance? What about the lack of knowledge about an emerging disease?
    (d) Is there a situation in which you feel the educational benefit does not outweigh the risks? If so, should students or other trainees be barred from patient contact in those situations?

6. When can physicians—ethically and legally—refuse to treat a patient? When are they obligated to treat patients?

7. Do physicians have special responsibilities in emergency situations?

Your attending physician resolves your colleague's concerns about nosocomial transmission of TB and asks you to partner with the infection control nurse to report this suspected case of tuberculosis to the health department. You learn that there have been three other people, all men, diagnosed with tuberculosis in the past month.

8. Is this an epidemic? How do you make that determination?

9. What is the role of the health department in this situation?

The infection control nurse and the attending physician then share an advisory from the health department that they had received earlier that month. In it, the health department details the results of an investigation of recent cases of tuberculosis. The investigation revealed that all of the men who had tuberculosis had been in homeless shelters over that past couple of months; two of them were known to be positive for human immunodeficiency virus (HIV). The advisory provided recommendations for healthcare facilities to help manage cases of tuberculosis, including a reminder for healthcare facilities to ensure that they have the capacity to rapidly identify and isolate any cases of suspected TB, as was accomplished in this case.

10. What is the difference between isolation and quarantine? Are either of these strategies likely to be useful in this case?

11. Would quarantine be an *ethical* tool in the case of tuberculosis, given what you know about its transmission?

12. How do you weigh individual autonomy against public protection when making decisions about quarantine or isolation? What diseases would you consider as requiring such interventions?

Contact case investigations play a critical role in controlling tuberculosis. Staff at the local health department (LHD) and the local shelters were able to compile a master list of individuals who were at risk for tuberculosis infection based on dates of possible exposure to tuberculosis. The list of individuals at risk consisted primarily of other homeless men who stayed at the shelters as well as staff who worked during this time period. Although the team worked diligently to identify and locate these individuals, not all were found.

13. During contact case investigations, despite efforts to protect privacy, it is common for contacts to figure out the identity of the case patient. How do you balance protecting the individual's confidentiality and protecting the public's health?

14. What factors should be taken into account when considering the target population in this particular contact investigation? What strategies would you employ to increase testing and screening in this vulnerable population?

All individuals who were determined to be at risk for tuberculosis were offered screening including an initial tuberculin skin test (TST), 3-month follow-up TST for those with an initial negative TST, chest X-ray, and clinical assessment. Any individual identified with active TB was isolated and treated under directly observed therapy (DOT).

15. Why do you think directly observed therapy (DOT) is recommended for active TB?

16. If a patient with active tuberculosis refuses DOT, legal action may be pursued. What are the potential risks and benefits of legal action, such as a court order, to mandate treatment?

17. Do you think court orders are an *ethical* tool? Why or why not?

18. Do you think court orders are an *effective* tool?

During this contact investigation, the health department also recommended DOT for many of the individuals identified with TB infection.

19. Do you agree with the health department's decision to provide DOT to individuals with latent TB infection? Why or why not?

Unfortunately, even though DOT was ordered for almost 90 men, very few individuals actually completed therapy. Over the following year, several more homeless men from this community were diagnosed with active TB. DNA fingerprinting demonstrated that the majority of cases were caused by an identical strain of Mycobacterium tuberculosis, indicating a common source. An isolate from one case was closely related and felt to be a variant of the same strain. Almost half of the men diagnosed with active TB had a history of a prior positive TST and had initiated therapy at some point in the past, but none was known to have completed a full course of therapy. It was determined that concurrent HIV infections had contributed to the epidemic. Over the ensuing years, the same strain of Mycobacterium tuberculosis was found in homeless individuals living in shelters in other states.

20. As a future physician or other healthcare practitioner practicing in this community, what public health policies would you recommend to address the prevention of tuberculosis in high-risk settings?

21. Do you think that physicians or other healthcare practitioners have a responsibility to advocate for such policies? If so, how would they do so and whom would they enlist to help?

4.3 Answers to Background Questions

1. Define an epidemic and describe essential elements of an outbreak. What is a contact case investigation?
   "Epidemic" refers to an increase in the number of cases of a disease above what is normally expected in a given population and geographic area (any activity above baseline). Typically, this implies a rapid increase above baseline; however, "epidemic" has also been used to describe a slower but consistent increase in baseline, such as the changes seen in prevalence of obesity or type 2 diabetes over the past several decades. If an epidemic spans several countries or continents and impacts a large number of people, it is often referred to as a "pandemic" [1].
   "Outbreak" is also defined as an increase in disease activity but is usually reserved

to describe infectious disease activity that occurs over a limited geographic area and over a shorter timeframe [1].

With respect to infectious diseases, epidemics or outbreaks can occur when there are adequate numbers of people who are susceptible to the agent. This can happen if the agent is new to the environment, if there is a change in the way it is transmitted, or if it becomes more virulent [1].

A contact case investigation involves a systematic approach to identify persons who may have been exposed to an infectious agent. According to the World Health Organization (WHO), "[c]ontacts are commonly investigated in high-income countries with low TB burdens and in settings in which a TB elimination policy is implemented, in order to identify persons with early active TB or who have recently been infected" [2]. Comprehensive guidelines for contact investigations for tuberculosis are published by the Centers for Disease Control and Prevention (CDC) and include detailed information about who should be screened for exposure to TB and how they should be screened [3].

2. **What is "directly observed therapy" for tuberculosis?**
All individuals with active TB should receive case management services because adherence to treatment for active TB can be challenging for many reasons, including the number of medications needed (multiple drugs are needed to improve cure rates and decrease likelihood of drug resistance), the side effects of these multiple drugs, the length of required treatment (minimum 6 months), and the fact that symptoms often resolve within a few weeks which can decrease motivation to continue medications. According to the CDC: "Directly observed therapy (DOT) may be defined as a course of treatment, or preventive treatment, for TB in which the prescribed course of medication is administered to the person or taken by the person under direct observation by a trained healthcare worker. DOT increases cure rates among patients with TB and is also effective in decreasing drug resistance, treatment failure, relapse, and mortality" [4]. The trained healthcare worker should then ideally record all of the treatment data into the health information system to optimize management of the case.

The CDC recommends that DOT be used for all patients with active TB [5]. In addition to recommending DOT as part of the treatment protocol for patients with active TB, DOT may be recommended as part of the treatment protocol for latent TB depending on the person being treated (e.g., children) or the treatment plan (e.g., once/twice weekly regimens) [6].

3. **What are known risk factors for tuberculosis infection and active disease?**
Most people who are exposed to infectious TB disease do not develop active TB disease. The most important risk factor for someone to have latent TB is to be exposed to someone with infectious TB disease. The following list of risk factors for exposure is from the CDC [7]:
   - Close contacts of a person with infectious TB disease
   - Persons who have lived in an area of the world with high rates of TB
   - Persons who work or reside with people who are at high risk for TB in facilities or institutions such as hospitals, homeless shelters, correctional facilities, nursing homes, and residential homes for those with HIV

Some people develop active TB disease soon after becoming infected (within weeks) before their immune system can fight the TB bacteria. Other people may get sick years later, when their immune system becomes weak for another reason.

Overall, about 5% to 10% of infected persons who do not receive treatment for latent TB infection will develop active TB disease at some time in their lives. For persons whose immune systems are weak, especially those with HIV infection, the risk of developing TB disease is much higher than for persons with normal immune systems.

Generally, persons at high risk for developing TB disease fall into two categories [7]:
- Persons, particularly young children, who have been *recently* infected with TB bacteria (see aforementioned risk factors for latent TB)
- Persons with medical conditions that weaken the immune system, including HIV infection, substance use disorders, silicosis, diabetes mellitus, severe kidney disease, or malnutrition, or conditions that require immunosuppressive treatments such as corticosteroids or chemotherapy (e.g., cancer or autoimmune diseases such as rheumatoid arthritis or Crohn disease)

4. **What are risk factors for becoming homeless?**
There are many known risk factors for being homeless including poverty or financial stress, mental health problems, substance use disorders, involvement with the criminal justice system, poor family functioning, low educational achievement, and poor housing conditions. Being young, gay, transgender, male, a veteran, and/or African-American are also independent risk factors for homelessness [8–10].

5. **What policies/procedures should hospitals have in place to prevent TB transmission in the hospital?**
Hospitals should have a comprehensive overarching policy, likely with many associated procedures, in place to prevent nosocomial transmission of TB. The policy should cover [11]:
- Risk assessment, including:
  - Early identification of individuals with infectious TB disease, for example, emergency department procedures for patients presenting with signs/symptoms of TB
  - Surveillance of hospital staff with annual testing and treatment for latent TB as indicated
- Exposure control with procedures to immediately place an individual with suspected infectious TB disease in an airborne infection isolation room
- Communication protocol for appropriate notification to local public health officials
- Personal protective equipment for all personnel involved in direct patient care of individuals with TB
- Appropriate laboratory safety precautions in place to ensure safe handling of patients' clinical specimens

6. **What legal authority, if any, is there to address non-compliance if a person with active pulmonary TB refuses to take medication?**
Laws authorize and obligate the government to protect the public's health within limits; for example, due process, equal protection, protection of privacy, and bodily integrity. The U.S. Supreme Court has repeatedly affirmed that the government may, again within limits, curtail individuals' rights to protect the public's health. School vaccination laws are a classic example of this. The balance of individual rights versus public protection should have a sound scientific basis.

With respect to communicable diseases, there are extensive laws addressing both surveillance (mandatory reporting of certain communicable diseases) and protection of those data. In most states, there are also laws that require testing for certain communicable diseases as a condition of employment or other privileges and benefits. In many jurisdictions, public health officials may require physical examinations and/or treatment of individuals who have or who are suspected of having select communicable diseases.

With respect to TB, three "fundamental interests support a state's use of compulsory examination or treatment in cases involving TB disease: (1) preserving an individual's own health or life, (2) preventing harm to others, and (3) avoiding the possible development of drug resistance", especially multidrug-resistant (MDR) TB and extensively drug-resistant (XDR) TB [12]. Persons with TB cannot be forced to undergo an exam or to take medication ("protection of bodily integrity"); however, they may be ordered by a court to remain isolated until no longer considered a threat to public health. Furthermore, "[p]ublic health laws may also authorize public health authorities to confine persons with contagious diseases to protect the community. Through detention laws, authorities may confine an individual to a health or other facility appropriate for his or her medical condition. Since detention presents a significant restriction on individual liberty, courts may generally require that procedural due process be satisfied regardless of whether existing detention statutes specifically delineate such due process" [12].

7. **Are there public health exemptions to HIPAA for contagious diseases? Why?**
   Yes. "The Health Insurance Portability and Accountability Act's privacy rule recognizes the legitimate need for public health authorities and others responsible for ensuring public health and safety to have access to protected health information to carry out their public health mission" [13]. Essentially, even sensitive health information that can be readily linked to an individual may be shared by health and public health workers and even with schools, employers, and others if necessary to protect the public's health. Any such sharing of information must be limited to only pertinent health information and must be handled with the greatest care to protect the individual's privacy to the extent possible. Other laws are in place to protect that individual from unwarranted discrimination [13].

4.4 Responses to Discussion Questions

Mr. Butler's symptoms and history lead you to consider TB in your differential diagnosis. In order to complete a thorough history, you apply the socioecological model to review known risk factors for TB.

1. **Intrapersonal: Considering what you should ask Mr. Butler, what individual factors would you take into account for this case?**
   The following topics should be explored as part of the history.
   - TB:
     - Prior history of TB
     - Known exposure to TB
     - TST (past and present)
     - CXR consistent with TB
   - Social history including alcohol, drug, and tobacco use:
     - Excessive alcohol use affects the immune system and increases risk of TB [14].
     - Drug use increases the likelihood of having a positive sputum smear and increases the length of contagiosity; this is likely due to treatment failure, possibly related to associated exposures (people who use drugs are more likely to experience incarceration and homelessness) [15].
     - Smoking increases risk of TB [16].
   - Medical history:
     - Other underlying medical problems, such as HIV positivity, diabetes, malnutrition, certain gastrointestinal (GI) conditions, certain malignancies, chronic steroid use, and general stress or poor health, are risk factors for reactivation.
   - Other risk factors:
     - Born in high-endemic country
     - Low socioeconomic status (especially in the urban population)
     - Age > 65
     - Occupation: healthcare workers

2. **Interpersonal: What role, if any, do family, friends, and social networks play in the transmission of tuberculosis? How would you assess this with Mr. Butler?**
   Family and friends may have been exposed to active tuberculosis; therefore, a full contact investigation is necessary. Consider that even though Mr. Butler is homeless, he may very well spend time with family members or friends outside the shelter.
   Who else is at risk? Include questions about:
   - Employment history.
   - Participation in religious organizations or other social organizations.

- Food insecurity (e.g., does he participate in meal programs?)
- Medical treatment (e.g., is he in any treatment programs?)

3. **Organizational: How do organizational factors impact the risk of tuberculosis within a community? What would you want to know about your community in this case?**
   - Organizational structures such as homeless shelters with strong TB prevention policies in place may decrease the likelihood of TB in a community.
   - Access to care for any individual at increased risk for TB infection can also impact the risk of TB. What role does the local health department play in the community? Is there a strong component of outreach? Does the health department provide comprehensive, free TB case management?
   - What are the support systems in place for persons with substance abuse?

You learn that Mr. Butler was born in the United States and has not travelled outside of the country. He is a smoker and states that in the past he was a heavy drinker and occasionally used intravenous drugs. He recalls being told that he had "a positive TB test" at a shelter in a different city but he did not take any medications. He denies any other pertinent past medical history.

Mr. Butler does not have any contact with family members. He "hangs out" with a few guys he knows from the local shelters. He usually, but not always, spends nights at one of two local homeless shelters. He shares that he has not been there for the past few weeks. He works day jobs when he can find them but that is not often.

Based on his clinical presentation and his history, you are concerned that Mr. Butler may have tuberculosis. You recommend a chest X-ray (CXR) and that he be placed in isolation. Your attending physician agrees. The CXR shows right upper lobe consolidation with mediastinal and hilar lymphadenopathy, suggestive of TB.

The patient is admitted to an airborne infection isolation room in the hospital. Sputum samples are sent to the lab. Your attending physician suggests that your fellow medical student also examine the patient, as a learning opportunity. However, when the student hears the complaint and suspicion of TB, he anxiously states, "I have young children at home, and my dad is getting chemotherapy. I don't want to go in there."

4. **How would you respond to your colleague's concerns?**
   As long as the patient is in an appropriate isolation room and the student adheres to airborne precautions (such as wearing the appropriate personal protection equipment), there is no significant risk to the student. This is an opportunity to educate the fellow student about some of the myths about the transmission of tuberculosis.

5. **Consider a more likely scenario in which a patient with AIDS is admitted for treatment of disseminated cytomegalovirus (CMV). Standard precautions are typically used for patients with CMV who are hospitalized. However, CMV infection in pregnancy can cause devastating birth defects; therefore, it is not unusual for pregnant healthcare practitioners to avoid providing direct patient care, if other providers are available.**

> **Box 4.1   Teaching Tip**
> The following questions (5a and 5b) should prompt an open-ended discussion with self-reflection. It would be helpful for facilitators to know if their teaching institutions have policies that (1) address students' or trainees' rights to refuse to be involved in a particular aspect of patient care, including the care of specific patients, and (2) address the role and protection of students and trainees during an outbreak of an emerging infectious disease.

   (a) Self-reflection: In this scenario, how would you feel about the examination if you or your colleague were pregnant?
   (b) Do you have the right, as a medical or health professions student or trainee, to refuse to examine, or be present during the examination of, a patient?

While medical and health professions students do not have the same legal obligations that physicians have with respect to treating patients without discrimination, they are still bound by the same ethical principles. Students should not be allowed to refuse to examine a patient or be present during the examination of a patient based on the patient's gender, race, ethnicity, income, education, sexuality, and so forth.

The discussion becomes more controversial when the key concern is about communicable disease transmission. There is an inherent risk of exposure to communicable diseases in health care. Do healthcare professionals assume that risk when they choose to enter the field? Primacy of patient welfare is a core principle for healthcare officials; however, it must be balanced with personal risk to the healthcare professional. The risk can and should be mitigated by strong infection control policies and procedure to minimize the risks. However, in situations of uncertainty, particularly with emerging infectious diseases, a common-sense risk reduction approach should be taken into account. For example, during the severe acute respiratory syndromes (SARS) epidemic in Toronto, Canada, 90% of the reported cases of SARS were associated with exposure at a single hospital. Healthcare workers, including students, comprised almost 40% of the cases. In a situation in which students have very little to offer with respect to patient welfare but are at significant personal risk, it is reasonable to question whether it is ethical to expose them to the risk [17].

(c) **How does lack of knowledge about an emerging disease affect our risk perception and risk tolerance?**
Lack of knowledge about an emerging disease can contribute to public fear. The perception that medical experts have a lack of knowledge about an emerging disease can exacerbate that fear. Recent examples of this fear include the 2009 H1N1 pandemic, the 2014 Ebola outbreaks, and more recently the Zika virus epidemic. In reference to the Ebola crisis, Lisa Rosenbaum states "when we face an uncertain prospect that we deeply fear, we evince what Cass Sunstein calls 'probability neglect': we tend to conflate the horror of what might happen with the likelihood that it will. Unless we can prove there's zero risk, the dreaded event feels exceedingly likely, and thus making probabilistic comparisons may not feel reassuring" [18]. Healthcare professionals, including medical students, are not immune to this phenomenon, which likely impacts both their risk perception and their risk tolerance [19].

> **Box 4.2   Teaching Tip**
> Consider sharing the following with learners to facilitate the discussion. In the early 1980s, when HIV and AIDS were first becoming recognized, it took more than 2 years for scientists to understand that a virus was the source of the disease. Many physicians refused to treat patients suspected of having the disease. One example taken from a *New York Times* article from 1987 is from the chief heart surgeon at Milwaukee hospital: "I've got to be selfish. I've got to think about myself; I've got to think about my family. That responsibility is greater than to the patient" [19].

(d) **Is there a situation in which you feel the educational benefit does not outweigh the risks? If so, should students or other trainees be barred from patient contact in those situations?**
Some factors to consider are:
- What are the *specific* educational benefits from seeing patients with these diseases?
- Is proper personal protective equipment (PPE) available for all members of the medical team?
- Are accepting these risks and seeing the patient part of the maturation process of a growing physician? Practicing physicians play a pivotal role as healthcare providers. Are these types of experiences necessary for that development?

✓ 6. **When can physicians—ethically and legally—refuse to treat a patient? When are they obligated to treat patients?**
In general, federal laws have protections to prevent discrimination, which prohibit physicians from refusing to treat patients for racial or religious reason. Furthermore, "[m]any states prohibit places of "public accommodation," including doctors' offices and hospitals, from discriminating on the basis of characteristics such as race, color, national origin, nationality, ancestry, religion, creed, age, marital status, familial status, sex, sexual orientation, gender identity, medical condition, disability, or other personal features — although, beyond the baseline federal protections, the grounds that are included vary by jurisdiction. Title VI of the federal Civil Rights Act of 1964 prohibits discrimination on the basis of race, color, and national origin in programs and activities that receive federal financial assistance, including Medicaid and Medicare. The Rehabilitation Act of 1973 adds disability to that list" [20].

Outside of the aforementioned parameters, it may not be illegal for a physician to refuse to treat a patient, but there are significant ethical concerns about a physician doing so. There is a complex balance between the physician's autonomy and the principles of beneficence, non-maleficence, and justice. For example, if another provider is not available to see the patient, a physician refusing to treat a patient for fear of contracting a communicable disease seems to be violating both non-maleficence and justice. The refusing physician may cause direct harm (by neglect) to the patient as well as to the community in which they live (by eroding trust). With respect to justice, consider justice both in relation to the patient who is not receiving timely care and the community to which the patient belongs. That community is now at risk if the case patient does not receive medical attention because there could be ongoing transmission of disease. In another scenario, in which a different provider is able and willing to see the patient, the patient is not directly harmed by a physician who refuses to treat. However, again, the physician's behavior may cause distrust (harm) in the healthcare system. Furthermore, is the physician who refuses to see the patient acting unjustly to the other physician present? Not only might this impact the patient load of the physician who picks up the extra case but now that individual is exposed to the communicable disease.

It is important to consider that the Hippocratic Oath outlines a responsibility to society as well as to individual patients.

✓ 7. **Do physicians have special responsibilities in emergency situations?**
The Emergency Medical Treatment and Labor Act (EMTALA) is a federal law that requires that anyone coming to an emergency department be stabilized and treated, regardless of their insurance status or ability to pay. They can only be transferred to another facility with the patient's request or if appropriate treatment cannot be provided at the original hospital. Although this only applies to hospitals that receive payment from the Centers for Medicare and Medicaid, that is the great majority of hospitals [21].

Your attending physician resolves your colleague's concerns about nosocomial transmission of TB and asks you to partner with the infection control nurse to report this suspected case of tuberculosis to the health department. You learn that there have been three other people, all men, diagnosed with tuberculosis in the past month.

✓ 8. **Is this an epidemic? How do you make that determination?**
Because the definition of epidemic includes "activity above baseline" and the baseline data are not shared here, it is difficult to definitively state that this is an outbreak. However, the temporal and geographic clustering strongly suggest that this is an epidemic and, more specifically, an outbreak.

> **Box 4.3 Teaching Tip**
> The student responsible for Background Question 1 should have the definitions for epidemic and outbreak.

9. **What is the role of the health department in this situation?**
   The local health department is responsible for establishing if there is a public health threat. Even a single case of infectious TB presents a potential public health threat and needs some level of intervention. In this case, there is an even more urgent concern given the consecutive appearance of cases, indicating the possibility of a larger outbreak. Every case of infectious TB should prompt a contact case investigation to identify other cases of latent or infectious TB. Rarely, a **source** case investigation is done to identify where the newly diagnosed person became infected. Since, in the United States, most cases of active TB are due to reactivation, years after initial infection, it is almost impossible to identify the source. A source case investigation may be indicated in a cluster of cases or in a young child who has primary TB.

   In addition to starting contact case investigations, the health department will report the cases to the state health department, which will then report to the CDC. The health department should also consider notification of healthcare providers, and potentially the public, depending on the specifics of the case. In this scenario, with several new cases of TB linked to homeless shelters, the health department would ideally work closely with the shelters and with other involved service agencies to ensure good communication and a coordinated approach to managing the contact investigation.

The infection control nurse and the attending physician then share an advisory from the health department that they had received earlier that month. In it, the health department details the results of an investigation of recent cases of tuberculosis. The investigation revealed that all of the men who had tuberculosis had been in homeless shelters over that past couple of months; two of them were known to be positive for human immunodeficiency virus (HIV). The advisory provided recommendations for healthcare facilities to help manage cases of tuberculosis, including a reminder for healthcare facilities to ensure that they have the capacity to rapidly identify and isolate any cases of suspected TB, as was accomplished in this case.

10. **What is the difference between isolation and quarantine? Is either of these strategies useful in this case?**
    When referring to the control of communicable diseases, the term "isolation" refers to the separation of **ill** persons from those who are not ill. In a hospital setting, this typically refers to placing a patient with a communicable disease, such as tuberculosis, in an appropriately equipped room. In the community setting, for an individual who is not ill enough to require hospitalization, this may refer to mandating restrictions of movement for the person with a communicable disease, often requiring that the individual stay in one place, removed from all others. For active infectious TB disease, the health department may consider random home visits or landline phone calls to ensure adherence to these requirements [22].

    The term "quarantine" refers to the separation and/or movement restrictions of **well** persons who are thought to have been exposed to a communicable disease but who are not yet ill. The timing and the duration of quarantine are directly related to the incubation period of the particular communicable disease [22].

11. **Would quarantine be an *ethical* tool in the case of tuberculosis, given what you know about its transmission?**
    A patient with latent TB does not have infectious TB disease and is not able to spread the disease and therefore poses no threat to others. In this case, it would not be an ethical tool.

    Only patients with untreated infectious TB disease are at risk of spreading TB. By definition, patients with untreated infectious TB are not considered "well"; therefore, quarantine is not applicable. The question then becomes whether isolation is an ethical tool. Most people argue that it is unethical *not* to isolate a person with infectious TB disease, as that individual is an imminent threat to others' health.

12. **How do you weigh individual autonomy against public protection when making decisions about quarantine or isolation? What diseases would you consider as requiring such interventions?**
    Individual autonomy must be respected to the extent possible; however, justice—in this case, obligation to protect others from a communicable disease—must be considered when an individual poses such a threat. Because infectious tuberculosis is a disease that is characterized by airborne transmission, any individual with infectious TB poses a public threat. In the United States, "states have police power functions to protect the health, safety, and welfare of persons within their borders. To control the spread of disease within their borders, states have laws to enforce the use of isolation and quarantine. These laws can vary from state to state and can be specific or broad. In some states, local health authorities implement state law. In most states, breaking a quarantine order is a criminal misdemeanor" [22].
    Diseases that are currently authorized for isolation or quarantine at the federal level based on an "Executive Order of the President" include cholera, diphtheria, infectious tuberculosis, plague, smallpox, yellow fever, viral hemorrhagic fevers, severe acute respiratory syndromes (SARS), and influenza that may cause a pandemic [22].

Contact case investigations play a critical role in controlling tuberculosis. Staff at the local health department (LHD) and the local shelters were able to compile a master list of individuals who were at risk for tuberculosis infection based on dates of possible exposure to tuberculosis. The list of individuals at risk consisted primarily of other homeless men who stayed at the shelters as well as staff who worked during this time period. Although the team worked diligently to identify and locate these individuals, not all were found.

13. **During contact case investigations, despite efforts to protect privacy, it is common for contacts to figure out the identity of the case patient. How do you balance protecting the individual's confidentiality and protecting the public's health?**
    Local health authorities should make every effort possible to protect the individual's identity, including using language such as "We are currently investigating a case of tuberculosis and have reason to believe that you may have been exposed to the disease" rather than identifying the individual who has tuberculosis.

> **Box 4.4 Teaching Tip**
> This question is slightly different from Discussion Question 12. For this question, it may be helpful to provide a specific scenario. For example, imagine that a case of TB is identified in a high school and only students who share the same classes need to be tested. When the infected student is suddenly absent from school, it will be very easy for all of the student's classmates to determine who has TB. Consider asking learners how they would feel if other people found out if they had a communicable disease. Similarly, consider a case in which someone had infectious TB but fails to adhere to the isolation requirements. In such a case health authorities may release the individual's name to the workplace, church, school, etc. as a means to protect others. Ask learners how they feel in that scenario, since it appears to be a violation of HIPAA.

14. **What factors should be taken into account when considering the target population (i.e., homeless individuals) in this particular contact investigation? What strategies would you employ to increase testing and screening in this vulnerable population?**
    Specific factors that should be considered when conducting a contact investigation in this situation include:
    – Accessibility: It is likely that it will be challenging to locate individuals who are homeless. Consider that there are likely:
        – Individuals who frequently stay at the shelters and therefore are more likely to be available for testing

- Individuals who only occasionally stay at the shelters and are more likely going to be very difficult to reach for testing
- Individuals for whom little information is available
- Mobility of the population outside of the community
- Possible distrust of public health officials

In terms of strategies, ideally a number of people, including medical providers for homeless individuals and managers of social programs, should be involved in planning efforts to locate the persons at risk. The team can then determine patterns of utilization of services to try to prioritize where the individuals at risk are likely to be found (such as places providing free meals, places offering daily employment, etc.). Other strategies include:

- Going door-to-door to try to locate individuals
- Contacting other agencies to assist with finding persons at risk (which would require the identities of those who may have been exposed)
- Developing informational social media posts
- Placing informational flyers/posters in strategic locations
- Providing small incentives (e.g., bus tokens, snacks, etc.)

All individuals who were determined to be at risk for tuberculosis were offered screening including an initial tuberculin skin test (TST), 3-month follow-up TST for those with an initial negative TST, chest X-ray, and clinical assessment. Any individual identified with active TB was isolated and treated under directly observed therapy (DOT).

15. Why do you think directly observed therapy (DOT) is recommended for active TB?

Treatment for active TB requires a lot of medications over a long period of time. Typically, treatment for infectious TB includes four-drug therapy for at least 2 months followed by an additional 4 months of two-drug treatment. Adherence is a significant challenge, and poor adherence is associated with development of multidrug-resistant (MDR) TB. Significant barriers for patients to adhere to the treatment regimen include the number and duration of medications, particularly once the individual is asymptomatic, as well as side effects. Having a person facilitate scheduling and delivery of medications may increase adherence and decrease the risk of drug resistance.

16. If a patient with infectious tuberculosis refuses DOT, legal action may be pursued under some circumstances. What are the potential risks and benefits of legal action, such as a court order, to mandate treatment?

Historically, court orders have been issued for involuntary detention/hospitalization if an individual with infectious tuberculosis does not comply with treatment recommendations. In many states, laws allow patients to be confined if they can be shown to be infectious (i.e., if they can be shown to be a present threat). In most states, this does not allow for future threat; in other words, a person cannot be forced to complete therapy once the sputa are negative. The key benefit for such legal action is increased adherence to the TB treatment protocol. In addition, such action may increase the public's trust of their local health authority. The key risk is that because there are typically limits on how long an individual can be detained or hospitalized, it is unlikely that treatment will be completed under these circumstances, and it is possible that the individual stops treatment and inadvertently contributes to drug resistance. Furthermore, on the part of the individual and the individual's family and friends, such aggressive tactics may increase distrust in the system and deter others from seeking care if they become ill.

> **Box 4.5 Teaching Tip**
> To encourage discussion, consider the following scenario as an alternative to the case with TB. If invasive group A strep (GAS) is identified in a patient shortly after a surgical procedure or if several patients in a facility develop GAS over a short period of time, healthcare workers (HCW) need to be considered as a potential source of the GAS. In this case, asymptomatic HCWs may need to provide their own swabs (throat, rectal, vaginal) for testing to ensure that they are not carriers. HCWs may be resistant but may be ordered to comply or risk their license. Ask learners how they would feel if asked to provide a rectal or vaginal swab.

17. **Do you think court orders are an *ethical* tool? Why or why not?**
    Yes. It has been well established that there are reasonable limits to a person's autonomy, but there are several caveats important to consider when ensuring that a court order is implemented appropriately. Is the person whose rights are being infringed upon a threat to the public? (Does the person have infectious TB disease?) Have all other measures to have that individual comply with isolation been exhausted? Will the individual's needs for food/shelter/medical care be met in a humane manner?

18. **Do you think court orders are an *effective* tool?**
    In the inpatient setting, adherence to isolation precautions for patients with communicable diseases can decrease nosocomial transmission of disease, but evidence for the impact of court orders is less well established. The effectiveness of court orders for containing the risk of tuberculosis is limited by the duration of treatment needed for tuberculosis and the interpretation of "a public health threat." For example, if a judge deems someone to be at risk only if he has positive sputa cultures, once the person is treated for 2 or more weeks, the sputa are likely to be negative, and the judge may deem that the individual no longer poses an imminent public health threat. If the individual is no longer compelled to adhere to the treatment regimen and discontinues treatment, he may develop drug-resistant TB, which in turn may lead to an even greater public threat. Given different laws across different jurisdictions, it is unlikely that the judicial system can address this issue consistently and comprehensively. Furthermore, it is conceivable that issuing a court order for one person with TB could lead other people suspected of having TB disease to go "underground," thereby potentially increasing the threat of transmission of TB within the community.

During this contact investigation, the health department also recommended DOT for many of the individuals identified with TB infection.

19. **Do you agree with the health department's decision to provide DOT to individuals with latent TB infection? Why or why not?**
    Individuals with risk factors for poor adherence (homelessness, substance use disorders, prior history of poor adherence, etc.) may be considered for DOT for latent TB. In addition, DOT is often recommended for children with latent TB. In some jurisdictions, the local health department works with school nurses to have the medications administered at school in order to increase adherence.

Unfortunately, even though DOT was ordered for almost 90 men, very few individuals actually completed therapy. Over the following year, several more homeless men from this community were diagnosed with active TB. DNA fingerprinting demonstrated that the majority of cases were caused by an identical strain of Mycobacterium tuberculosis, indicating a common source. An isolate from one case was closely related and felt to be a variant of the same strain related. Almost half of the men diagnosed with active TB had a history of a prior positive TST and had initiated therapy at some point in the past, but none was known to have completed a full course of therapy. It was determined that concurrent HIV infections had contributed to the epidemic. Over the ensuing years, the same strain of Mycobacterium tuberculosis was found in homeless individuals living in shelters in other states.

20. **As a future physician or other healthcare practitioner practicing in this community, what public health policies would you recommend to address the prevention of tuberculosis in high-risk settings?**

    Examples of potential policies to decrease the risk of TB in high-risk settings include:
    - Policy for homeless shelters to ensure identification of infectious and latent TB, for example, requiring overnight visitors/residents to have a card that documents their TST results (and treatment status if applicable)
    - Improved ventilation, ultraviolet (UV) lights in homeless shelters
    - Improvement in housing policies that decrease the number of people in shelters
    - Incentives for treatment of latent tuberculosis infection (LTBI) (fast-food coupons, transportation tokens, small gifts, drinks, pudding, etc. for the children)

    > **Box 4.6 Teaching Tip**
    >
    > To encourage discussion, have learners consider the relative effectiveness, feasibility, social will, risks, etc. of the different strategies. For example, the personal incentives for testing are likely to be less effective than a strict policy requiring documented testing. However, the strict policy might have the adverse effect that some homeless people without documentation have nowhere to stay.

21. **Do you think that physicians or other healthcare practitioners have a responsibility to advocate for such policies? If so, how would they do so and whom would they enlist to help?**

    One aspect of physicians' professional obligations is to society. Physicians and other healthcare practitioners are uniquely situated to understand the individual consequences of certain policies. In this case, they can ask the local health department what they are doing and advocate for them to get involved if they have not enacted all of these types of policies already. In most areas, the local health department will be the subject matter expert in this; physicians can work with the health department to understand barriers to implementation and identify ways that local physicians can better support the department.

Acknowledgments   This case was adapted from an earlier case written by Cynthia Morrow, Don Cibula, and Lloyd Novick that was published in the *American Journal of Preventive Medicine* in 2003.

References

1. Centers for Disease Control and Prevention. Principles of epidemiology in public health practice: introduction to applied epidemiology and biostatistics. 3rd edition [Internet]. Atlanta (GA): CDC; 2012 May 18 [cited 2017 Nov 1]. Available from: https://www.cdc.gov/ophss/csels/dsepd/ss1978/lesson1/section11.html.
2. World Health Organization. Tuberculosis (TB) Programme contact investigations [Internet]. Geneva: WHO. [cited 2017 Nov 1]. Available from: http://www.who.int/tb/areas-of-work/laboratory/contact-investigation/en/.
3. Taylor Z. Guidelines for the investigation of contacts of persons with infectious tuberculosis: recommendations of the recommendations from the National Tuberculosis Controllers Association and CDC [Internet]. Atlanta (GA): CDC; 2005 Dec 16 [cited 2017 Nov 1]. Available from: https://www.cdc.gov/mmwr/preview/mmwrhtml/rr5415a1.htm.
4. Centers for Disease Control and Prevention. Menu of suggested provisions for state tuberculosis prevention and control laws, case management, treatment guidelines, and required treatment [Internet]. Atlanta (GA): CDC; 2010 Oct 8 [cited 2017 Nov 1]. Available from: https://www.cdc.gov/tb/programs/laws/menu/treatment.htm.
5. Centers for Disease Control and Prevention. TB 101 for health care workers [Internet]. Atlanta (GA): CDC. [cited by 2017 Nov 1]. Available from: https://www.cdc.gov/tb/webcourses/tb101/page3832.html.
6. Centers for Disease Control and Prevention. Treatment options for latent tuberculosis infection [Internet]. Atlanta (GA): CDC; 2016 August 31 [cited 2017 Nov 1]. Available from: https://www.cdc.gov/tb/publications/factsheets/treatment/ltbitreatmentoptions.htm.
7. Centers for Disease Control and Prevention. TB risk factors [Internet]. Atlanta (GA): CDC; 2016 Mar 18 [cited 2017 Nov 1]. Available from: https://www.cdc.gov/tb/topic/basics/risk.htm.
8. Shelton K, Taylor P, Bonner A, van den Bree M. Risk factors for homelessness: evidence from a population-based study. Psychiatr Serv. 2009;60(4):465–72.
9. To MJ, Palepu A, Aubry T, Nisenbaum R, Gogosis E, Gadermann A, Cherner R, Farrell S, Misir V, Wang S. Predictors of homelessness among vulnerably housed

adults in 3 Canadian cities: a prospective cohort study. BMC Public Health. 2016;16:1041.
10. Tsai J, Rosenheck R. Risk factors for homelessness among US veterans. Epidemiol Rev. 2015;37:177–95.
11. The Johns Hopkins University/The Johns Hopkins Hospital. Health, safety, and environment manual. The Airborne Pathogen Control Program. Policy HSE 601 [Internet]. Baltimore (MD). [cited 2017 Nov 1]. Available from: https://hpo.johnshopkins.edu/hse/policies/156/10979/policy_10979.pdf.
12. The Centers for Law and the Public's Health: A Collaborative at Johns Hopkins and Georgetown Universities. Tuberculosis control laws and policies: a handbook for public health and legal practitioners [Internet]. 2009 Oct 1 [cited 2017 Nov 1]. Available from: https://www.cdc.gov/tb/programs/tblawpolicyhandbook.pdf.
13. United States Health and Human Services. Special topics in health information privacy [Internet]. Washington, DC: HHS; 2002 Dec 3 [updated 2003 Apr 3; cited 2017 Nov 1]. Available from: https://www.hhs.gov/hipaa/for-professionals/special-topics/public-health/index.html.
14. Lönnroth K, Williams BG, Stadlin S, Jaramillo E, Dye C. Alcohol use as a risk factor for tuberculosis - a systematic review. BMC Public Health. 2008;8:289.
15. Oeltmann JE, Kammerer JS, Pevzner ES, Moonan PK. Tuberculosis and substance abuse in the United States, 1997-2006. Arch Intern Med. 2009;169(2):189–97.
16. Slama K, Chiang CY, Enarson DA, Hassmiller K, Fanning A, Gupta P, et al. Tobacco and tuberculosis: a qualitative systematic review and meta-analysis. Int J Tuberc Lung Dis. 2007;11(10):1049–61.
17. Wallington T, Berger L, Henry B, Shahin R, Yaffe B, Mederski B, et al. Update: severe acute respiratory syndrome -Toronto, Canada, 2003. MMWR. 2003;52(23):547–50.
18. Rosenbaum L. Communicating uncertainty - Ebola, public health, and the scientific process. N Engl J Med. 2015;372:7–9.
19. When Doctors Refuse to Treat AIDS. The New York Times [Internet]. 1987 August 3 [cited 2017 Nov 29]. Available from: http://www.nytimes.com/1987/08/03/opinion/when-doctors-refuse-to-treat-aids.html.
20. Lynch HF. Discrimination at the doctor's office. N Engl J Med. 2013;368:1668–70.
21. United States Government. Centers for Medicare & Medicaid. Emergency medical treatment & labor act (EMTALA) [Internet]. Baltimore (MD): CMS; 2012 Mar 26 [cited 2017 Nov 29]. Available from: https://www.cms.gov/Regulations-and-Guidance/Legislation/EMTALA/.
22. Centers for Disease Control and Prevention. Legal authorities for isolation and quarantine [Internet]. Atlanta (GA): CDC; 2014 Oct 8 [cited 2017 Nov 1]. Available from: https://www.cdc.gov/quarantine/about-lawsregulationsquarantineisolation.html.

Further Reading on this Topic

Centers for Disease Control and Prevention. Fact sheet, tuberculosis: an overview [Internet]. Atlanta (GA): CDC; 2014 March. Available from: https://www.cdc.gov/nchhstp/newsroom/docs/factsheets/tb-overview-factsheet.pdf.

Gupta V, Sugg N, Butners M, Allen-White G, Molnar A. Tuberculosis among the homeless—preventing another outbreak through community action. N Engl J Med. 2015;372(16):1483–5.

Community Perspectives

Contents

Chapter 5   "I Think I'm in Labor" – 79

Chapter 6   "Why Does My Son Have Lead in His Blood?" – 99

"I Think I'm in Labor"

Jordana L. Gilman and Sarah Cumbie Reckess

5.1 Background Questions – 80

5.2 Additional Case Information and Questions for Discussion – 81

5.3 Answers to Background Questions – 83

5.4 Responses to Discussion Questions – 87

References – 95

■ **Glossary**

Jail - Refers to a locally operated correctional facility where people who are awaiting trial and sentencing (i.e., they have been charged but not yet tried for a crime) or who have been sentenced and are serving a short sentence (usually for a misdemeanor or low-level offense) are held. Jails also hold inmates awaiting transfer to state or felony prisons [1].

Prison - Refers to a correctional facility where people serve longer sentences once they have been sentenced to a crime. Prisons tend to be reserved for higher-level offenses such as felony convictions, and they are usually operated by the state government or the U.S. Federal Bureau of Prisons. Private prisons are for-profit correctional facilities that are operated by a third party through a contract with a state or federal agency [1].

Conspiracy - Although it varies by jurisdiction, the criminal act of conspiracy is generally defined as an agreement by two or more people to commit a crime, with at least one person proceeding to carry out an overt act of criminal activity. Each person is punishable and held to the same standard as the one who may have engaged in the activity, even if they do not know the identities of their co-conspirators. Broadening of the conspiracy statutes at the federal level, as well as the U.S. Supreme Court's decision in the *United States vs. Shabani, 513 U.S. 10 (1994)*, has led to a significant increase in the number of incarcerated women, particularly in federal drug crimes, because it allows every participant in a crime to be held liable for the actions of every other participant, regardless of what role they played in the crime.

Preterm labor - Regular contractions of the uterus resulting in changes in the cervix that start before 37 weeks of pregnancy. Changes in the cervix include effacement (the cervix thins out) and dilation (the cervix opens so that the fetus can enter the birth canal).

Learning Objectives

1. Understand the complexity of treating inmates, including working with jail staff to ensure safety for the inmate, staff, health care practitioners, and any others involved in providing care.
2. Develop strategies for working with nonmedical partners to provide optimal health care and improved outcomes.
3. Understand risk factors for poor maternal and neonatal outcomes, considering the role of race and ethnicity.
4. Develop community-wide recommendations to address a population health issue.
5. Explain how justice is relevant in both an individual patient encounter and in population health.

Case Background

You are an obstetrician in a private practice affiliated with a hospital. Your practice has a contract to provide correctional health services for inmates at a local jail. Jail staff has contacted your office to request that you provide prenatal care services to an inmate, as their on-site medical staff consists of a part-time physician assistant (PA) and a registered nurse (RN). The inmate, Nia Jones, is a 17-year-old girl who is 33 weeks pregnant, G2P0010, and is experiencing symptoms of preterm labor. She has been held for the past 2 months in the county jail awaiting trial because she cannot post $8000 in bail.

5.1 Background Questions

1. What federal laws protect inmate health? What are the policies in your state that govern inmate healthcare access?

2. Do individual jail facilities have policies governing pregnant inmate health care? Who oversees these policies?

3. How many pregnant women are incarcerated in the United States? Is this number increasing or decreasing?

4. Do the American Medical Association (AMA) and American College of Obstetricians and Gynecologists (ACOG) have best practices or guidelines for treating pregnant inmates? What are the medical risks of being physically restrained (e.g., shackled) during labor and delivery? Is this practice legal in your state?

5. How is confidentiality protected when providing health care for an inmate? Does the Health Insurance Portability and Accountability Act (HIPAA) apply to inmates?

6. How does incarceration impact maternal and infant health outcomes?

7. What is the incidence of teen pregnancy in the United States? Is teen pregnancy increasing or decreasing in the United States? What demographic groups are most at risk for teen pregnancy?

8. What are risk factors for preterm labor?

"I Think I'm in Labor"

5.2 Additional Case Information and Questions for Discussion

During a phone call with the jail's physician assistant, you learn that Nia's contractions have been going on for at least a day and that she believes they are getting closer together. Nia did not receive any prenatal care prior to incarceration. She has experienced poor weight gain during this pregnancy, currently appears to be underweight, and admits to smoking cigarettes during the early stages of her pregnancy. When Nia entered the justice system, her initial health screen revealed chlamydial cervicitis.

1. Self-reflection: Did you assume anything about the patient in this case? If so, what did you assume?
    (a) Why do you think you made those assumptions?

2. What else do you want to know about this patient?
    (a) What risk factors were present and likely contributed to her preterm labor?

3. Why do you think the patient did not receive prenatal care until entering the justice system?
    (a) Even when all logistical and structural barriers to care are removed, some women still do not present for timely medical care during their pregnancies. What other sociocultural factors may influence seeking care?
    (b) How can we as healthcare practitioners combat medical mistrust?

Some county jail facilities contract with a private healthcare practitioner, local hospital, or teaching hospital to provide care, while other jail facilities hire and staff medical personnel in-house. In your area, the county jail contracts with a private healthcare practitioner—your employer. While the majority of your patients are not incarcerated, a small minority are incarcerated women. Incarcerated patients do not have a choice of practitioner in this situation, and there are no "consumer rights" in jails and prisons.

4. What are some ethical concerns that physicians and other healthcare practitioners may face when providing inmate health care?
    (a) How does a person's background, criminal or otherwise, affect the care provided?
    (b) How does the lack of patient autonomy or privacy affect the care provided? Does it pose an ethical dilemma for practitioners?

5. How do you hold yourself accountable to ensure that all your patients are receiving the best level of care when their situations are so different?
    (a) How do you hold your medical group accountable for incarcerated patient care?
    (b) Creative problem-solving: Currently, one mechanism of accountability for private companies that provide correctional health services is lawsuits. The company can be sued if there are poor health outcomes under their care. Can you think of more effective ways to ensure quality of care for incarcerated patients?

6. Self-reflection: What does it mean to provide care in a punishing environment?

7. Is it ethical to treat health care as a reward or a benefit while incarcerated?

You meet Nia in the medical exam room at the local community health center. A male jail deputy stands in the room with you, and Nia is handcuffed for the examination. When you ask if she can be uncuffed for the remainder of the examination to allow better access, you are told no.

8. Self-reflection: Do you have any fears or safety concerns about caring for an incarcerated patient?

9. Creative problem-solving: How can you work with the jail deputy to address his concern (safety) with your needs (accessing the patient)?

Since entering the jail, Nia has been seen twice by the PA—once at intake and then a month later. Last week, she submitted a medical ticket asking to be seen by the nurse because she was experiencing severe cramping and light bleeding. Several days later, the nurse who examined her then made the referral to your office. Based on Nia's recall, it has been 5 days since the cramping began, rather than the 1 day reported in the referral. The jail deputy interrupts to say that it has in fact been 4 days since the cramping began, because it started in the middle of the night.

10. **How is the healthcare practitioner-patient relationship impacted by the presence of the jail deputy?**
    (a) How do you establish rapport with this patient given the circumstances of the visit?

11. **Should incarcerated patients legally have the right to a second opinion or an appeal process?**
    (a) If incarcerated patients do not have the right to a second opinion from a healthcare practitioner, is that ethical? Why or why not?

During your examination, Nia states that she is feeling hopeless and has recurrent dreams of dying during childbirth. Her eyes get teary when you ask if the father is involved. She states that he is also in the county jail and that they were arrested together. She says, "He's going away for a long time." Her tone becomes hard as she explains that she is being charged with conspiracy even though she did not know he was selling drugs. Nia believes that she is being charged by the prosecutor in order to convince her boyfriend to take a plea deal. As she starts to sob, she tells you that her boyfriend is not going to take the plea deal and that he does not care that their baby will be placed in foster care if she gives birth while in jail.

12. **As a healthcare practitioner, you have a unique and intimate view of the impact of these laws upon the lives of your patients. Should practitioners advocate for changes in penal law that impact patients?**
    (a) Creative problem-solving: What changes to laws and/or policies might you suggest in this case?

Based on the examination, you determine that Nia should be on bed rest for the remainder of her pregnancy. You tell the jail deputy that she needs to be monitored carefully and admitted to the hospital at the first sign of labor. The jail deputy asks, "How can we be sure she is in labor? We're not going to transport her just because she says she is having the baby."

13. **How do you respond?**

Three days later, you see Nia for her follow-up appointment. You ask her how she is feeling, and she says that she is still sad and has many unanswered questions. She asks you where her baby will go if she delivers while she is in jail. You ask if she has any family that can take the baby. She says, "No one wants the burden," and starts to cry. She asks if she will be able to hold the baby after giving birth. You say, "I don't know the answer to your question, but I can find out." She then asks if anyone can attend the delivery besides the jail staff; she would like to ask her aunt. You tell her that the jail does not alert family when she is in labor in order to maintain public safety at the hospital but that it is possible her aunt can visit after the baby is born, if the jail staff approves the request. She wonders if she will be able to breastfeed at the hospital. You tell her it depends on whether the jail will provide a nursery accommodation for the baby. Finally, she asks if she will be handcuffed during birth; other women in the jail have warned her that pregnant women are shackled during labor and delivery. You assure her that you will do your best to prevent her from being restrained while she delivers. You feel inept at not knowing the answers to these questions, concerned for Nia and her baby, and frustrated that so much of what will happen is dependent on jail staff and not medical guidelines.

14. **Self-reflection: What does a humane birth look like to you?**

15. **Do you think that the public safety issues that have led to the shackling of pregnant inmates (e.g., potential harm to staff and the inmate herself due to violent behavior, escape of inmates) are legitimate concerns?**
    (a) What about the restriction on having family members present at the birth?

16. **What is the healthcare practitioner's role in advocating for a safe delivery for a pregnant patient who is incarcerated?**
    (a) **There is a concept called "zealous representation" in law that assumes that a lawyer is doing everything possible to advocate for the client's goals within the boundaries of the law. Does this standard apply in medicine? Should it?**
    (b) **Self-reflection: Do you think that your treatment of Nia and her baby would change if the circumstances were different? What if Nia had been found guilty of murder and was facing 25 years to life for her crime?**

17. **What are Nia's options once the baby is born?**

5.3 Answers to Background Questions

1. **What federal laws protect inmate health? What are the policies in your state that govern inmate healthcare access?**

   In the U.S. Constitution, the Eighth Amendment states "Excessive bail shall not be required, nor excessive fines imposed, nor cruel and unusual punishments inflicted." The prohibition on unduly harsh penalties on incarcerated individuals is the basis for providing adequate health care to inmates [2].

   The 1976 Supreme Court Case *Estelle v. Gamble, 429 U.S. 97*, established that doctors must provide appropriate care for prisoners' illnesses and injuries under the Eighth Amendment and that deliberate indifference to an inmate's medical illness constituted cruel and unusual punishment [3].

   Despite these federal laws, states offer differing levels of access to quality health care for inmates. In 35 states, inmates have co-payments for medical treatment, which is taken from commissary accounts, although inmates may only earn 12 cents per hour [4]. For inmates who are chronically ill, elderly, or otherwise unable to work due to their health, this policy presents a major challenge as they have higher medical costs and less money in wages [5].

   Healthcare access varies widely by state. As of 2012, more than 20 states contracted with private healthcare companies as a way to reduce healthcare costs for inmates, in contrast to hiring individual medical professionals [6]. States that adopted the Medicaid expansion allow incarcerated individuals to stay enrolled in their Medicaid coverage while in prison or jail, making it easier to receive care once released [7, 8]. Increasing privatization of prisons also has implications for healthcare access, as companies may sacrifice care to keep costs low and maximize profit [6].

2. **Do individual jail facilities have policies governing pregnant inmate health care? Who oversees these policies?**

   Health care for pregnant inmates varies widely by facility and state and whether the facility is private or public. Individual states set their own policies regarding inmate health care; some states require all jail and prison facilities to follow similar rules, while others allow local jurisdictions to set their own policies and standards. All states have a state-run department that oversees corrections. There is no federal legislation that provides a standard of care for female reproductive care while in state custody. Many jails and prisons lack healthcare practitioners who are trained in obstetrics and gynecology, and gynecological exams are not required on an annual basis or at admission [9]. Among women who are pregnant while in state prison, only 54% reported receiving some type of pregnancy care [10].

   New York State's law ("Births to inmates of correctional institutions and care of children of inmates of correctional institutions") sets out the rights of mother and child to live together in a state penitentiary until 1 year of age, unless the mother is physically unfit to care for the child, and states that women who are nursing a child under age 1 when they

are sentenced to a state penitentiary may bring the child with them [11]. Another New York State regulation, "Prenatal and Infant Care Services," states that the Department of Corrections and Community Supervision is required to provide "comprehensive prenatal care delivered by qualified specialists… which shall include regular medical examinations, advice on appropriate levels of activity and safety precautions, nutritional guidance, and HIV education" [12]. Twenty-two states and the District of Columbia outlaw shackling during childbirth [13].

3. **How many pregnant women are incarcerated in the United States? Is this number increasing or decreasing?**
The most recent available data indicates that there are more than 200,000 incarcerated women in the United States (205,400 as of 2015). This number constitutes roughly 9% of the total incarcerated population, and the number of incarcerated women is increasing [14]. The population of incarcerated women is increasing largely due to increased criminality of nonviolent crimes, such as minor property offenses, drug charges, and conspiracy [15]. Estimates show that 6–10% of incarcerated females are pregnant. In 1998, the last year for which data is available, 1,400 women gave birth while incarcerated [16]. Reporting of this data is not mandatory, so estimates of incarcerated pregnant females are likely to be lower than the true number. As the number of incarcerated women increases and the rate of pregnancy among incarcerated women remains the same, the number of pregnant women in correctional facilities is increasing.

4. **Do the American Medical Association (AMA) and American College of Obstetricians and Gynecologists (ACOG) have any best practices or guidelines for treating pregnant inmates? What are the medical risks of being physically restrained (e.g., shackled) during labor and delivery? Is this practice legal in your state?**
The AMA embarked on a large-scale national effort to address inmate health care in the 1970s, and their work led to the development of the National Commission on Correctional Health Care (NCCHC). The NCCHC develops national standards of care for jails, prisons, and juvenile detention facilities, including guidelines and standards for pregnant inmates. Standards of care for pregnant women include comprehensive counseling and assistance in family planning, prenatal care, and nonuse of restraints (shackling) during labor and delivery [9]. The AMA also developed an advocacy tool kit that supports legislation to ban shackling during labor and delivery [13].

In 2012, ACOG issued a Committee Opinion on Reproductive Health Care for Incarcerated Women and Adolescent Females. The Committee Opinion emphasizes the need for incarcerated women to have access to prenatal care, referrals for continuing care, counseling and mental health care, and abortion services. The ACOG Committee Opinion is strongly against shackling of incarcerated females during labor and delivery, citing medical risks as well as the demeaning nature of the practice. The document also emphasizes the responsibility of physicians "to ensure that medical needs of incarcerated women and adolescent females are being addressed appropriately, such as by providing training or consultation to healthcare practitioners and correctional officers in prison settings" [17].

As of 2017, 22 states (Arizona, California, Colorado, Delaware, Florida, Hawaii, Idaho, Illinois, Louisiana, Maine, Maryland, Massachusetts, Minnesota, Nevada, New Mexico, New York, Pennsylvania, Rhode Island, Texas, Vermont, Washington, and West Virginia) and the District of Columbia restrict the use of shackles during childbirth [13, 18]. Even among states that restrict use of restraints during labor and delivery, laws vary widely regarding physician authority to remove restraints, whether the restriction applies to juveniles, whether restraints may be used during medical transport, and whether restraints may be used during the postpartum recovery period [18].

Even in states that have outlawed the practice of shackling inmates during labor, violations are common because there is ineffective enforcement and policing. For instance, New York State enacted the Anti-Shackling Law in 2009, but 23 out of 27 incarcerated women who were interviewed for a study by the Women In Prison Project reported that they were shackled at least once in violation of the statute [19].

> **Box 5.1 Patient Perspective**
> One woman who was shackled in violation of the 2009 law told the Correctional Association: "My ankles were shackled during the whole trip to the hospital when I was in labor. They pushed me in a wheelchair from the van to the hospital, and at one point the wheelchair almost tipped over. I would not have been able to catch myself very well… I was shackled until I got to the delivery room, but even then, they kept one of my ankles shackled to the bed. [They] only took it off when it was time to start pushing… I couldn't rotate the way I needed to and I had to sit in one spot the whole time I was in labor. The baby was pushing and I was going through contractions and I wanted to lie on my side but I couldn't because I couldn't move my leg."

There are many medical and psychological risks of shackling during labor. Shackling heightens the risk of blood clots, limits mobility during delivery, and increases the risk of falling or being able to safely break one's fall to protect the fetus. Shackles interfere with healthcare practitioners' abilities to deliver care, especially in emergency situations such as seizures due to eclampsia and hemorrhage. Shackling during the postpartum period prevents women from healing and bonding with their newborns [17].

Shackling during labor is also considered to be unnecessary based on the evidence. The stated purpose of shackling is to prevent harm to the staff and the inmate herself from violent behavior, but 63% of women are incarcerated for non-violent crimes; in fact, only 14% of all violent offenders are women [20]. Proponents of the use of shackles in childbirth sometimes argue that they are necessary. The use of shackles in childbirth has been cited as a way to prevent inmates from escaping, but no escape attempts have been reported among pregnant incarcerated women who were not shackled during childbirth [17].

5. **How is confidentiality protected when providing health care for an inmate? Does HIPAA apply to inmates?**
   HIPAA applies to any entity that performs the functions of a health plan or a healthcare provider, so a jail or prison system that performs those functions is subject to HIPAA [21]. If there are situations that require the healthcare provider to disclose personal patient information, the provider should inform inmates at the beginning of the healthcare encounter [22]. The American Public Health Association and the American Bar Association both state that inmate patients should be provided the same privacy of healthcare information as patients in the community, but this standard falls upon the healthcare provider to enforce [23, 24]. The U.S. Constitution does not expressly provide the right to health information privacy, but the U.S. Supreme Court has upheld that right in regard to information held in government databases [25].

   HIPAA allows correctional facilities to obtain or use protected health information if necessary for providing health care to an inmate; for the health and safety of inmates, officers, or staff; and for administration and maintenance of the safety, security, and good order of the correctional institution [26]. For instance, officers may need to know about an inmate's seizure history before placing the inmate in isolation, disabilities may require accommodation and assistive devices, and inmate medication management may warrant recurring medical appointments. Medical conditions or certain medications may affect the tasks an inmate can perform safely while working [27]. While HIPAA does protect inmates' right to health information privacy, there are exceptions that are intended to protect the health and safety of all.

6. **Does incarceration impact maternal and infant health outcomes?**
Incarceration may impact maternal and infant health outcomes either positively or negatively, depending on the woman's previous context and community. For many women who lack reliable meals, access to prenatal health care, access to drug cessation programs, and shelter, incarceration can lead to higher birth weight babies and improved health outcomes [28–30]. The initial health screening in jail may be the first time an incarcerated woman learns she is pregnant, allowing for prenatal care to commence [9].

Conversely, pregnancy care in correctional facilities may be substandard, and access may be limited, leading to negative birth outcomes [31]. In 2008, only 54% of pregnant inmates received any type of prenatal health care [10]. Incarcerated women may be forced to perform manual labor that puts the pregnancy at risk or may be shackled during transportation or other activities, which limits mobility and increases the risk for dangerous falls [19]. Shackling during labor and delivery is harmful for health outcomes, as discussed in Background Question 4. The correctional facility environment may cause extreme stress or exacerbate anxiety and depression, which also increase the risk of the pregnancy [32]. Incarceration is associated with higher rates of substance abuse, cigarette smoking, and sexually transmitted infections, which may lead to high-risk pregnancies [33]. The combination of poor conditions or access to care while incarcerated and risk factors that are frequently associated with incarceration lead to low birth weight babies born to incarcerated mothers [34].

Maternal-infant attachment is important for both mother and baby after birth. Since maternal-infant attachment is not a priority for correctional facilities, many women are separated from their infant 1–2 days after giving birth [9]. Even if the pair is not separated, breastfeeding is not easily facilitated in the correctional environment. However, breastfeeding is important to incarcerated women and contributes to their psychosocial well-being and self-worth as a mother [35]. Creating barriers to breastfeeding or prohibiting it altogether may have deleterious effects on inmates' well-being and physical health [36]. Breast milk also has important health benefits for newborns [37].

7. **What is the incidence of teen pregnancy in the United States? Is teen pregnancy increasing or decreasing in the United States? What demographic groups are most at risk for teen pregnancy?**
The most recent available data indicates a teen birth rate of 20.3 per 1000 females ages 15–19 in the United States or 209,809 babies born to women in this age group in 2016. This teen birth rate is down 67% from 1991 when it was at a record high rate of 61.8 births per 1000 females ages 15–19 [38].

Teen pregnancy rates (which include pregnancies that end in a live birth and those that end in fetal loss or elective termination) have also declined continuously over the past three decades. The most recent available data shows a teen pregnancy rate of 43.4 pregnancies per 1000 females ages 15–19 in 2013, compared to 117.6 pregnancies per 1000 females ages 15–19 in 1990 [39]. These declines are attributed to increasing numbers of teens who are waiting to have sexual intercourse and who are using more effective contraception when they do choose to have sexual intercourse [40]. An estimated 77% of teen pregnancies are unplanned [41], and 89%

of live births to teens occur outside of marriage [38].

Demographic factors such as age, race, and ethnicity influence rates of teen childbearing. Most adolescents who give birth are 18 or 19 years old. Birth rates are highest among Hispanic teens, followed closely by black teens. Data from 2016 shows that there were 31.9 births per 1000 Hispanic females ages 15–19, while there were 29.3 births per 1000 black females of the same ages and 14.3 births per white females in the same age range [38].

8. **What are risk factors for preterm labor?**
Women who have a history of previous preterm labor or those who have had a preterm delivery in the past are high risk for preterm labor [42]. Short interval and multiple gestation pregnancies are also associated with a higher risk of preterm labor [43]. Pregnancies conceived using in vitro fertilization or assisted reproductive technologies are at higher risk for preterm labor because they are more likely to be multiple gestations [44].

Certain demographic factors increase the risk of preterm labor. Women younger than 18 and older than 35 are more likely to experience preterm labor. For those over 35, this is because they are more likely to have other health conditions that cause preterm labor, such as hypertension.

Preterm labor and birth also occur more often among African American mothers [45]. A growing body of evidence suggests that this is due to chronic stress from exposure to racism, known as the "weathering hypothesis" [46]. Several studies have found that African immigrants have rates of preterm birth comparable to white American women; however, their daughters have rates that are comparable to black American women in general [47, 48].

Lifestyle and environmental factors can also increase the risk for preterm labor and delivery. Late presentation to prenatal care or lack of prenatal care, smoking, alcohol use, illicit drug use, domestic violence and abuse, lack of social support, high stress, physically demanding work or manual labor, and exposure to certain environmental pollutants are all risk factors for preterm labor [44].

Mothers who have health conditions prior to the pregnancy may also be at risk. Being underweight or obese before conception can increase the risk of preterm labor. A history of diabetes, hypertension, and coagulation disorders also increase the risk [44].

Conditions acquired during pregnancy, if untreated, can increase the risk of preterm labor. Urinary tract infections, sexually transmitted infections, bacterial vaginosis, hypertension, and gestational diabetes have been shown to increase the risk of preterm labor and delivery. Certain developmental abnormalities in the fetus and placenta previa can also increase the risk [44].

5.4 Responses to Discussion Questions

During a phone call with the jail's PA, you learn that Nia's contractions have been occurring for at least a day and she believes they are getting closer together. Nia did not receive any prenatal care prior to incarceration. She has experienced poor weight gain during this pregnancy, appears to be underweight and admits to smoking cigarettes during the pregnancy prior to incarceration. When Nia entered the justice system, her initial health screen revealed chlamydial cervicitis.

1. **Self-reflection: Did you assume anything about the patient in this case? If so, what did you assume?**
   (a) **Why do you think you made those assumptions?**

> **Box 5.2 Teaching Tip**
> Learners may assume that the patient is guilty of whatever crime she has been arrested for. It is important to note that people in jail are typically *awaiting* trial and have not yet been found guilty of a crime. It may be useful to explore how arrests

> and criminal charges are perceived in different communities—some learners may come from communities in which it is widely accepted that people may be falsely accused.
> More importantly, however, is that the healthcare practitioner in this case does not need to know the patient's alleged crime or guilt status. The healthcare practitioner should not ask the patient about the crime unless the patient begins discussing it first.

2. **What else do you want to know about this patient?**
   It is important to obtain the rest of the relevant medical history, including information about previous health care, other drugs or alcohol use, previous pregnancies, and blood type [17]. The initial vignette states that Nia is G2P0010, indicating a previous pregnancy that ended in abortion; it would be helpful when providing prenatal care to know whether that was the result of spontaneous abortion or elective termination and at what gestational age. Nia should be treated as any other prenatal patient with late presentation to care, with special attention paid to risk factors that are more prevalent among incarcerated women.

   (a) **What risk factors were present and likely contributed to her preterm labor?**
   The patient's race has not been disclosed in this case, but if the patient is black, her race may have an association with preterm labor [45]. The patient's age, chlamydial cervicitis, history of cigarette smoking, poor weight gain, psychosocial stress, and late access to care are all risk factors for preterm labor [44].

3. **Why do you think the patient did not receive prenatal care until entering the justice system?**
   Structural factors, such as inconvenient appointment times or availability of transportation, likely played a role. Nia may have been unable to miss school or work, or unable to afford the consequences—such as lost income—of doing so. She may also have had difficulty finding a healthcare practitioner willing to accept her insurance provider, especially if she is a Medicaid recipient. Finally, she may not have been aware of the importance of seeking prenatal care early.

   (a) **Even when all logistical and structural barriers to care are removed, some women still do not present for timely medical care during their pregnancies. What other sociocultural factors may influence seeking care?**
   Gynecological examinations by a doctor have the potential to re-traumatize women who have experienced trauma and sexual abuse. A study of Bedford Hills Correctional Facility inmates in 1999 found that 94% of women had experienced physical or sexual abuse during their lives. Survey respondents who reported feeling "bad" after gynecological appointments identified past abuse as the cause of this feeling. Fear of being re-traumatized leads some survivors to avoid seeking medical care [49].

   In addition to fear of invasive exams, patients with knowledge of unethical medical experimentation may harbor mistrust of healthcare practitioners and the medical profession. Research has long suggested that the ill effects of the Tuskegee Study extend beyond those men and their families to the greater whole of black culture. Black patients consistently express less trust in their physicians and the medical system than white patients, are more likely to believe medical conspiracies, and are much less likely to have common, positive experiences in healthcare settings [50].

   Even among patients who do not have an awareness of these

experiments, patients may avoid seeking care because of racism and classism, which they personally experience when visiting a healthcare provider [51].

Patients may also have cultural differences in seeking care or a lack of awareness that one should seek care during the early stages of a pregnancy. In rare cases, women may have denial of pregnancy, an inappropriate reaction to pregnancy that also prevents women from seeking appropriate prenatal care [52].

Substance abuse was identified as the reason for not seeking prenatal care by 30% of participants in one study [53]. Some states require medical staff to report suspected substance use by pregnant women to law enforcement. Additionally, other states mandate drug testing of all pregnant patients by healthcare professionals. The Association of Women's Health, Obstetric and Neonatal Nurses has stated that "laws that criminalize drug use during pregnancy have the potential to deter women from seeking prenatal care that can provide them access to appropriate counseling, referral, and monitoring" and to diminish trust and reduce disclosure by patients to healthcare professionals [54].

- Honor patient autonomy and emphasize informed consent in all encounters.
- Fully explain all procedures or exam maneuvers before initiating them, and ask patients for permission to begin the exam or procedure.
- Emphasize patient privacy policies in ways that are easy for patients to understand.
- Involve the patient in all care decisions.
- Involve case management as quickly as possible, when available, which has been shown to improve birth weight outcomes for incarcerated mothers [34].
- Invite an advocate for the patient into the provider-patient relationship, as a way of empowering patients and combating mistrust.
- Acknowledge the history of racism and classism in medical care environments and in medical studies.
- Assure office staff and providers are adequately trained in cultural competency [55].

Some county jail facilities contract with a private healthcare practitioner, local hospital, or teaching hospital to provide care, while other jail facilities hire and staff medical personnel in-house. In your area, the county jail contracts with a private healthcare practitioner—your employer. While the majority of your patients are not incarcerated, a small minority are incarcerated women. Incarcerated patients do not have a choice of practitioner in this situation, and there are no "consumer rights" in jails and prisons.

> **Box 5.3 Teaching Tip**
> If learners are struggling to generate ideas in response to this question, try offering the following words and asking what comes to mind: trust, fear, racism, bias, trauma, and culture.

(b) **How can we as healthcare practitioners combat medical mistrust?**
There are many ways to foster a trusting patient-practitioner relationship:

✓ 4. **What are some ethical concerns that physicians and other health care practitioners may face when providing inmate health care?**
Physicians and other health care practitioners may be concerned about:
- The welfare and rights of the incarcerated patient
  - Is the patient's consent truly voluntary as an inmate?
  - How will confidentiality be protected if a corrections officer is present?
  - Will the patient encounter biases that negatively impact the quality of care?

- The welfare and rights of other patients
  - Could the patient present a physical danger to other patients?
  - If the patient requires additional resources (e.g., one-on-one nursing), could this limit resources available to others?
- The welfare and rights of the healthcare practitioner and other staff
  - Could the patient present a physical danger to staff?

(a) **How does a person's background, criminal or otherwise, affect the care provided?**
Healthcare practitioners may feel personally conflicted about providing care to an incarcerated patient because that person has been accused of a violent crime or one they find morally reprehensible.

(b) **How does the lack of patient autonomy or privacy affect the care provided? Does it pose an ethical dilemma for practitioners?**
Confidentiality and privacy are important concerns when treating inmates; in a jail environment, it may be difficult to prevent staff or other inmates from hearing information. HIPAA may be waived for the safety or health of the patient and staff [27]. Additionally, incarcerated patients may be unable to follow medical recommendations due to jail rules and regulations. For example, inmates may be prohibited from staying in their bed during daytime hours despite being placed on bed rest, or postpartum inmates may sleep in an upper bunk that requires them to climb up and down during the healing process. Also, patients may be forced to comply with physician recommendations that they disagree with or feel are not in their best interests.

Many correctional facilities lack adequate resources devoted to inmate health or uniform policies regarding medical care [56]. Healthcare practitioners may find substandard exam rooms, scheduling constraints, inadequate space to examine the patient and conduct a private interview, and a general lack of professional autonomy if working for a contracted healthcare company. All of these factors may contribute to unequal care being provided to incarcerated patients.

5. **How do you hold yourself accountable that all your patients are receiving the best level of care when their situations are so different?**
One way to ensure that patients receive fair and equitable health care is by relying on guidelines, such as those provided by organizations such as ACOG, to standardize care. Healthcare practitioners should perform the same standard tests and exams on all patients, if clinically indicated, regardless of the ability to pay.

(a) **How do you hold your medical group accountable for incarcerated patient care?**
In large part, it falls on the individual practitioner to ensure that patients' feedback is heard and that every incarcerated patient receives the standard of care. Sharing case notes and establishing mentorships between senior and junior practitioners in your practice is one way to ensure accountability and equity.

(b) **Creative problem-solving:** Currently, one mechanism of accountability for private companies that provide correctional health services is lawsuits. The company can be sued if there are poor health outcomes under their care. Can you think of more effective ways to ensure quality of care for incarcerated patients?

6. **Self-reflection: What does it mean to provide care in a punishing environment?**

7. **Is it ethical to treat health care as a reward or a benefit while incarcerated?**
   American prisoners have a constitutional right to health care through the Eighth Amendment's prohibition of "cruel and unusual" punishment, but many still view health care and treatment as a reward, not a right [3]. The "principle of equivalence" states that prisoners are entitled to a standard of health care equivalent to that available outside of prisons [57, 58]. When health care for the incarcerated is guided ethically by the concept of a "right to health" regardless of a person's status (legal, criminal, or otherwise), a need arises beyond the principle of equivalence. Because incarcerated patients often present with more complicated medical needs than those found in the community setting, equivalent levels of health care are not sufficient to achieve the right to health. In those cases, it is appropriate and necessary to provide health care beyond the minimum standards set by the principle of equivalence [59].

You meet Nia in the medical exam room at the local community health center. A male jail deputy stands in the room with you, and Nia is handcuffed for the examination. When you ask if the patient can be uncuffed for the remainder of the examination to allow better access to the patient, you are told no.

8. **Self-reflection: Do you have any fears or safety concerns about caring for an incarcerated patient?**

9. **Creative problem-solving: How can you work with the jail deputy to address his concern (safety) with your needs (accessing the patient)?**

Since entering the jail, Nia has been seen twice by the PA—once at intake and then about a month later. Last week, she submitted a medical ticket asking to be seen by the nurse because she was experiencing severe cramping and light bleeding. Several days later, the nurse who examined her then made the referral to your office. Based on Nia's recall, it has been 5 days since the cramping began, rather than the 1 day reported in the referral. The jail deputy interrupts to say that it has in fact been 4 days since the cramping began, because it started in the middle of the night.

10. **How is the healthcare practitioner-patient relationship impacted by the presence of the jail deputy?**
    The patient may be unwilling to share or feel uncomfortable sharing certain aspects of her medical history, which could lead to an incorrect assessment. The patient may also feel unsafe or threatened by the jail deputy depending on how she has been treated by that jail deputy in the past. The presence of the jail deputy creates a definite hierarchy in the room that prevents a more intimate, trusting, and collaborative dynamic from forming between the healthcare practitioner and the patient.
    (a) **How do you establish rapport with this patient given the circumstances of the visit?**
    The healthcare practitioner should:
    - Focus all the attention on the patient.
    - State clearly that the practitioner is there for the patient, not for the jail deputy.
    - Make good eye contact.
    - Listen carefully.
    - Summarize what the patient has said.
    - Trust the patient's timeline of events. (The jail deputy may interrupt, but the jail deputy will not be the one providing the medical history. Practitioners should treat the jail deputy with respect and de-escalate any tensions that arise but maintain the focus on the patient.)
    - Ask if the patient has questions or concerns and address them.
    - Demonstrate that the practitioner is in control of the encounter.

✓ **11. Should incarcerated patients legally have the right to a second opinion or an appeal process?**
Incarcerated patients should have the right to a second opinion or an appeal process if they feel they are not receiving adequate care, but generally the systems are not in place for them to exercise that right. They may bring a lawsuit against the correctional facility or healthcare company when they are no longer incarcerated, but options are very limited while they are still in jail or prison. Medical practices can address this issue internally by identifying a process for incarcerated patients to receive a second opinion or the opportunity to switch practitioners within the practice.
  (a) **If incarcerated patients do not have the right to a second opinion from a healthcare practitioner, is that ethical? Why or why not?**
  All patients should feel that the care they are given is fair and adequate, regardless of their incarceration status [60]. When patients do not feel they are being cared for appropriately, they are less likely to follow instructions, comply with recommendations, or seek continuing care. This may lead to poorer health outcomes or a further breakdown in trust between patients and practitioners.

During your examination, Nia states that she is feeling hopeless and has recurrent dreams of dying during childbirth. Her eyes get teary when you ask if the father is involved. She states that he is also in the county jail and that they were arrested together. She says, "He's going away for a long time." Her tone becomes hard as she explains that she is being charged with conspiracy even though she did not know he was selling drugs. Nia believes that she is being charged by the prosecutor in order to convince her boyfriend to take a plea deal. As she starts to sob, she tells you that her boyfriend is not going to take the plea deal and that he does not care that their baby will be placed in foster care if she gives birth while in jail.

✓ **12. As a healthcare practitioner, you have a unique and intimate view of the impact of these laws upon the lives of your patients. Should practitioners advocate for changes in penal law that impact patients?**
Healthcare providers are typically held in high esteem by their communities, and with that respect comes power and influence [61, 62]. Advocating against unfair laws may be one way that providers exercise that power. Conspiracy is one such law that disproportionately targets women who would not otherwise be involved in the justice system. Changing such a law could promote reproductive justice, which is closely tied to reproductive health and well-being [63]. Reproductive justice is a movement that emphasizes how different social identities, such as race, gender, and class, interact with each other and contribute to inequalities. The reproductive justice movement expands upon a woman's autonomy in health decisions to include the right to have children when desired and the right to parent with dignity in the case of incarcerated women [64].
  (a) **Creative problem-solving: What changes to laws and/or policies might you suggest in this case?**

Based on the examination, you determine that Nia should be on bed rest for the remainder of her pregnancy. You tell the jail deputy that the client needs to be monitored carefully and admitted to the hospital at the first sign of labor. The jail deputy asks, "How can we be sure she is in labor? We're not going to transport her just because she says she is having the baby."

✓ **13. How do you respond?**
First, the healthcare practitioner should try to understand the deputy's perspective. Many correctional facilities, including jails, are overcrowded and understaffed. The deputy's response may be based on his knowledge that transporting an inmate to the hospital requires additional manpower and resources, such as a transport van. He may foresee the complications that Nia's

preterm labor may have on the safety and security of the jail population. Understanding the deputy's frustration and finding solutions to keep Nia and her fetus safe is imperative [65].

Using empathy and understanding, healthcare practitioners can make recommendations for the health of their patients. For example, they can:
- Provide the jail deputy with brief instructions on the signs of preterm labor
- Distribute brochures or "cheat sheets" on the signs of preterm labor to the jail facility for quick reference
- Provide an emergency phone number for healthcare practitioners and support staff
- Respectfully emphasize to the jail deputy the risks of not admitting the patient and how those risks could become liabilities to the jail facility

Three days later, you see Nia for her follow-up appointment. You ask her how she is feeling, and she says that she is still sad and has many unanswered questions. She asks you where her baby will go if she delivers while she is in jail. You ask if she has any family that can take the baby. She says, "No one wants the burden," and starts to cry. She asks if she will be able to hold the baby after giving birth. You say, "I don't know the answer to your question, but I can find out." She then asks if anyone can attend the delivery besides the jail staff; she would like to ask her aunt. You tell her that the jail does not alert family when she is in labor in order to maintain public safety at the hospital but that it is possible her aunt can visit after the baby is born, if the jail staff approves the request. She wonders if she will be able to breastfeed at the hospital. You tell her it depends on whether the jail will provide a nursery accommodation for the baby. Finally, she asks if she will be handcuffed during birth; other women in the jail have warned her that pregnant women are shackled during labor and delivery. You assure her that you will do your best to prevent her from being restrained while she delivers. You feel inept at not knowing the answers to these questions, concerned for Nia and her baby, and frustrated that so much of what will happen is dependent on jail staff and not medical guidelines.

> **Box 5.4 Teaching Tip**
> This is a great opportunity for learners to reflect on how they may view health care differently from others—even in their class and profession. Birth plans typically address elements of the birth experience such as preferred methods of pain control, choice of delivery posture, skin-to-skin contact with the infant after delivery, delayed clamping of the umbilical cord, and feeding of the infant. A key theme in humane childbirth is patient choice and healthcare practitioner communication when patient choice cannot be safely implemented [67]. Medicalization of childbirth may be discussed directly or indirectly; it is explored in detail in ▶ Chapter 7.

✓ 14. **Self-reflection: What does a humane birth look like to you?** [66].

✓ 15. **Do you think that the public safety issues that have led to the shackling of pregnant inmates (e.g., potential harm to staff and the inmate herself due to violent behavior, escape of inmates) are legitimate concerns?**
Public safety issues include escape, harming of self or the fetus, harming the medical staff or corrections officers, and destruction of hospital property [67]. Fear of harm from violent action, however, is less likely given that the majority of incarcerated women are nonviolent offenders [20]. As noted in Answer to Background Question 4, there are no reported escape attempts by incarcerated pregnant women who were *not* shackled during labor [18, 68].

> **Box 5.5 Personal Perspective**
> Carolyn Sufrin, AM, MD, PhD, explains her experience as a healthcare practitioner in this situation in her book, *Jailcare: Finding the Safety Net for Women Behind Bars* [69]:
>
> ❯❯ I first encountered an incarcerated patient in 2004, when I was a first-year Ob/Gyn resident in Pennsylvania. A woman from the local jail was in labor, and I was the doctor delivering her baby—as she remained shackled to the hospital bed. I was deeply troubled by this moment, and horrified at my own complicity

> in the act. A scripted authority figure in the birth room, I did not demand that the guard unshackle the mother. Among many disbelieving thoughts, the simplest was this: there were women—pregnant women—behind bars.

(a) **What about the restriction on having family members present at the birth?**
Having family members present at the birth may present additional safety issues. Public safety issues include family members smuggling drugs, weapons, or other contraband to the inmate and family members threatening or harming the inmate or correctional staff. These concerns may not be relevant to certain family members, or the support the family members provide to the patient may outweigh these concerns.

There may be other jail staff or professionals who provide services inside the jail who can accompany the inmate during delivery. Healthcare practitioners can ask about the possibility of a counselor, mental health professional, reentry specialist, doula, or other support person who has clearance with the jail and could be called for assistance.

16. **What is the healthcare practitioner's role in advocating for a safe delivery for a pregnant patient who is incarcerated?**
The practitioner must strive to give the incarcerated patient a safe and calm delivery experience for the health of the mother and the baby [17]. As an authority figure, the practitioner has the right and responsibility to advocate for safe conditions, for healthy positioning (no handcuffs), and for equal access to healthcare resources [62].

(a) There is a concept called "zealous representation" in law that assumes that a lawyer is doing everything possible to advocate for the client's goals within the boundaries of the law. Does this standard apply in medicine? Should it?
The medical parallels are beneficence (act in a way that benefits others) and non-maleficence (do no harm). The term "zealous" lacks bioethical meaning but can and should be applied to the care of patients and advocacy on their behalf. In legal matters, the lawyer acting on behalf of their client is aware that there is a lawyer acting in opposition with the same level of zealous representation, so when an opposing party is well represented, a lawyer can be a zealous advocate on behalf of a client and at the same time assume that justice is being done [69].

(b) **Self-reflection: Do you think that your treatment of Nia and her baby would change if the circumstances were different? What if Nia had been found guilty of murder and was facing 25 years to life for her crime?**

17. **What are Nia's options once the baby is born?**
Correctional facilities offer different resources and have different policies for mothers and newborns. This varies not only by state but by facility. If possible, healthcare practitioners can offer suggestions that can facilitate healthy bonding and development for the newborn and an improved postpartum recovery period for Nia.

To address the situation at hand, the healthcare practitioner might suggest that the newborn goes with Nia's aunt or other relatives, who can then set up visits between Nia and her baby at the jail. Nia may be able to pump breastmilk and send it to her newborn. Doula programs, case management, and "holistic defense" may also be able to facilitate creative solutions to this problem. Holistic defense involves social workers and lawyers working together for the inmate. See ▶ Box 5.6 for case conclusion.

Box 5.6 Case Conclusion

Nia gives birth in the hospital to a premature infant born at 34 weeks. She is able to deliver without handcuffs thanks to her obstetrician's advocacy. Her aunt is called after she delivers the baby. The jail where Nia is incarcerated does not have a nursery, so she is not able to care for the baby while she is still in jail and is unable to breastfeed. Ultimately Nia's aunt cares for the infant, while Nia awaits trial.

References

1. Freudenberg N. Jails, prisons, and the health of urban populations: a review of the impact of the correctional system on community health. J Urban Heal Bull New York Acad Med. 2001;78(2):214–35.
2. Stevenson BA, Stinneford JF. Amendment VIII – The United States Constitution [Internet]. Philadelphia (PA): National Constitution Center. [cited 2018 May 22]. Available from: https://constitutioncenter.org/interactive-constitution/amendments/amendment-viii.
3. *Estelle v Gamble* [1976] 429 U.S. 97 (U.S.). Available from: https://www.law.cornell.edu/supremecourt/text/429/97.
4. Awofeso N. Making prison health care more efficient. BMJ. 2005;331(7511):248–9.
5. Morrison Piehl A, Useem B, Dilulio JJ. Right-sizing justice: a cost-benefit analysis of imprisonment in three states [Internet]. New York: Manhattan Institute; 1999 [cited 2018 May 22]. Available from: https://www.manhattan-institute.org/pdf/cr_08.pdf.
6. Penn Wharton Public Policy Initiative. The current state of public and private prison healthcare [Internet]. New York: Manhattan Institute, Center for Civic Innovation; 2017; [cited 2018 May 22]. Available from: https://www.manhattan-institute.org/pdf/cr_08.pdf.
7. Centers for Medicare & Medicaid Services. Health coverage options for incarcerated people [Internet]. Baltimore (MD): CMS. [cited 2018 May 22]. Available from: https://www.healthcare.gov/incarcerated-people/.
8. Gates A, Artiga S, Rudowitz R. Health coverage and care for the adult criminal justice-involved population [Internet]. San Francisco (CA): The Henry J. Kaiser Family Foundation; 2014 Sep 5 [cited 2018 May 22]. Available from: https://www.kff.org/uninsured/issue-brief/health-coverage-and-care-for-the-adult-criminal-justice-involved-population/.
9. National Commission on Correctional Health Care Board of Directors. Women's health care in correctional settings [Internet]. Chicago (IL): NCCHC; 1994 Sep 25 [updated 2014 Oct 19; cited 2018 May 22]. Available from: https://www.ncchc.org/womens-health-care.
10. Maruschak LM. Medical problems of prisoners [Internet]. Bureau of Justice Statistics. Washington, DC: US Department of Justice; 2008 [cited 2018 May 22]. Available from: https://www.bjs.gov/content/pub/pdf/mpp.pdf.
11. New York Consolidated Laws, COR § 611 [correction law on the Internet]. Available from: https://codes.findlaw.com/ny/correction-law/cor-sect-611.html
12. Prenatal and infant care services, CRR-NY 7651.17 [New York codes, rules and regulations on the Internet]. Available from: https://govt.westlaw.com/nycrr/Document/I4fb49723cd1711dda432a117e6e0f345?viewType=FullText&originationContext=documenttoc&transitionType=CategoryPageItem&contextData=(sc.Default).
13. American Medical Association. An act to prohibit the shackling of pregnant prisoners model state legislation [Internet]. Chicago (IL): Advocacy Resource Center; 2015 [cited 2018 May 22]. Available from: https://www.ama-assn.org/sites/default/files/media-browser/specialtygroup/arc/shackling-pregnant-prisoners-issue-brief.pdf.
14. Walmsley R. World female imprisonment list [Internet]. London (UK): Institute for Criminal Policy Research; 2015 [cited 2018 May 22]. Available from: http://www.prisonstudies.org/sites/default/files/resources/download/world_female_imprisonment_list_third_edition_0.pdf.
15. Braithwaite RL, Treadwell HM, Arriola KRJ. Health disparities and incarcerated women: a population ignored. Am J Public Health. 2005;95(10):1679–81.
16. Clarke JG, Hebert MR, Rosengard C, Rose JS, DaSilva KM, Stein MD. Reproductive health care and family planning needs among incarcerated women. Am J Public Health. 2006;96(5):834–9.
17. Committee on Health Care for Underserved Women. Committee opinion: health care for pregnant and postpartum incarcerated women and adolescent females. Washington, D.C.: American College of Obstetricians and Gynecologists; 2011 Nov. [reaffirmed 2016; cited 2018 May 22]. Available from: https://www.acog.org/Clinical-Guidance-and-Publications/Committee-Opinions/Committee-on-Health-Care-for-Underserved-Women/Health-Care-for-Pregnant-and-Postpartum-Incarcerated-Women-and-Adolescent-Females.
18. American College of Obstetricians and Gynecologists. 2017 State legislation tally: incarcerated women: limiting use of restraints [Internet]. Washington, D.C.: ACOG; 2017 [cited 2017 Dec 18]. Available from: https://www.acog.org/-/media/Departments/State-Legislative-Activities/2017ShacklingTally.pdf?dmc=1&ts=20171218T1843221836.
19. Kraft-Solar T. Reproductive injustice: the state of reproductive health care for women in New York state prisons [Women in Prison Project report on the Internet]. Northampton (MA): Prison Policy Initiative; 2015 [cited 2018 May 22]. Available from: https://static.prisonpolicy.org/scans/Reproductive-Injustice-FULL-REPORT-FINAL-211-15.pdf.
20. Irwin J, Schiraldi V, Ziedenberg J. America's one million nonviolent prisoners [policy report on the Internet]. Washington, D.C.: Justice Policy Institute; 1999 Mar [cited 2018 May 22]. Available from: http://www.justicepolicy.org/images/upload/99-03_rep_onemillionnonviolentprisoners_ac.pdf
21. National Governors Association. The Privacy Rule: Corrections, law enforcement and the courts [publication

on the Internet]. Washington, D.C.: NGA; 2002 [cited by 2018 May 22]. Available from: https://www.nga.org/files/live/sites/NGA/files/pdf/FACTSHIPAACORRECT.pdf.
22. Goldstein MM. Health information privacy and health information technology in the US correctional setting. Am J Public Health. 2014;104(5):803–9.
23. American Bar Association. ABA standards for criminal justice: treatment of prisoners. 3rd ed. Washington, D.C.: ABA; 2011. Available from: https://www.americanbar.org/content/dam/aba/publications/criminal_justice_standards/Treatment_of_Prisoners.authcheckdam.pdf.
24. American Public Health Association Task Force on Correctional Health Care Standards. Correctional health care standards and accreditation [policy statement on the Internet]. Washington, D.C.: APHA; 2004 Nov 9 [cited 2018 May 22]. Available from: https://www.apha.org/policies-and-advocacy/public-health-policy-statements/policy-database/2014/07/02/12/07/correctional-health-care-standards-and-accreditation.
25. Rothstein MA. Currents in contemporary bioethics: constitutional right to informational health privacy in critical condition. J Law Med Ethics. 2011;39(2):280–284.
26. Uses and disclosures for which an authorization or opportunity to agree or object is not required, 45 CFR § 164.512 (2004).
27. Standards for privacy of individually identifiable health information. Office of the Assistant Secretary for Planning and Evaluation, DHHS. Final rule. Fed Regist. 2000;65(250):82462–829.
28. Martin SL, Rieger RH, Kupper LL, Meyer RE, Qaqish BF. The effect of incarceration during pregnancy on birth outcomes. Public Health Rep. 112(4):340–6.
29. Kyei-Aboagye K, Vragovic O, Chong D. Birth outcome in incarcerated, high-risk pregnant women. J Reprod Med. 2000;45(3):190–4.
30. Howard DL, Strobino D, Sherman SG, Crum RM. Maternal incarceration during pregnancy and infant birthweight. Matern Child Health J. 2011;15(4):478–86.
31. Mertens DJ. Pregnancy outcomes of inmates in a large county jail setting. Public Health Nurs. 18(1):45–53. Available from: http://www.ncbi.nlm.nih.gov/pubmed/11251873.
32. Fogel CI. Pregnant inmates: risk factors and pregnancy outcomes. J Obstet Gynecol Neonatal Nurs JOGNN. 1993;22(1):33–9.
33. Egley CC, Miller DE, Granados JL, Ingram-Fogel C. Outcome of pregnancy during imprisonment. J Reprod Med. 1992;37(2):131–4.
34. Bell JF, Zimmerman FJ, Cawthon ML, Huebner CE, Ward DH, Schroeder CA. Jail incarceration and birth outcomes. J Urban Heal Bull New York Acad Med. 2004;81(4):630–44.
35. Huang K, Atlas R, Parvez F. The significance of breastfeeding to incarcerated pregnant women: an exploratory study. Birth. 2012;39(2):145–55.
36. Schwarz EB, Nothnagle M. The maternal health benefits of breastfeeding. Am Fam Physician. 2015;91(9):603–4.
37. Mathur NB, Breastfeeding DD. Indian J Pediatr. 2014;81(2):143–9.
38. Martin JA, Hamilton BE, Osterman MJK, Driscoll AK, Mathews TJ. Births: final data for 2015. Natl Vital Stat Rep. 2017;66(1):1.
39. Kost K, Maddow-Zimet I, Arpaia A. Pregnancies, births and abortions among adolescents and young women in the United States, 2013: National and state trends by age, race and ethnicity [Internet]. New York: Guttmacher Institute; 2017 [cited 2018 May 22]. Available from: https://www.guttmacher.org/report/us-adolescent-pregnancy-trends-2013.
40. Santelli JS, Lindberg LD, Finer LB, Singh S. Explaining recent declines in adolescent pregnancy in the United States: the contribution of abstinence and improved contraceptive use. Am J Public Health. 2007;97(1):150–6.
41. Mosher WD, Jones J, Abma JC. Intended and unintended births in the United States: 1982–2010. Natl Heal Stat Rep. 2012;55:1–28. Available from: https://www.cdc.gov/nchs/data/nhsr/nhsr055.pdf.
42. Ekwo EE, Gosselink CA, Moawad A. Unfavorable outcome in penultimate pregnancy and premature rupture of membranes in successive pregnancy. Obstet Gynecol. 1992;80(2):166–72.
43. Gardner M, Goldenberg R, Cliver S, Tucker J, Nelson K, Copper R. The origin and outcome of preterm twin pregnancies. Obstet Gynecol. 1995;85(4):553–7.
44. Eunice Kennedy Shriver National Institute of Child Health and Human Development. What are the risk factors for preterm labor and birth? [Internet]. Rockville (MD): NICHD Information Resource Center; 2017 Jan 31 [cited 2018 May 22]. Available from: https://www.nichd.nih.gov/health/topics/preterm/conditioninfo/who_risk.
45. Centers for Disease Control and Prevention. Preterm birth [Internet]. Atlanta (GA): CDC; 2018 [cited 2018 May 22]. Available from: https://www.cdc.gov/reproductivehealth/maternalinfanthealth/PretermBirth.htm.
46. Geronimus AT. The weathering hypothesis and the health of African-American women and infants: evidence and speculations. Ethn Dis. 1992;2(3):207–21.
47. David RJ, Collins JW. Differing birth weight among infants of U.S.-born blacks, African-born blacks, and U.S.-born whites. N Engl J Med. 1997;337(17):1209–14.
48. David R, Collins J Jr. Disparities in infant mortality: what's genetics got to do with it? Am J Public Health. 2007;97(7):1191–7.
49. Testimony by Gail T. Smith, Director, Women in Prison Project, The Correctional Association of New York, Before Assembly Committees on Health and Corrections [Internet]. CANY; 2017 Oct 30. Available from: https://www.correctionalassociation.org/wp-content/uploads/2017/10/Oct-30-17-Health-care-hearing-gts-testimony-draft-2-21.pdf.
50. Alsan M, Wanamaker M. Tuskegee and the health of black men [appendix for online publication]. 2017 June [cited 2018 May 23]. Available from: http://www.nber.org/data-appendix/w22323.
51. Arnett MJ, Thorpe RJ, Gaskin DJ, Bowie JV, LaVeist TA. Race, medical mistrust, and segregation in primary care as usual source of care: findings from the exploring health disparities in integrated communities study. J Urban Heal. 2016;93(3):456–67.

52. Schultz MJ, Bushati T. Maternal physical morbidity associated with denial of pregnancy. Aust New Zeal J Obstet Gynaecol. 2015;55(6):559–64.
53. Friedman SH, Heneghan A, Rosenthal M. Characteristics of women who do not seek prenatal care and implications for prevention. J Obstet Gynecol Neonatal Nurs. 2009;38(2):174–81.
54. Criminalization of pregnant women with substance use disorders. J Obstet Gynecol Neonatal Nurs. 2015;44(1):155–7.
55. Jongen CS, McCalman J, Bainbridge RG. The implementation and evaluation of health promotion services and programs to improve cultural competency: a systematic scoping review. Front Public Heal. 2017;5:24.
56. McKillop M. Prison health care spending varies dramatically by state [Internet]. Philadelphia (PA):The Pew Charitable Trusts; 2017 [cited 2018 May 23]. Available from: http://www.pewtrusts.org/en/research-and-analysis/analysis/2017/12/15/prison-health-care-spending-varies-dramatically-by-state.
57. Lines R. From equivalence of standards to equivalence of objectives: the entitlement of prisoners to health care standards higher than those outside prisons. Int J Prison Health. 2006;2(4):269–80.
58. United Nations. Basic principles for the treatment of prisoners. 1990 [cited 2018 Jul 5]. Available from: https://www.un.org/ruleoflaw/blog/document/basic-principles-for-the-treatment-of-prisoners/
59. Exworthy T, Samele C, Urquía NFA. Asserting prisoners' right to health: progressing beyond equivalence. Psychiatr Serv. 2012;63(3):270–5.
60. Axon A, Hassan M, Niv Y, Beglinger C, Rokkas T. Ethical and legal implications in seeking and providing a second medical opinion. Dig Dis. 2008;26(1):11–7.
61. Gruen RL, Campbell EG, Blumenthal D. Public roles of US physicians. JAMA. 2006;296(20):2467.
62. Gruen RL, Pearson SD, Brennan TA. Physician-citizens—public roles and professional obligations. JAMA. 2004;291(1):94.
63. Sufrin C, Kolbi-Molinas A, Roth R. Reproductive justice, health disparities and incarcerated women in the United States. Perspect Sex Reprod Health. 2015;47(4):213–9.
64. Luna Z, Luker K. Reproductive justice. Annu Rev Law Soc Sci. 2013;9(1):327–52.
65. Galvin G. Understaffed and overcrowded: state prisons crippled by budget constraints, bad leadership [Internet]. US News and World Report; 2017 Jul 26 [cited 2018 May 27]. Available from: https://www.usnews.com/news/best-states/articles/2017-07-26/understaffed-and-overcrowded-state-prisons-crippled-by-budget-constraints-bad-leadership.
66. Suárez-Cortés M, Armero-Barranco D, Canteras-Jordana M, Martínez-Roche ME. Use and influence of delivery and birth plans in the humanizing delivery process. Rev Lat Am Enfermagem. 2015;23(3): 520–6.
67. Redding H. The components of prison security [report on the Internet]. Naples (FL): International Foundation for Protection Officers; 2004 Sep [cited 2018 May 27]. Available from: http://www.ifpo.org/resource-links/articles-and-reports/protection-of-specific-environments/the-components-of-prison-security/
68. Sufrin C. Jailcare: finding the safety net for women behind bars. Oakland: University of California Press; 2017.
69. American Bar Association, House of Delegates. Model rules of professional conduct. American Bar Association; 2008.

Further Reading on this Topic

Dubler N. Ethical dilemmas in prison and jail health care [Internet]. Bethesda (MD): Health Affairs; 2014 Mar 10. Available from: https://www.healthaffairs.org/do/10.1377/hblog20140310.037605/full/.

Inside Edition. Inmate gave birth on jail floor claiming staff wouldn't take her to hospital [web streaming video]. 2017 February 9. Available from: https://www.youtube.com/watch?v=TqH323ioHmg.

Quinn A. In Labor, In Chains. The New York Times [Internet]. 2014 Jul 27. Available from: https://www.nytimes.com/2014/07/27/opinion/sunday/the-outrageous-shackling-of-pregnant-inmates.html.

"Why Does My Son Have Lead in His Blood?"

Travis R. Hobart

6.1 Background Questions – 100

6.2 Additional Case Information and Questions for Discussion – 100

6.3 Answers to Background Questions – 103

6.4 Responses to Discussion Questions – 108

References – 120

© Springer Nature Switzerland AG 2019
A. E. Caruso Brown et al. (eds.), *Bioethics, Public Health, and the Social Sciences for the Medical Professions*, https://doi.org/10.1007/978-3-030-03544-0_6

Glossary

Early intervention - A federally funded, locally run program that provides evaluations and services for children suspected to have developmental delay.

Free erythrocyte protoporphyrin or FEP - An enzyme in the heme pathway that becomes elevated in the presence of prolonged high levels of lead in the blood.

Lead exposure or lead poisoning - Any detectable elevation of blood lead level. The U.S. Centers for Disease Control and Prevention (CDC) also sets a reference level of 5 μ(mu)g/dL, above which public health actions should be undertaken.

Learning Objectives

1. Understand the importance of a complete social and environmental history and of the need to partner with local public health agencies in the management of certain illnesses.
2. Begin to form independent views on the extent to which the physician has a broader social responsibility beyond the individual patient-physician relationship and defend your views persuasively.
3. Develop the skill to identify the need for change in your own community, for instance, within your medical school, clinic, hospital, neighborhood, or city.
4. Identify ways in which physicians can advocate for change at the local, regional, state, and national level.

> **Case Background**
>
> *You are a pediatrician in an impoverished urban area. You routinely screen children for lead poisoning. You just saw a happy, active 2-year-old boy named Calvin Wilson for his well-child exam. Your nurse gives you the results of his lead test, which shows that his capillary blood lead level (BLL) is elevated (25 μg/dL).*
>
> *You note this is the fifth case this month of a child with an elevated lead level (greater than 5 μg/dL), though he is your first patient with a level greater than 20 μg/dL. Typically, about 5% of your patients have an elevated level. You are alarmed because there seems to be a trend toward more children in your practice having elevated BLLs and you know that lead is a potent neurotoxin.*

6.1 Background Questions

1. What is the major source of lead exposure in the United States? Identify a data source used to determine the incidence of lead exposure. Based on this source, what is the incidence?

2. What are the adverse health outcomes associated with acute, subacute, and chronic lead exposure?

3. In the United States, which groups of children are most likely to have high lead levels? Why? What are the current U.S. Centers for Disease Control and Prevention (CDC) guidelines regarding screening for lead exposure? What are the requirements for screening in your state?

4. How could you figure out where this child was exposed to lead? Whom could you call for help? What public health systems are in place to address such issues?

5. What action(s) do we take for primary prevention of lead exposure? Secondary prevention? Tertiary prevention?

6. If a child is suspected to have developmental delay, what public health resources are available for families in the United States?

7. The Kennedy Krieger Institute at Johns Hopkins University performed a public health study involving lead and lead abatement, which was ethically controversial. How was the study done and why was this ethically problematic? What were the findings of this study?

8. What were the findings of the study in Flint, Michigan, done by Dr. Mona Hanna-Attisha and colleagues? What kind of study was it? How did the water crisis there affect the lead levels in the local population?

6.2 Additional Case Information and Questions for Discussion

Because of the high lead level, you send Calvin to have the blood lead level retested with a venous sample. Capillary samples are known to sometimes be falsely elevated. The venous sample confirms the lead level of 25 μ(mu)g/dL. You assess for anemia, which is both a consequence of lead exposure and a commonly occurring comorbidity, in the form of iron deficiency anemia. He is mildly anemic. You also test for free erythrocyte porphyrin (FEP), a blood enzyme that becomes elevated with long-term exposure to high lead levels in the body (analogous to hemoglobin A1C elevation in diabetes). The results show that the exposure has been fairly recent.

When you call Ms. Wilson to let her know about the results and your plan to retest Calvin in 1 month, she asks you where you think the lead came from. She mentions that her tap water has a funny taste and smell but that they continue to drink it because they cannot afford to buy bottled water. You tell her you are not sure about the source and that you will be looking into it. This is not the first time you have heard your patients' families complain about the taste, smell, and color of the water. After some initial research, you learn that the city recently switched its source of water. You also learn that city officials have decided not to do "corrosion control" to prevent lead from leaching out of pipes into the water, though you believe that it is probably recommended. You are worried about the risk to children and, after Calvin's results, you decide to do a study to determine whether the change in water source is causing higher rates of lead exposure.

1. What types of studies could you perform to answer this question?
   (a) What are the ethical concerns with each?
   (b) How do the ethical concerns raised in the Kennedy Krieger study guide your thinking?

You recognize that this situation is similar to the events occurring in Flint, MI, in 2014–2015. In Flint, a pediatrician recognized the health risks and did a study to assess the risk to her patients. The results of that study were famously announced at a press conference and only later released in a medical journal.

2. What are healthcare practitioners' ethical and legal obligations when conducting research that might have immediate health implications?
   (a) Do healthcare practitioners have an obligation to share study results without first going through the peer review process of a medical journal?
      (i) If so, under what circumstances?
      (ii) If not, why not?

3. If healthcare practitioners suspect that they are witnessing a health concern that may affect the health of more than one patient, what is their obligation to investigate further?
   (a) If an issue of concern is validated and a solution to address the concern is identified, what is the healthcare practitioner's obligation to advocate for changes to address the problem?
   (b) When you see such a pattern, whom would you call?

You see Calvin for follow-up of his elevated BLL. He is and has been growing well without any medical problems. When he was screened for lead at age 1, his level was less than 2 µg/dL (lower limit of detection). At his visit today, you note that he is very mobile and constantly exploring. His motor development is on track. However, he only says five words so far, and he is not putting any words together (which indicates speech delay for a 2-year-old). His mother is also concerned that he throws a tantrum whenever she uses the vacuum cleaner or a loud truck passes the apartment, and that he does not seem interested in other children. His mother is 19 years old and is 25 weeks into her second pregnancy.

4. What would you do now with respect to the lead exposure?
   (a) What services are available in the community?
   (b) How can you determine whether the water is the source of his lead exposure or if it is something else in part or in total?

5. Developmental delay is both a consequence of lead exposure and a risk factor for lead exposure. Why do you think that is?
   (a) What would you do now to assess his possible developmental delay?

You call the county health department lead poisoning program. They have been receiving an increased number of calls recently due to people's concerns about the water. They state that they are currently coordinating with the public water utility to assess the possible dangers associated with the recent water source change. Because Calvin's level is so high, they plan to visit his home to assess it for lead hazards.

6. Calvin may have lifelong cognitive impairments because of his lead exposure (which can cause irreversible neurological damage). Who should be responsible for paying for his treatment and the treatment of other people affected by the change in water source?

7. The usual method for identifying lead exposure is through testing children's lead levels. What level of prevention is this?
   (a) Is this form of prevention just?
   (b) What would be more just?

The county lead poisoning program investigates Ms. Wilson's rented apartment over the next few weeks. You are surprised to learn that Calvin and his mother live with Calvin's aunt and her five children, ranging in age from 15 months to 12 years. In addition to lead in the tap water, the health department inspector finds multiple lead hazards in the home, including paint chipping in the bedroom where Calvin sleeps with two of his older cousins. The older cousins report that Calvin sometimes eats the paint chips.

8. What would you do now?
   (a) Are you worried about his pregnant mother and the fetus? Why or why not?
   (b) Does it matter whether the lead in Calvin's blood comes from the paint or the water? Why or why not?

9. Do you have any obligations to Calvin's cousins?

The health department notifies the landlord that the lead hazards exist in the apartment. He works to mitigate the hazards by repainting the bedroom and some of the windows. As Calvin's physician, you are concerned that this will not be enough, but that is all the law requires of the landlord. Over the next several months, you check Calvin's lead level monthly. You learn that his cousins, his mother, and his new baby sister have been tested. All were negative for lead exposure. Calvin's lead level initially goes down steadily, and you are reassured that the landlord's fixes have helped. However, about 6 months later, in early summer, Calvin's BLL suddenly spikes up to 60 µg/dL. This level of lead is likely to affect his brain, leading to learning and attention problems in the long run (although we can never be sure of the cause in individual cases). You recommended hospitalization for 5 days of chelation therapy. When you talk to Ms. Wilson, she tells you that they have recently had to open the windows to cool the house down, including some of the windows with old lead paint that were recently repainted. The water pipes still contain lead as well.

10. Self-reflection: How does this make you feel?

11. What would you do now?
    (a) Are you comfortable sending him back to the same apartment after he is discharged?

The health department has been working with the family for months to find alternative housing. They have a Section 8 voucher, which gives them government assistance for housing, but only certain landlords are willing to take Section 8 tenants. Most Section 8 options will not accommodate such a large family, with two adults and seven children. The process, which normally takes years, has been expedited due to the lead hazards, but no suitable housing has been found yet. Because of the new exposure, you now tell the health department that you think Calvin's immediate family should be separated from his aunt and cousins in order to get him into a lead-free environment.

12. What are the risks and benefits of separating the families? Is it worth it?

The research from Flint, MI, showed an increase in the average BLL for the cohort of children tested, and the effect was worse on children in impoverished neighborhoods, in part because they happened to be near the "end of the line" of the water system, which allowed more time for lead to accumulate in the water prior to being consumed.

13. Thinking about the socio-ecologic model, what institutional level factors affect the rates of lead exposure? What can be done about them?
    (a) Community level factors?
    (b) Policy level factors?

"Why Does My Son Have Lead in His Blood?"

Structural violence "is one way of describing social arrangements that put individuals and populations in harm's way." [1, 2] As an example, in the 1930s and 1940s in the United States, the practice of "redlining" was the norm in many cities, which made it impossible for African Americans, Jewish, and foreign-born residents to receive mortgages for homes in certain desirable neighborhoods. Many of the cities continue to have geographic disparities that originated with or were perpetuated by these practices [3].

14. **In what ways can the events in Flint, MI, be considered a recent example of structural violence?**
    (a) **Self-reflection: What does that mean to you, as a future healthcare practitioner?**

Recent research from Flint, MI, showed that the percentage of children with a BLL above 5 µg/dL increased from approximately 3% to approximately 4–5% of children before and after the water switch [4–6]. However, the geometric mean BLL in the community only went up from 1.19 µg/dL to 1.30 µg/dL, and lead levels since 2006 have in fact decreased in the community [6]. Exposure from water is less likely than exposure from paint to cause higher levels above 10 or 15 µg/dL that we see in Calvin. In many places, there is lead in the water and also in old homes from the paint and soil.

15. **Based on the long-term trend that has recently been further characterized in the peer-reviewed literature, do you think that Dr. Hanna-Attisha and her team were premature in their announcement to the press?**

16. **Which do you think is more concerning: exposure of most of the population to low levels of lead (those drinking tap water) or exposure of a smaller portion of the population to higher levels of lead (those exposed to paint hazards)?**

17. **Creative problem-solving: Since we have limited resources as a society, should we focus our efforts on lead pipes and make a small impact on a large population, or should we focus on lead paint to make a larger impact on a small population? How can we achieve both as a society?**

6.3 Answers to Background Questions

1. **What is the major source of lead exposure in the United States? Identify a data source used to determine the incidence of lead exposure. Based on this source, what is the incidence?**

The major source of lead exposure in the United States is lead-based paint found in homes built before 1978 (when such paint was banned). Other sources include soil, water (through corrosion of pipes containing lead), poor-quality toys, spices and cosmetics imported from other countries, and ammunition [7]. In addition, some people have exposure in the workplace, while making car batteries, soldering, or painting certain infrastructure (such as bridges, for which lead-based paint is still used) [8].

The lead typically enters the body through ingestion or inhalation. Ingestion of lead paint continues to be common. As old paint deteriorates over time, the dust or flakes from the paint end up on the floor of the home, where children are exposed. Toddlers, in particular, are prone to exposure because of their proximity to the ground and frequent, developmentally typical hand-to-mouth behavior; that is, they put everything in their mouths. Ingestion of lead-contaminated water is another source. Inhalation has traditionally occurred due to burning of leaded gasoline, but this fortunately no longer occurs since leaded gasoline has been banned. However, the lead from that process is still in the soil, especially in communities near a highway or in urban centers, where it can be ingested through hand-to-mouth behavior. In addition, some activities, such as melting car batteries or elemental lead for hobby use, can still produce respirable lead [9]. A major data source that reports the incidence of lead exposure is the CDC. The CDC funds lead surveillance programs in the health departments in 29 states and 6 cities (including Washington, DC). These states and cities must report their data to the CDC on a quarterly basis. Other states and cities are not funded by the CDC but can voluntarily report their data. The CDC is explicit in the fact that these data are incomplete. These funded programs target children at highest risk and do not reliably test all children. Therefore, the data are not comprehensive and cannot be used to estimate incidence in states or to compare states [10].

2. **What are the adverse health outcomes associated with acute, subacute, and chronic lead exposure?**
The major effect of lead exposure in childhood seen today is the long-term effect on intelligence, learning, and attention due to unrecognized subclinical exposures. The strength of the inverse correlation of BLL and IQ has been consistently shown in studies, even at low levels of exposure [9, 11]. Most have found that "as blood lead levels in 2-year-olds increase by 10 µg/dL, the IQ at age 4 years and older declines by 2 to 3 points" on average [9]. One study found that at 10 µg/dL the increase caused IQ to decrease by 4.6 points and that the decrease in IQ was faster in those whose BLL did not go above 10 µg/dL [12]. Some studies have pointed out caveats to this. First, an average decrease of IQ of 2–3 points will mean that some children have above average decreases that are more significant. Second, and more importantly, IQ is not a very nuanced measure of learning in children, and significant attention and learning problems can result from lead exposure and coexist with a relatively high IQ [11].

At levels of 25 µg/dL or greater, anemia is a consequence. As alluded to in the case background, the elevation of FEP can indicate elevation above this level for a prolonged period [9]. It should also be noted that concurrent iron deficiency anemia is also often present.

Blood lead levels above 60 µg/dL might lead to symptoms of headaches, abdominal pain, and constipation [9]. It is worthwhile to note that these non-specific symptoms are commonly experienced by children not exposed to lead, making them difficult to use as diagnostic indicators. Neurologic symptoms are also seen, such as clumsiness, agitation, and somnolence, leading to vomiting, stupor, and convulsions [9]. In high enough doses, lead exposure can lead to death. In 2006, a child died after ingesting a single metal charm from a bracelet that came for free with a pair of Reebok shoes [13].

Children are particularly vulnerable to the toxic effects of lead for several reasons. They absorb a higher percent of ingested lead from their intestine than adults. In addition, the lead more readily crosses the blood-brain barrier. Lastly, the rapidly developing neurologic system of a child is more susceptible to toxins than the relatively more stable brain of an adult [11].

Though the vulnerability of adults is lower, exposure in adulthood can also have lasting effects. As in children, it can have neurologic and hematologic effects. In addition, effects that have been primarily identified after occupational exposure are hypertension and kidney toxicity [14].

3. **In the United States, which groups of children are most likely to have high lead levels? Why? What are the current U.S. Centers for Disease Control and Prevention (CDC) guidelines regarding screening for lead exposure? What are the requirements for screening in your state?**
Groups of children at higher risk of elevated BLL are those who are poor, members of racial-ethnic minority groups, and/or recent immigrants. The major reason for this is that they are more likely to live in older, poorly maintained rental homes [7, 14]. In older housing stock, lead paint often remains under subsequent paint layers and, if poorly maintained, sheds contaminated paint dust and flakes. Young children exhibiting developmentally typical hand-to-mouth behavior are at particular risk in such an environment [9]. Another group at risk are children whose parents have

occupational exposure [7]. In these families, the exposed worker can carry lead particles home on their clothing, skin, or hair and expose other family members. Additionally, lead does cross the placental barrier, so pregnant women can inadvertently expose the fetus through occupational or other exposure [14].

The CDC guidelines are derived from an expert report from the Advisory Committee on Childhood Lead Poisoning Prevention (ACCLPP). Up until 1997, the recommendations from the ACCLPP were to screen all children at 12 and 24 months of age and older children at higher risk. Due to decreasing prevalence of lead exposure, the 1997 updated guidelines recommended more targeted screening and encouraged state and/or local agencies to formulate their own screening plan based on local housing conditions. Absent a local plan, the CDC continues to recommend universal screening. In addition, due to higher risk among Medicaid recipients, that program requires that all enrolled children have screening at 12 and 24 months, or by 72 months if not done prior [9, 15]. The American Academy of Pediatrics (AAP) concurs with the CDC recommendation [9]. Interestingly, the United States Preventive Services Task Force (USPSTF) does not agree. The new draft recommendation, released in late 2018, found insufficient evidence to recommend for or against screening for lead in asymptomatic children, essentially reaffirming the previous recommendation [16].

The state legislature and/or health department would set legal requirements for lead screening, and these differ by state. As an example, New York state law requires screening for all children at ages 12 and 24 months, [17] as previously recommended by the CDC.

4. **How could you figure out where this child was exposed to lead? Whom could you call for help? What public health systems are in place to address such issues?**

A careful history might determine one or more possible sources, but it is unlikely that the healthcare practitioner would be able to identify the definitive source of lead in this way. Instead, providers will likely rely on services available in the community to investigate the home. The local health department is the first place to start. It depends on the state and county, but most counties in the United States have a lead poisoning program. The CDC funds such programs in 29 states, 5 cities, and the District of Columbia [10]. Many other states have their own programs as well. Depending on the lead level and what the health department offers, the health department will likely send an inspector and/or public health nurse to the home to assess for lead hazards. The inspector typically assesses all surfaces inside and outside of the home for lead and looks for certain household items that are commonly found to have lead in them (e.g., certain toys, spices or cultural remedies or cosmetics sent from foreign countries, talcum powder) [7]. All water utilities are required by federal law to test the water supply for lead and copper at customer taps. If there is lead above a level of 15 parts per billion in the water, then the utility must take steps to notify the public about the problem and fix it using better corrosion control [18]. Many water utilities can direct consumers to water testing options, though they are not required to test on demand.

5. **What action(s) do we take for primary prevention of lead exposure? Secondary prevention? Tertiary prevention?**
Primary prevention is preventing lead exposure before it ever takes place. To do this, we would primarily have to move children out of old houses with lead paint and/or remediate housing stock so it is lead-free. We would also have to assure that the water supply and pipes were free of lead. The U.S. Department of Housing and Urban Development (HUD) oversees and funds multiple strategies for primary prevention of lead exposure, including making grant money available

to homeowners and landlords for repairs and enforcing certain lead regulations related to housing [19]. As another example, the Green and Healthy Homes Initiative, a primary prevention program, focuses on renovations of homes to address hazards such as lead. In addition, in their strategic plan, they propose many strategies to prevent lead exposure, including several legal strategies that require more comprehensive assessment of homes occupied by children [20].

Secondary prevention is screening for disease before it has major consequences. The CDC and AAP recommendation for screening at the ages of 12 and 24 months [15] is secondary prevention. These screening tests for lead in the blood provide an opportunity for early identification of elevated BLLs and subsequent interventions to stop further exposure or treat the patient with chelation therapy if indicated. However, such screening cannot prevent all consequences of lead exposure because we know that effects on the brain can begin at any level of exposure. Furthermore, screening does not indicate for how long the exposure has been ongoing.

Tertiary prevention is preventing long-term consequences after having developed a disease. In the case of lead exposure, an important example is the provision of early intervention and special education to affected children. We know these services can help a child with developmental delay progress and become more successful in school and the workplace [21]. These services are available to all children in the United States with a developmental disability (see Answer to Background Question #6 below). Lead poisoning is a diagnosis that qualifies a child for such services if it causes impairment of the child's learning ability [22].

6. **If a child is suspected to have developmental delay, what public health resources are available for families in the United States?**
In the United States, any child under age 3 with suspected developmental delay is eligible for early intervention services. This publicly funded program, mandated by the Individuals with Disabilities Education Act (IDEA), provides for evaluation and treatment of any child suspected of having developmental delay. If a child qualifies for services (speech, physical, and occupational therapy and/or special education) based on the initial evaluation, the program provides a tailored approach to meet the child's particular needs [23]. For children aged 3 and older, the IDEA also mandates appropriate therapy through the local school district. This is true even if the child is not enrolled in the local school district. Parents do not need a referral from a healthcare practitioner to be evaluated for these services [24].

Each state may set the developmental milestones used as a benchmark for these programs. If a child has not met certain milestones by a certain age, then they may qualify for services. In addition, certain conditions automatically qualify a child for these services. These typically include hearing or vision loss and certain diseases such as Down syndrome in which developmental delay is expected [23, 24]. As noted previously, the original law singled out several conditions that are likely to cause difficulty with school, including lead poisoning [22].

7. **The Kennedy Krieger Institute at Johns Hopkins University performed a public health study involving lead and lead abatement, which was ethically controversial. How was the study done and why was this ethically problematic? What were the findings of this study?**
The investigators in this study were assessing whether there could be an affordable lead abatement strategy for houses that could prevent lead exposure in the occupants. They identified 108 houses in the city of Baltimore, MD, in which children were living. They divided the houses into five groups. Two control groups consisted of lead-safe houses: one group was built after 1978 and therefore did not contain lead paint, and the other group had been

previously remediated of lead hazards. Three study groups were divided into different lead paint remediation plans costing $1,650, $3,500, and $7,000. They had previously studied these abatement plans in unoccupied homes and had proven that they were effective at reducing lead hazards, though admittedly not eliminating them. As a comparison, the cost of totally eliminating lead hazards in a home was roughly $20,000 at the time and well out of reach of the average homeowner or landlord. The results of the study showed that the lead values of almost all the children decreased, regardless of the remediation plan used. That is, all remediation plans worked equally well in the short term [25, 26].

The study was deemed ethically problematic afterward when two mothers who were enrolled in the study sued the Kennedy Krieger Institute (KKI). They contended that they were not fully informed of the risks in the study and that the researchers had not notified them in a timely manner after identifying lead hazards in their homes. Lower courts initially dismissed the case, but the appeals court in Maryland eventually ruled against the KKI. The judge ruled that the informed consent had been inadequate because parents could not consent to a study that would have risk to their children without any hope of personal benefit. The other issues were that the researchers had a duty to warn about the hazards found and that the Institutional Review Board (IRB) process at Johns Hopkins had been too concerned with producing successful research at the expense of the research subjects [25, 27].

The ethical issues raised center around the idea of justice. Critics of the study contend that the money and time spent researching the less effective remediation plans should have been used instead on fully remediating the houses and, further, that researching alternatives undermines any efforts to improve housing conditions in society. They argue that these remediation plans treat the participants unfairly and perhaps even exploit them by giving them less effective remediation in order to benefit others [25].

The counter-argument is that the researchers themselves were not treating anyone unfairly. Rather, society has been treating the participants unfairly by unjustly allowing some citizens to live in unsafe living conditions. Working to address a problem now, if imperfectly, is justifiable when those children otherwise would not likely see any benefit from a potential future change in societal norms [25].

8. **What were the findings of the study in Flint, Michigan, done by Dr. Mona Hanna-Attisha and colleagues? What kind of study was it? How did the water crisis there affect the lead levels in the local population?**

The city of Flint, MI, switched the source of the drinking water from Lake Huron to the Flint River in April 2014, primarily to save money. Initially, there were complaints about its taste, smell, and color. The water was found to be contaminated with coliform bacteria in August–September 2014. Subsequently, the water was found to have high levels of trihalomethanes, a chlorine byproduct (a result of treating the bacteria with high levels of chlorine). In addition, the water utility failed to use corrosion control to prevent lead from leaching out of pipes into the water, which was legally required given the water source and the size of the city. Despite numerous complaints and significant evidence of both infection risk and contaminants such as lead and trihalomethanes, the government did not address the problem [28, 29]. In September, an outside study done by Virginia Tech was released, in which scientists found high levels of lead in drinking water samples taken from many homes. In addition, the study by Dr. Hanna-Attisha was publicized. With this publicity, the government finally responded [4, 28].

Worried about the risk to children, Mona Hanna-Attisha, a local pediatrician (and pediatric residency program director) at the major children's health clinic in Flint, decided to do a study to determine whether the change in water source was causing higher rates of lead exposure. The study was a retrospective cohort study looking at BLLs in children who had their blood drawn in the lab at Hurley Medical Center in Flint. It compared levels before and after the water source change. The findings indicated that the percentage of children in Flint who had elevated BLL (above 5 μg/dL) increased from 2.4% before the water source change to 4.9% afterward. Further, they found that areas with high measured lead in the drinking water had an even higher percentage of children with elevated BLL, increasing from 4% to 10.6% [4]. More recent research has further characterized the extent of the problem and has even called into question the severity of the problem. A 2018 study found somewhat similar changes in the percentage of children with BLLs above 5 μg/dL (from 2.2% to 3.7%, though their more conservative statistical methods did not show significance). They also calculated the geometric mean BLL in the community and found it only went up from 1.19 μg/dL to 1.30 μg/dL for those years. Furthermore, they noted that the percentage of children with BLL above 5 μg/dL in 2006 was 11.8% and that the mean BLL at that time was 2.33 μg/dL. Thus, while it was certainly a failure of local and state government that the water was not being treated appropriately, the lead exposure that resulted was still lower than just a few years prior [6].

6.4 Responses to Discussion Questions

Because of the high lead level, you send Calvin to have the blood lead level retested with a venous sample. Capillary samples are known to sometimes be falsely elevated. The venous sample confirms the lead level of 25 μg/dL. You assess for anemia, which is both a consequence of lead poisoning and a commonly occurring comorbidity, in the form of iron deficiency anemia. He is mildly anemic. You also test for free erythrocyte porphyrin (FEP), a blood enzyme that becomes elevated with long-term exposure to high lead levels in the body (analogous to hemoglobin A1C elevation in diabetes). The results show that the exposure has been fairly recent.

When you call Ms. Wilson to let her know about the results and your plan to retest Calvin in 1 month, she asks you where you think the lead came from. She mentions that her tap water has a funny taste and smell but that they continue to drink it because they cannot afford to buy bottled water. You tell her you are not sure about the source and that you will be looking into it. This is not the first time you have heard your patients' families complain about the taste, smell, and color of the water. After some initial research, you learn that the city recently switched its source of water. You also learn that city officials have decided not to do "corrosion control" to prevent lead from leaching out of pipes into the water, though you believe that it is probably recommended. You are worried about the risk to children and, after Calvin's results, you decide to do a study to determine whether the change in water source is causing higher rates of lead exposure.

1. **What types of studies could you perform to answer this question?**
   The original Flint study was a retrospective cohort study looking at blood lead levels in children tested in their lab [4]. It compared levels before and after the water source change. The healthcare practitioners in this case could do a similar study in their clinic. They could also do a prospective cohort study of the children in the city. They could study their lead levels now and compare them over time once the water problem is fixed or they have bottled water. There are two main problems with a prospective study: it would be ethically flawed if it allowed patients or subjects to continue to drink water known to be contaminated, and it involves a significant input of time and money to follow a population for a prolonged period. A retrospective study would be not only better at showing a change due to the water source change but also less ethically problematic, less expensive, and less time-consuming. An alternative would be to study two

different but similar communities retrospectively, one that had lead in the water and one that did not.

A case control study, comparing patients with elevated blood lead levels to matched controls with normal blood lead levels and then comparing their water source or other sources of exposure, is also an option. This would not be as helpful as a cohort study because the data would be incomplete and purely correlative. It would merely show that exposure to lead in the water increases the odds of having a high level of lead in blood. We already know that exposure to lead causes high blood lead levels, so the only benefit is showing that the specific water source is a problem.

(a) **What are the ethical concerns with each?**
Any study allowing patients to drink contaminated water if other options were available would not be ethically sound. It would not be ethical to randomize children for any study like this because it would expose them to lead.

(b) **How do the ethical concerns raised in the Kennedy Krieger study guide your thinking?**
The judge in the Kennedy Krieger case had three main critiques: (1) that it was not possible for the families to give informed consent for a study without direct benefit to the children, (2) that the researchers failed to adequately warn the participants, and (3) that the IRB at Johns Hopkins was more concerned with having a valuable study than in the safety of the participants. In addition, others have raised related ethical concerns (mentioned in the response to Background Question 7 above), primarily that the study allowed children to remain in risky environments and that the study undermined any attempts to fully remediate homes with lead hazards because it was searching for a cheaper and potentially less effective alternative. This allowed an unjust situation to continue [25].

These concerns would guide researchers to assure that no child was put at risk without any potential benefit. In addition, researchers must ensure good communication with the participants in the study about any potential risks arising before or during the study. The judge's ruling also guides institutions to assure the independence of their IRB; it must always act in the best interest of the research participants even if that is not in the best interest of the institution. Lastly, researchers must value justice in their work and strive to align their research questions with the most just course for society. If, as in the Kennedy Krieger study, the research being proposed does not line up with the best-case scenario for the participants, there must be sound reasoning to support something less. Buchanon and Miller argue that such public health research may be ethically justifiable if four conditions are met, namely, (1) that there is a large population in need, (2) that a cheaper though less effective solution to a problem is likely to exist, (3) that some constraint makes use of the expensive solution unlikely, and (4) that the less expensive option, if proven effective, is highly likely to be implemented [25]. In that case, the benefits of finding a cheaper yet implementable solution outweigh the risks of doing the research or continuing to do nothing. This is certainly still up for debate, however.

You recognize that this situation is similar to the events occurring in Flint, MI, in 2014–2015. In Flint, a pediatrician recognized the health risks and did a study to assess the risk to her patients. The results of that study were famously announced at a press conference and only later released in a medical journal.

✓ 2. **What are healthcare practitioners' ethical and legal obligations when conducting research that might have immediate health implications?**

> **Box 6.1 Teaching Tip**
> This question is meant to provoke learners to think about whether physicians and other healthcare providers have an obligation to assure rapid dissemination of findings to the public. Encourage them to think about the risks and benefits of such approaches. The risk, for instance, of creating panic or promulgating conclusions that have not been rigorously tested, replicated, or evaluated through the peer review process versus the benefit of protecting people from continued exposure.

Most interventional studies have pre-defined stopping rules and will have data safety monitoring boards in place. If at any point the researchers recognize that the study poses a danger to participants that is more substantial than expected, the researchers are ethically obligated to stop the study. Sometimes studies can be reopened after modifications are made. It is acceptable to have some risk from study procedures and interventions, but if risk is deemed "more than minimal," then the interventions must have expected benefit as well. If the knowledge of the risk changes over the course of the study (due to study findings *or* other published findings), that can be a reason to act. Research ethics are explored in much more depth in ▶ Chapters 19 and 20.

(a) **Do healthcare practitioners have an obligation to share study results without first going through the peer review process of a medical journal?**
Yes, in some circumstances.

(i) **If so, under what circumstances?**
If the healthcare practitioner believes there is an imminent risk to the public posed by not sharing their data, then they would have an ethical obligation to share. In many cases, the research involved is epidemiologic; that is, a pattern of risk has been noticed in a population. In the case of Flint, MI, they believed that sharing the data would help to address the water issue and lessen further exposure to the population. They were right, and though it still took time, it probably happened faster than it would have otherwise. It should also be noted that it was not without risks. The researcher who completed a more recent study about Flint also wrote an Op-Ed in *The New York Times* concerned about the negative impacts of labeling the citizens of Flint. He notes that the people living there face many challenges already and their lead levels have actually been decreasing over the past 10 years. There may be unintended harms that arise from calling the city "poisoned" [30].

In randomized controlled clinical studies, a study may be stopped early if clear benefit is demonstrated for the new intervention; however, such study should still go through the peer review process before publication.

(ii) **If not, why not?**
The reasons not to share include concerns that the data are not conclusive or that the threat is not imminent and concrete. That is, one would have to assess the risks and benefits of sharing or not sharing. A researcher sharing data too soon could be harmful to public health if it induces an unnecessary panic or causes labeling or stigma. If the researcher lacks enough data to be statistically conclusive, it is a mistake to share the data. The peer review process is meant to help assure that research is sound before it is publicized. In order to skip that process,

✓ 3. **If healthcare practitioners suspect that they are witnessing a health concern that may affect the health of more than one patient, what is their obligation to investigate further?**

In some cases, such as with reportable diseases, there is a legally defined obligation to report to appropriate authorities. These diseases are chosen by the state health department as potential public health threats and, once reported by a healthcare practitioner, are usually investigated by the local health department. There is no legal obligation for a healthcare practitioner to do the same with regard to an environmental hazard such as lead in the water, but many would see doing so as an ethical obligation. The principles of justice and nonmaleficence would steer us to further investigation. It is not a new idea to have an obligation to society as well as our patients. Many modern physician pledges, including the Modern Hippocratic Oath by Dr. Louis Lasagna and the American Medical Association (AMA) Code of Ethics, have clauses directed to a physician's obligation to the society or community [31].

(a) **If an issue of concern is validated and a solution to address the concern is identified, what is the healthcare practitioner's obligation to advocate for changes to address the problem?**

The healthcare practitioner has an ethical obligation to help address the problem. In ethical terms, the principles of beneficence and nonmaleficence compel the practitioner to improve the health or limit the deterioration of health to his or her patients. In addition, the principle of justice compels the professional to assure that all patients have access to the conditions in which they can live a healthy life. In the case of excess lead in the water, these conditions have been compromised. As noted previously, there is a special obligation that physicians have to society. There are many ways that healthcare practitioners can address such societal problems, and different practitioners have differing views on how far to take this responsibility. Some feel compelled to actively seek out such health risks and address them, where others might take a less active approach. Some examples are noted below.

(b) **When you see such a pattern, whom would you call?**

Two issues arise from this question. The first is how to address an issue that poses an imminent threat to the public's health. This could be considered direct advocacy for patients and for the community. In this case, a healthcare practitioner may want to call the appropriate division of the health department, which would typically handle a number of public health concerns. In the case of lead exposure, a lead inspector can examine the house for lead hazards. If rabies is identified in an animal, health department personnel will help identify others who have been exposed to the animal and recommend rabies post-exposure prophylaxis if indicated. If an outbreak of foodborne illness is suspected in the community, an epidemiologist can assist with identifying others at risk. In other cases, it may be more appropriate to contact the hospital infection control team first. For example, in the event of an increase in hospital-acquired infections, then the hospital is the first point of contact.

The second issue regards how to help enact long-term changes related to the health concern. A healthcare provider practicing this kind of advocacy can work with various stakeholders to assess problems with laws and policies and find solutions to help fix them. One possibility is connecting with elected officials at the local, state, or federal level in order to educate them about the health issue at hand and encourage them to consider it when making policy decisions. Another avenue for advocacy is to work with a local or national advocacy organization, such as the American Academy of Pediatrics. These organizations often involve themselves with politicians and the policy-making process in order to further health goals. At a local level, one can work with the health department on a longer-term basis. Often, local health departments have advisory boards made of local healthcare providers and experts who help craft their public health agenda. Lastly, one method for advocacy is to use the media to send a health-related message.

You see Calvin for follow-up of his elevated BLL. He is and has been growing well without any medical problems. When he was screened for lead at age 1, his level was less than 2 μg/dL (lower limit of detection). At his visit today, you note that he is very mobile and constantly exploring. His motor development is on track. However, he only says five words so far, and he is not putting any words together (which indicates speech delay for a 2-year-old). His mother is also concerned that he throws a tantrum whenever she uses the vacuum cleaner or a loud truck passes the apartment, and that he does not seem interested in other children. His mother is 19 years old and is 25 weeks into her second pregnancy.

4. **What would you do now with respect to the lead exposure?**
   It is important to find the source of exposure to prevent worsening of the problem. This can begin with a careful history of Calvin's behaviors and environment but is primarily accomplished by the local public health colleagues in the health department and/or water treatment facilities.
   (a) **What services are available in the community?**
   Given his high lead level, Calvin's apartment and any other place where he visits regularly should be assessed for lead hazards. This inspection should include interior and exterior surfaces in the home and potentially water testing as well. Though lead in water is a problem in some places, lead in paint is still the major cause of lead exposure in the United States. Lead paint usually chips off the wall and forms dust as it ages. As children crawl or play in the dust and have (developmentally typical) hand-to-mouth behavior, they ingest the lead dust or paint chips causing elevated levels. In addition, there are services available for his developmental delay (see Response to Discussion Question #5a).
   (b) **How can you determine whether the water is the source of his lead exposure or if it is something else in part or in total?**
   Taking a careful history can help determine the source. In some cases, siblings or parents will report that they have seen the child eating paint chips or dirt that would raise concerns for those as sources. However, this would not be conclusive without also having someone examine his home environment for hazards and test the water. Even then, the results are often speculative unless specific behaviors are observed. Furthermore, finding one source (such as the water) does not eliminate the possibility of other sources.

5. **Developmental delay is both a consequence of lead exposure and a risk factor for lead exposure. Why do you think that is?**
   Patients with certain types of developmental delay may be more likely to eat things they should not and to have continued hand-to-mouth behavior after an age where it is no longer developmentally typical. With these behaviors, some children with developmental delays are more likely to be exposed to lead. As discussed in Background Question 2, lead exposure can cause children to have learning and attention problems, which is a form of developmental delay [9].

"Why Does My Son Have Lead in His Blood?"

(a) **What would you do now to assess his possible developmental delay?**
Calvin certainly needs an assessment from early intervention for his developmental delay. This federally funded program will send physical, occupational, and speech therapists to the home, daycare, or school to assess the child for delays; it is available in every county in the United States and free regardless of family income—although individual states set the eligibility requirements. Each state and county run the program locally, and sometimes also hire subcontractors to help, such as private occupational and physical therapy practices. If delays are found, program staff will create and implement an individualized plan for the child [23].

You call the county health department lead poisoning program. They have been receiving an increased number of calls recently due to people's concerns about the water. They state that they are currently coordinating with the public water utility to assess the possible dangers associated with the recent water source change. Because Calvin's level is so high, they plan to visit his home to assess it for lead hazards.

✅ 6. **Calvin may have lifelong cognitive impairments because of his lead exposure (which can cause irreversible neurological damage). Who should be responsible for paying for his treatment and the treatment of other people affected by the change in water source?**
A recent Pew report estimates that the hypothetical prevention of any lead exposure to the group of children born in 2018 would save $84 billion over their lifetime. That would be approximately $21,000 for each child. These savings would mostly accrue to the family in the form of increased lifetime earnings ($77 billion). In addition, there would be savings of $1.9 billion associated with education due to avoided effects on learning and cognition, as well as $1.7 billion due to avoided healthcare costs. In addition, there are estimated savings of $3.1 billion due to the increased number of quality-adjusted life years (QALYs) from limited lead exposure. In addition, this report found that total avoidance of lead exposure would lead to an additional 14,600 students earning a high school diploma and 15,200 earning a 4-year college degree, as well as the avoidance of 6000 teens becoming parents and 15,500 children being convicted of a crime [32].

Currently, the costs to address learning deficits are borne by the county early intervention program (for children under age 3) or the school district (for children 3 and older) in which the child resides. These programs are funded by a combination of local, state, and federal funds, meaning that all U.S. taxpayers have a stake in the impacts of local and state policies. The cost of any health effects would be borne by health insurance but is not likely to be needed until well in the future. The health insurance could be private or public, again spreading the costs to many people who might have no ability to affect changes in local or state policies. This distribution of costs is just in that we all share the burden [33].

Box 6.2 Patient Perspective

Resident Leanne Walters, a mother of 4, testified that her then 4-year-old twins asked her if they were going to die from lead poisoning after learning of the crisis. Now 5 years old, one twin "hasn't grown in a year," Walters said, adding that he still weighs 35 pounds, while his brother now weighs 56 pounds. Walters also said doctors "didn't know what was wrong" with her 14-year-old son despite multiple ultrasounds, a colonoscopy, and an emergency CAT scan. "They could not figure out where his pains were coming from," Walters recalled, adding that he suffered from dizziness, headaches, and rashes. Walters said her older son's rashes were so bad that when he would take a bath, "He would scream and cry about how bad his skin burned… Yes, it keeps me up at night. Yes, it makes me emotional," she said. "These are my kids. These are everybody's kids… As I was showing them my water, I was told I was a liar and I was stupid by showing these bottles of water," she said.

Julia Jacobo reporting for ABC news, March 29, 2016.
▶ https://abcnews.go.com/US/flint-mother-emotional-testimony-water-crisis-affected-childrens/story?id=38008707 [33].

7. **The usual method for identifying lead exposure is through testing children for elevated levels. What level of prevention is this?**
This is secondary prevention because it is a screening test with the goal of identifying a disease before it becomes symptomatic or more severe. The goal is to reduce the risk of adverse consequences arising from the disease. Tertiary prevention is an intervention for a disease in order to rehabilitate or limit the worst effects of the disease. Unfortunately, secondary prevention of lead exposure is not very effective because a relatively low BLL can do damage to a child's brain and development long before causing outward symptoms. Identification of an elevated BLL through screening does usually result in prevention of further exposure and sometimes treatment with chelation therapy, but in many cases, long-term damage has already occurred. Thus, the main approach to someone with elevated BLL identified through screening is to assure adequate services are in place for developmental and school assistance (which is tertiary prevention).

Primary prevention is complete prevention of exposure to lead. This can be achieved by keeping lead out of the water and fixing lead paint and soil hazards in the home before anyone is exposed.

(a) **Is this form of prevention just?**
The screening test is administered justly, in that it is equally available to all people at risk without significant cost (that cost is borne by all U.S. taxpayers). However, the screening program must be viewed in its entirety, which includes not only the screening test administration but also the response when an elevated BLL is found. The current system largely puts the costs and work of remediating lead hazards on the homeowner or landlord. This often results in inadequate repairs due to lack of resources, especially when the homeowner is impoverished or the renter has limited ability to force the landlord to comply. Furthermore, many impoverished families are afraid to bring housing problems to the attention of their landlord out of fears of eviction. It is illegal to evict someone for this, but it does happen, unfortunately. For these reasons, the current system of secondary prevention is not just. Some children are still exposed to dangerous levels of lead, and those children are more likely to be poor, racial/ethnic minorities, or recent immigrants [7, 9, 15].

(b) **What would be more just?**
Primary prevention. A more just form of prevention would be to eliminate the lead hazards as best as possible, assuming that the costs to those unexposed to lead are not overly burdensome. If the lead paint, dust, and pipes could be removed and replaced at reasonable cost, then the risk would be vastly reduced for all members of society, especially those currently at higher risk. This would eliminate the likelihood of any children having long-term effects from lead exposure. The recent Pew analysis on the costs of primary prevention found that a few measures produced long-term cost savings in addition to the health benefits of protecting children from lead hazards. They estimate that replacing lead service pipes would save $1.33 for every dollar spent over 18 years. Similarly, replacing the lead paint in older houses would save $1.39 for every dollar spent. Simply enforcing rules around renovation and repair would save $3.10 per dollar spent [32].

The county lead poisoning program investigates Ms. Wilson's rented apartment over the next few weeks. You are surprised to learn that Calvin and his mother live with Calvin's aunt and her five children, ranging in age from 15 months to 12 years. In addition to lead in the tap water, the health department inspector finds multiple lead hazards in the home, including paint chipping in the bedroom where Calvin sleeps with two of his older cousins. The older cousins report that Calvin sometimes eats the paint chips.

8. **What would you do now?**

   There are a number of children in the home who are also at risk. It is important to assure that the other children have been tested for lead, whether or not they are your patients. It also emphasizes the importance of having the health department do an inspection, both to find the hazards in the environment and to identify other kids at risk by completing a more thorough social and environmental history than often occurs in the office.

   (a) **Are you worried about his pregnant mother and the fetus? Why or why not?**

   It would be quite dangerous for a pregnant woman to be exposed to lead, because it does cross the placenta and expose the fetus. Fortunately, most pregnant women do not have exposure to lead because they do not usually do the kinds of things that young children do (e.g., crawl around on the ground and put things in their mouths). There are, however, some pregnant women who are exposed through pica behavior, occupational exposure, or performance of renovations in homes contaminated with lead paint. In addition, lead in the water supply is a source of exposure. The American College of Obstetricians and Gynecologists does recommend testing for lead during pregnancy when risk factors are present [34].

   Lead in the water is of great concern if the newborn is drinking formula. Powdered formula must be mixed with water before feeding, and if the family uses contaminated water, that would be a large source of exposure for the infant. Because of the small size of the infant and a diet exclusively of formula mixed with water, this exposure would be relatively much greater than an older child or adult drinking tap water.

   (b) **Does it matter whether the lead in Calvin's blood comes from the paint or the water? Why or why not?**

   It matters in terms of finding and mitigating the source, but it does not have any differing effects on the body. If from the paint, then the interior of the house needs to be renovated or repainted. Often, windows and doors are sources of exposure, so they might need replacement. If it is found in the water, then the source is most likely the pipes within the home. In that case, those would need to be replaced, or the water would need to be filtered with a certain type of filter before drinking.

9. **Do you have any obligations to Calvin's cousins?**

   Assuming they are not patients of the practitioner in the vignette, there is no physician-patient relationship with them, so there is no legal obligation to do anything. However, knowing that they are at high risk of lead exposure, there is an ethical obligation to them. One would want to make sure that they are tested for lead as well, either by contacting their doctor or by encouraging them to get testing done through the health department. The health department will usually follow up with all the residents in the home to make sure testing is done.

The health department notifies the landlord that the lead hazards exist in the apartment. He works to mitigate the hazards by repainting the bedroom and some of the windows. As Calvin's physician, you are concerned that this will not be enough, but that is all the law requires of the landlord. Over the next several months, you check Calvin's lead level monthly. You learn that his cousins, his mother, and his new baby sister have been tested. All were negative for lead exposure. Calvin's lead level initially goes down steadily, and you are reassured that the landlord's fixes have helped. However, about 6 months later, in early summer, Calvin's BLL suddenly spikes up to 60 μ(mu)g/dL. This level of lead is likely to affect his brain, leading to learning and attention problems in the long run (although we can never be sure of the cause in individual cases). You recommended hospitalization for 5 days of chelation therapy. When you talk to Ms. Wilson, she tells you that they have recently had to open the windows to cool the house down, including some of the windows with old lead paint that were recently repainted. The water pipes still contain lead as well.

✓ 10. **Self-reflection: How does this make you feel?**

✓ 11. **What would you do now?**
   (a) **Are you comfortable sending him back to the same house after he is discharged?**
   No. It would be risky to send him back to the same house after having such a high level, unless the home can be thoroughly risk reduced. The usual options are either to send the child/family to a friend or relative's home or to find a new apartment/house that can be moved to. Clearly, the latter takes a long time. Occasionally, the health department would find a hotel for the family for a short time until an alternative can be arranged.

The health department has been working with the family for months to find alternative housing. They have a Section 8 voucher, which gives them government assistance for housing, but only certain landlords are willing to take Section 8 tenants. Most Section 8 options will not accommodate such a large family, with two adults and seven children. The process, which normally takes years, has been expedited due to the lead hazards, but no suitable housing has been found yet. Because of the new exposure, you now tell the health department that you think Calvin's immediate family should be separated from his aunt and cousins in order to get him into a lead-free environment.

✓ 12. **What are the risks and benefits of separating the families? Is it worth it?**
   Separating the families has the benefit of moving Calvin to a safe place, free of lead hazards. However, there are likely social and economic reasons that the families live together, which should be considered. It may be that some aspect of daily life will be significantly affected. The mother may not be able to afford rent on her own, she may need help from her sister for child care while she works, or a host of other reasons may make things difficult after separation. If the separation leads to financial troubles or job loss for the mother, then the benefits may not be worth the risks.

The research from Flint, MI, showed an increase in the average BLL for the cohort of children tested, and the effect was worse on children in impoverished neighborhoods, in part because they happened to be near the "end of the line" of the water system, which allowed more time for lead to accumulate in the water prior to being consumed.

❓ 13. **Thinking about the socio-ecologic model, what institutional level factors affect the rates of lead exposure? What can be done about them?**
   There are several institutional level factors involved, including the landlord or apartment rental institution, the water system, the state and local health departments, the housing department, and even the doctor's office where the child is seen. For example, many urban apartment rentals are run by landlords who live in another place and have many apartments. They might not be as responsive to the needs of a tenant as a landlord who lives in the same building or has a property manager who lives there. As would be expected, cheaper apartments are less likely to have landlords or managers who respond quickly to requests for repairs or repainting. Avoiding or delaying these repairs is a way that landlords keep their costs down. Many housing agencies tasked with helping to assure adequate housing are struggling to keep up with their caseloads and may not be able to help someone fight the landlord or find alternate housing.

   Local health departments vary in their ability to respond quickly to lead poisoning concerns, both based on policy (discussed below) and based on the size of the department and caseload for the various concerns that they manage. Some might have dedicated staff to deal only with lead exposure, while others might have staff that deal with multiple public health issues, giving them less time to manage the lead issue.

The child's healthcare provider's adherence to particular screening guidelines affects when and whether an elevated BLL is identified. Also, whether the healthcare provider has a laboratory on site to draw the blood or has to rely on the patient/family getting to another location is an important institutional factor. Relying on families to take multiple steps to get the screening done means that some kids will be missed. This is not necessarily due to negligence on the family's part, but rather because the institution has not made it easy for the family amidst their competing demands.

(a) **Community level factors?**

These factors show how the institutions interact within the particular community. These interactions often have significant historical context. For instance, if the health and housing departments do not typically work together and are not located in the same place, then they are less likely to coordinate their efforts well.

In Flint, MI, people working in the state and local government failed to protect the residents from lead hazards. While the city of Flint was under state control due to fiscal problems, the state-appointed emergency manager chose to change the source of drinking water in order to save money, despite the fact that the Flint water treatment facility was not adequately prepared to provide safe drinking water. Later, officials showed "intransigent disregard of compelling evidence of water quality problems and associated health effects," "callous and dismissive responses to citizens' expressed concerns," and "persistent delays in coordinating appropriate responses," all of which perpetuated the crisis [29]. These institutional and community level factors led to a public health crisis.

On the other hand, if a community is particularly attuned to lead exposure and takes steps to address it, that could have significant positive effects. For example, in Rochester, NY, a group of concerned individuals and organizations formed the Coalition to Prevent Lead Poisoning in 2000. This group included community members, local government agencies, schools, healthcare providers, researchers, and others. They have since been able to reduce the number of children with BLL above 10 µg/dL by 84% [35], which was 2.4 times faster than NY state over the same period [36]. The community really coordinated their efforts in order to achieve a particular goal.

(b) **Policy level factors?**

These factors are based on the laws and policies of the city, state, or country that affect the issue. Three major changes to U.S. law in the last 50 years have greatly decreased the population's exposure to lead. First, lead was phased out of gasoline for use in automobiles starting in 1973 [37]. Second, it was banned from use in house paint in 1978 [38]. Third, water systems have been required to monitor and control for lead contamination in people's tap water since 1991 [18]. This is where the government failed to follow the law in Flint.

Other policies that affect this issue are related to what level of lead is considered dangerous for humans. The CDC currently uses a reference level of 5 µg/dL, over which a public health response is recommended. However, it is also clear there is no safe level of exposure, and even levels below 5 µg/dL can cause lasting effects [39]. These guidelines have changed over time, meaning that a child with an elevated BLL today might not have been considered in the past to have had lead poisoning. On a more local level, each state and county decide when they are going to inspect the home and help the family with public health nursing, depending both on the BLL of the child and the resources available in the community. In some places, the county health department may initiate an in-person investigation for levels over 5 µg/dL, while in others they may do so only when the level reaches a higher threshold because of lack of adequate resources or use of an outdated recommendation.

Other policy level factors include the screening guidelines set forth by various groups. In the case of lead exposure, these guidelines differ in key ways. The United States Preventive Services Task Force (USPSTF) has found insufficient evidence to recommend for or against screening for lead in asymptomatic children in its 2018 Draft Recommendation [16]. In the state of New York, however, public health law requires testing of all 1- and 2-year-olds because of concerns about older housing stock in the state [17]. These differing screening policies can clearly lead to different outcomes.

Other examples are policies around housing. In this case, lead exposure has been identified by testing a child. However, that means that the child was only identified after the exposure occurred. Enforceable housing policy that requires landlords and owners to identify and remediate lead hazards before a child lives in the home would achieve primary prevention [15].

Structural violence "is one way of describing social arrangements that put individuals and populations in harm's way." [2] *As an example, in the 1930s and 1940s in the United States, the practice of "redlining" was the norm in many cities, which made it impossible for African Americans, Jewish, and foreign-born residents to receive mortgages for homes in certain desirable neighborhoods. Many of the cities continue to have geographic disparities that originated with or were perpetuated by these practices* [3].

14. **In what ways can the events in Flint, MI, be considered a recent example of structural violence?**
    As described by Galtung [1] and Farmer [2], among others, this would be any social structure that places some members of society at risk. Lead risks in the environment have arisen through a failure of society to mitigate these hazards, either by assuring clean drinking water or by assuring lead paint is removed. These hazards do harm to certain members of society, often those who are most marginalized, thus limiting their potential. The official Michigan governor's Advisory Task Force came to the "inescapable conclusion" that the events in Flint were a case of environmental injustice. The residents of Flint, predominantly African American and with a high rate of poverty, were not afforded the same level of protection from environmental hazards available to other state residents. In addition, because the Flint local government power had been ceded to a state-appointed city manager, the residents had no say in their local decisions [29].
    (a) **Self-reflection: What does that mean to you, as a future healthcare practitioner?**

Box 6.3   Teaching Tip
This question is meant to encourage reflection about how social structures and customs sometimes lead to worse health outcomes. Depending on the level of student, they may not have thought much about this before. This should also provoke some discussion about the healthcare practitioner's responsibility to patients and to society. Encourage them to discuss how there are different levels of response to this type of information. A practitioner could ignore it entirely and only see patients who have means and/or private insurance. Another practitioner could choose to advocate strongly for changes (such as Mona Hanna-Attisha) or work in a health department on exactly these issues. Most practitioners will fall somewhere in the middle; for those, it is crucially important for them to recognize that their medical advice may not be feasible for some patients and that recommending something without an understanding of the patient's environment is detrimental to the relationship with the patient.

Recent research from Flint, MI, showed that the percentage of children with a BLL above 5 µg/dL increased from approximately 3% up to approximately 4–5% of children before and after the water switch [4–6]. However, the geometric mean BLL in the community only went up from 1.19 µg/dL to 1.30 µg/dL, and lead levels since 2006 have in fact decreased in the community [6]. Exposure from water is less likely than exposure from paint to cause higher levels above 10 or 15 µg/dL that we see in Calvin. In many places, there is lead in the water and also in old homes from the paint and soil.

15. **Based on the long-term trend that has recently been further characterized in the peer-reviewed literature, do you think that Dr. Hanna-Attisha and her team were premature in their announcement to the press?**
    This question gets back to some of the original questions of the case. The later research has definitely raised questions about how significant the increase was considering that the levels were higher only a few years prior. It is also worth noting that many cities, including several in Michigan, have higher rates of elevated BLLs [30]. Perhaps more of this would have been raised had the study gone through the peer review process first. There is also an argument that the levels in the past are moot. There was a failure of the government to provide clean water, and it did affect the lead levels in the children in the city. Their research brought this to light in a way that could no longer be ignored.

16. **Which do you think is more concerning: exposure of most of the population to low levels of lead (those drinking poorly treated tap water) or exposure of a smaller portion of the population to higher levels of lead (those exposed to paint hazards)?**
    Lead in water is measured in parts per billion, or micrograms per liter of water, with the U.S. Environmental Protection Agency (EPA) action level of 15 ppb (µg/L). It is usually measured when the water has been sitting in the pipes for 6–8 hours (usually first thing in the morning). That is not the only time that people are drinking the water. Several homes in Flint were above 100 ppb but most were below that [40]. Lead from paint is usually measured in the dust that falls off the wall as the paint deteriorates. It is measured in micrograms per square foot. A small study of older homes found amounts of more than 100 µg/ft^2 on the floors and more than 10,000 for window wells [41]. The lead dust is available on the floor at all times except just after cleaning. These data indicate there is more lead available in the dust than in the water. However, fewer members of the household are ingesting the dust than the water. The available data show that about 70% of the children with elevated BLL (over 10 µg/dL in that study) have been exposed to lead paint and dust and that 30% have another source, including water (at 10–20%) [42].

    House paint seems to be more responsible for BLLs above 10 µg/dL. However, we know there are likely effects from exposure to lower levels [12]. Some have argued that the chronic low-level exposure in the 2–5 µg/dL range actually has more effect on long-term collective brain function in a community than the relatively few children with higher BLLs—though those individuals are likely to be worse off on an individual level. This argument supposes that the widespread consumption of lead-contaminated water affects everyone a small amount and that the community as a whole would be worse off.

17. **Creative problem-solving: Since we have limited resources as a society, should we focus our efforts on lead pipes and make a small impact on a large population, or should we focus on lead paint to make a larger impact on a small population? How can we achieve both as a society? How can you, as a future healthcare practitioner, advocate for changes such as these? Do you have any obligation to do so?**

Box 6.4 Teaching Tip

It may be helpful for facilitators to explore different concepts of distributive justice that can be applied to this case, such as egalitarian (everyone gets the same level of service), welfare-based (people's level of service is dependent on their needs with people who are most disadvantaged getting the most services), utilitarian (services are distributed simply to achieve greatest "good"), and so forth.

To engage the students, facilitators can split the group in two, give them a hypothetical scenario, and have them argue to a conclusion. For example:

You are a physician consultant for the county health department. They received X dollars to reduce lead exposure in the community. They need to choose to spend it on lead pipes or lead paint remediation. The data show that 10% of children are being exposed to lead in water and the exposed children's mean BLL has increased to 3 µg/dL. The data also show that 3% of children in the community are exposed to lead paint hazards and their BLL has increased to 10 µg/dL on average. How would you advise the county to spend the money?

Box 6.5 Case Conclusion

Because of the advocacy of physicians and others in Flint, public outcry grew, and the city and state were forced to change back to the safe water source. They also restarted corrosion control, but that takes time to be effective. In theory the water is now safe, but many are still wary.

Michigan state and local officials have been indicted on charges that they withheld information about the contamination and also that they doctored reports to hide the problems. The governor's Flint Water Advisory Task Force found the events in Flint to be a "story of government failure, intransigence, unpreparedness, delay, inaction, and environmental injustice." [29] Fortunately, recent research has shown that the trajectory of BLLs in Flint is downward since 2006 and the events produced only a minor, transient elevation in average BLL [6].

References

1. Galtung J. Violence, peace, and peace research. J Peace Res. 1969;6(3):167–91.
2. Farmer P, Nizeye B, Stulac S, Keshavjee S. Structural violence and clinical. PLoS Med. 2006;3(10):e449.
3. Gross T. A 'Forgotten history' of how the US government segregated America [Internet]. NPR; 2017 May 3 [cited 2018 Aug 2]. Available from: https://www.npr.org/2017/05/03/526655831/a-forgotten-history-of-how-the-u-s-government-segregated-america.
4. Hanna-Attisha M, LaChance J, Sadler RC, Champney SA. Elevated blood lead levels in children associated with the Flint drinking water crisis: a spatial analysis of risk and public health response. Am J Public Health. 2016;106(2):283–90.
5. Kennedy C, Yard E, Dignam T, Buchanan S, Condon S, Brown MJ, et al. Blood lead levels among children aged <6 years: Flint, Michigan, 2013–2016. MMWR. 2016;65(25). Available from: https://doi.org/10.15585/mmwr.mm6525e1.
6. Gomez HF, Borgialli DA, Sharman M, Shah K, Scolpino AJ, Oleske JM, et al. Blood lead levels of children in Flint, Michigan: 2006–2016. J Pediatr. 2018;197:158–64. Available from: https://www.jpeds.com/article/S0022-3476(17)31758-4/fulltext.
7. Centers for Disease Control and Prevention. Lead: information for parents [Internet]. Atlanta (GA): CDC; 2018 [updated 2015 Dec 8; cited 2018 Jul 17]. Available from: https://www.cdc.gov/nceh/lead/parents.htm.
8. US Department of Labor. Lead [Internet]. Washington, D.C.: Occupational Safety and Health Administration; [cited 2018 Aug 2]. Available from: https://www.osha.gov/SLTC/lead/.
9. Committee on Environmental Health, American Academy of Pediatrics. Lead. In: Etzel RA, Balk SJ, editors. Pediatric environmental health. 2nd ed. Elk Grove Village: American Academy of Pediatrics; 2003. p. 249–66.
10. Centers for Disease Control and Prevention. Lead: learn more about CDCs childhood lead poisoning data [Internet]. Atlanta (GA): CDC; updated 2016 May 4; [cited 2018 July 18]. Available from: https://www.cdc.gov/nceh/lead/data/learnmore.htm.
11. Lidsky TI, Schneider JS. Lead neurotoxicity in children: basic mechanisms and clinical correlates. Brain. 2003;126(1):5–19.
12. Canfield RL, Henderson CR, Cory-Slechta DA, Cox C, Jusko TA, Lanphear BP. Intellectual impairment in children with blood lead concentration below 10 µg per deciliter. N Engl J Med. 2003;348:1517–26.
13. Berg KK, Hull HF, Zabel EW, Staley PK, Brown MJ, Homa DM. Death of a child after ingestion of a metallic charm – Minnesota, 2006. Morbidity and Mortality Weekly Report [Internet]. Atlanta (GA): CDC; 2006 Mar 23 [cited 2018 Jul 19]; 55(Dispatch):1–2. Available from: https://www.cdc.gov/mmwr/preview/mmwrhtml/mm55d323a1.htm.
14. Agency for Toxic Substances and Disease Registry (ATSDR). Toxicological profile for lead [Internet]. Washington, D.C.: U.S. Department of Health and Human Services, Public Health Service; 2007 Aug [cited 2018 Jul 20]. p. 528. Available from: https://www.atsdr.cdc.gov/ToxProfiles/tp13.pdf.

15. Advisory Committee on Childhood Lead Poisoning Prevention of the Centers for Disease Control and Prevention. Low level lead exposure harms children: a renewed call for primary prevention [Internet]. Washington, D.C.: U.S. Department of Health and Human Services; 2012. [cited 2018 Jul 20]. 54 p. Available from: https://www.cdc.gov/nceh/lead/acclpp/final_document_030712.pdf.
16. U.S. Preventive Services Task Force. Draft recommendation statement; elevated blood lead levels in children and pregnant women: screening. Rockville (MD): USPSTF Program Office; updated 2018 Oct [cited 2019 Jan 25]. Available from: https://www.uspreventiveservicestaskforce.org/Page/Document/draft-recommendation-statement/elevated-blood-lead-levels-in-childhood-and-pregnancy-screening.
17. NYS Regulations for Lead Poisoning Prevention and Control – NYCRR Title X, Part 67, Amended 2009 [cited 2018 Aug 13]. Available from: https://www.health.ny.gov/regulations/nycrr/title_10/part_67/index.htm.
18. US Environmental Protection Agency. Drinking water requirements for states and public water systems [Internet]. Washington, D.C.: US Environmental Protection Agency; updated 2017 Mar 15 [cited 2018 Jul 15]. Available from: https://www.epa.gov/dwreginfo/lead-and-copper-rule.
19. Office of Lead Hazard Control and Healthy Homes (OLHCHH) [Internet]. Washington, D.C.: U.S. Department of Housing and Urban Development. [cited 2018 Aug 2]. Available from: https://www.hud.gov/healthyhomes.
20. Green & Healthy Homes Initiative. Strategic plan to end childhood lead poisoning – a blueprint for action. Baltimore (MD): Green & Healthy Homes Initiative; 2016 [cited 2018 Aug 13]. 19 p. Available from: https://www.greenandhealthyhomes.org/wp-content/uploads/strategic-plan.pdf.
21. Educational Services for Children Affected by Lead Expert Panel. Educational interventions for children affected by lead. Atlanta (GA): CDC; 2015 [cited 2018 Aug 13]. 69 p. Available from: https://www.cdc.gov/nceh/lead/publications/Educational_Interventions_Children_Affected_by_Lead.pdf.
22. Individuals with Disabilities Education Act, 34 CFR § 300.8-Child with a disability. [cited 2018 Aug 13]. Available from: https://www.ecfr.gov/cgi-bin/text-idx?SID=50cdeae64d97568f13e26ceab8ea690c&mc=true&node=pt34.2.300&rgn=div5#se34.2.300_18.
23. Center for Parent Information and Resources. Overview of early intervention [Internet]. Newark (NJ): US Department of Education, Office of Special Education Programs; 2017 [cited 2018 Jul 22]. Available from: https://www.parentcenterhuborg/ei-overview/.
24. Center for Parent Information and Resources [Internet]. Newark (NJ): US Department of Education, Office of Special Education Programs; 2010 Sept 9; Special education services for preschoolers with disabilities; 2016 [cited 2018 Jul 22]. Available from: https://www.parentcenterhub.org/preschoolers/.
25. Buchanan DR, Miller FG. Justice and fairness in the Kennedy Krieger institute lead paint study: the ethics of public health research on less expensive, less effective interventions. Am J Public Health. 2006;96(5): 781–7.
26. Lead-based paint abatement and repair and maintenance study in Baltimore: findings based on two years of follow-up [Internet]. Washington, D.C.: US Environmental Protection Agency; 1997 Dec [cited 2018 Aug 13]. Available from: https://www.epa.gov/sites/production/files/documents/24folup.pdf.
27. Kaiser J. Court rebukes Hopkins for lead paint study. Science. 2001;293(5535):1567–9.
28. Kennedy M. Lead-laced water in Flint: a step-by-step look at the makings of a crisis. NPR [Internet]. 2016 April 20 [cited 2018 Aug 13]. Available from: https://www.npr.org/sections/thetwo-way/2016/04/20/465545378/lead-laced-water-in-flint-a-step-by-step-look-at-the-makings-of-a-crisis.
29. Davis M, Kolb C, Reynolds L, Rothstein E, Sikkema K. Flint water advisory task force: final report. Lansing (MI): Office of the Governor; 2016 [cited 2018 Jul 25]. 62 p. Available from: https://www.michigan.gov/documents/snyder/FWATF_FINAL_REPORT_21March2016_517805_7.pdf.
30. Gómez H, Dietrich K. The children of Flint were not 'poisoned.' The New York Times [Internet]. 2018 Jul 22 [cited 2018 Aug 2]. Available from: https://www.nytimes.com/2018/07/22/opinion/flint-lead-poisoning-water.html.
31. Association of American Physicians and Surgeons. Physician oaths [Internet]. Tucson (AZ): AAPS [cited 2018 Aug 2]. Available from: http://www.aapsonline.org/ethics/oaths.htm.
32. A report from the Health Impact Project. 10 policies to prevent and respond to childhood lead exposure. Philadelphia (PA): Pew Charitable Trusts; 2017 [cited 2018 Aug 4]. Available from: http://www.pewtrusts.org/-/media/assets/2017/08/hip_childhood_lead_poisoning_report.pdf.
33. Jacobo J. Flint mother gives emotional testimony of how water crisis affected her children's health. ABC News [Internet]. 2016 Mar 29 [cited 2018 Aug 13]. Available from: https://abcnews.go.com/US/flint-mother-emotional-testimony-water-crisis-affected-childrens/story?id=38008707.
34. American College of Obstetricians and Gynecologists. Committee Opinion No. 533. Lead screening during pregnancy and lactation. Obstet Gynecol; 2012 [reaffirmed 2016; cited 2018 Aug 4];120:416–20. Available from: https://www.acog.org/-/media/Committee-Opinions/Committee-on-Obstetric-Practice/co533.pdf?dmc=1&ts=20180804T1350405152.
35. Coalition to Prevent Lead Poisoning History. Rochester (NY): Coalition to Prevent Lead Poisoning; 2018 [cited 2018 Jul 11]. Available from: http://www.theleadcoalition.org/wp-content/uploads/2017/09/About-CPLP-Katrina-Article-1.pdf.
36. Kennedy BS, Doniger AS, Painting S, Houston L, Slaunwhite M, Mirabella F, et al. Declines in elevated blood Lead levels among children 1997-2011. Am J Prev Med. 2014;46(3):259–64.
37. US Environmental Protection Agency. EPA takes final step in phaseout of leaded gasoline [Internet]. Washington, D.C.: US Environmental Protection Agency; 1996 [updated 2016 Aug 11; cited 2018

38. US Environmental Protection Agency. Lead: protect your family from exposures to lead [Internet]. Washington, D.C.: US Environmental Protection Agency; updated 2017 Aug 30 [cited 2018 Jul 15]. Available from: https://www.epagov/lead/protect-your-family-exposures-lead.
39. Centers for Disease Control and Prevention. Lead [Internet]. Atlanta (GA): CDC; updated 2018 May 4 [cited 2018 Jul 1]. Available from: https://www.cdc.gov/nceh/lead/default.htm.
40. Edwards M. Our sampling of 252 homes demonstrates a high lead in water risk: Flint should be failing to meet the EPA lead and copper rule [Internet]. Flint Water Study; 2015 Sep 8 [cited 2018 Aug 4]. Available from: http://flintwaterstudy.org/2015/09/our-sampling-of-252-homes-demonstrates-a-high-lead-in-water-risk-flint-should-be-failing-to-meet-the-epa-lead-and-copper-rule/.
41. Vijayalakshmi P, Hegarty-Steck M, Hu H. Blood lead levels in relation to paint and dust lead levels: the lead-safe Cambridge program. Am J Public Health. 2001;91(12):1973–4.
42. Levin R, Brown MJ, Kashtock ME, Jacobs DE, Whelan EA, Rodman J, et al. Lead Exposures in U.S. Children, 2008: implications for prevention. Environ Health Perspect. 2008;116(10):1285–93.

Further Reading on this Topic

Flint Study

Bellinger DC. Lead contamination in Flint—an abject failure to protect public health. N Engl J Med. 2016;374(12):1101–3.

Hanna-Attisha M. The future for Flint's children. The New York Times [Internet]. 2016 Mar 26. Available from: https://www.nytimes.com/2016/03/27/opinion/sunday/the-future-for-flints-children.html.

Rosner D. Flint, Michigan: a century of environmental injustice. Am J Public Health. 2016;106(2):200–1.

Rosner D. A lead poisoning crisis enters its second century. Health Affairs. 2016;35(5):756.

Sadler RC, LaChance J, Hanna-Attisha M. Social and built environmental correlates of predicted blood Lead levels in the Flint water crisis. Am J Public Health. 2017;107(5):763–9.

Kennedy Krieger

Glantz LH. Nontherapeutic research with children: Grimes v Kennedy Krieger Institute. Am J Public Health. 2002;92(7):1070–3.

Mastroianni AC, Kahn JP. Risk and responsibility: ethics, Grimes v Kennedy Krieger, and public health research involving children. Am J Public Health. 2002;92(7):1073–6.

Nelson RM. Nontherapeutic research, minimal risk, and the Kennedy Krieger lead abatement study. IRB Ethics Hum Res. 2001;23(6):7–11.

Ross LF. In defense of the Hopkins lead abatement studies. J Law Med Ethics. 2002;30(1):50–7.

General

Earnest MA, Wong SL, Federico SG. Perspective: physician advocacy: what is it and how do we do it? Acad Med. 2010;85(1):63–7.

Office of Disease Prevention and Health Promotion. Environmental health [Internet]. Rockville (MD): ODPHP. [cited 2018 Aug 13]. Available from: https://www.healthypeople.gov/2020/topics-objectives/topic/environmental-health.

Office of the Surgeon General (US). The Surgeon General's Call to Action to Promote Healthy Homes [Internet]. 2: The connection between health and homes. Rockville (MD): Office of the Surgeon General (US); 2009 [cited 2018 Aug 13]. Available from: https://www.ncbi.nlm.nih.gov/books/NBK44199/.

Srinivasan S, O'Fallon LR, Dearry A. Creating healthy communities, healthy homes, healthy people: initiating a research agenda on the built environment and public health. Am J Public Health. 2003;93(9):1446–50.

The Rachel Maddow Show. Kids' toxic test results raised alarm: interview with Dr. Mona Hanna-Attisha [web streaming video]. MSNBC; 2015 December 18 [cited 2018 Aug 14]. Available from: http://www.msnbc.com/rachel-maddow/watch/kids-toxic-test-results-raised-alarm%2D%2Ddoctor-588639299776.

Healthcare Systems and Comparative Approaches

Contents

Chapter 7 "Our Baby Is Turning Blue" – 125

Chapter 8 "I Have a Touch of Sugar but I Can't Afford My Meds" – 145

"Our Baby Is Turning Blue"

Caitlin M. Nye

7.1 Background Questions – 126

7.2 Additional Case Information and Questions for Discussion – 126

7.3 Answers to Background Questions – 128

7.4 Responses to Discussion Questions – 133

References – 143

© Springer Nature Switzerland AG 2019
A. E. Caruso Brown et al. (eds.), *Bioethics, Public Health, and the Social Sciences for the Medical Professions*, https://doi.org/10.1007/978-3-030-03544-0_7

■ **Glossary**

Hypoplastic left heart syndrome - A rare congenital cardiac defect in which the left side of the heart is underdeveloped or absent

Medicalization - The process by which common physical processes and states come to be pathologized within the medical model and become understood as potential or actual illness

Learning Objectives
1. Describe the ethical and professional responsibilities of maternal healthcare practitioners, including obstetricians, nurse-midwives, and direct-entry midwives.
2. Describe how health policy can be used to improve maternal and infant health outcomes.
3. Describe cognitive bias and its influence on medical decision-making for both practitioners and patients.

> **Case Background**
>
> *You are working in a rural emergency department when the Nichols family rushes in carrying their son, Kaden, who was born at home 3 days earlier. The infant is blue, limp, and without respiratory effort. Despite a vigorous resuscitation, he is pronounced dead after 30 minutes.*

7.1 Background Questions

1. What are the trends in home births in the United States? Consider both rates and demographics of those choosing home births. How do outcomes compare between home and hospital births?

2. How do rates of infant and maternal mortality in the United States compare to other countries? What are the most common causes of infant mortality in the United States? What are the most common causes of maternal mortality?

3. In the United States, what types of healthcare practitioners are typically involved in deliveries? What training and credentialing options exist for midwives? Where do midwives practice?

4. What is the American College of Obstetricians and Gynecologists (ACOG) policy on home birth? What is the American College of Nurse-Midwives (ACNM) position on home birth?

5. What is the United Kingdom's National Institute for Health and Care Excellence (NICE) recommendation for low-risk pregnant patients who have already given birth?

6. Identify a legal case in which a pregnant patient with decisional capacity was compelled to undergo an intervention without her consent. How was the decision justified? Has it been upheld or overturned?

7. How many states have legislation mandating pulse oximetry testing at birth hospitals? What is the evidence behind such policies?

7.2 Additional Case Information and Questions for Discussion

In the aftermath of Kaden's death, you ask the family about the events leading up to their infant's cardiorespiratory arrest. They explain to you that he is their third child and that their other babies were delivered in their home under the supervision of a midwife. Both previous deliveries were uneventful, and those children are thriving. This delivery, however, appeared different right away. The baby was blue after birth and never turned pink like his siblings. At the time, the midwife reassured the family that everything was fine. Over the past 3 days, Kaden became increasingly lethargic and gradually developed labored breathing, which prompted the family to come to the hospital.

1. **Self-reflection: What is your first reaction to this vignette?**

2. **Whose voices are missing in this story? What would you like to know about them?**

In most states, an investigation is required for cases of unexpected, unattended infant deaths. In this case, an autopsy was performed and revealed that the newborn had a complex congenital cyanotic heart lesion called hypoplastic left heart syndrome (HLHS). Every healthcare decision carries potential risks as well as potential benefits. How patients and clinicians weigh those risks and benefits—which they consider acceptable and which they do not—has a significant impact on the decision-making process and thus the outcome.

In this case, without treatment, HLHS is uniformly fatal. With treatment, infants who survive the three-stage surgical repair for HLHS are often not considered cured because they may have lifelong complications. For this reason, some families choose to forgo surgical repair and focus on comfort care for the infant. This decision is considered within the zone of parental discretion.

3. Why do you think that most states require investigations of unexpected deaths?

4. What does this information add to your initial impression of the case?

5. Information such as this likely influences families' and healthcare practitioners' perception of risk and thus is likely to influence decision-making. What cognitive biases are involved in risk assessment?

6. What are the ethical obligations of any maternal healthcare practitioner, midwife or physician, when it comes to describing the risks and benefits of home birth vs. hospital birth?

7. How might you talk to the family about their experience afterward, given the circumstances of their child's birth and death?
   (a) Would you contact the midwife to discuss the situation with her? Why or why not?

You consult a colleague in the newborn nursery who tells you that currently the standard of care in newborn nurseries is to perform pulse oximetry screening in the first 48 hours of life to detect cyanotic heart defects; many but not all such defects are diagnosed even earlier, during a detailed ultrasound examination around the twentieth week of pregnancy.

8. Does the home birth setting itself preclude either of these interventions: pulse oximetry screening or ultrasound?
   (a) Does that change your assessment of responsibility in this case?
   (b) Should the midwife be informed of this standard of care? If so, whose responsibility is it to do so?

9. Overwhelmingly, current state laws mandating newborn pulse oximetry apply only to hospitals. Is it just to exempt home births from these laws? Are there good reasons to do so?

A few weeks later, your colleague sends you an interesting case report that describes a scenario in which a cardiac anomaly of uncertain severity was detected on a routine 20-week ultrasound [1]. The pregnant patient had already planned for a birth at home and was followed by a midwife; she declined further testing, in part because of concerns that the midwife might refuse to participate in the home birth. A maternal-fetal medicine specialist and a palliative care physician met with the patient to discuss possible outcomes, which included death at home. Both the patient and her partner stated that they accepted that their child may die during or shortly after birth.

10. Why do you think some mothers and families choose home birth?
    (a) What other lifecycle events might stimulate a conversation about hospital vs. in-home care or "medicalization" versus a "natural" process of care?

11. Are there *any* situations in which patient autonomy during pregnancy or childbirth should be overridden to protect the fetus?
    (a) Is drug use during pregnancy morally different? Why or why not? What about other health behaviors: eating well, exercising, etc.?
    (b) Self-reflection: Some ethicists have argued that although there is no legal obligation for pregnant patients to put the interests of the fetus first, there is an ethical obligation. Do you agree or disagree? Why?

In the aforementioned published case, the midwife struggled, ethically, with whether to be present at the birth because it was, in some respects, outside her scope of practice, and because she was concerned that her presence could be misconstrued as unilateral support for the family's decision. Eventually, she decided that she was obligated to be present.

12. Why do you think this midwife decided to participate in the birth?

13. Healthcare practitioners weigh risks differently, from each other and from patients. What are the ethical obligations of healthcare practitioners when it comes to describing the risks and benefits of home birth vs. hospital birth?

After this experience, you lead a community task force seeking to establish transfer agreements between local hospitals and midwifery practices that provide care for home births. The first meeting is somewhat contentious. The obstetricians and pediatricians representing the hospitals feel that the choice to give birth at home is irresponsible and likely to end badly; they cite examples of other cases with poor outcomes to support their claims. The midwives and patient advocates point to the NICE recommendation in the United Kingdom and to studies around the world demonstrating equivalent or better outcomes for low-risk patients giving birth at home [2–8].

14. What do you think might account for the different recommendations in the United States and the United Kingdom?
    (a) What is different about the healthcare systems in the United States and the United Kingdom?

15. Many studies around the world have looked at outcomes in home births and compared them to hospital births [2–8]. What types of study design do you think are used? Why?
    (a) What are challenges and limitations to these study designs?

16. Describe the interprofessional communication you would want to have between hospitals and midwives to ensure best possible outcomes.

    (a) Creative problem-solving: What local policies could you propose to achieve these outcomes?

17. Self-reflection: At the beginning of the discussion, you were asked about your emotional response to this case. Has that changed through the discussion or because of what you've learned?

7.3 Answers to Background Questions

1. **What are the trends in home births in the United States? Consider both rates and demographics of those choosing home births. How do outcomes compare between home and hospital births?**
   According to the National Center for Health Statistics, in 2012, 1.36% of births took place outside of hospitals, up from 1.26% in 2011 and representing a nearly 15-year-long steady increase in the United States. Two-thirds of these out-of-hospital births occurred at home, with most of the remaining third occurring at freestanding birth centers. The percentage of planned home births increased from 0.56% in 2004 to 0.89% in 2012 [9]. Several studies have noted that of people who chose home birth, they tend to overwhelmingly be white, married, college educated, and employed outside the home and tend to be more affluent than pregnant patients in general [9, 10].
   Studies from the Netherlands, Norway, and Canada have reported that low-risk women who experienced planned home births had reduced rates of maternal morbidity, including anal sphincter tears, postpartum hemorrhage, and manual extraction of the placenta [5–7]. These studies, however, analyzed outcomes in populations that are more ethnically and socioeconomically homogeneous than in the United States and in countries in which most women have access to state-supported healthcare systems providing universal coverage. Prevalence of planned home birth is also much greater in these countries than in the United States (for

instance, in the Netherlands, 29% of women give birth outside an obstetric unit), and midwifery care is well integrated with the larger healthcare system, resulting in good access to emergency services, consultation, and transportation for transfer [7].

In the United States, studies have generally supported the finding that women experiencing home birth receive fewer interventions during labor and delivery (e.g., augmentation, episiotomy, assisted delivery) than women experiencing hospital birth, but they have been inconclusive or contradictory in findings related to maternal and neonatal morbidity. For example:

- Cheyney and colleagues analyzed data from the Midwives Alliance of North American Statistics Project registry, a Web-based data collection system created for collecting information primarily on home and birth centers, and concluded that low-risk women experienced high rates of normal physiologic birth at home, with no increase in adverse events [3].
- A 2008 retrospective cohort study evaluated birth certificate data from 27 states [11]. They found that home births had a higher rate of 5-minute Apgar scores (which alone are not a useful marker of adverse neonatal outcomes) of less than 4 and of neonatal seizures but were otherwise comparable in outcomes [12]. Additionally, the population included some women at higher risk for certain adverse outcomes, including those choosing vaginal birth after cesarean section (VBAC) and with certain medical conditions. They also included births occurring after 42 weeks, which disproportionately applied to home births; evidence strongly suggests that the risk of intrauterine fetal demise and stillbirth begins to increase at 41 weeks and increases dramatically after 42 weeks (the medical definition of postmaturity) [13].
- A study of U.S. birth and death certificate data from 2006 to 2009 found that midwife-attended home births had a nearly fourfold increase in total neonatal mortality compared with those attended by hospital midwives, leading to an excess 15.42 deaths per 10,000 births. A significant proportion of those deaths occurred in women with risk factors that have been previously described but that were not considered high-risk exclusion criteria for home birth: specifically, nulliparous women, those over age 35, and those delivering after 41 weeks completed gestation [14].

2. **How do rates of infant and maternal mortality in the United States compare to other countries? What are the most common causes of infant mortality in the United States? What are the most common causes of maternal mortality?**

In 2015, using the WHO definition, the rate of maternal mortality in the United States was 14 maternal deaths per 100,000 live births. This rate is comparable to Nicaragua (13.3), Qatar (13), and Uruguay (15). In contrast, for most Western European countries, Japan, and a few others, the rate is around 5 maternal deaths per 100,000 live births. In the United Kingdom, it is 9. The highest rates of maternal mortality are found in countries such as Zimbabwe (443), South Sudan (789), Nigeria (814), and Sierra Leone (1360) [15]. Using the broader CDC definition, the rate of maternal mortality in the United States in 2013 was 17.3 maternal deaths per 100,000 live births and apparently rising over the previous 25 years. However, this rate belies a significant disparity: For white women, it is 12.7 maternal deaths per 100,000 live births, while for black women, it is 43.5 maternal deaths per 100,000 live births [16].

In 2016, the rate of infant mortality in the United States was 5.9 infant deaths per 1000 live births, or 587 per 100,000 live births [17, 18]. Among non-Hispanic whites this rate was 4.9; among non-Hispanic blacks it was 11.3 [18]. More than 40 countries have lower rates of infant mortality than the United States (including

Greece, Poland, Bosnia, and Cuba); nearly 70 have lower rates than the rate of mortality for black American infants. The latter rate is comparable to Thailand, Turkey, Libya, Tunisia, and Peru. The highest rates of infant mortality are found in Chad (75), Somalia (83), Sierra Leone (83), and the Central African Republic (89) [19].

The most common causes of maternal mortality in the United States are cardiovascular disease, infection, hemorrhage, cardiomyopathy, pulmonary embolism, and preeclampsia and eclampsia—conditions that are potentially treatable if identified early [16]. A synthesis of data from maternal mortality review committees showed that, among deaths that occurred during pregnancy, hemorrhage and cardiovascular conditions were the most common causes of death; among deaths that occurred within 42 days of the end of pregnancy, infection and hemorrhage were the most common causes; and among deaths occurring 43 days to 1 year after end of pregnancy, cardiomyopathy, mental health conditions, and embolism were the most common causes of death [20]. Higher rates of cesarean delivery may also contribute to maternal mortality [21]. It is important to better understand the contribution of differential access to health care (e.g., insurance status, transportation, availability of maternity care), standards of care (e.g., routine postpartum visits within 6 weeks after delivery), and bias (experiences of racial, ethnic, and other forms of discrimination as they impact healthcare delivery) when considering how to decrease maternal mortality in the United States.

Age, race, ethnicity, socioeconomic status, maternal education, marital status, body mass index, and parity are all associated with maternal and infant morbidity and mortality. In general, women who are more affluent, more highly educated, married, and multiparous have better birth outcomes, both for themselves and for their children, as do women who are younger but post-adolescent and those who are not overweight [22–24].

The most common causes of infant mortality are birth defects, preterm and low-birthweight births, sudden infant death syndrome (SIDS), maternal pregnancy complications, and injuries. Of these, SIDS and injury are most likely to be preventable; many preterm births and maternal pregnancy complications may also be preventable [17, 18].

Rates vary significantly by state. For example, New Hampshire experienced 3.7 infant deaths per 1000 live births in 2016, while California had 4.2, and Alabama had 9.1. In general, infant mortality rates are higher in Southern and some Midwestern states. This may be due in part to access to care, as well as higher rates of maternal obesity [17, 18]. There are also differences between rural and urban areas with regard to the most common causes of neonatal death. The infant mortality rate for congenital defects, SIDS, and unintentional injuries is higher in rural areas than in urban areas; the infant mortality rate due to low birth weight and maternal complications is higher in urban areas than rural areas [25].

3. **In the United States, what types of healthcare practitioners are typically involved in deliveries? What training and credentialing options exist for midwives? Where do midwives practice?**
Obstetricians, certified nurse-midwives, and registered nurses are the most commonly found healthcare practitioners in hospital delivery rooms. Pediatric staff, including neonatal nurse practitioners and neonatologists, may attend higher-risk and complicated deliveries, to provide immediate care to the infant(s).

In freestanding birth centers, typically certified nurse-midwives and registered nurses are present. In home births, depending upon state regulations, a certified nurse-midwife (CNM), a certified midwife (CM), or a direct-entry midwife may be in attendance.

The American College of Nurse-Midwives (ACNM) and Accreditation Commission for Midwifery Education accredit academic and training programs that prepare nurses to complete either a master's or doctoral degree (DNP) in nurse-midwifery, and these nurse-midwives are licensed as CNMs. They also accredit programs that prepare nurses and non-nurses to practice as CMs.

All states allow CNMs to practice. These midwives have master's or doctoral degree level education, have a nursing background, and must pass licensure requirements. All training programs include clinical hours in both hospital and out-of-hospital births, though opportunities for practice settings vary by market [26]. A few states allow CMs to practice; they also obtain a master's degree and can be licensed and certified, though a background in nursing is not required. Additionally, there are educational programs that are accredited to train certified professional midwives (CPM), who are licensed and regulated in 31 states. In some states, professionally attended home births are effectively illegal [27].

Last, so-called direct-entry, or lay, midwives do not, in general, go through any kind of formally accredited education into the practice of midwifery, learning instead through apprenticeship and mentoring, and many states do not allow them to be licensed [26, 28]. Were they to deliver a baby at home in those states, they would be breaking the law (i.e., practicing medicine without a license). Some states, such as New York, require licensure of midwives, though, notably, they do not require continuing education, which is required for physicians [29].

4. **What is the American College of Obstetricians and Gynecologists (ACOG) policy on home birth? What is the American College of Nurse-Midwives (ACNM) position on home birth?**
It is the position of the ACNM that "every family has a right to experience childbirth in an environment where human dignity, self-determination, and the family's cultural context are respected" and that "every woman has a right to an informed choice regarding place of birth and access to safe home birth services" [30].

The ACNM notes that "certified nurse-midwives (CNM) and certified midwives (CM) are maternity care professionals who are qualified to provide ongoing assessment of appropriate birth site selection over the course of the antepartum, intrapartum and postpartum periods" including in the home. The group also posits that an "integrated system of health care that includes collaboration among all health care providers is essential and fundamental to supporting a safe, seamless, transfer of care from home and/or out of the hospital setting when necessary" and promotes research and evaluation in "normal birth, including the influence of birth setting" [30].

In contrast, ACOG has stated that:

> Women inquiring about planned home birth should be informed of its risks and benefits based on recent evidence. Specifically, they should be informed that although planned home birth is associated with fewer maternal interventions than planned hospital birth, it also is associated with a more than twofold increased risk of perinatal death (1–2 in 1,000) and a threefold increased risk of neonatal seizures or serious neurologic dysfunction (0.4–0.6 in 1,000). These observations may reflect fewer obstetric risk factors among women planning home birth compared with those planning hospital birth. Although the American College of Obstetricians and Gynecologists (the College) believes that hospitals and accredited birth centers are the safest settings for birth, each woman has the right to make a medically informed decision about delivery [31].

The ACOG statement goes on to note the importance of recognizing risk factors that make home birth particularly dangerous and that fetal malpresentation, multiple gestation, or prior cesarean delivery are absolute contraindications [31]. Consistent with the ACNM statement, they agree on the importance of qualified healthcare practitioners and "an integrated and regulated health system; ready access to consultation; and access to safe and timely transport to nearby hospitals."

5. **What is the United Kingdom's National Institute for Health and Care Excellence (NICE) recommendation for low-risk pregnant patients who have already given birth?**
The NICE recommendations note that for woman at low risk of complications, giving birth is generally quite safe. It encourages healthcare practitioners to support patients in their choice of birthplace and explicitly states that low-risk multiparous patients should be advised that "planning to give birth at home or in a midwifery-led unit (freestanding or alongside) is particularly suitable for them because the rate of interventions is lower and the outcome for the baby is no different compared with an obstetric unit" [32]. Benefits of birth outside the obstetric unit include higher rates of spontaneous vaginal delivery and lower rates of interventions, such as instrumental vaginal delivery (e.g., vacuum, forceps), cesarean section, and episiotomy, compared with planning birth in other settings [3, 5, 8].

6. **Identify a legal case in which a pregnant patient with decisional capacity was compelled to undergo an intervention without her consent. How was the decision justified? Has it been upheld or overturned?**
The landmark case on this issue is *In the matter of A.C* [33]. Angela Stoner Carder was a 27-year-old woman with a history of recurrent Ewing sarcoma, who was found to have relapsed again when she was 25 weeks pregnant. Although her disease was thought to be incurable, she was encouraged to undergo treatment until the 28th week of pregnancy in order to give the fetus a chance to survive. In the 26th week of pregnancy, the hospital where she was admitted sought a court order for a cesarean section, over the objections of her husband, parents, and personal physicians, who opposed it on grounds that it would hasten her death and that the baby was unlikely to survive. After the deaths of both Carder and her child, her family sought to have the court order, and legal precedent, vacated. In 1990, the District of Columbia Court of Appeals did so. Subsequent decisions in 1994 (*In re Baby Boy Doe,* 632 N.E.2d 326, Ill. App. Ct. 1994) and 1997 (*In re Fetus Brown,* 689 N.E.2d 397, 400, Ill. App. Ct. 1997) supported the idea that the rights of a competent pregnant person superseded the rights of the fetus in the eyes of the court, even when the patient's decision might result in harm to the fetus [33–35]. Despite these precedents, a recent review identified 10 such cases between 1990 and 2014. Cases primarily involved situations in which physicians felt that cesarean sections were indicated to preserve the life or health of the fetus but patients refused to provide consent. The authors noted that many other similar cases involved the use of coercive tactics rather than judicial orders [36].

7. **How many states have legislation mandating pulse oximetry testing at birth hospitals? What is the evidence behind such policies?**
Approximately 20% of critical congenital heart defects (CCHD) are not identified on prenatal ultrasounds or by routine physical examination shortly after birth but can be identified by routine pulse oximetry screening 24–48 hours after birth. Early diagnosis may prevent death or neurological injury from hypoxia. New Jersey was the first state to pass legislation mandating such screening, in 2011. Soon after, the U.S. Department of Health and Human Services recommended that all states incorporate this screening into newborn screening programs (generally a panel of blood tests for diseases that can be detected early in life and benefit from early diagnosis and treatment) [37]. By 2014, 43 states had taken steps to require pulse oximetry screening through legislation, regulations, or hospital guidelines [38].

In the first 9 months after screening was mandated in New Jersey, 73,320

"Our Baby Is Turning Blue"

infants underwent screening. Forty-nine were referred for further evaluation; 3 were diagnosed with CCHD and thought to have benefitted from early diagnosis (before they became symptomatic), while 17 had other diagnoses (non-critical congenital heart defects or systemic disease). Six years earlier, in Tennessee, a task force was convened on the same issue; in that state and at that time, however, the task force recommended that routine screening *not* be mandated. The rationale was that the sensitivity and specificity of pulse oximetry was too low, giving a high rate of false-positive results [39]. Vermont conducted its own study of CCHD diagnoses in a 10-year period and concluded that a screening program would have identified five patients and possibly avoided the need for resuscitation in three; the authors did not state that these small numbers proved such screening should not be mandated but raised questions about cost-benefit considerations in settings with limited resources [40]. Although the screening itself is relatively low cost—hospitals and birth centers need pulse oximeters available to evaluate infants who are ill-appearing—effective implementation requires education of practitioners regarding screening protocols, including referral, evaluation, and follow-up of infants who do not pass screening [41].

7.4 Responses to Discussion Questions

In the aftermath of Kaden's death, you ask the family about the events leading up to their infant's cardiorespiratory arrest. They explain to you that he is their third child and that their other babies were delivered in their home under the supervision of a midwife. Both previous deliveries were uneventful, and those children are thriving. This delivery, however, appeared different right away. The baby was blue after birth and never turned pink like his siblings. At the time, the midwife reassured the family that everything was fine. Over the past 3 days, Kaden became increasingly lethargic and gradually developed labored breathing, which prompted the family to come to the hospital.

1. **Self-reflection: What is your first reaction to this vignette?**

> **Box 7.1   Teaching Tip**
> Encourage learners to recognize that healthcare practitioners can and do react emotionally in many situations, particularly to the unexpected loss of a patient and perhaps especially—for some—to the death of someone who should be just beginning to live. Some learners may make quick judgments of the family's or midwife's culpability in the infant's death. Encourage them to recognize a snap judgment for what it is, and then put it aside to explore the case with an open mind.

2. **Whose voices are missing in this story? What would you like to know about them?**

The Nichols' version of events has been shared briefly, but still, relatively little is known about them. What do they understand about their child's life and death? How are they grieving? Do they have other family members, friends, or faith leaders to help support them at this difficult time? Who is caring for their other children right now? What do they want to happen next? The emergency practitioner, in this situation, is almost certainly concerned about the role of home birth in the baby's outcome. Is the family thinking about this too? What did they understand about the risks and benefits of their birth choices?

The midwife is completely silent at this point in the narrative. What was her training and level of experience? What was her relationship to the local obstetrical community or hospital itself? Is the family communicating with her? Did they inform her that they were taking Kaden to the hospital? If so, what was her reaction to this information?

In most states, an investigation is required for cases of unexpected, unattended infant deaths. In this case, an autopsy was performed and revealed that the newborn had a complex congenital cyanotic heart lesion called hypoplastic left heart syndrome (HLHS). Every healthcare decision carries potential risks as well as potential benefits. How patients and clinicians weigh those risks and benefits—which they consider acceptable and which they do not—has a significant impact on the decision-making process and thus the outcome.

In this case, without treatment, HLHS is uniformly fatal. With treatment, infants who survive the three-stage surgical repair for HLHS are often not considered cured because they may have lifelong complications. For this reason, some families choose to forgo surgical repair and focus on comfort care for the infant. This decision is considered within the zone of parental discretion.

3. **Why do you think that most states require investigations of unexpected deaths?**
   There are several reasons for this. First, in many cases of sudden death involving children, non-accidental trauma (child abuse) is suspected and must be ruled out, both to protect other children in the home and to provide justice for the deceased child. In other cases, without an investigation and autopsy, this child's death might be mistakenly classified as due to SIDS or simply unexplained. Many families derive comfort from knowing the reason for their child's death. Furthermore, while HLHS is not hereditary, other birth defects and causes of early neonatal death are, and that information may be beneficial to the health of the child's parents, siblings, or future siblings.

   However, there is also an important public health rationale for requiring investigations. Investigations and autopsies, along with other surveillance systems and databases tracking causes of death, provide crucial information to inform public policy as it relates to health care. For example, when a state is considering mandating pulse oximetry, it would be useful to know how many infant deaths are attributable to undiagnosed cardiac lesions—information that would be unavailable without such a requirement.

4. **What does this information add to your initial impression of the case?**
   While there is not a single answer to this question, had the Nichols family received a prenatal diagnosis of HLHS, they might have—legally, ethically, morally—planned to deliver at home and to receive palliative care only, keeping the baby comfortable until he died.

   On the other hand, after receiving education about the diagnosis, interventions, and possible outcomes, they might also have decided to deliver in a tertiary care hospital with pediatric cardiology, cardiothoracic surgery, and neonatal intensive care readily available; or they might have chosen to terminate the pregnancy. It should not be assumed that any one of these options is more likely simply because of the family's preference for childbirth at home.

   It is also interesting to note that the decisions that people make are affected by how the data regarding treatment options and interventions are presented to them. For instance, for HLHS cardiac surgeons tend to focus on short-term surgical survival data, while cardiologists focus more on long-term outcomes and quality of life. In situations like this, parents are more likely to agree to surgical intervention when it is initially presented by a surgeon.

5. **Information such as this likely influences families' and healthcare practitioners' perception of risk and thus is likely to influence decision-making. What cognitive biases are involved in risk assessment?**
   In shared decision-making, patients and healthcare practitioners approach decision-making regarding the patient's health care as a collaborative endeavor. Patients have widely varying preferences for how collaborative this process truly is. On one end of the spectrum, some will want to be given all the available information and then will take time to privately consider and evaluate it, perhaps soliciting opinions from family and friends and checking the veracity of the practitioner's recommendations. On the other end are patients who want to be well-informed but prefer to delegate responsibility for the final decision to the clinician [42]. Thus, cognitive biases are likely to impact both patients and practitioners.

Cognitive biases emerge when people rely on heuristic frameworks for decision-making, leading to errors in judgment compared with so-called rational decision-making. There are several cognitive biases that commonly influence the processes of diagnosis and intervention by healthcare practitioners [43]. Some of these biases are particularly impactful in the arena of risk assessment. Most obvious is the tendency to simplify probabilities in order to make them easier to process, as in when diseases, interventions, or even patients themselves are classified in binary groups as "high-risk" or "low-risk" [44]. There is some evidence that ability to understand and use numerical information (numeracy) affects reasoning about risk and increases susceptibility to cognitive biases, including effects of mood or how information is presented [45].

However, there are several others. Commission biases prompt clinicians, presented with an uncertain or ambiguous situation, to err on the side of action or intervention, regardless of the clinical evidence. Although patients can also experience this bias—a desire to do *something*—there is evidence that clinicians are more likely to want to act in the face of uncertainty. Omission bias, on the other hand, is a preference to avoid intervention; this bias leads one to weigh the risks of action more heavily than inaction, even when they are objectively comparable. Confirmation bias occurs when people place undue emphasis on information that corroborates previous decisions or beliefs while possibly minimizing information that does not support those decisions or beliefs.

In the essay "Risk and the Pregnant Body," Lyerly and colleagues describe the ways in which risks are systematically distorted in the care provided to pregnant and laboring patients [44]. They write:

» In the first two sections, we explore the dual nature of attitudes toward medical intervention during pregnancy and birth. When treating pregnant women's non-obstetrical medical needs, it turns out, there is a tendency to notice the risks of intervening without adequately noting the risks of failing to intervene. In contrast, when we turn from management of pregnancy to management of birth, we note a tendency to intervene without due regard for the burdens to both fetus and woman that such interventions may bring. If risk perception is often distorted, the nature of the distortion changes markedly depending on the circumstance of a pregnant woman's health needs. In the third section, we move outside the clinic, noting patterns of advice about decisions faced in everyday life during pregnancy—what to eat, how to sleep, which activities to engage in, and which to avoid. Here we see a retreat from evidence-based advice to the mantra of 'better safe than sorry,' even in the face of reassuring data. [44]

6. **What are the ethical obligations of any maternal healthcare practitioner, midwife or physician, when it comes to describing the risks and benefits of home birth vs. hospital birth?**
It is the responsibility of any maternal healthcare practitioner to be well versed in the most recent evidence about the risks and benefits of home birth as well as the risks and benefits of hospital birth and to share information about these risks and benefits in a neutral manner. In addition, the clinician needs to carefully delineate the definition of "low-risk" pregnancy and childbirth with regard to home birth, as guidelines described by both ACNM and ACOG regarding home birth specify the eligibility of mothers with low-risk pregnancy.

7. **How might you talk to the family about their experience afterward, given the circumstances of their child's birth and death?**
Although there are many approaches to communicating with grieving families, critiques of other clinicians, assignment

of blame, or questioning the family's choices do not have a place in the initial conversation about the kind of terrible outcome described in this case.

(a) **Would you contact the midwife to discuss the situation with her? Why or why not?**
The emergency practitioner should discuss this with the family and proceed in a sensitive matter. It is important that the midwife be aware of what has happened and why. This is true for any healthcare practitioner who has had a patient die or experience significant morbidity.

You consult a colleague in the newborn nursery who tells you that currently the standard of care in newborn nurseries is to perform pulse oximetry screening in the first 48 hours of life to detect cyanotic heart defects; many but not all such defects are diagnosed even earlier, during a detailed ultrasound examination around the 20th week of pregnancy.

8. **Does the home birth setting itself preclude either of these interventions: pulse oximetry screening or ultrasound?**
No. Midwives use both technologies, and pulse oximetry is easily performed at home with inexpensive and easily portable equipment. In many cases, whether interventions and treatments are perceived as "inpatient," "outpatient," or "home" care is dependent on healthcare systems and services. There is more flexibility in such designations than patients are led to believe.

(a) **Does that change your assessment of responsibility in this case?**
It should. The failure to diagnose the heart condition earlier was not the inevitable result of the family's preference for home birth. A contrasting example might help. If this mother had a precipitous birth at home with postpartum hemorrhage and developed hypovolemic shock from blood loss because she did not have rapid access to a blood transfusion, that would be a risk that she knowingly (with proper informed consent) took by choosing to give birth in a home setting. Many patients who choose home birth weigh such risks against the risks of receiving an unnecessary intervention in a hospital setting (with the attendant risks of iatrogenic complications) and choose the former.

In the case of pulse oximetry, it can and should have been performed at the time of the home delivery; this was possibly a mistake on the part of the midwife. A trained, licensed midwife would be aware that this is a recommended screening procedure for the health of the baby, and she should have the equipment and knowledge to do it. If she fails, having all that, then it is her mistake.

It is also important to consider this case as representing a systems-based problem. If it were required as part of licensure or as part of the midwife's agreement with the local "backup" birthing hospital, then the system would make it much harder for any midwife to fail to do this simple screening test. Often, however, such systems are not in place in the United States, and where they are, there is great state-to-state variation.

> **Box 7.2 Teaching Tip**
> Learners may want to discuss the pros and cons of allowing direct-entry midwives to practice. They may also have some confusion about the multiple entry points into the practice of midwifery. It is important to clarify any factual matters that hinder accurate discussion and debate. If learners have access to computers, tablets, or smartphones, they may be invited to look up the requirements on their state's licensing website.

(b) **Should the midwife be informed of this standard of care? If so, whose responsibility is it to do so?**
The midwife should be made aware of the outcome of the case. In addition, the midwife should be advised

of standard practices that might have prevented the outcome or at least allowed the family to have made more informed choices in their child's life and death. Only by recognizing such errors and analyzing the events that led to them can practitioners and healthcare systems improve practices.

Unfortunately, there are few systems or structures in place to assure that direct-entry midwives have opportunities for quality, safety, and process improvement and for continuing education. A hospital might, for instance, present this case at a morbidity and mortality conference or undertake a root-cause analysis to understand why pulse oximetry was not performed. Including community healthcare practitioners who care for the same patients but are not affiliated with hospitals (such as direct-entry midwives) in such discussions can improve patient care. Given the realities of fragmented health care, in this situation the emergency department practitioner may be the only person in a position to inform the midwife.

9. **Overwhelmingly, current state laws mandating newborn pulse oximetry apply only to hospitals. Is it just to exempt home births from these laws? Are there good reasons to do so?**
Such laws are often driven by advocacy on the part of those affected by the absence of screening, as much or more so than the evidence supporting screening. However, it is worth considering: Does the law in question intend to regulate practice by attempting to assure that those who seek services receive appropriate care (a sort of consumer protection intent), or does it intend to protect infants directly (like a law requiring the use of an appropriate car seat)? In either case, there are good arguments that home birth should not be treated differently since infants are equally deserving of early diagnosis regardless of where they are born. Depending on the state, implementing and enforcing regulations for health care that is delivered outside of a hospital can be far more difficult than implementing and enforcing them in the already tightly regulated hospital system.

Attempts to improve the quality of medical care via legislation have not been consistently successful, and many professional groups have favored self-regulation of practice over external regulation (as in the form of legislation).

A few weeks later, your colleague sends you an interesting case report that describes a scenario in which a cardiac anomaly of uncertain severity was detected on a routine 20-week ultrasound [1]. The pregnant patient had already planned for a birth at home and was followed by a midwife; she declined further testing, in part because of concerns that the midwife might refuse to participate in the home birth. A maternal-fetal medicine specialist and a palliative care physician met with the patient to discuss possible outcomes, which included death at home. Both the patient and her partner stated that they accepted that their child may die during or shortly after birth.

10. **Why do you think some mothers and families choose home birth?**
In the context of childbirth, "medicalization" has driven the adoption of interventions with little evidence to support their use, such as routine episiotomies and continuous electronic fetal monitoring during labor [46, 47]. Concerns about medicalization and unnecessary interventions, as well as beliefs that home birth is safer and has better outcomes, were the top two reasons described in a qualitative study of Americans who chose home birth [10].

Families may favor home birth for a variety of reasons [48]. Some have had prior negative experiences, personally or involving friends or relatives, with the hospital birth environment. They may have experienced or fear that they will experience the loss of autonomy during labor and delivery [48, 49]. They may be aware that patients who give birth in hospitals are more likely to receive medical interventions (such as synthetic oxytocin [Pitocin] to accelerate labor, episiotomies, blood transfusions, vacuum extraction and forceps deliveries, and cesarean sections). They may find their homes to be more comfortable and safe [50, 51]. They may also wish to avoid interventions for their infants, such as the hepatitis B vaccine and intramuscular vitamin K—although they could refuse these in the hospital, and in some states, midwives can administer vitamin K at home [52].

> **Box 7.3 Personal Perspective**
> Dana Lathrope, a labor and delivery nurse, talked about the birth of her second child at home, after her first was born in the hospital setting [52]:
>
> » Experiencing home birth with a midwife solidified my belief that we are doing it so wrong in hospital. We are not treating birth as the natural event that it should be. At home, birth was so beautiful, peaceful and comfortable. In the hospital, it's not that way. Even when birth is at its simplest and easiest, the many hospital routines and interventions conspire to make it otherwise. I think having the home birth made me realize that birth could be a beautiful, comfortable event. It didn't have to be this scary thing that so many people perceive it as being.

(a) **What other lifecycle events might stimulate a conversation about hospital vs. in-home care or "medicalization" versus a "natural" process of care?**
There is a growing body of research on people's preference for comfort, dignity, and the home setting at the end of life—a "natural" death away from the hospital setting. However, there are a variety of complicated factors that can contribute to patients having a "medicalized" dying process. These factors can include patient and family beliefs about death and dying, cognitive biases on the part of healthcare practitioners, the patient's disease process(es) at the end of life, and other environmental factors [43, 53]. Cognitive biases on the part of clinicians and patients (or their loved ones) can cloud judgment regarding futile care, symptom management, and either escalating or avoiding interventions at the end of life, as well as the feasibility of, for example, transfer home from an intensive care unit [54]. The comparison between care of patients at these two momentous lifecycle events—childbirth and death—reveals some obvious differences. Nonetheless, clinicians may wish to compare the two to examine them for similar ethical considerations regarding bias, patient and family desires and beliefs, and clinical decision-making.

11. **Are there *any* situations in which patient autonomy during pregnancy or childbirth should be overridden to protect the fetus?**
This question is addressing the moral status of a fetus. That is, under what circumstances, if any, should the fetus be considered to be valuable enough to be considered separately from the mother. This is controversial and the subject of much scholarly analysis. It is also important to recognize that fetal interests can never be known directly—they are always imposed by others, whether by the parents or healthcare practitioners [44, 55, 56].
Some of the factors ethicists have considered include:
- *Viability.* Might the moral status of a fetus after the point of viability, when it could conceivably survive independent of the pregnant person, be different than before viability and deserves more consideration?
- *The individual's intent regarding continuation of the pregnancy.* Assuming that the individual has access to contraception and abortion and has chosen to become pregnant and/or to continue the pregnancy (whether planned or unplanned), does that decision confer moral status to the fetus?

- *The burden of the action or inaction.* Is asking a pregnant individual to abstain from an activity different from asking the same individual to undertake a certain action? Does it matter if the burdens of either are minimal? For example, is there a difference between avoiding alcohol consumption in moderation and being on bedrest? What about choosing to give birth at home versus declining fetal surgery?
- *The relative risks and benefits to each party.* This is controversial as well since, as noted earlier, people seldom agree on risk assessment. For example, is taking an anticonvulsant during pregnancy (clear benefit to the pregnant individual with a seizure disorder but known risks to the fetus) acceptable while taking an antidepressant (some benefit to the pregnant individual, infrequent but real risk to the fetus) is not? Such assessments are further complicated by the dynamic nature of medical science: A study published in the late 1980s, when court-ordered obstetrical interventions were still supported by case law, suggested that the medical recommendation was retrospectively wrong in one-third of cases [57]. This is explored further in the next question.

For each of these, there are arguments on the extreme ends, with some scholars arguing that the autonomy of the living person always supersedes the potential rights of a future person (the fetus) and others arguing that pregnant individuals and fetuses have equal rights. It must be acknowledged, however, that the latter position, no matter how strongly articulated, cannot resolve disputes in which maternal and fetal interests are clearly in conflict. One must choose in such cases, and it is difficult to justify placing the rights of an unborn child above those of an adolescent or adult who has identity and agency, as well as relationships and responsibilities [58]. As Harris notes, "A pregnant woman can represent her own autonomy interests, but a fetus obviously is unable to articulate its beneficence-based needs. Fetal needs exist only as a projection of a physician's or another party's determination of what is thought to be in the best interest of a fetus" [56]. She goes on to suggest that clinicians approach perceived ethical conflicts between pregnant individuals and their unborn children in three ways:

1. "[A]ttempt to understand the pregnant woman and her decisions within her broad social networks and communities."
2. "[A]sk how the standpoint of clinicians framing the ethical dilemma is related to the outcome judged to be ethical, and recognize that opinions from a diverse group of people might best accomplish this."
3. "[A]ttend to issues of [gender] and race equality by asking, 'Does this approach reduce or enhance existing conditions of advantage and disadvantage?'" [56].

(a) **Is drug use during pregnancy morally different? Why or why not? What about other health behaviors: eating well, exercising, etc.?** Drug use presents an ethical challenge, but fortunately one with solutions. Most people agree that intrauterine drug exposure is not a good thing, both on an individual level and on a population level. However, substance use disorder is now increasingly recognized as a disease (see ▶ Chapter 18 for further discussion of opioid use disorders), one that requires treatment and that may involve relapse. Nevertheless, perceptions of substance use as a moral failure persist, and nowhere is this so true as when it involves pregnancy [59, 60].

Tennessee was the first state to pass a "fetal assault" law, explicitly criminalizing the use of illegal substances during pregnancy (distinct from laws regarding possession or sale) [60]. Although that law expired in 2016 and was not renewed, one-third of states still use pre-existing laws, such as those related to child abuse, in order to prosecute pregnant individuals for drug use [60]. Are such laws just? It seems that they cannot be if one truly accepts that substance use is a disease.

More broadly, ethical criticism of attempts to regulate other behaviors by pregnant women has included many of the concerns addressed previously, namely, that recommendations regarding such behaviors and the evidence supporting them are often weak and are rapidly evolving (examples abound: women were once told to avoid exercise but are now encouraged to exercise; more recently ACOG dropped recommendations to avoid consumption of sushi and sashimi during pregnancy); that women, particularly those who are poor or belong to racial or ethnic minority groups, are disproportionately burdened, contributing to inequities; and, ultimately, that the consequences of infringed upon bodily autonomy—to individuals, societies, and populations—are far greater than any potential benefits [44, 56].

There are also important consequentialist concerns regarding attempts to regulate the behavior of pregnant people by courts or healthcare practitioners acting paternalistically: patients might avoid prenatal care, compromising outcomes for both mothers and infants [56].

(b) **Self-reflection: Some ethicists have argued that although there is no *legal* obligation for pregnant patients to put the interests of the fetus first, there is an *ethical* obligation. Do you agree or disagree? Why?**

> **Box 7.4 Teaching Tip**
> Consider asking learners who agree whether they feel this obligation extends to interventions such as fetal surgery that carry significant risks to the pregnant patient (including potential difficulties with future childbearing) and relatively intangible benefits (potential prevention of emotional pain and suffering and possibly financial burdens if the surgery produces the desired outcome for the unborn child).

In the aforementioned published case, the midwife struggled, ethically, with whether to be present at the birth because it was, in some respects, outside her scope of practice and because she was concerned that her presence could be misconstrued as unilateral support for the family's decision. Eventually, she decided that she was obligated to be present.

✓ 12. **Why do you think this midwife decided to participate in the birth?**
The midwife has conflicting professional obligations. On one hand, she likely recognizes that she was not qualified to assist with a high-risk delivery, such as the one in this case. She probably wants to avoid potentially inflicting harm by practicing outside her scope. She might hope that, without any healthcare practitioner willing to attend the birth, the family will reconsider their decisions. On the other hand, she may have both ethical and pragmatic considerations. She may feel it is unethical to try to coerce the family into a different choice (by withdrawing her services), or she may realize that such a tactic simply will not work. In either case, she might then feel that she has a continued duty to care for the patient, in the absence of transfer to another healthcare practitioner.

✓ 13. **Healthcare practitioners weigh risks differently, from each other and from patients. What are the ethical obligations of healthcare practitioners when it comes to describing the risks and benefits of home birth vs. hospital birth?**
All healthcare practitioners have an obligation to be truthful with their patients, which means taking care to present

"Our Baby Is Turning Blue"

information as objectively as possible—that is, they avoid conveying their own biases to their patients.

After this experience, you lead a community task force seeking to establish transfer agreements between local hospitals and midwifery practices that provide care for home births. The first meeting is somewhat contentious. The obstetricians and pediatricians representing the hospitals feel that the choice to give birth at home is irresponsible and likely to end badly; they cite examples of other cases with poor outcomes to support their claims. The midwives and patient advocates point to the NICE recommendation in the United Kingdom and to studies around the world demonstrating equivalent or better outcomes for low-risk patients giving birth at home [2–8].

14. **What do you think might account for the different recommendations in the United States and United Kingdom?**
    There are many possibilities. The midwifery profession is more established and respected in the United Kingdom and many other countries in Europe; consequently, interprofessional communication and collaboration between midwives and obstetricians is likely to be better. Low-risk pregnant patients are steered toward receiving care from a midwife, rather than an obstetrician; obstetricians are viewed not as first-line providers but as specialists who care for higher-risk patients. This allows for the rapid and coordinated transfer of patients who begin labor at home but develop complications. In the United States, many physicians and hospitals have no system in place for communication during a transfer; the midwife may not speak with a physician at all, with all medical information being relayed through emergency medical services, after 911 is called [61]. In the United Kingdom, an intrapartum or postpartum transfer from home to hospital is not viewed as a serious mistake or failure but rather as a sign that the system has worked well.
    (a) **What is different about the healthcare systems in the United States and the United Kingdom?**
    One major difference is the existence of the National Health Service (NHS). There is a significantly greater focus on the value of primary care, as well as cost-effectiveness. The dominant cultural expectation is that "normal" pregnancies will be managed by midwives. Similarly, most people, including children, receive routine primary care from general practitioners. Pediatric consultants, in this setting, are mostly hospital-based specialists who care for children who are ill.

15. **Many studies around the world have looked at outcomes in home births and compared them to hospital births [2–8]. What types of study design do you think are used? Why?**
    Most studies have been retrospective cohort or case-control studies using birth certificate data. Randomized controlled trials are not typically feasible because few, if any, patients would be willing to be randomized to home or hospital births. Most patients have strong feelings regarding their preferred location for childbirth. Increasingly, systematic reviews and meta-analyses have been performed, combining data from multiple case-control or cohort studies.
    (a) **What are challenges and limitations to these study designs?**
        Limitations include:
        – *Selection bias*: Matching controls can be difficult; the population of patients choosing home birth tends to be quite different in measurable, and likely nonmeasurable, ways compared to the population choosing hospital birth.
        – *Design bias*: Failure to adequately control for the many possible confounders and effect modifiers, such as outcomes of previous births, maternal age, etc., would affect validity.
        – *Measurement bias*: Birth certificate data is itself limited. It does not typically identify whether

a home birth was planned or unplanned (the results of preterm or precipitous delivery without time to seek medical attention—these are more likely to be complicated) nor whether a hospital birth occurred after labor at home and transfer for complications. It cannot, by design, include assessments of quality of care.
- *Reporting bias*: Large cohort studies done by obstetricians-gynecologists have tended to show small but statistically significant increases in negative outcomes in home births; similar studies done by midwives have tended to show no differences. One interpretation is that the researchers are, perhaps inadvertently, allowing their own views to influence the study design and data interpretation; another is that they are only publishing studies that support their previously held positions. Case-control and cohort studies can demonstrate associations but not prove causation. Cohort studies generally provide stronger evidence than case-control because they are less likely to suffer from selection bias.

16. **Describe the interprofessional communication you would want to have between hospitals and midwives to ensure best possible outcomes.**
A starting point would be mutual respect between healthcare practitioners with different strengths and scopes of expertise. Such respect might start with each stakeholder understanding the other's educational and professional background. Midwives might try to understand the perspectives of obstetricians and neonatologists, who see "worst-case scenarios," while obstetricians and other hospital-based practitioners might try to appreciate that such scenarios are relatively uncommon.

Once professional relationships are established, then effective communication can proceed as in any other situation when a patient requires higher-level or more specialized care. Such relationships would also facilitate earlier dialogue and transfer, since midwives might feel more comfortable consulting with an obstetrician-gynecologist at a local hospital to discuss potential concerns before emergencies arise. This is not unlike ordinary interprofessional communication that occurs every day in medicine. A hospitalist might make a critical care specialist aware of a patient who is not responding as quickly as hoped to treatment but does not yet require intensive care. A pediatric nurse practitioner seeing a child with iron-deficiency anemia might call a pediatric hematologist to confirm that he is not missing something more serious.

Notably, a team of researchers recently constructed a 50-state database, using regulatory data, to describe the environment for midwifery practice and interprofessional collaboration, and then developed a scoring system (the Midwifery Integration Scoring System or MISS) [62]. They ranked states according to this system and then compared states' maternal and infant outcomes with their assessed integration of midwives into health care, after controlling for race (see ► Chapter 5 for further discussion of the association between race, racism, and birth outcomes in the United States). They found that greater integration of midwives, as well as higher density of midwives, showed "significantly higher rates of spontaneous vaginal delivery, vaginal birth after cesarean, and breast-feeding, and significantly lower rates of cesarean, preterm birth, low birth weight infants, and neonatal death" [62].
(a) **Creative problem-solving: What local policies could you propose to achieve these outcomes?**

17. **Self-reflection: At the beginning of the discussion, you were asked about your emotional response to this case. Has that changed through the discussion or because of what you've learned?**

Acknowledgments This case was adapted from an earlier case written by Thomas Curran and Amy Caruso Brown.

References

1. Jankowski J, Burcher P. Home birth of infants with anticipated congenital anomalies: a case study and ethical analysis of care providers' obligations. J Clin Ethics. 2015;26(1):27–35.
2. Janssen PA, Saxell L, Page LA, Klein MC, Liston RM, Lee SK. Outcomes of planned home birth with registered midwife versus planned hospital birth with midwife or physician. Can Med Assoc J. 2009;181(6–7):377–83.
3. Cheyney M, Bovbjerg M, Everson C, Gordon W, Hannibal D, Vedam S. Outcomes of care for 16,924 planned home births in the United States: the midwives Alliance of North America statistics project, 2004 to 2009. J Midwifery Womens Health. 2014;59:17–27.
4. Olsen O, Clausen JA. Planned hospital birth versus planned home birth. Cochrane Database Syst Rev. 2012;(9):CD000352.
5. Blix E, Huitfeldt AS, Øian P, Straume B, Kumle M. Outcomes of planned home births and planned hospital births in low-risk women in Norway between 1990 and 2007: a retrospective cohort study. Sex Reprod Healthc. 2012;3(4):147–53.
6. Hutton EK, Reitsma AH, Kaufman K. Outcomes associated with planned home and planned hospital births in low-risk women attended by midwives in Ontario, Canada, 2003-2006: a retrospective cohort study. Birth. 2009;36(3):180–9.
7. de Jonge A, Mesman JA, Manniën J, Zwart JJ, van Dillen J, van Roosmalen J. Severe adverse maternal outcomes among low risk women with planned home versus hospital births in the Netherlands: Nationwide cohort study. Br Med J. 2013;346:f3263.
8. Rossi A, Prefumo F. Planned home versus planned hospital births in women at low-risk pregnancy: a systematic review with meta-analysis. Eur J Obstet Gynecol Reprod Biol. 2018;222:102–8.
9. MacDorman MF, Mathews TJ, Declerq E. Trends in out-of-hospital births in the United States, 1990-2012. NCHS Data Brief. 2014;144:1–8.
10. Boucher D, Bennett C, McFarlin B, Freeze R. Staying home to give birth: why women in the United States choose home birth. J Midwifery Womens Health. 2009;54(2):119–26.
11. Cheng YW, Snowden JM, King TL, Caughey AB. Selected perinatal outcomes associated with planned home births in the United States. Am J Obstet Gynecol. 2013;209(4):325.e1–8.
12. American Academy of Pediatrics, Committee on Fetus and Newborn; American College of Obstetricians and Gynecologists, Committee on Obstetric Practice. The Apgar score. Pediatrics. 2006;117(4):1444–7.
13. Divon MY, Haglund B, Nisell H, Otterblad PO, Westgren M. Fetal and neonatal mortality in the postterm pregnancy: the impact of gestational age and fetal growth restriction. Am J Obstet Gynecol. 1998;178(4):726–31.
14. Grünebaum A, McCullough LB, Sapra KJ, Brent RL, Levene MI, Arabin B, et al. Early and total neonatal mortality in relation to birth setting in the United States, 2006–2009. Am J Obstet Gynecol. 2014;211(4):390.e1–7.
15. WHO, UNICEF, UNFPA, World Bank Group and the United Nations Population Division. Trends in maternal mortality, 1990–2015. New York: United Nations Population Fund; 2015 [cited 2018 July 22]. Available from: https://www.unfpa.org/publications/trends-maternal-mortality-1990-2015.
16. Centers for Disease Control and Prevention. Pregnancy mortality surveillance system [Internet]. Atlanta (GA): CDC. [cited 2018 July 22]. Available from: https://www.cdc.gov/reproductivehealth/maternalinfanthealth/pmss.html.
17. Kochanek KD, Murphy SL, Xu JQ, Arias E. Mortality in the United States, 2016. NCHS data brief, no 293. Hyattsville (MD): National Center for Health Statistics; 2017.
18. Centers for Disease Control and Prevention. Infant mortality [Internet]. Atlanta (GA): CDC. [cited 2018 July 22]. Available from: https://www.cdc.gov/reproductivehealth/maternalinfanthealth/infantmortality.htm.
19. UN Inter-agency Group for Child Mortality Estimation. Mortality rate, infant (per 1,000 live births). Washington, D.C.: The World Bank. Available from: https://data.worldbank.org/indicator/SP.DYN.IMRT.IN?year_high_desc=false.
20. Building U.S. Capacity to Review and Prevent Maternal Deaths. Report from nine maternal mortality review committees. Maternal Mortality Review Information Application [Internet]; 2018 [cited 2018 Jul 22]. Available from: http://reviewtoaction.org/Report_from_Nine_MMRCs.
21. Amnesty International. Deadly delivery: the maternal health care crisis in the USA. London: Amnesty International Publications; 2010.
22. Waldenström U, Aasheim V, Nilsen AB, Rasmussen S, Pettersson HJ, Schytt E. Adverse pregnancy outcomes related to advanced maternal age compared with smoking and being overweight. Obstet Gynecol. 2014;123(1):104–12.
23. Bai J, Wong FW, Bauman A, Mohsin M. Parity and pregnancy outcomes. Am J Obstet Gynecol. 2002;186(2):274–8.
24. Kent ST, McClure LA, Zaitchik BF, Gohlke JM. Area-level risk factors for adverse birth outcomes: trends in urban and rural settings. BMC Pregnancy Childbirth. 2013;13:129.
25. Ely DM, Hoyert DL. Differences between rural and urban areas in mortality rates for the leading causes of infant death: United States, 2013-2015. NCHS Data Brief. 2018;300:1–8.
26. Carr KC. Midwifery education and accreditation in the U.S. [PowerPoint presentation]. Silver Spring [MD]: ACNM; 2014. Available from: http://www.midwife.org/acme.
27. Midwives Alliance of North America. Legal status of U.S. midwives [Internet]. Montvale (NJ): MANA. Available from: http://www.mana.org/about-midwives/legal-status-of-us-midwives.
28. American College of Nurse-Midwives. Comparison of certified nurse-midwives, certified midwives, certified professional midwives clarifying the distinctions among professional midwifery credentials in the U.S. Silver Spring (MD): ACNM; 2017. Available from

http://www.midwife.org/acnm/files/ccLibraryFiles/FILENAME/000000006807/FINAL-ComparisonChart-Oct2017.pdf.
29. New York State Office of the Professions. Midwifery – questions and answers [Internet]. Albany: New York State Education Department; [cited 2018 Jul 22]. Available from: http://www.op.nysed.gov/prof/midwife/midwifeqa.htm.
30. American College of Nurse-Midwives. Position statement: home birth. Silver Spring (MD): ACNM; 2011. Available from: http://www.midwife.org/ACNM/files/ACNMLibraryData/UPLOADFILENAME/000000000251/Home%20Birth%20Aug%202011.pdf.
31. American College of Obstetricians and Gynecologists. Planned home birth. Committee Opinion No. 697. Obstet Gynecol. 2017;129:e117–22.
32. National Institute for Health and Care Excellence. Clinical guideline CG190: intrapartum care for healthy women and babies. UK: National Institute for Health and Care Excellence; 2017. Available from: https://www.nice.org.uk/guidance/cg190/chapter/Recommendations.
33. Curran WJ. Court-ordered caesarean sections receive judicial defeat. NEJM. 1990;323:489–92.
34. In re Fetus Brown, 689 N.E.2d 397, 400, Ill. App. Ct. 1997.
35. In re Baby Boy Doe, 632 N.E.2d 326, Ill. App. Ct. 1994.
36. Morris T, Robinson JH. Forced and coerced cesarean sections in the United States. Contexts. 2017;16(2):24–9.
37. Garg LF, Van Naarden BK, Knapp MM, Anderson TM, Koppel RI, Hirsch D, et al. Results from the New Jersey statewide critical congenital heart defects screening program. Pediatrics. 2013;132(2):e314–23.
38. Glidewell J, Olney RS, Hinton C, Pawelski J, Sontag M, Wood T, et al. State legislation, regulations, and hospital guidelines for newborn screening for critical congenital heart defects-United States, 2011-2014. Morbidity and Mortality Weekly Report (MMWR). 2015;64(23):625–30.
39. Liske MR, Greeley CS, Law DJ, Reich JD, Morrow WR, Baldwin HS. Et al; Tennessee task force on screening newborn infants for critical congenital heart disease. Report of the Tennessee task force on screening newborn infants for critical congenital heart disease. Pediatrics. 2006;118(4):e1250–6.
40. Good RJ, Canale SK, Goodman RL, Yeager SB. Identification of critical congenital heart disease in Vermont: the role of universal pulse oximetry screening in a rural state. Clin Pediatr. 2015;54(6):570–4.
41. Oster ME, Aucott SW, Glidewell J, Hackell J, Kochilas L, Martin GR, et al. Lessons learned from newborn screening for critical congenital heart defects. Pediatrics. 2016;137(5). pii: e20154573.
42. Charles C, Gafni A, Whelan T. Shared decision-making in the medical encounter: what does it mean? (or it takes at least two to tango). Soc Sci Med. 1997;44(5):681–92.
43. Yuen T, Derenge D, Kalman N. Cognitive bias: its influence on clinical diagnosis. J Family Practice. 2018;67(6):366.
44. Lyerly AD, Mitchell LM, Armstrong EM, Harris LH, Kukla R, Kuppermann M, et al. Risk and the pregnant body. Hastings Cent Rep. 2009;39(6):34–42.
45. Reyna VF, Nelson WL, Han PK, Dieckmann NF. How numeracy influences risk comprehension and medical decision making. Psychol Bull. 2009;135(6):943.
46. Carroli G, Belizan J. Episiotomy for vaginal birth. Cochrane Database Syst Rev. 2000;(2):CD000081.
47. Thacker SB, Stroup DF. Continuous electronic heart rate monitoring for fetal assessment during labor. Cochrane Database Syst Rev. 2000;(2):CD000063.
48. Hazen H. "The first intervention is leaving home": reasons for electing an out-of-hospital birth among Minnesotan mothers. Med Anthropol Q. 2017;31(4):555–71.
49. Newnham E, McKellar L, Pincombe J. "It's your body, but…" Mixed messages in childbirth education: Findings from a hospital ethnography. Midwifery. 2017;55:53–9.
50. Parry D. "We wanted a birth experience, not a medical experience": exploring Canadian women's use of midwifery. Health Care for Women Intl. 2008;29(8–9):784–806.
51. Fleming S, Donovan-Batson C, Burduli E, Barbosa-Leiker C, Hollins Martin C, Martin C. Birth satisfaction scale/birth satisfaction scale-revised (BSS/BSS-R): a large scale United States planned home birth and birth centre survey. Midwifery. 2018;41:9–15.
52. Smith M. Why choose home birth. Midwifery Today Int Midwife. 2017;122:9–12.
53. Ko MC, Huang SJ, Chen CC, Chang YP, Lien HY, Lin JY, et al. Factors predicting a home death among home palliative care recipients. Medicine (Baltimore). 2017;96(41):e8210.
54. Lin Y, Myall M, Jarrett N. Uncovering the decision-making work of transferring dying patients home from critical care units: an integrative review. J Adv Nurs. 2017;73(12):2779–3233.
55. Chervenak FA, McCullough LB. The fetus as a patient: an essential ethical concept for maternal-fetal medicine. Journal of Maternal-Fetal Medicine. 1996;5(3):115–9.
56. Harris LH. Rethinking maternal-fetal conflict: gender and equality in perinatal ethics. Obstet Gynecol. 2000;96(5):786–91.
57. Kolder VEB, Gallagher J, Parsons MT. Court-ordered obstetrical interventions. N Engl J Med. 1987;316:1192–6.
58. Tan ML. Fetal discourses and the politics of the womb. Reprod Health Matters. 2004;12(Suppl 24):157–66.
59. Patrick SW, Schiff DM. A public health response to opioid use in pregnancy. Pediatrics. 2017;139:e20164070.
60. Hui K, Angelotta C, Fisher CE. Criminalizing substance use in pregnancy: misplaced priorities. Addiction. 2017;112(7):1123–5.
61. Cheyney M, Everson C, Burcher P. Homebirth transfers in the United States: narratives of risk, fear and mutual accommodation. Qual Health Res. 2014;24(4):443–56.
62. Vedam S, Stoll K, MacDorman M, Declercq E, Cramer R, Cheyney M, et al. Mapping integration of midwives across the United States: impact on access, equity, and outcomes. PLoS One. 2018;13(2):e0192523.

Further Reading on this Topic

Davis-Floyd RE, Sargent C. Childbirth and authoritative knowledge: cross-cultural perspectives. Berkeley/Los Angeles: University of California Press; 1997.

Martin N. Lost mothers: maternal care and preventable deaths. ProPublica [Internet]. 2017 July 17. Available from: https://www.propublica.org/series/lost-mothers.

United Nations Population Fund. The state of the world's midwifery. New York: UNFPA; 2014. Available from: https://www.unfpa.org/sowmy.

"I Have a Touch of Sugar but I Can't Afford My Meds"

Martha A. Wojtowycz and Ahmed A. Malik

8.1 Background Questions – 146

8.2 Additional Case Information and Questions for Discussion – 146

8.3 Answers to Background Questions – 147

8.4 Responses to Discussion Questions – 154

References – 160

Learning Objectives

1. Describe the ways in which socioeconomic status and health are interrelated.
2. Analyze the ways that race affects health, including the role of unconscious (implicit) bias from health care providers.
3. Explain how the current system of health care finance impacts health outcomes.

> **Case Background**
>
> *You are caring for Jacob Messenger, a 54-year-old man with foot pain in the emergency department (ED). He says he had "high sugars" in the past, but he has not seen his primary care physician recently. His physical exam reveals cellulitis. You also find that his blood glucose level is 200 mg/dL and his hemoglobin A1C is 9.5%. When you ask about his medications, he says that he has not been taking them regularly. He states, "I just can't afford them."*

8.1 Background Questions

1. Identify at least two financial and two non-financial barriers to (a) accessing health care and (b) achieving optimal health.

2. Using a source that is no more than 5 years old, find average life expectancy of impoverished and affluent Americans (e.g., by lowest income quartile vs. highest quartile) as well as for white and black Americans by sex.

3. Choose a chronic disease (e.g., diabetes, heart disease, chronic obstructive pulmonary disease [COPD]), and describe the prevalence by income, education, and race.

4. Adjusting for income and education, what is the evidence that race is an independent risk factor for poor health outcomes? What is the role of racism? Choose at least one scholarly article addressing this, and describe its methods and limitations.

5. The American Academy of Pediatrics (AAP) recommends screening for poverty. What was the rationale behind this recommendation, and what is the recommended screening question? What is the evidence supporting this approach?

6. What is an accountable care organization (ACO)? What is the U.S. Centers for Medicaid and Medicare Services' (CMS) "Accountable Health Communities" initiative?

7. Since the passage of the Affordable Care Act (ACA), what do current data show with respect to rates of uninsured individuals and with respect to those enrolled in Medicaid in the United States?

8. Describe some of the myths about poverty in the United States today.

8.2 Additional Case Information and Questions for Discussion

Mr. Messenger lives in a state that expanded Medicaid, but he has not enrolled in Medicaid yet. He has a part-time minimum wage job that does not provide health insurance. His car recently broke down and he cannot afford to repair it. He has not been able to afford his medications, and he has stopped seeing his doctor because of the costs of co-pays and the difficulty with transportation.

1. What does this mean for his ability to access health care?
   (a) What if Mr. Messenger lived in a state that did not expand Medicaid?

2. Why do nonprofit and teaching hospitals take all patients?
   (a) What are the legal reasons?
   (b) How is this financially feasible?

Until a couple of months ago, Mr. Messenger worked at a large brewery as a mechanical technician. Through that job, he was able to get employer-sponsored insurance. However, he has had gradually worsening intermittent foot pain over the past year, and there were days that he could not go to work because of the pain. He quickly exhausted his (unpaid) sick leave, and his manager felt he was unreliable. When his car broke down, he missed work 3 days in a row, and he was let go. Since then, he has held several consecutive part-time jobs, but none has provided health insurance, and he has not been able to keep one position long enough to accumulate sick leave.

"I Have a Touch of Sugar but I Can't Afford My Meds"

❓ 3. Poverty can clearly impact health outcomes, but how can poor health impact a patient's financial situation?

❓ 4. Self-reflection: In thinking about CMS's Accountable Health Communities, how do you feel about physicians being held accountable for their patients' transportation, housing, food security, etc.?

❓ 5. What can you do to help Mr. Messenger get his medications?

Mr. Messenger mentions to you that he was in a different ED about 6 months ago with similar pain in his foot. He had a lot of pain at the time but was not offered any pain medication or symptomatic relief, so he decided to leave. The following day, when the pain became unbearable, he went back to the ED. He was admitted to the hospital. You note that his medical records show that he had a deep abscess that formed in his lower leg, and he required 10 days of intravenous antibiotics.

❓ 6. Creative problem-solving: It is not unusual for patients' accounts to differ in significant ways from the written medical record. How would you "fill in the gaps" between Mr. Messenger's account and the notes you are able to see? What do you think happened?

❓ 7. What does implicit bias mean?
   (a) Self-reflection: What does it mean to you as a future physician?
   (b) How do you think implicit bias might have impacted Mr. Messenger's care in the ED?
   (c) What can we do to address our own biases in order to provide the best possible care for our patients?

You determine that Mr. Messenger needs admission for treatment of his cellulitis. Upon learning this, Mr. Messenger becomes distraught, expressing concern that an admission will jeopardize his current job.

❓ 8. Self-reflection: Before entering the health professions, how did you think about poverty? Have your experiences as a health care professional or future health care professional changed your perspective? How so?

❓ 9. How can you, as future physicians, help to improve the lives of those in poverty?

8.3 Answers to Background Questions

✅ 1. Identify at least two financial and two nonfinancial barriers to (a) accessing health care and (b) achieving optimal health.

One way to better understand these barriers is to use a classification system such as the one proposed by the health care access barriers model (HCAB) described by Carrillo et al. [1]. The HCAB addresses barriers that are modifiable and categorizes them into financial, structural, and cognitive. The latter two can be considered to be nonfinancial barriers for purposes of this question.

The most commonly known financial barrier to accessing health care is the cost of accessing health care, which many readily identify as a lack of insurance. Without insurance, the cost of a simple visit to a primary care doctor can be prohibitive. A study published in *Health Affairs* found that the average cost for one primary care visit, across 10 different states in different regions of the United States, ranged from $128 to $188 [2]. Costs for laboratory and other tests were not included in this estimate, and therefore the actual cost of a visit could potentially be significantly greater. Having insurance by itself, however, is not a complete solution to this financial barrier.

Even with insurance, people face financial barriers in terms of the cost of their contribution to the health insurance, co-pays, and deductibles. Co-pays are a fixed amount paid out of pocket for services, usually upon receipt of the service. A person with diabetes, who needs to have regular visits with his or her primary care doctor, may not be able to afford the frequent visits that would result in recurring co-pays. For other individuals, having deductibles may serve as another financial barrier to accessing health care. Deductibles require insurance enrollees to pay out of pocket for their medical care until they have spent a certain amount. It is only after this amount is reached that the insurance coverage begins. Even after that point, however, insured individuals are usually responsible to pay a certain percentage of the cost. These cost considerations can serve as barriers to accessing health care.

A third layer of financial barrier to accessing health care is the cost of transportation. It is important to note that many people living below the poverty level may not be able to afford an extra trip to the doctor's office. Even with the availability of public transportation, a person who is living paycheck to paycheck may have to carefully plan his or her transportation. Ensuring that there are enough funds to get to and from work may be of paramount importance. As a consequence, the inability to afford transportation to and from a medical appointment may become a barrier to accessing health care.

A fourth layer of financial barrier (in addition to insurance status, co-pays/deductibles, and transportation) is the cost associated with purchasing the medications that the physician prescribes. If a person has limited or no prescription coverage, he or she may feel that visiting the physician may be an exercise in futility if the medications that the physician prescribes are not affordable.

Similarly, the financial barriers to achieving optimal health start with limited or complete lack of insurance. For people who have diabetes, the same barriers explored above may still prevent them from maintaining necessary frequent visits. If they are unable to access the health care services they need, they are not likely to achieve optimal health (e.g., they may experience preventable diabetic retinal complications, peripheral vascular disease, amputations, etc.). Furthermore, if they have limited or no prescription drug coverage and are not able to afford their medications, they still may not be able to achieve their optimal health despite being able to visit a physician.

In following with the HCAB classification, nonfinancial barriers to accessing health care can be structural or cognitive. One key structural barrier is the refusal of providers to accept new Medicaid patients or limits to the number of Medicaid patients in their practices [3]. Another structural barrier is the lack of providers of the same gender as the patient, which is an important cultural and religious consideration, and may result in delayed care-seeking behavior [4]. Other examples of structural barriers include the limited hours of operation of a health care facility; lack of accessible entrances to health care facilities; lack of transportation, in general; and, specifically, modes appropriate to the individual needs of a person or something as seemingly simple as street safety [1]. These can be represented by hours of operation that do not account for an individual's work hours, the lack of ramps at facility entrances for wheelchair-bound individuals, and the lack of accessible transport vans or busses. The lack of access to health care can then compound the inability to achieve optimal health. In addition to the structural nonfinancial barriers, cognitive barriers can also preclude a person from accessing health care and from achieving optimal health.

Nonfinancial cognitive barriers to accessing health care include limited or no English proficiency, the lack of interpretive services, and/or lack of translated materials [1]. If patients cannot communicate their needs to the provider and vice versa, and they are not able to find a way to do so, then they may not seek the needed care. It is also possible that even if they do seek medical care, they may not be able to understand information or appointment reminders that arrive in the mail or via phone calls. In addition to language barriers, not being able to understand the medical jargon can be a barrier for patients to access health care.

The inability to understand the directions to a health care facility can be another cognitive barrier to accessing health care. A person may be unable to navigate the public transportation system, for example, or may not be able to understand printed directions to the facility. This could potentially prevent a person from accessing needed care. If a patient is given a referral to a different location for special diagnostic testing or a different location for therapeutics—requiring visits to multiple locations—the logistics involved and the potential of not being able to locate the facilities may further prevent a person from accessing health care [1].

Similarly, nonfinancial barriers to achieving optimal health include limited or no English proficiency, lack of interpretive services, and limited health literacy. If a person with limited or no English proficiency is given medication directions in English, he or she will not be able to follow them and thus will be unable to achieve optimal health. Depending on the reading level or health literacy of an individual (even one for whom English is the primary language), he or she may not be able to follow the prescription directions. This could lead to underdosing or overdosing and, thus, could prevent an individual from achieving optimal health.

In addition to all of the above, we must consider lost days of work and the lack of childcare support as significant barriers to accessing and achieving optimal health care. Depending on how one wants to qualify the "lost-days-of-work barrier," it can be considered a financial barrier—as it may lead to loss of pay, which translates to a financial consideration—or a nonfinancial barrier as it may lead to limited availability and thus becomes a logistical consideration. Similarly, if a person cannot find childcare for his or her child for the duration of an appointment, this will serve as a barrier to accessing health care. In the case of a diabetic person, the inability to find affordable, and even available, childcare would prevent a person from scheduling regular visits and thus would serve as a barrier to achieving optimal health.

Finally, it is essential for health care practitioners to consider community resources. For example, patients with type 2 diabetes will be encouraged to eat a healthy diet and incorporate physical activity into their daily life. Access to affordable healthful food and safe places to be physically active can dramatically impact individuals' ability to optimize their health.

✅ 2. **Using a source that is no more than 5 years old, find average life expectancy of impoverished and affluent Americans (e.g., by lowest income quartile vs. highest quartile) as well as for white and black Americans by sex.**

According to the Centers for Disease Control and Prevention (CDC), in 2016, the average life expectancy of Americans was 78.7 years—76.1 years for men and 81.1 years for women—a slight decrease from 2015 [5].

According to an article published in the *Journal of the American Medical Association*, analyzing data from 2001 to 2014, the life expectancy in the affluent 1% was 14.6 years and 10.1 years higher than the life expectancy in the impoverished 1% of men and women, respectively [6]. The expected age of death for men in the top 1% of the income distribution was 87.3 years, while the expected age of death for men in the lowest 1% of the income distribution was 72.7 years. Similarly, women in the top 1% of the income distribution had a life expectancy of 88.9 years compared to a life expectancy of 78.8 years for women in the lowest 1% of the income distribution [6].

In 2015, the life expectancy at birth for the white population was 79.0 years compared to 75.5 years for the black. White males had a life expectancy of 76.6 years, while black males had a life expectancy of 72.2 years. Life expectancy was 81.3 years for white females and 78.5 years for black females [7].

3. **Choose a chronic disease (e.g., diabetes, heart disease, chronic obstructive pulmonary disease), and describe the prevalence by income, education, and race.**

   The CDC's National Center for Health Statistics is a good resource for finding information on chronic diseases, such as chronic obstructive pulmonary disease (COPD). The data in ◘ Table 8.1 were extracted from the FastStats section of the National Center for Health Statistics website [8] and from the *Health, United States, 2016* report [7]. Caution should be used when interpreting these findings as these data, in some cases, rely on self-reporting by respondents to the National Health Interview Survey. The questions used to ascertain disease status, however, were structured based on validated questions.

4. **Adjusting for income and education, what is the evidence that race is an independent risk factor for poor health outcomes? What is the role of racism? Choose at least one scholarly article addressing this, and describe its methods and limitations.**

   There is a large body of research on the topic of racial disparities in health outcomes, the role of racism, and the effect of socioeconomic status (SES) on the racial disparities seen in health outcomes. Although there are a number of studies highlighting the role that race plays in health outcome disparities, the question of to what extent socioeconomic factors, such as education and income, are confounding these findings, or even acting as effect modifiers, is not readily answered. The role that racism plays

◘ **Table 8.1** Prevalence of chronic diseases in the United States [7, 8]

	Diabetes	Heart disease	COPD	Cancer	Stroke
Total prevalence[a]	20.6%	10.7%	1.3%	8.5%	2.8%
Prevalence by poverty level[a]					
Below 100%	27.3%	14.8%[b]	3.3%[b]	6.4%[b]	5.6%[b]
100%–199%	21.7%	12.5%[b]	2.3%[b]	6.7%[b]	3.8%[b]
200% or more	16.6%	9.9%[b]	0.8%[b]	9.3%[b]	2.2%[b]
Prevalence by education[a]					
Less than high school	15.3%	13.2%	2.9%	7.0%	5.1%
High school or GED	10.9%	12.0%	2.0%	9.0%	3.4%
Some college	10.6%	13.0%	1.4%	10.6%	3.7%
Bachelor's degree or higher	6.6%	10.3%	0.5%	10.9%	1.8%
Prevalence by race/ethnicity[a]					
White non-Hispanic	16.6%	11.0%	1.4%	9.3%	2.7%
Black non-Hispanic	23.9%	10.1%	0.9%	5.2%	4.0%
Asian	17.3%	5.5%	0.4%	3.9%	1.1%
Hispanic or Latino	29.8%	8.1%	0.9%	4.5%	2.9%

Reported in the National Health Interview Survey summary report
COPD chronic obstructive pulmonary disease, *GED* general equivalency development/diploma
[a]Percentages are reported as age-adjusted percentages of adults, age 18 years and older, living with the specific chronic condition in 2016
[b]For these percentages, poverty level was defined as "poor," "near poor," and "not poor"

is also challenging to answer, but as Dr. David Williams from Harvard School of Public Health states, a "large and growing body of research shows that day-to-day experiences of African-Americans create physiological responses that lead to premature aging (meaning that people are biologically older than their chronological age)… and experiences of racial discrimination are an important type of psychosocial stressor that can lead to adverse changes in health status and altered behavioral patterns that increase health risks" [9].

A study by Franks and colleagues examined racial disparities in life expectancy and quality-adjusted life expectancy (QALE), using quality-adjusted life years (QALY), a subjective measure based on health-related quality of life (HRQL) [10]. The authors analyzed the data before and after adjusting for SES and considered both morbidity and mortality in their assessment of the differences in health between the black and the white populations in the study. HRQL was obtained from the 2000 Medical Expenditure Panel Survey (MEPS). Mortality ratios were obtained from the National Health Interview Surveys (NHIS) and linked to the National Death Index (NDI). The MEPS is a subset of the NHIS, providing a nationally representative sample of the adult non-institutionalized, civilian population. The researchers found that African-Americans had 67,000 more deaths annually than white Americans, translating to 2.2 million years of life lost (YLL). After adjusting for SES, this dropped to 1.1 million YLL. There were no significant differences in HRQL after adjusting for SES. However, some of the limitations of this study were the reliance on self-reported data, use of subjective measures that could vary depending on an individual's perception of quality of life measures, and a limitation on the range of variables that define SES.

Another study, using regression models, showed that after adjusting for SES, black Americans were still 50% more likely than white Americans to report poor health [11]. Using propensity models, the study showed that this dropped to an odds ratio of 1.3. All models were adjusted for gender, age, labor force status, marital status, health insurance status, and female-headed household. The SES measure in each of the models included educational attainment. When additional elements were added to the SES measure (wealth, long-term family income, and long-term neighborhood poverty), the odds ratio (95% confidence interval) of reporting poor health in the black population changed from 1.21 (1.00–1.46) in the regression model to 1.04 (0.80–1.35) in the propensity model. From these findings, the authors concluded that long-term measures of SES, such as long-term family income and long-term neighborhood poverty, could better explain the disparity in health outcomes than race. With fewer measures of SES, there was still a statistically significant difference in the reporting of poor health between African-Americans and white Americans. This study, however, again, relied on self-reported health outcome perceptions and created mathematical models to extrapolate the possibility of SES confounding the seemingly independent effect of race on health outcomes.

A third study looking at racial disparities in health among non-poor African-Americans and Hispanic Americans found that while Hispanic Americans, who were upwardly mobile in terms of SES, faced less discrimination as their income increased, African-Americans did not experience a similar drop in discrimination with upward financial mobility [12]. Colen et al. used data from two cohorts of the National Longitudinal Study of Youth: the original 1979 adult cohort and the young adult cohort. The first cohort consisted of men and women from ages of 14 to 21 years, and the second cohort consisted of men and women from the ages of 14 to 41 years. Both cohorts were from a nationally representative sample. Health was assessed using self-rated health. Experiences of discrimination were ascertained using the Major

Experiences of Discrimination Scale and the Everyday Discrimination Scale, which have been used widely in the literature. Again, one of the prominent limitations of this study was its reliance on self-reported health.

As a fourth example, in 2012, Duro et al. published an analysis of data from more than 4500 people from the third National Health and Nutrition Examination Survey (1988–1994) to look at racial disparities in mortality. One of the variables that they included was the allostatic load score, which refers "to the accumulation of physiological perturbations as a result of repeated or chronic stressors in daily life." The authors used 10 biomarkers (including metabolic, renal, and cardiovascular markers) in their analysis. The authors conclude that "(a)llostatic load burden partially explains higher mortality among blacks, independent of SES and health behaviors. These findings underscore the importance of chronic physiologic stressors as a negative influence on the health and lifespan of blacks in the United States" [13]. One limitation to this study is the use of biomarkers as a proxy for chronic physiologic stress, as there are many other factors that contribute to these biomarkers.

The degree to which any of the aforementioned findings can be attributed to structural racism is not well studied; however there is growing attention to better understand its role in health disparities. Refer to ▶ Chapter 6 for a related discussion on structural violence.

5. **The American Academy of Pediatrics (AAP) recommends screening for poverty. What was the rationale and what is the recommended screening question? What is the evidence supporting this approach?**
Research has shown that poverty is linked to many health problems in children. Persistent and severe poverty has been shown to lead to higher infant mortality, higher rates of asthma and obesity, and increased risk of injuries [14].

There are a number of services available to help families with children living in poverty such as the earned income tax credit; the child tax credit; Temporary Assistance for Needy Families (TANF); Medicaid and the Children's Health Insurance Program (CHIP); early childhood education; the Supplemental Assistance Program for Women, Infants, and Children (WIC); Supplemental Nutrition Assistance Program (SNAP, formerly Food Stamps); the National School Lunch Program; home visitation services; programs for early childhood education; and many others [15]. Early identification of children and their families who may need these services can facilitate preventive health services and support a healthy cognitive development [15].

The AAP policy statement "Poverty and Child Health in the United States" provides numerous recommendations, including screening for social determinants (based on the evidence that poverty impacts health and the proven interventions to mitigate against poverty). The policy reports that using a single question "Do you have difficulty making ends meet at the end of the month?" will allow a pediatrician to identify a family needing to be linked to services, with 98% sensitivity [15].

6. **What is an accountable care organization (ACO)? What is the Centers for Medicaid and Medicare Services' (CMS) "Accountable Health Communities" initiative?**
The CMS defines accountable care organizations (ACOs) as the voluntary coming together of a group of doctors, hospitals, and other health care providers to provide "coordinated high-quality care" to Medicare patients. The goal of this coordination is to decrease medical errors, to decrease duplication of medical services, and to provide patients with the appropriate care at the appropriate time [16]. While CMS defines ACOs as being unique to Medicare patients, the concept has been widely adopted by

commercial insurers. In 2017, there were close to 1000 commercial and public ACOs covering more than 32 million Americans [17].

An accountable health community, formed by the Innovation Center of CMS, is a model bridging community services with clinical services to identify Medicaid and Medicare beneficiaries who have social needs, such as unstable or nonexisting housing or food insecurity, which are not being met. Research has shown that these needs can increase the risk of developing chronic conditions, increase health care costs, and lead to avoidable health care utilization. As of May 2018, there were 31 organizations participating in this model [18]. The mechanism in which this model will work is described by the CMS as follows:
- Screening of community-dwelling beneficiaries to identify certain unmet health-related social needs
- Referral of community-dwelling beneficiaries to increase awareness of community services
- Provision of navigation services to assist high-risk community-dwelling beneficiaries with accessing community services
- Encouragement of alignment between clinical and community services to ensure that community services are available and responsive to the needs of community-dwelling beneficiaries

The Accountable Health Community Initiative of the CMS provides funding to organizations that want to implement this model. The model will be supported over a 5-year period and will test the effectiveness of various approaches to linking beneficiaries to resources for their unmet social needs. Each of the approaches will be implemented in the following way, described by CMS [18]:
- Identify and partner with clinical delivery sites (e.g., physician practices, behavioral health providers, clinics, hospitals) to conduct systematic health-related social needs screenings of all beneficiaries and make referrals to community services that may be able to address the identified health-related social needs
- Coordinate and connect beneficiaries to community service providers through community service navigation
- Align model partners to optimize community capacity to address health-related social needs

7. **Since the passage of the Affordable Care Act (ACA), what do current data show with respect to rates of uninsured individuals and with respect to those enrolled in Medicaid in the United States?**
The Affordable Care Act contained several key provisions aimed at reducing the number of uninsured. These included insurance reforms such as allowing young adults to remain on their parents' insurance up to age 26, the expansion of Medicaid eligibility to 138% of the Federal Poverty Level (FPL), creation of health insurance marketplaces through which individuals could purchase affordable insurance and receive premium subsidies, and the requirement that most individuals have health insurance.

A large portion of these provisions of the ACA went into effect in 2014. The number of nonelderly Americans who were uninsured fell from 44 million in 2013 to less than 28 million at the end of 2016, decreasing the rate of the uninsured by approximately 6.8 percentage points [19]. Based on information from the 2014, 2015, 2016, and 2017 Current Population Survey Annual Social and Economic Supplements and the American Community Survey, the U.S. Census Bureau estimates that the number of Americans who were enrolled in Medicaid increased from 54.9 million in 2013 to 62.3 million in 2016, increasing the rate of people covered by Medicaid by 1.9 percentage points [20]. These numbers show that the passage of the ACA allowed for more individuals to get health insurance coverage and thus overcome one of the financial barriers to accessing health care.

✓ 8. **Describe some of the myths about poverty in the United States today.**

> **Box 8.1   Teaching Tip**
> One narrative is that people in poverty are victims of a cruel system and generational poverty. Another narrative is that people in poverty have become dependent on the welfare system, consequently losing motivation to work and improve their situation. Students should recognize that no single narrative is accurate but that perceptions of social mobility are part of the American myth.

There are a number of myths about poverty in the United States today. Although they are fraught with inaccuracies or incomplete depictions of the situation, the miniscule amount of fact upon which they may have been based can perpetuate policy decisions about services for those living in poverty. One perspective of poverty in the United States is that systemic processes and generational practices created a hostile environment and victimized certain segments of the society into poverty. Such forces impair social mobility. Such a concept is contrary to the American idealized concept of the "rags to riches story" and the possibility that, with some luck and a lot of hard work, anybody with gumption can change his or her socioeconomic standing.

A study by Ferrie et al., which examined multigenerational socioeconomic mobility, found that the correlation in education over a number of generations was 20% higher than previously known. In other words, a person's social mobility, as measured by educational attainment, was more heavily influenced by their grandfather's or great-grandfather's educational attainment than previously thought. Unfortunately, this study did not look into what might be contributing to this limitation in mobility [21].

Another perspective is that by providing generous support, the U.S. government has contributed to people in poverty being dependent on government funding. Fortifying this myth is the unfounded notion that people living in poverty who are dependent on government services are not working, are not actively looking for jobs, or are not working hard enough at their jobs to make a livable income. The facts present a different picture, however. Close to one-third of the people on public assistance are actually from working families [22]. Additionally, many of the top 10 occupations in the United States—secretaries, retail sales, and cashiers—pay anywhere from $9.08 to $15.25 an hour, leaving these employees and their families well below the federal poverty line, despite holding steady jobs [23]. In 2015, 8.6 million people were among the working poor, and 4.6 million families were living below the federal poverty line despite having at least one member in the work force for at least half the year [24]. As these data demonstrate, no one conceptualization of the nature of poverty in the United States accurately describes the nature of the problem. Poverty is a multifaceted issue that has been correlated with lower access to health care and poorer health outcomes.

8.4   Responses to Discussion Questions

Mr. Messenger lives in a state that expanded Medicaid but he has not enrolled in Medicaid yet. He has a part-time minimum wage job that does not provide health insurance. His car recently broke down and he cannot afford to repair it. He has not been able to afford his medications, and he has stopped seeing his doctor because of the costs of co-pays and the difficulty with transportation.

✓ 1. **What does this mean for his ability to access health care?**
   Medicaid has traditionally covered low-income children, pregnant women, and parents/guardians of children, some elderly adults, and persons with disability. The Medicaid expansion allowed states that adopted it to provide coverage for most individuals based on income requirements alone (less than 138%

federal poverty level). With a part-time minimum wage job, it is likely that Mr. Messenger would meet his state's Medicaid income eligibility of 138% of the federal poverty level: $16,753 for a single person in 2018 [25].

Despite being eligible for Medicaid, Mr. Messenger might face barriers to enrollment, including lack of knowledge on how and where to apply, misunderstanding the eligibility rules, and difficulty completing the enrollment process, including the required documentation [26].

Even with Medicaid, Mr. Messenger might have difficulty finding a regular provider who accepts Medicaid. As a self-pay patient with multiple medical problems, he would typically need to come up with $250 or more for an outpatient visit. Some hospitals and clinics charge self-pay patients more because insurance companies have negotiating power while individual patients do not. On the other hand, some private physicians may discount self-pay patients slightly (charge them less) because they do not have to bill, wait for reimbursement, or contest denials of coverage. Some clinics geared toward indigent care, such as federally qualified health centers, have a sliding scale to bill patients based on income. However, such a clinic may not be available.

Lack of transportation is another factor that will limit Mr. Messenger's access to care since his car is not working. Mr. Messenger may also avoid following up with physicians if he is ashamed about his inability to comply with treatment.

Some people have argued that everyone has access because they can go to the emergency room, as this patient has. Should a visit to the emergency room be considered "access to health care"? On the one hand, he will likely receive care at a hospital ED because of the federal Emergency Medical Treatment and Labor Act (EMTALA), which requires all hospital EDs to medically screen every patient and, when indicated, stabilize or transfer the patient [27] (unlike a primary care office or urgent care clinic, where insurance is typically checked prior to the visit). On the other hand, he will likely get a bill that is much more expensive than he would have at a clinic or urgent care center. Furthermore, Mr. Messenger will also not have access to most recommended screenings and health maintenance services—things that might prevent his condition from worsening and that could lower his expenses in the long run.

(a) **What if Mr. Messenger lived in a state that did *not* expand Medicaid?**

In states that did not expand Medicaid, the income eligibility for Medicaid ranges from 18% to 105% of the FPL for parents, and only one state provides coverage for childless adults [28].

If Mr. Messenger was a parent caring for a child, he might qualify for Medicaid, depending on the state he resided in. For example, the states of Maine and Texas have not expanded Medicaid; in Maine, a 54-year-old man who is a parent would be eligible up to 105% of the FPL, but in Texas the same man would be eligible only up to 18% of FPL [28].

If he was childless, Mr. Messenger would not be eligible for Medicaid if his state did not expand Medicaid. He would most likely be out of luck for obtaining coverage since his employer does not provide coverage and insurance purchased through the ACA Marketplace is likely going to be too expensive for him. At this age, he would have to wait 11 years until he became eligible for Medicare—if he survives that long (a legitimate concern given the severity of his disease and inability to access consistent care).

The ACA created subsidies to help low- and moderate-income individuals cover costs of health insurance, particularly those individuals without access to employer-based insurance or Medicaid or Medicare insurance. Because the ACA subsidies were created with the assumption that Medicaid would expand, the subsidies are only available to those who make between 100% and 400% of the FPL. Thus, if Mr. Messenger earned less than 100% FPL in a state that did not expand Medicaid, he will fall into a "coverage gap," unable to get assistance through subsidies [29]. For adults with dependent children in a state that did not expand Medicaid, it is possible that they fall into a "coverage gap" if they have an income above the Medicaid eligibility limits but below the limit for Marketplace subsidies [29]. People in these situations cannot realistically afford health insurance on their own (considering a health plan could cost as much as his/her annual earnings).

2. **Why do nonprofit and teaching hospitals take all patients?**
   Nonprofit hospitals are exempt from federal, state, and local taxes and in exchange for this are expected to provide charity care. Since the passage of the ACA, nonprofit hospitals are required to provide and report on their services that benefit the community in order to maintain tax-exempt status. Many teaching hospitals are "safety net providers," which are providers of health care services to Medicaid, uninsured, and other vulnerable populations. These hospitals, either by mandate or mission, provide medical and enabling ("wrap around") services regardless of the patient's ability to pay.

   (a) **What are the legal reasons?**
   Claiming legal nonprofit status as a hospital exempts the institution from federal, state, and local taxes, which saves quite a bit of money. Since the introduction of Medicare and Medicaid in the 1960s, nonprofit hospitals have also been required to provide a "community benefit" consisting of the following requirements [30]:
   1. Operate a 24-hour emergency room
   2. Provide charity care to the extent of the hospital's financial ability
   3. Extend medical staff privileges to all qualified physicians in the area, consistent with the size and nature of the facility
   4. Accept payment from Medicare and Medicaid programs on a nondiscriminatory basis
   5. Maintain a community-controlled board (i.e., a governing board with membership, by appointment, primarily from the local community)

   These measures of community benefit have always been controversial and are probably not adequate for the current state of our medical system. In many states, there are additional requirements or waivers to these criteria. In addition, the ACA added the following criteria [31]:
   - Conducting a community health needs assessment with an accompanying implementation strategy
   - Establishing a written financial assistance policy for medically necessary and emergency care
   - Complying with specified limitations on hospital charges for those eligible for financial assistance
   - Complying with specified billing and collections requirements

   As seen in Discussion Question 1, all hospitals (profit and nonprofit) have a legal obligation to provide emergency care under EMTALA legislation. The hospital must stabilize any patient with life-threatening illness, regardless of ability to pay. They may then transfer to an appropriate facility. The upshot is that no hospital can let someone die on their front steps because they do not have insurance.

   (b) **How is this financially feasible?**
   Medicaid is the largest funder of uncompensated care. Hospitals

receive additional funding from states to offset some of their uncompensated care costs, which include the difference between Medicaid payments and cost of providing services to the Medicaid population, plus the cost of care for uninsured patients. By law, state Medicaid programs make disproportionate-share hospital (DSH) payments to hospitals with a large proportion of Medicaid and low-income patients. These DSH payments vary by state and are limited by federal allotments. Approximately half of all hospitals in the United States receive DSH payments. However, with the implementation of the ACA and the anticipated reduction in the number of uninsured, federal DSH allotments will be reduced, resulting in a reduction in DSH payments [32]. Other sources of uncompensated care funding include state and local appropriations for indigent health programs and services, private sector funding, the Veterans Administration, community health centers, and Medicare [33].

Teaching hospitals also receive funds from Medicare to help pay for the costs of graduate medical education (i.e., residency). Medicare pays a percentage extra based on the number of residents per hospital bed that are training in the hospital. This incentivizes a hospital to train medical residents, which is an added cost to the hospital. It should be noted that Medicare, on average, pays only about $0.87 on the dollar for what the hospital services cost. Hospitals make up the difference with payments from private insurance. This is part of the reason why an incentive for more than Medicare typically pays would be appealing to a hospital. It would limit the losses from Medicare patients [34].

Until a couple of months ago, Mr. Messenger worked at a large brewery as a mechanical technician. Through that job, he was able to get employer-sponsored insurance. However, he has had gradually worsening intermittent foot pain over the past year, and there were days that he could not go to work because of the pain. He quickly exhausted his (unpaid) sick leave and his manager felt he was unreliable. When his car broke down, he missed work 3 days in a row, and he was let go. Since then, he has held several consecutive part-time jobs, but none has provided health insurance, and he has not been able to keep one position long enough to accumulate sick leave.

✅ 3. **Poverty can clearly impact health outcomes, but how can poor health impact a patient's financial situation? How so?**
Poverty is both a cause and consequence of poor health. Patients who are poor are more likely to postpone seeking costly care until they are sicker. They are less likely to have benefits through their employers, such as generous paid sick leave policies or flexible workplace environments. Patients who are in poor health are more likely to be less productive and to lose their jobs, which may result in a loss of income and other assets, and ultimately poverty. In addition, as a result of their poor health, they are likely to face increased costs, both direct health-related costs such as co-pays, deductibles, and medications even if they are insured, and indirect costs such as increased transportation-associated costs and childcare.

Patients in poverty are often operating at the brink of disaster. That is, they are "making ends meets" most of the time, but a small increase in expenses, like a car breaking down (or a major medical bill), will put them over the financial edge. For upper-middle-class professionals who hold salaried employment with generous sick leave (as most health care professionals do), a car breakdown or medical issue is much less likely to affect employment status. This is not true for less well-off individuals.

✅ 4. **Self-reflection: In thinking about CMS's Accountable Health Communities, how do you feel about physicians being held accountable for their patients' transportation, housing, food security, etc.?**

✅ 5. **What can you do to help Mr. Messenger get his medications?**
This physician, like all health care practitioners, is part of a health care team. The

first thing to do is to enlist the assistance of a social worker or case manager. Most hospitals make sure that this happens for uninsured patients staying in the hospital. A social worker or case manager will help him sign up for insurance, if at all possible (Medicaid or Marketplace plan if eligible and affordable). Doing so will help him pay for both the hospital admission (if needed) and medications for discharge.

In some states, such as New York, Medicaid enrolls some patients in the Health Home program. This program provides a "care manager" who helps the patient find medication, get to appointments, and access other pertinent support services. To qualify, patients must be eligible and active in Medicaid and must have two identified chronic conditions or human immunodeficiency virus (HIV) infection or serious mental illness. In addition, patients must be shown to have some risk factors that indicate that they will likely receive benefit from the program [35].

The social worker also may be able to connect Mr. Messenger to local programs to offset costs. Some states and many health systems have "patient assistance programs" to address unmet needs. Many pharmaceutical companies offer financial assistance, and some community-based organizations might provide such support.

Mr. Messenger mentions to you that he was in a different ED about 6 months ago with similar pain in his foot. He had a lot of pain at the time but was not offered any pain medication or symptomatic relief, so he decided to leave. The following day, when the pain became unbearable, he went back to the ED. He was admitted to the hospital. You note that his medical records show that he had a deep abscess that formed in his lower leg, and he required 10 days of intravenous antibiotics.

6. **Creative problem-solving: It is not unusual for patients' accounts to differ in significant ways from the written medical record. How would you "fill in the gaps" between Mr. Messenger's account and the notes you are able to see? What do you think happened?**

> **Box 8.2 Teaching Tip**
> Encourage learners to discuss the possibility that Mr. Messenger was undertreated during his first ED visit and to recognize that both the patient's and the provider's accounts may be accurate; people can interpret the same events differently.

7. **What does implicit bias mean?**
Implicit biases refer to attitudes or stereotypes that affect people's understanding, decisions, and actions without conscious knowledge. They lead to an assumption, usually negative, about an individual based on characteristics that are not relevant to the decision-making process, e.g., race, ethnicity, gender, and sexual orientation. Although many people may consciously reject negative assumption of these groups, they are constantly exposed to negative and stereotypic depictions of these groups and may not be aware of their unconscious biases.

Implicit bias is of great concern to health care practitioners who should provide impartial care to their patients. For example, a provider may unconsciously perceive an ethnic minority patient as less competent and decide against prescribing a particular treatment regimen. Some of the groups affected by implicit bias may already be vulnerable and disadvantaged on many different levels: as examples, patients who are impoverished, of a racial or ethnic minority, transgender, lesbian, or gay; patients who have a mental health problem or who are disabled; and patients who do not speak English proficiently or who are refugees. These are just some of the examples. Implicit bias is of particular concern when it affects the quality of health care and contributes to suboptimal health outcomes. In fact, implicit race bias has been identified as one possible explanation for the racial health care disparities [36].

> **Box 8.3 Teaching Tip**
> Ask other learners if they agree or disagree with the student who first responds.

(a) **Self-reflection: What does it mean to you as a future physician?**

> **Box 8.4   Teaching Tip**
> Consider having learners take an implicit association test before or during the discussion, and discuss their results.

(b) **How do you think implicit bias might have impacted Mr. Messenger's care in the ED?**
FitzGerald and Hurst conducted a recent systematic review of implicit bias in health care professionals and reported that most studies found evidence for the same levels of implicit biases among physicians and nurses as in the general population. These biases are correlated to the characteristics of these providers, e.g., race, gender, years of practice, place of medical training, and type of health care setting. Patient characteristics that were associated with implicit bias included age, gender, race/ethnicity, socioeconomic status, disability, weight, mental illness, having acquired immunodeficiency syndrome (AIDS), and intravenous drug use. Implicit bias is a complex phenomenon as patients may possess two or more of these characteristics. These biases manifest themselves through the number of questions asked of the patient, diagnoses, number of tests ordered, and treatment recommendations [36].

Based on his socioeconomic status, age, race, ethnicity, and weight, Mr. Messenger may have been asked fewer questions, had fewer tests ordered, been diagnosed with cellulitis and diabetes later in his medical assessment, received inferior pain management, and been less likely to be referred to a specialist than patients with more "favorable" characteristics.

(c) **What can we do to address our own biases in order to provide the best possible care for our patients?**
Implicit bias can be reduced through increased awareness and taking actions to change our responses. The first step is recognition of our own biases. A common tool is Harvard's Project Implicit®. If health care practitioners recognize implicit bias and are motivated to change, they can employ different strategies to reduce or eliminate those biases. One important method is to have frequent human interactions (such as thoughtful doctor-patient interactions or, better yet, frequent "every day" interaction) with people who are different from us. Over time, these interactions with people decrease bias. Other strategies to reduce implicit bias include consciously adjusting a stereotypic response, imagining the patient as opposite of the stereotype, focusing on the patient as an individual rather than a stereotype, putting yourself in the patient's shoes, and viewing the patient as an equal [37].

You determine that Mr. Messenger needs admission for treatment of his cellulitis. Upon learning this, Mr. Messenger becomes distraught, expressing concern that an admission will jeopardize his current job.

✓ 8. **Self-reflection: Before entering the health professions, how did you think about poverty? Have your experiences as a health care professional or future health care professional changed your perspective? How so?**

✓ 9. **How can you, as future health care practitioners, help to improve the lives of those in poverty?**
Some possibilities that future health care practitioners can employ include:
- Working in a medically underserved area and/or accepting all patients, including patients with Medicaid
- Offering some nontraditional office hours that allow patients who cannot take time away from work during the day to seek care
- Screening patients for poverty, food insecurity, housing insecurity, etc. and appropriately referring them to resources in the community (government programs or charitable organizations)
- Recognizing the role that poverty might play in undesirable health

behaviors (such as nonadherence to treatment or unhealthy lifestyle) and attempting to address the root causes of those behaviors, rather than assuming patients are not interested or willing to improve their own health

- Advocating for local, state, or national laws, policies, and programs that directly address poverty
- Educating future students and trainees to understand the relationship between poverty and health [38].

> **Box 8.5 Patient Perspective**
>
> "My name is Tianna Gaines-Turner. I am married to the father of our 3 biological children… I also have 3 teenage stepchildren, whom I love dearly…I work part time for a childcare provider at a recreation center making about $10 dollars an hour and my husband works behind the deli counter at a grocery store making $8 dollars an hour. We haven't been able to find full-time jobs. With the part-time jobs, our incomes go up and down. Not only do we have incomes that are inadequate, but they are also unstable and unpredictable. When programs like SNAP (food stamps) rely on stable income reports, it makes it harder to keep this nutrition support steady. So we may lose food stamps one month because we make too much, and then a few months later, when our companies choose to reduce our hours at their own convenience, we make less money and we need to turn to food stamps again to feed our kids healthy meals. …My 3 children have medical issues that concern me every day. All of my children suffer from epilepsy and have to take life-saving medication every day. All 3 of my children also have moderate to severe asthma. I worry about a day that might come where my children won't be able to see a specialist because I can't afford the co-pay. In addition, neither my husband nor I qualify for medical assistance because we make too much money in our part-time jobs, but our jobs do not offer health insurance. I, too, suffer from asthma and epilepsy. I currently can't afford to get an inhaler. The thought that my own children may not be able to get the care they need worries me every day. …There have been times when my oldest son was sick or having seizures, and of course, as would you, I wanted to be at the hospital to care for him and help him get the care he needed, and to comfort him. But that meant my husband had to stay home to take care of the twins. And then we were both unable to work, so we lost money that month, and ultimately had to make a choice—do we pay the rent or do we pay the light bill? Not to mention, how do we buy food? No family should have to choose between paying a bill or putting food on the table. …We are whole human beings. Poverty has to do with a whole person who is in a family, in a neighborhood, in a community, and our country. The policies you work on in the House of Representatives affect me and my family in very deep and important ways. I am living out your policy-making and I see how your decisions affect both physical and mental health, especially the physical and mental health of my children and me."
>
> Testimony of Tianna Gaines-Turner Chair, Witnesses to Hunger before the Committee on the Budget United States House of Representatives Hearing The War on Poverty: A Progress Report July 31, 2013 [38].

References

1. Carrillo JE, Carrillo VA, Perez HR, Salas-Lopez D, Natale-Pereira A, Byron AT. Defining and targeting health care access barriers. J Health Care Poor Underserved. 2011;22(2):562–75.
2. Saloner B, Polsky D, Kenney GM, Hempstead K, Rhodes KV. Most uninsured adults could schedule primary care appointments before the ACA, but average price was $160. Health Aff. 2015;34(5):773–80.
3. Hing E, Decker SL, Jamoom E. Acceptance of new patients with public and private insurance by office-based physicians: United States, 2013. Atlanta (GA): Centers for Disease Control and Prevention. NCHS Data Brief. 2015;195:1–8. [cited 2018 July 23]. Available from: https://www.cdc.gov/nchs/data/databriefs/db195.pdf.
4. Padela AI, Gunter K, Killawi A, Heisler M. Religious values and healthcare accommodations: voices from the American Muslim community. J Gen Intern Med. 2012;27(6):707–15.
5. CDC National Center for Health Statistics. Mortality in the United States. CHS Data Brief No. 293. Atlanta (GA): CDC; December 2017. [cited 2018 Aug 5]. Available from: https://www.cdc.gov/nchs/products/databriefs/db293.htm.
6. Chetty R, Stepner M, Abraham S, Lin S, Scuderi B, Turner N, et al. The association between income and life expectancy in the United States, 2001-2014. JAMA. 2016;315(16):1750–66.

7. U.S. Department of Health and Human Services. CDC National Center for Health Statistics Health, United States, 2016 Report. Atlanta (GA): CDC; 2016 [cited 2018 Jul 24]. Available from: https://www.cdc.gov/nchs/data/hus/hus16.pdf#015.
8. CDC National Center for Health Statistics. FastStats – diseases and conditions [Internet]. Atlanta (GA): CDC; 2016 July 6 [cited 2018 July 24]. Available from: https://www.cdc.gov/nchs/fastats/diseases-and-conditions.htm.
9. Williams DR. Why discrimination is a health issue. Robert Wood Johnson Foundation Culture of Health Blog [Internet]. Princeton (NJ): Robert Wood Johnson Foundation; 2017 Oct [cited 2018 Aug 5]. Available from: https://www.rwjf.org/en/blog/2017/10/discrimination-is-a-health-issue.html.
10. Franks P, Muennig P, Lubetkin E, Jia H. The burden of disease associated with being African-American in the United States and the contribution of socio-economic status. Soc Sci Med. 2006;62(10):2469–78.
11. Do DP, Frank R, Finch BK. Does SES explain more of the black/white health gap than we thought? Revisiting our approach toward understanding racial disparities in health. Soc Sci Med. 2012;74(9):1385–93.
12. Colen CG, Ramey DM, Cooksey EC, Williams DR. Racial disparities in health among nonpoor African-Americans and Hispanics: the role of acute and chronic discrimination. Soc Sci Med. 2018;199:167–80.
13. Duru OK, Harawa NT, Kermah D, Norris KC. Allostatic load burden and racial disparities in mortality. J Natl Med Assoc. 2012;104(1–2):89–95.
14. Poverty and Child Health in the United States. Pediatrics. 2016;137(4):e20160339.
15. Council On Community Pediatrics COC. Poverty and child health in the United States. Pediatrics. 2016;137(4):e20160339.
16. Centers for Medicare and Medicaid Services. Accountable Care Organizations (ACOs) [Internet]. Baltimore (MD): CMS; 2018 May 3 [cited 2018 July 25]. Available from: https://www.cms.gov/Medicare/Medicare-Fee-for-Service Payment/ACO/.
17. Pierce-Wrobel C, Micklos J. How the most successful ACOs act as factories of innovation. Health Affairs Blog [Internet]. 2018 Jan 29 [cited 2018 July 25]. Available from: https://www.healthaffairs.org/do/10.1377/hblog20180124.514403/full/.
18. Alley DE, Asomugha CN, Conway PH, Sanghavi DM. Accountable health communities — addressing social needs through Medicare and Medicaid. N Engl J Med. 2016;374(1):8–11.
19. The Henry J. Kaiser Family Foundation. Key facts about the uninsured population [Internet]. San Francisco (CA): Kaiser Family Foundation; 2017 Nov 29 [cited 2018 Jul 25. Available from: https://www.kff.org/uninsured/fact-sheet/key-facts-about-the-uninsured-population/.
20. Barnett JC, Berchick ER. Health insurance coverage in the United States: 2016. United States Census Bureau. 2017 Sep [cited 2018 Jul 25]. Available from: https://www.census.gov/library/publications/2017/demo/p60-260.html.
21. Ferrie J, Massey C, Rothbaum J. Do grandparents and great-grandparents matter?: multigenerational mobility in the US, 1910–2013. Cambridge (MA): NBER; 2016 Sep [cited 2018 Jul 25]. Available from: http://www.nber.org/papers/w22635.pdf.
22. Gladstone B. #2: Who deserves to be poor? On The Media [web streaming audio]. WNYC Studios; 2016 Aug 7 [cited 2018 Jul 25]. Available from: https://www.wnycstudios.org/story/who-deserves-to-be-poor.
23. Babic M. 5 myths about the working poor in America [Internet]. Oxfam America; 2016 Sep [cited 2018 July 25]. Available from: https://politicsofpoverty.oxfamamerica.org/2016/09/5-myths-about-the-working-poor-in-america/.
24. U.S. Bureau of Labor Statistics. BLS Reports. A Profile of the Working Poor, 2015 [Internet]. Washington, D.C.; 2017 Apr [cited 2018 Jul 25]. Available from: https://www.bls.gov/opub/reports/working-poor/2015/home.htm.
25. Centers for Medicare and Medicaid Services. Federal Policy Level (FPL) [Internet]. Baltimore (MD): CMS. [cited 2018 Aug 1]. Available from: https://www.healthcare.gov/glossary/federal-poverty-level-fpl/.
26. Kaiser Commission on Medicaid and the Uninsured. Issue brief, key lessons from medicaid and chip for outreach and enrollment under the affordable care act: getting into gear for 2014. San Francisco (CA): Kaiser Family Foundation; 2013 [cited 2018 Aug 1]. Available from: https://kaiserfamilyfoundation.files.wordpress.com/2013/06/8445-key-lessons-from-medicaid-and-chip.pdf.
27. Centers for Medicare and Medicaid Services. Emergency Medical Treatment and Labor Act. Baltimore (MD): CMS; 2012 Mar 26 [cited 2018 Aug 5]. Available from: https://www.cms.gov/Regulations-and-Guidance/Legislation/EMTALA/.
28. The Henry J. Kaiser Family Foundation. Where are the states today? Medicaid and CHIP eligibility levels for children, pregnant women and adults [Internet]. Kaiser Family Foundation: San Francisco (CA); 2018 Mar 28 [cited 2018 Aug 1]. Available from: https://www.kff.org/medicaid/fact-sheet/where-are-states-today-medicaid-and-chip/.
29. Garfield R, Damico A, Orgera K. The coverage gap: uninsured poor adults in states that do not expand medicaid [Internet]. San Francisco: The Henry J. Kaiser Family Foundation; 2018 Jul 12 [cited 2018 Aug 2]. Available from: https://www.kff.org/medicaid/issue-brief/the-coverage-gap-uninsured-poor-adults-in-states-that-do-not-expand-medicaid/.
30. Rubin DB, Singh SR, Young GJ. Tax-exempt hospitals and community benefit: new directions in policy and practice. Annu Rev Public Health. 2015;36:545–57.
31. Robert Wood Johnson Foundation. Health policy brief: nonprofit hospitals' community benefit requirements. Health Affairs [Internet]. Princeton (NJ); 2016 Feb 25 [cited 2018 Aug 5]. Available from: https://www.rwjf.org/content/dam/farm/reports/issue_briefs/2016/rwjf426962.
32. Medicaid and CHIP Payment and Assessment Commission. Report to Congress on Medicaid and CHIP. Overview of Medicaid policy on disproportionate share hospital payments. Washington, D.C.: MACPAC; 2016 [cited 2018 Aug 2]. Available from:

33. https://www.macpac.gov/wp-content/uploads/2016/03/Overview-of-Medicaid-Policy-on-Disproportionate-Share-Hospital-Payments.pdf.
33. Coughlin TA, Holahan J, Caswell K, McGrath M. Uncompensated care for the uninsured in 2013: a detailed examination [Internet]. San Francisco (CA): Kaiser Family Foundation; 2014 May 30 [cited 2018 Aug 2]. Available from: https://www.kff.org/report-section/uncompensated-care-for-the-uninsured-in-2013-a-detailed-examination-sources-of-funding-for-uncompensated-care/.
34. American Hospital Association. Trends in hospital financing. Trendwatch chartbook [Internet]. Dallas (TX): AHA; 2018 [cited 2018 Aug 2]. Available from: https://www.aha.org/system/files/2018-05/2018-chartbook-table-4-4.pdf.
35. New York State Department of Health. Eligibility requirements: identifying potential members for health home services [Internet]. Albany (NY): NYSDOH; 2017 Mar 2 [cited 2018 Aug 5]. Available from: https://www.health.ny.gov/health_care/medicaid/program/medicaid:health_homes/docs/health_home_chronic_conditions.pdf.
36. FitzGerald C, Hurst S. Implicit Bias in healthcare professionals: a systematic review. BMC Med Ethics. 2017;18(19):1–18.
37. Institute for Healthcare Improvement. How to rieduce implicit bias? [Internet]. Boston (MA): IHI; 2017 Sep 28 [cited 2018 Aug 2]. Available from: http://www.ihi.org/communities/blogs/how-to-reduce-implicit-bias.
38. U.S. Government Printing Office. The war on poverty: a progress report hearing before the committee on the budget house of representatives one hundred thirteenth congress first session hearing held in Washington, DC, July 31, 2013. Prepared statement of Tianna Gaines-Turner, p. 64. [cited 2018 Aug 5]. Available from: https://www.gpo.gov/fdsys/pkg/CHRG-113hhrg81981/pdf/CHRG-113hhrg81981.pdf.

Further Reading on this Topic

Coates T. Between the world and me. New York: Spiegel & Grau; 2015.

Hoffman KM, Trawalter S, Axt JR, Oliver MN. Racial bias in pain assessment and treatment recommendations, and false beliefs about biological differences between blacks and whites. Proceedings of the National Academy of Sciences. 2016:201516047.

Onie R. TED talk: What if our healthcare system kept us healthy? [web streaming video]. TEDMED; 2012 Apr. Available from: https://www.ted.com/talks/rebecca_onie_what_if_our_healthcare_system_kept_us_healthy?language=en.

Tweedy D. Black man in a white coat: a doctor's reflections on race and medicine. New York: Picador; 2016.

Decision-Making for Children and Adolescents

Contents

Chapter 9 "I Don't Want My Child to Get Vaccines" – 165

Chapter 10 "Our Son's Cancer Is Gone. Why Can't We Stop Treatment?" – 185

"I Don't Want My Child to Get Vaccines"

Manika Suryadevara and Joseph B. Domachowske

9.1 Background Questions – 166

9.2 Additional Case Information and Questions for Discussion – 166

9.3 Answers to Background Questions – 168

9.4 Responses to Discussion Questions – 173

References – 183

© Springer Nature Switzerland AG 2019
A. E. Caruso Brown et al. (eds.), *Bioethics, Public Health, and the Social Sciences for the Medical Professions*, https://doi.org/10.1007/978-3-030-03544-0_9

Learning Objectives
1. Apply the best interest standard and the harm principle, two approaches to surrogate decision-making for children, to vaccination.
2. Analyze relationships between medicine, ethics, law, and culture, using vaccination as a case study.
3. Explore ethical, social, and public health issues that are unique to certain developmental periods, such as infancy and adolescence.
4. Describe healthcare practitioners' legal and ethical responsibilities when patients and families decline to follow medical advice.

> **Case Background**
>
> *You are a primary care pediatrician practicing in an affluent, highly educated community with a moderately high rate of vaccine refusal. Your practice has traditionally accepted new patients regardless of their vaccination philosophies, preferring to establish relationships with vaccine-hesitant parents in a concerted effort to build rapport and provide education on the benefits of vaccination over time.*

9.1 Background Questions

1. What is the process in your state for a family to obtain a medical exemption from vaccination for their child? If your state permits religious or personal belief exemptions (all except three states), what is the process for a family to obtain such an exemption from vaccination?

2. Is there evidence that state or local policies on school vaccine requirements impact vaccination rates? Do they impact rates of vaccine-preventable diseases?

3. Describe the spectrum of attitudes toward vaccination. Define and describe vaccine hesitancy. In the USA, which groups of children are more likely to be unvaccinated? Which groups are more likely to be under-vaccinated?

4. Are healthcare practitioners or medical practices allowed to dismiss or to refuse to accept families who choose not to immunize their children? What guidance do professional groups, such as the American Academy of Pediatrics, provide regarding this?

5. What communication approaches can healthcare practitioners and medical practices take when discussing vaccines with families? Which approaches are most effective in achieving vaccine acceptance?

6. What is the Vaccine Adverse Event Reporting System (VAERS)? How does it work? What other vaccine safety policies or surveillance processes are in place in the USA?

7. Why was the original rotavirus vaccine (1998) withdrawn from the market less than a year after its licensure by the U.S. Food and Drug Administration (FDA)? Identify another historical event important to public perceptions of vaccines. (Do not use the events associated with Andrew Wakefield referred to in Background Question 8.)

8. Describe the circumstances regarding Andrew Wakefield's research and the reasons for its retraction.

9.2 Additional Case Information and Questions for Discussion

Recently, there has been an outbreak of measles in a neighboring county. You are now concerned because you have two patients who are currently undergoing chemotherapy for leukemia and another with an immunodeficiency who cannot receive the live viral measles vaccine. Your immunocompromised patients have medical contraindications to receiving measles vaccine, are at greater risk of contracting measles, and are more likely than others to develop severe infection or complications. You discuss this concern with your partners in the practice.

1. **Self-reflection: In this situation, how do you feel about accepting families who refuse vaccinations, knowing there are at-risk children who are unable to be immunized in the practice?**

"I Don't Want My Child to Get Vaccines"

2. What are the potential benefits and risks of refusing to care for families who do not immunize their children? Consider ethical, legal, and public health implications.
   (a) What are the risks to the patients themselves?
   (b) What are the risks to the public?
   (c) What might happen if all healthcare practitioners in the area refused to accept unvaccinated patients?

3. Is the potential impact of vaccine refusal the same for all vaccines? How do vaccines differ with regard to impact of vaccine refusal?

4. Is there an ethical difference between a religious exemption and a personal belief exemption? If so, what is the difference?
   (a) Should there be a legal difference? Why or why not?

Your partners meet to discuss the benefits and risks of caring for families who do not immunize their children. After intense debate, the decision is made by all partners to no longer accept families who refuse to immunize their children. You then ask what the practice should do about parents who refuse vaccines but are already in the practice or parents who have accepted some vaccines but refuse others.

5. If you opt to implement a policy to not accept patients who refuse immunizations, what do you do with families who refuse vaccines but are already established in your practice?
   (a) What are your legal and ethical commitments to these families? Can you dismiss them from the practice? If so, how do you do that?

6. Self-reflection: How do you feel about vaccines personally? Have you experienced any hesitancy over the decision to consent to vaccines for yourself (such as the influenza vaccine)?
   (a) Would you consider participating in a vaccine trial yourself? Why or why not?
   (b) If you have or plan to have children, do you think you will choose to vaccinate your children with all recommended vaccines? Do you feel any hesitancy over this decision?

7. In some institutions, healthcare professionals are required to either be immunized against influenza or wear a mask during influenza season. What is the ethical justification for such policies?

Two days after your meeting, you are seeing patients on a typically busy afternoon. Your first patient is a 13-year-old girl with a sore throat. You know her well and recall that the family has refused the human papillomavirus (HPV) vaccine in the past, expressing concerns about "a new vaccine that is meant to protect against a sexually transmitted infection." The parents have stated that their daughter is not having sex so is not at risk for acquiring infection. The mother did state at the last visit that she would consider the vaccine for her daughter in the future, "when she was older."

8. Should you continue to recommend that she receives the HPV vaccine at each visit? Why or why not?

9. Should HPV vaccination be required for entry into middle school? Why or why not?

Your next patients are a 2-year-old girl and her 6-month-old brother, both of whom are in your office for routine visits. The children are receiving vaccines according to an alternative schedule requested by their parents. During the visit, the parents mention that they have friends who are expecting a child soon and wanted to bring the baby to your practice. However, they have heard this will not be possible under the practice's new policy. They are disappointed. They feel that you have both "learned from each other" since their daughter's birth. You do not know if that is how you would describe your relationship: Instead, you feel that you have reluctantly accepted their alternative schedule because you worry that they would refuse all vaccinations if you did not.

10. Is it possible to have trust between a family and a healthcare practitioner when they disagree about something so fundamental?

11. **Strongly recommending vaccines, in a presumptive manner, is known to increase vaccine uptake. How do you balance the potential positive public health impact of this approach with concerns of paternalism and infringement upon patient autonomy or parental authority?**

12. **Reflecting on the story of Andrew Wakefield, how do you think the dissemination of misinformation affects vaccine confidence within the community?**
    (a) How do you think real safety concerns have affected public trust in vaccines?
    (b) Creative Problem-Solving: A recent study in the journal *Science* demonstrated that false information posted to social media spread much more quickly and widely than true information, a finding that applied to science news as well as other types of stories. As a future healthcare leader, what would you propose to combat scientific misinformation?

9.3 Answers to Background Questions

1. **What is the process in your state for a family to obtain a medical exemption from vaccination for their child? If your state permits religious or personal belief exemptions, what is the process for a family to obtain such an exemption from vaccination?**
   Medical exemptions from vaccinations mandated for school entry are allowed in all 50 states. The process for obtaining this exemption, however, varies by location. State-specific criteria for receiving a medical exemption can include a combination of the following: a written statement from a physician; a completed state-specific medical exemption form with or without notarization; approvals by the state's health department, with or without a requirement for an annual review; and reapproval of the medical exemption [1].

   In New York State, for example, to obtain a medical exemption from school vaccination, a state-licensed physician must complete a medical exemption form certifying which vaccine is contraindicated, the reason for the contraindication, and the duration of the medical exemption. This form is then presented to the school, followed by notification of exemption approval or denial provided in writing to the parents [2]. The process is somewhat different in New Mexico, where obtaining a medical exemption requires that a licensed physician complete a medical exemption form attesting that the required vaccines would be harmful to the child's life. The original form must be notarized and filed with the New Mexico Department of Health, which then notifies the parent or guardian of the request's approval or denial [3].

   Currently, all but three states (California, Mississippi, and West Virginia) allow religious exemptions to vaccinations mandated for school entry. In many other states, obtaining such an exemption is a simple process: A parent or guardian need only to sign and return a letter to the school attesting that it is the family's religious belief that the child should not be immunized. However, the process in other states is more labor-intensive. In New York, parents or guardians seeking religious exemptions must complete and submit to the school a form providing a detailed explanation of the reasons for the exemption, including the specific religious principles that guide their objections to vaccination and the religious basis for proscribing any specific immunizations. Furthermore, if the school has additional questions regarding the existence of the religious belief, the school principal may request supporting documents, such as a letter from an authorized representative or literature materials of the religious affiliation [2]. In almost all states, medical exemptions are decided and overseen by healthcare practitioners, while religious exemptions are under the jurisdiction of the Department of Education.

2. **Is there evidence that state or local policies on school vaccine requirements impact vaccination rates? Do they impact rates of vaccine-preventable diseases?**
   Laws regarding immunization mandates in the USA date back to the nineteenth century when the State of Massachusetts (1809) passed the first such law, which required smallpox vaccination. In 1922, the U.S. Supreme Court affirmed the rights of states to pass laws requiring vaccinations prior to school entry (*Zucht v. King*). In the 1960s, school vaccine laws were geared more for the control of disease outbreaks, particularly measles, because schools remained the primary site for disease transmission. As evidence grew that schools in states with enforcement of strict immunization laws had lower rates of measles infection, the focus shifted from enforcing immunization mandates during emergency outbreak situations to preventing disease. This ultimately led to public health codes that mandate vaccinations prior to school entry for children in all states [4].

   Immunization mandates for school entry or ongoing attendance are consistently associated with increases in vaccination coverage for both elementary and middle school children [5–9]. For example, one study showed that within 1 year of enacting a law to mandate meningococcal and hepatitis B vaccines for entry into seventh grade, adolescent vaccine completion rates in San Diego, CA, increased from 13% to 67%. For students who were not fully vaccinated, 90% were found to be in the process of completing the vaccine series [8]. Strictly enforced state laws requiring vaccinations prior to school entry result in both higher vaccination coverage rates in the community and in a reduction in the incidence of vaccine-preventable diseases, best illustrated for measles and pertussis [10–13].

   In the last decade or so, there has been a steady increase in nonmedical (religious or personal belief) vaccine exemptions among the families of school-aged children around the country [14, 15]. Exemptions from mandated vaccinations leave individuals at an increased risk for contracting disease and the community at an increased risk for a disease outbreak, even among fairly well-vaccinated populations [13, 16]. For example, unvaccinated children in Colorado were 22 times as likely to contract measles and 6 times as likely to acquire pertussis than fully immunized children [13]. During the 2010 pertussis outbreak in California, pertussis cases occurred in higher numbers within geographic areas with the highest vaccine exemption rates [12]. Similarly, investigations of measles outbreaks have found that more than half of measles cases occurred in individuals who were unimmunized, the majority of whom had claimed a nonmedical exemption to vaccination [11].

3. **Describe the spectrum of attitudes toward vaccination. Define and describe vaccine hesitancy. In the USA, which groups of children are more likely to be unvaccinated? Which are more likely to be under-vaccinated?**
   The Strategic Advisory Group of Experts (SAGE) on Immunizations Working Group on Vaccine Hesitancy defines vaccine hesitancy as the "delay in acceptance or refusal of vaccination despite the availability of vaccination services. Vaccine hesitancy is complex and context specific, varying across time, place, and vaccines. It is influenced by factors such as complacency, convenience, and confidence" [17]. The vaccine attitude spectrum spans from complete rejection of all vaccines (vaccine refusers) to active acceptance of all vaccines (vaccine acceptors). Vaccine hesitancy comprises the group in between, including those who accept vaccines but with concern, those who accept some vaccines and delay or refuse others, and those who accept vaccines but elect to receive them on an alternative schedule (ones not supported by evidence-based guidelines).

   Under-immunization refers to patients who have not received all the recommended vaccines for age for any reason, including reasons of vaccine hesitancy as described above. Under-immunization also occurs for a variety of social factors that interfere with access to health care independent of concerns over vaccine hesitancy. Clustering of under-immunized and vaccine refusers has been well-described, for a variety of reasons including local community culture, healthcare practitioner attitudes, and school policies [18]. Furthermore, the consequences of low vaccination coverage rates are more significant within these clusters than outside of them [18]. When compared to under-immunized children, unvaccinated children were more likely to be white or male; to have married, college-educated mothers; and to be living in households with higher incomes or with four or more children. Furthermore, unvaccinated children are more likely to have parents or guardians who deliberately refuse vaccines, while under-immunized children tend to be vaccine delayed due to healthcare system or sociodemographic factors [13, 19].

4. **Are healthcare practitioners or medical practices allowed to dismiss or to refuse to accept families who choose not to immunize their children? What guidance do professional groups, such as the American Academy of Pediatrics, provide regarding this?**
Physicians and other independently practicing healthcare practitioners have the legal right to dismiss or to refuse to accept families who will not vaccinate their children. If no physician-patient relationship has yet been established, they can choose to accept or refuse to care for patients for any reason that does not involve discrimination by gender, race, ethnicity, religion, age, or sexual orientation. After the establishment of the physician-patient relationship, dismissal must occur in concordance with state laws regarding patient abandonment, for example with official written notification of the patient or parent, delivery of information to find a new healthcare practitioner, and the continuation of medical care for a set period, typically 30 days.

The American Academy of Pediatrics (AAP) is an organization whose goal is to support their pediatrician members in the attainment of "optimal physical, mental, and social health and well-being for all infants, children, and young adults." [20] While there are many other medical organizations, including the American Medical Association (AMA) and the American College of Physicians (ACP), the AAP is one of few organizations focused on ensuring a standard of care for pediatric patients.

In the past, the AAP has strongly discouraged discontinuing medical care for families who refuse to vaccinate their children [21]. Recently, however, it issued an updated policy statement maintaining that individual pediatricians and pediatric group practices may consider dismissal of families who do not immunize their children after all efforts for parent education have been undertaken and after the parents receive notification of practice policy of dismissal of unvaccinated children. It is important to ensure that dismissal of these families from the practice occurs in concordance with the state health laws to avoid real or perceived patient abandonment [22].

5. **What communication approaches can healthcare practitioners and medical practices take when discussing vaccines with families? Which approaches are most effective in achieving vaccine acceptance?**
Healthcare practitioners play a significant role in parents' vaccine decision-making process, including for those parents who are vaccine-hesitant [22]. Acknowledging parental concerns regarding vaccines while correcting vaccine misperceptions in a nonconfrontational manner are key components of the relationship. Taking the opportunity to discuss the benefits of immunizations and to address new or ongoing parental concerns at each visit helps to optimize healthcare practitioner-parent vaccine discussions [21, 22]. How healthcare practitioners initiate vaccine discussion influences vaccine acceptance [23]. Specifically, the use of presumptive language when discussing vaccines

is associated with high rates of vaccine acceptance when compared to the use of participatory wording. Presumptive language refers to the presentation of vaccines as recommended and necessary. Patients of healthcare practitioners who maintain the recommendation even with initial parental resistance are more likely to be vaccinated [23]. In contrast, patients of healthcare practitioners who use participatory language (e.g., "Which vaccines do you want your child to receive today?") have lower immunization rates. Other productive strategies include the use of parent-centered motivational interviewing, personalizing vaccine acceptance, and conveying experiences and anecdotes about vaccine successes [22].

6. **What is the Vaccine Adverse Event Reporting System (VAERS)? How does it work? What other vaccine safety surveillance processes are in place in the USA?**

An adverse event refers to a new medical problem that occurs during or after the administration of a vaccine that may or may not be caused by the vaccine itself [24]. Clinically significant severe adverse events caused by immunizations are quite rare, while low-grade fevers or local injection site reactions are commonplace. On a case-by-case basis, it is difficult to determine whether an uncommon adverse event is causal or is simply temporally associated with having received a vaccination. Rare events, given the low numbers, can even be difficult to assign causality in a population until a very large number of individuals are immunized and followed systematically [25]. Pre-licensure clinical vaccine trials identify common vaccine-related adverse events; however, they may not be large enough to detect such rare events. Thus, post-licensure surveillance systems are needed to ensure the safety of these vaccines at the population level.

The Vaccine Adverse Event Reporting System (VAERS), established in 1990, is a national, passive surveillance system for monitoring signals that may indicate problems with vaccine safety. Managed by the U.S. Centers for Disease Control and Prevention (CDC) and the FDA, the VAERS was designed to identify rare adverse events following vaccinations; monitor increases in known adverse events following vaccinations; identify potential patient risk factors for adverse events following vaccinations; assess the safety of newly licensed vaccines; determine and address possible reporting clusters; recognize persistent administration errors; and provide a national safety monitoring system to enhance the response to public health crises [24].

Reporting an adverse event following vaccination to VAERS can be done by anyone, including healthcare professionals, patients, parents, or caregivers. Online reports and reporting forms can be accessed and submitted through the VAERS site and should be completed for all clinically relevant and important post-vaccination events, including vaccination errors, regardless of whether it is known if the vaccine caused the adverse events. VAERS data are submitted to a central facility for analysis to detect vaccine safety signals [24]. As VAERS is a passive surveillance system, no effort is made to contact individuals who have experienced an adverse event for active data collection, and no determination of causality can be assigned.

In addition to the VAERS, the CDC created two more vaccine safety surveillance systems intended to determine whether a potential safety signal identified through VAERS is valid or causal: (1) Vaccine Safety Datalink and (2) Clinical Immunization Safety Assessment Project. Created in 1990 by the CDC National Immunization Program to conduct post-marketing vaccine safety surveillance, the Vaccine Safety Datalink is a collaborative project between the CDC Immunization Safety Office and eight large healthcare organizations around the country. Together, they rely on electronic health data from participating sites to collect information on vaccines administered and medical visits of those who are immunized. This data system uses rapid cycle analysis to detect adverse events following vaccination in near-real-time to inform the public about vaccine safety risks as soon as possible. It is also used to design case-control studies when needed to assess the validity of potential signals detected from the VAERS. Therefore, the main objectives of the Vaccine Safety Datalink are to conduct large population-based vaccine safety studies of questions arising in the literature or from other vaccine safety systems and to monitor possible adverse events after new vaccines are licensed or new vaccine recommendations are made. These data are then used to inform organizations and national advisory committees as vaccine recommendations are being developed or revised [26].

The Clinical Immunization Safety Assessment Project was created to improve the scientific understanding of vaccine safety at the individual level. Comprised of a national network of vaccine safety experts, the objective of this program is to serve as a vaccine safety resource for consultation regarding clinical vaccine safety questions; to develop strategies to determine who may be at increased risk for adverse events following immunizations; and to conduct studies to identify risk factors and strategies to prevent these adverse events [25].

7. **Why was the original rotavirus vaccine (1998) withdrawn from the market less than a year after its licensure by the U.S. Food and Drug Administration? Identify another historical event important to public perceptions of vaccines. (Do not use the events associated with Andrew Wakefield referred to in another background question.)**

In August 1998, the FDA licensed an oral tetravalent rotavirus vaccine (RotaShield, Wyeth Laboratories, based in Philadelphia, Pennsylvania) for vaccination of infants for the prevention of severe rotavirus gastroenteritis. The Advisory Committee of Immunization Practices (ACIP) recommended the routine use of the tetravalent rotavirus vaccine for healthy infants at 2, 4, and 6 months of age, with the 3-dose series completed by the first birthday [27]. Vaccine uptake was robust, and pediatricians noted an almost immediate reduction in the number of infants and young children in need of hospitalization for rotavirus-associated dehydration. Less than 1 year after its approval and widespread use, the VAERS identified 15 cases of intussusception temporally associated with receipt of the vaccine. Most cases occurred within 1 week of receiving a dose of vaccine, especially after the first dose of the vaccine series. In response to this alert, post-licensure data were analyzed and a multi-state investigation was conducted to determine whether the temporal association between rotavirus vaccination and the development of intussusception could be causal. In July 1999, the CDC recommended postponing the use of rotavirus vaccine until the data analysis was complete. Just a few months later, in October 1999, based on the scientific evidence revealing that intussusceptions occurred more frequently in the first 2 weeks following rotavirus vaccination than would be expected to occur by chance, the ACIP withdrew its recommendation for the use of tetravalent rotavirus vaccine in infants in the USA [28, 29]. The FDA subsequently withdrew licensure.

In the mid-1950s, following the approval and licensure of the polio vaccine, Cutter Laboratories, one of the few companies producing vaccine, unintentionally had live poliovirus contaminating 200,000 doses of their inactivated virus vaccine. This resulted in illness in more than 40,000 individuals, paralysis in 200, and death in 10 [30].

Not quite a decade later, the first formalin-inactivated respiratory syncytial virus (RSV) vaccine was studied in infants and young children during a randomized placebo controlled clinical trial. Infants who were immunized with the investigational vaccine were more likely to become infected with RSV and experienced more severe disease than those in the placebo group. Ultimately more than 80% of infants who received the study vaccine were hospitalized with RSV disease. Two infants died. It was nearly 50 years before investigational RSV vaccines would be included again in clinical trials. A safe and effective vaccine for this ubiquitous infection remains elusive [31].

8. **Describe the circumstances regarding Andrew Wakefield's research and the reason for its retraction.**
   In 1998, Andrew Wakefield, a British gastroenterologist, published a case series in the *Lancet*, suggesting that the measles-mumps-rubella (MMR) vaccine may predispose individuals to neuropsychiatric disease, particularly autism. Wakefield's study, which included only 12 children and made vague claims connecting the measles virus from the vaccine to behavioral regression and chronic enterocolitis, resulted in a significant decline in MMR vaccination rates. Following the publication of this report, many scientists tried to replicate these findings with no success. Subsequently, co-authors of the paper claimed that there was insufficient evidence to state that the MMR vaccine leads to autism. Around the same time, it came to light that Wakefield had financial conflicts that were not disclosed and that authors had conducted invasive procedures on the enrolled children without consent and, furthermore, selected out the data they chose to report. With the accumulation of this information, the *Lancet* finally retracted the original article in 2010 [32]. However, concerns that MMR vaccination causes autism persist, despite both the public retraction of the study and numerous well-designed studies demonstrating no association.

9.4 Responses to Discussion Questions

Recently, there has been an outbreak of measles in a neighboring county. You are now concerned because you have two patients who are currently undergoing chemotherapy for leukemia and another with an immunodeficiency who cannot receive the live viral measles vaccine. Your immunocompromised patients have medical contraindications to receiving measles vaccine, are at greater risk of contracting measles, and are more likely than others to develop severe infection or complications. You discuss this concern with your partners in the practice.

1. **Self-reflection: In this situation, how do you feel about accepting families who refuse vaccinations, knowing there are at-risk children who are unable to be immunized in the practice?**

   > **Box 9.1 Teaching Tip**
   > This question is intended to elicit learners' immediate, emotional response to the dilemma. Do they feel judgmental toward the vaccine-refusing families? Protective of the immunocompromised children? By acknowledging these sentiments and recognizing that they have duties to children from both families, they will be better able to reason logically during the remainder of the discussion.

2. **What are the potential benefits and risks of refusing to care for families who do not immunize their children? Consider ethical, legal, and public health implications.**
   Healthcare practitioners have obligations to act in the best interests of their patients (beneficence) while minimizing harm to their patients (non-maleficence). In pediatrics, they fulfill this duty with the understanding that parents are ideal surrogate decision-makers for their children in most circumstances—assuming that they are not putting their child at risk for harm. The determination of how much harm comes to these unvaccinated children and the decision to accept or refuse to care for these families has been widely debated within the pediatric community.

   On one hand, by refusing to provide care to families who choose to not immunize their children, healthcare practitioners send a strong message regarding the importance of vaccinations in disease prevention in the pediatric population. Furthermore, these healthcare practitioners are protecting the other patients in the practice by reducing the risk of transmitting and acquiring vaccine-preventable

diseases within the pediatric office [33]. On the other hand, by refusing to accept these patients, healthcare practitioners may push families to see other healthcare practitioners who do not believe strongly in vaccines, losing the opportunity for future vaccine discussions [34]. Maintaining relationships with families who do not vaccinate their children allows for ongoing, continued discussions with the potential for parental reconsideration and vaccination as the children grow older [35].

The ethical dilemma. The central ethical dilemma for healthcare practitioners who are deciding whether to accept patients who decline immunizations is the ability to maintain mutual trust and respect with a family who disagrees with the basic value of vaccinations in pediatric disease prevention and health promotion. Many healthcare practitioners argue that if families cannot agree with such an essential practice so early in the therapeutic relationship, it may be difficult for them to move forward with other medical decision plans. However, families who are dismissed may opt for care outside of traditional pediatrics, potentially compromising the quality of care the children receive beyond the issues of immunizations.

The ethics behind accepting and refusing patients who decline immunizations can be further divided at the individual and community level [36]. At the individual level, the surrogate decision-makers (parents and guardians) have autonomy in their medical decisions, which should not impact the quality of care received [36]. Only accepting immunized patients into the practice may show a lack of respect for parental decision-making rights or may make parents feel forced into choosing to vaccinate.

At the population level, high immunization uptake is needed to protect the community health, particularly those who cannot be immunized. Accepting patients who refuse immunizations into the practice has the potential to expose others to life-threatening, vaccine-preventable diseases. Furthermore, parents who do not immunize children, stating low disease risk because of the already established herd immunity, may be thought of as taking advantage of others who accepted the minimal risks associated with immunization for disease protection.

The legal dilemma. There are several different legal considerations at play when determining whether or not to care for patients who refuse vaccinations. First, it is important to remember that parents have the legal authority to refuse vaccinations for their children. This right has been contested and supported in case law for decades.

By accepting vaccine-refusing families into the practice, healthcare practitioners may face legal action by families who vaccinate their children claiming that the practice puts their child at risk for life-threatening illness by also allowing unvaccinated children into the practice. There could also be legal action taken by the families who have chosen not to vaccinate should their child develop a vaccine-preventable disease. Detailed documentation of the discussions regarding the risks of remaining unimmunized and collecting parent-signed vaccine refusal waivers may reduce this particular medicolegal concern, but does not guarantee immunity to such liability [37].

Physicians and other healthcare practitioners who dismiss families from the practice because of vaccine refusal may be accused of negligence or patient abandonment. In negligence, a physician is responsible if the physician fails to meet the obligation to a patient and patient harm occurs. The physician who is found to violate the professional code of ethics, in this manner, can also be found guilty of professional misconduct [36]. Specifically, as per the American Medical Association Code of Medical Ethics, physicians have an obligation to ensure continuity of care for their established patients and alert these patients if there is an inability to maintain the clinical relationship. In this scenario, a physician could provide the family with sufficient notice of dismissal and facilitate seamless transfer of care to another community healthcare practice without the occurrence of patient harm as a result. However, further legal questions arise when there are no other available healthcare practitioners willing or able to accept the care of the dismissed patient.

The public health dilemma. As more healthcare practitioners decline to care for unimmunized patients, there may be communities where no healthcare practitioners exist to care for those patients. Alternatively, in communities with practices that do accept unimmunized patients, there may be large clusters of unimmunized children within a given practice. This situation results in a concentrated cluster of unimmunized individuals, a circumstance already known to increase the risk of acquiring and transmitting vaccine-preventable diseases [21].

Physicians play a crucial public health role in disseminating information regarding vaccine importance to the community. If physicians refuse to care for patients who are unvaccinated, the ability to continue to provide this information is lost, and the chance for future vaccination opportunities is significantly reduced.

(a) **What are the risks to the patients themselves?**
Unvaccinated populations are at risk for acquiring and transmitting life-threatening, vaccine-preventable diseases, including diseases that were at one time eliminated from the USA, such as endemic measles. A review of the published U.S. literature describes the association between vaccine refusal and vaccine-preventable diseases. Upon review of measles epidemics, more than half the cases described occurred in individuals with no history of vaccination. Even more noteworthy, most unvaccinated children had a nonmedical vaccine exemption. Unvaccinated children were a large portion of the total cases early in the outbreaks, with epidemic curves starting a year earlier among the unvaccinated population compared to those who have received vaccine. More specifically, unvaccinated children were 22–35 times more likely to acquire measles than vaccinated children [11].

(b) **What are the risks to the public?**
The risk of vaccine-preventable diseases extends beyond the unvaccinated individual. Vaccines are not 100% effective in disease prevention; thus it is important to maintain high vaccine coverage rates in the community to provide optimal protection. Thus, in a measles epidemic, it is possible for a significant proportion of vaccinated individuals to contract disease from a single unvaccinated individual [11]. When reviewing outbreaks of a disease with a less effective vaccine, such as pertussis, schools and communities with high rates of vaccine exemptions had higher rates of pertussis disease, even among the vaccinated population [11].

(c) **What might happen if all the healthcare practitioners in the area refused to accept unvaccinated patients?**
If all the healthcare practitioners within a community refused to accept patients who were unimmunized, it could lead to a large population of vulnerable, unvaccinated children without access to medical care. These children lose the opportunity for future vaccinations through continued discussion as well as all the preventive care that comes with having a continuity healthcare practitioner. If a small number of healthcare practitioners accepted these families, there would then be large clusters of unvaccinated children within a single practice, increasing the risk of acquiring and transmitting vaccine-preventable diseases through the community even further.

3. **Is the potential impact of vaccine refusal the same for all vaccines? How do vaccines differ with regard to impact of vaccine refusal?**

The potential public health impact of vaccine refusal depends on the prevalence, transmissibility, and severity of the infection(s) the rejected vaccines are meant to prevent; the overall vaccine effectiveness in the immunized population; and vaccination coverage rates. Attack rates are highest for infections that are transmitted by the airborne route, such as tuberculosis, smallpox, varicella, and measles. Infections that are transmitted by respiratory droplets, such as influenza and pertussis, are still quite contagious, but with attack rates substantially lower than those seen from airborne infections. Infections that are transmitted by direct contact have lower transmission rates because of the requirement for direct contact between the infected and susceptible individual.

The potential public health impact on refusal of measles vaccination, for example, can be dramatic even though the prevalence of measles infection is lower than the prevalence of pertussis or influenza. Measles is spread by the airborne route with the virus surviving suspended in the air for up to 2 hours [38]. Ninety percent of nonimmune individuals who are exposed to the measles virus will become infected. Since the virus remains in the air for such a long period of time, an individual may not be aware of the exposure. Complications of measles infection include otitis media, bacterial pneumonia, and encephalitis. Out of every 1000 infected individuals, one or two will die from the infection. When administered as recommended, the measles vaccine is highly effective: the 2-dose series, given at 12–15 months of age and at 4–6 years, has an efficacy of 93% and 97% for 1 and 2 doses, respectively [39]. Due to intense efforts by the public health system, in pursuit of universal MMR vaccination, endemic measles was declared eliminated from the USA in 2000 [40].

Overall, rates of measles vaccination in the USA have remained excellent, but it is concerning that increasing numbers of unimmunized children tend to be found in clusters. These clusters represent very high-risk areas when a single case of measles infection is identified, because there is a general lack of "community immunity." As vaccine refusals increase, the risk of acquiring measles increases, particularly at a time when travel to measles-endemic countries happens frequently. For example, when unvaccinated children expose other nonimmune children to the measles virus in waiting rooms of healthcare facilities, where the virus can be transmitted for at least 2 hours after the infected patient has already left, the risk of transmission of infection is approximately 90%. Considering that infants are not routinely vaccinated until 12 months of age, there are many young patients who are vulnerable. Thus, the potential for measles transmission within the pediatric practice described may directly impact the healthcare practitioners' decisions to accept or refuse to care for unimmunized children.

In contrast, refusal of human papillomavirus (HPV) vaccine among members of the pediatric practice would not place other individuals seeking care at the office at increased risk for disease. HPV infection is transmitted by direct contact, usually but not always sexual in nature. HPV infects the oropharynx and genitourinary tracts and can predispose to both oropharyngeal and genitourinary cancers. Each year, in the USA, there are approximately 14 million new HPV infections, more than 31,000 new HPV-associated cancer diagnoses, and more than 6000 HPV-associated cancer deaths in both males and females. The HPV vaccine is effective in the prevention of HPV-associated cancers. It is recommended for universal administration starting at the 11- to 12-year-old well-child visit. Despite this longstanding recommendation, adolescent HPV vaccination rates remain low. Reasons for suboptimal vaccination rates include the lack of healthcare practitioner recommendation; parent and healthcare practitioner concerns regarding the association between HPV and sex; and the lack of understanding of the importance of HPV vaccine as cancer prevention. While this continues to remain a significant public health issue, the risk of HPV transmission to the unimmunized population in the medical practice is low and may be less of a factor in the decision to accept or refuse to care for patients declining HPV vaccination [41].

4. **Is there an ethical difference between a religious exemption and a personal belief exemption? If so, what is the difference?**

    Religious exemptions refer to vaccine refusals based on immunizations being contrary to the patient's or family's religious beliefs. Personal belief exemptions refer more generally to vaccine refusals based on immunizations being contrary to the parental beliefs, including religious beliefs but encompassing beliefs beyond the conventional scope of religion [42]. Religious exemptions effectively represent a subset of personal belief exemptions. Some states permit religious exemptions but not personal belief exemptions, and this is ethically controversial.

    In some cases, these exemptions may be separated by the presence or absence of a sincere faith-based conviction that prohibits vaccination. For example, one person might believe that seeking medical intervention, rather than prayer, shows a lack of faith in God and is therefore a sin. Another person might also believe in God (and perhaps even belong to a different denomination of the same faith tradition) but instead reject vaccines because of equally ardent belief that the human body is healthier when "naturally" exposed to disease and that this "purity" is a central part of a "living a good life." A third person might endorse, with the same degree of conviction as the previous two, a belief that vaccination is part of a larger government conspiracy. It can be argued that what all three beliefs share is (1) a lack of acceptance by mainstream Western medicine; (2) a deep concern that permitting vaccination of their children will lead to real and lasting harm, spiritual or physical; and (3) the inability of modern science to disprove such beliefs. Given these examples, it is controversial as to whether the first case is truly ethically "different" from the other two.

    > **Box 9.2   Teaching Tip**
    > Prompt learners to consider whether a patient's values are qualitatively different if they are supported by a formal religious institution, as opposed to developed independently of any organized group. All beliefs, religious and otherwise, are unique in certain ways. Does it change the learners' approach or willingness to tolerate values different from their own?

    (a) **Should there be a legal difference? Why or why not?**

    The First Amendment to the U.S. Constitution guarantees freedom of religion and forbids the restriction of an individual's religious practices. However, the state government is ultimately responsible for the protection of public health, including the use of mandated vaccination laws. In 1905, the U.S. Supreme Court upheld the constitutionality of a state-mandated smallpox vaccination program during a smallpox epidemic in the case of *Jacobson v. Massachusetts*. Similarly, school-mandated vaccination laws have resulted in many legal battles, yet the U.S. Supreme Court has maintained that states have the authority to require vaccinations for school entry to ensure protection of public health [43].

    All 50 states have their own legislation regarding mandated vaccinations for school entry and the exemption process. Currently, the majority of states continue to offer religious exemptions, while only a few states still allow personal belief exemptions. In practice, many vaccine-refusing parents learn to make religious cases for their refusal, and many websites exist to help guide such parents through the process. Over the past few years, however, several states have passed legislation making religious exemptions more difficult, or in the case of California, no longer legal to obtain [44].

Your partners meet to discuss the benefits and risks of caring for families who do not immunize their children. After intense debate, the decision was made by all partners to no longer accept families who refuse to immunize their children. You then ask what the practice should do about parents who refuse some or all vaccines but are already in the practice.

5. **If you opt to implement a policy to not accept patients who refuse immunizations, what do you do with families who refuse vaccines but are already established in your practice?**
   The practice may decide to keep existing patients, with whom they already have a relationship, or to inform these patients of the change in policy. Information clearly outlining how the change of policy affects the existing clinical relationship is necessary. The options available to the patient should be described and include reasonable timelines for changes to occur. Some patients may choose to immunize when faced with such a decision. It is prudent to list the specific vaccines required to remain in the practice, along with an acceptable time frame to begin and to complete those immunizations. Some patients will choose to leave the practice. Medical care should be provided to those patients until a new healthcare practitioner has been secured. Some practices have changed their policy for new patients only, electing to keep established patients without requiring them to be immunized. In these circumstances, some healthcare practitioners elect to implement additional measures, such as adding a requirement that parents sign a waiver indicating that they acknowledge the risks of not immunizing their children. Others arrange to have a separate waiting room for unimmunized children or establish an agreement that sick, unimmunized children enter the practice through an alternate entrance designed to minimize exposure to vaccinated children [45].

(a) **What are your legal and ethical commitments to these families? Can you dismiss them from the practice? If so, how do you do that?**
    The legal and ethical commitments to families with established clinical relationships overlap. The American Medical Association Code of Ethics states that healthcare practitioners are obligated to care for their established patients. If the healthcare practitioner can no longer care for the patient, advance notification needs to be given to the patient regarding this obstacle to care, and the healthcare practitioner is required to facilitate transfer of care to another healthcare practitioner. During this time of transition, the healthcare practitioner is ethically responsible for delivering medical care to the established patient. Failing to follow this code of ethics leaves the healthcare practitioner open to charges of professional misconduct as well as claims of medical malpractice for negligence. Laws in effect to protect patients from abandonment by their healthcare practitioner vary not only by state but also by jurisdiction. Existing laws typically outline the necessary steps and timelines used to dismiss a patient. It is important for healthcare practitioners to know their local laws to be sure that any practice policies that include the possibility of dismissing the patient are legal.

> **Box 9.3   Teaching Tip**
> Ask learners to consider how they would feel about dismissing a family they had known for several years.

"I Don't Want My Child to Get Vaccines"

✅ 6. **Self-reflection: How do you feel about vaccines personally? Have you experienced any hesitancy over the decision to consent to vaccines for yourself (such as the influenza vaccine)?**
   (a) Would you consider participating in a vaccine trial yourself? Why or why not?
   (b) If you have or plan to have children, do you think you will choose to vaccinate your children with all recommended vaccines? Do you feel any hesitancy over this decision?

✅ 7. **In some institutions, healthcare professionals are required to either be immunized against influenza or wear a mask during influenza season. What is the ethical justification for such policies?**
The Advisory Committee on Immunization Practices recommends that all U.S. healthcare personnel—including physicians, nurses, and other workers or students in inpatient and outpatient settings, medical emergency response workers, and employees of nursing home and long-term care facilities—who have contact with patients be immunized against influenza to reduce staff absenteeism, nosocomial acquisition of influenza infection, and influenza-related mortality among persons at increased risk for severe influenza illness [46, 47]. Strategies that include promotion of the influenza vaccine and on-site administration of influenza vaccine at no cost over multiple days are associated with an increase in healthcare personnel influenza vaccine uptake but may still result in suboptimal vaccine uptake among this population [48]. Mandating influenza vaccination for healthcare personnel with alternative infection control measures for personnel who do not get vaccinated, such as wearing a mask during influenza season, consistently results in higher influenza vaccination rates for this group [48].

Ethical considerations for these vaccine mandates have been widely discussed, balancing one's medical decision-making autonomy with the duty to protect the health of the community. Healthcare professionals have ethical obligations to their patients. Mandating that healthcare personnel be immunized against influenza is supported by the ethical principles of beneficence (protecting self to protect others), non-maleficence (reducing the risk of nosocomial transmission of influenza to patients who may be at high risk for developing influenza-related complications), and justice (ensuring that patients who cannot be immunized are being cared for by immunized healthcare professionals), particularly since influenza vaccination programs are safe and effective in the reduction of severe influenza infection and the protection of vulnerable, high-risk patients. Ideally, less invasive measures, such as voluntary immunization programs with education, should be instituted. However, when these methods fail to result in high vaccination rates among healthcare personnel, mandatory influenza immunization programs may be needed to maximize health and promote the larger good, guiding principles of utilitarian ethics [49–52].

Two days after your meeting, you are seeing patients on a typically busy afternoon. Your first patient is a 13-year-old girl with a sore throat. You know her well and recall that the family has refused the HPV vaccine in the past, expressing concerns about "a new vaccine that is meant to protect against a sexually transmitted infection." The parents have stated that their daughter is not having sex so is not at risk for acquiring infection. The mother did state at the last visit that she would consider the vaccine for her daughter in the future, "when she was older."

✅ 8. **Should you continue to recommend that she receive the HPV vaccine at each visit? Why or why not?**
While it may seem to be burdensome and time-consuming to discuss vaccines with a family who has refused them in the past, it is important that healthcare practitioners continue the vaccine discussion at each medical visit in an effort to reduce missed opportunities, defined as visits with patients who are due to be vaccinated but leave the

visit without receiving vaccines. Parents of unvaccinated girls have commonly cited concerns regarding vaccine safety and perceived lack of need (due to beliefs that their daughters are not sexually active) as reasons for non-vaccination [53]. It is the duty of the healthcare practitioner to continue to recommend HPV vaccine to their eligible patients for the prevention of HPV-associated cancers and to discuss the reasons for early vaccination, including vaccine efficacy in younger adolescents and the benefits of being immunized before sexual debut. The practice of discussing such issues at each encounter offers ongoing opportunities for the patient to ask questions and provides a consistent message that the recommendation is important. There is some risk that this approach will be perceived as "badgering" even if great care is taken to deliver the information objectively each time. One technique that can help circumvent being perceived as badgering is to inform the patient of the intention to discuss the recommendation at each visit. Subsequently, each time the topic is discussed, the interview can close with a statement that it will be discussed again at the next visit.

Healthcare practitioners often underestimate the level of importance parents place on vaccines and tend to overestimate parental vaccine concerns [54]. The anticipation of hesitancy also discourages healthcare practitioners from continuing to make vaccine recommendations [55]. Furthermore, parents who initially refuse vaccines may change their mind and immunize children after recurring discussions with the healthcare practitioner or with others in their lives, outside the clinic. Ultimately, healthcare practitioners need to make strong vaccine recommendations at each medical visit in order to capture those adolescents who may not make it in for a well-child exam, to reach families who have changed their minds about vaccinations, and to convey the importance of optimizing health through immunizations.

9. **Should HPV vaccination be required for entry into middle school? Why or why not?**
Immunizations are effective in the reduction of vaccine-preventable diseases. When vaccination coverage rates in a community are high, herd immunity develops and the incidence of disease tends to remain low. One measure to ensure high vaccination rates within a community are to require vaccines for school entry. School vaccine mandates are successful in reaching a wide population within the community and ultimately protect children from vaccine-preventable diseases in an environment where infections are readily transmissible. This is efficient and cost-effective in achieving high vaccination rates, ensuring herd immunity, and optimizing the health of both the individual and the public. This is particularly true for contagious infections that could be easily transmitted within the school population, such as measles and varicella. While concerns for impingement on personal liberty have been voiced, the federal government has given the states the power and responsibility to protect the health of the public. All states require immunizations for school entry with varying degrees of exemptions. Only a few states require HPV vaccination for middle school entry.

HPV is currently the most common sexually transmitted infection in the USA, resulting in significant morbidity and mortality, particularly with the development of HPV-associated genitourinary and oropharyngeal cancers among men and women. This vaccine is safe and effective in the prevention of HPV-associated cancers, yet requiring HPV vaccine for middle school entry has been widely debated. Those against HPV vaccine mandates express concerns regarding parental autonomy, particularly for a vaccine preventing an infection that is not acquired through casual contact at school. However, the hepatitis B vaccine, also preventing a sexually transmitted infection that may lead to cancer, has been required for school entry by most states for more than 15 years. Those who are in favor of HPV vaccine mandates describe the importance of this vaccine in cancer prevention and the role of school immunization requirements in maintaining high community-wide vaccination rates, ultimately protecting the health of the community. Mandates, however, can backfire, as groups opposed to vaccines and/or vaccine mandates for school entry can be highly organized, well-funded, and skilled in engaging social and traditional media to advance their agenda. More consistent and widespread anti-vaccine messages may prevail in areas that saw little such activity prior to sponsoring a legislative bill to mandate a vaccine for school attendance.

Your next patients are a 2-year-old girl and her 6-month-old brother, both of whom are in your office for routine visits. The children are receiving vaccines according to an alternative schedule requested by their parents. During the visit, the parents mention that they have friends who are expecting a child soon and wanted to bring the baby to your practice. However, they have heard this will not be possible under the practice's new policy. They are disappointed. They feel that you have both "learned from each other" since their daughter's birth. You do not know if that is how you would describe your relationship: Instead, you feel that you have reluctantly accepted their alternative schedule because you worry that they would refuse all vaccinations if you did not.

10. **Is it possible to have trust between a family and a healthcare practitioner when they disagree about something so fundamental?**
Trust between families and healthcare practitioners can be characterized by (1) parental confidence that the healthcare practitioner is recommending what is best for their child and (2) healthcare practitioner confidence that the parents will adhere to their recommendations. This trust leads to improved patient health outcomes. While agreement on all aspects of the medical care is not necessary to maintain trust, the principle of immunizations in disease prevention is fundamental to pediatric practice. One of the first discussions pediatricians have with their new parents surrounds vaccines. When agreement cannot be made on a medical practice that healthcare practitioners so strongly believe in, it is difficult to imagine parental adherence to future medical recommendations. The concerns for breach of trust, in this situation, may be so strong that the healthcare practitioner asks the family to find a new practitioner. There are other healthcare practitioners, however, who may maintain the relationship with the family to keep the communications open and continue to build trust to allow for ongoing discussions regarding the importance of vaccines in the health of the child.

11. **Strongly recommending vaccines, in a presumptive manner, is known to increase vaccine uptake. How do you balance the potential positive public health impact of this approach with concerns of paternalism and infringement upon patient autonomy or parental authority?**
Presumptive vaccine recommendations are stated as a declaration of the vaccines to be given. An example of a presumptive recommendation is as follows: "There are 3 vaccines due to be given today." This style is in contrast to the participatory vaccine recommendations, which are formatted as "Are we giving shots today?" It is the duty of the healthcare practitioner to do good and protect the health of his/her patients. Immunizations are one of medicine's greatest successes in disease prevention and public health. Vaccinating patients protects the individual as well as the community. Presumptive vaccine recommendations show the parents how much importance the healthcare practitioner places on immunization as central to health and well-being. Furthermore, parents are more likely to immunize their children when healthcare practitioners make strong, presumptive recommendations to keep them

protected from life-threatening vaccine-preventable diseases [23]. Ultimately, vaccination of the patient is still the decision of the parents. Paternalism is defined as the limiting of a person's autonomy in their supposed best interest. The presumptive vaccine recommendation format does not limit the parent's autonomy in medical decision-making, but instead provides the parents with the understanding of the importance of immunizations to the healthcare practitioner.

12. **Reflecting on the story of Andrew Wakefield, how do you think the dissemination of misinformation affects vaccine confidence within the community?**
Following the widespread publicizing of Wakefield's results, the MMR vaccination rates in England dropped from 91% to below 80% in the following 5 years, and the previously eradicated measles began to resurface and spread [56]. This vaccine hesitancy was not limited to England. The 2009 U.S. National Immunization Survey of parents with children ages 24–35 months revealed that 26% and 8% delayed and refused vaccines for their children, respectively [57]. The spread of vaccine misinformation, now facilitated by social media, plays a major role in community-wide vaccine hesitancy, despite the fact that misinformation is based primarily on parental anecdotes, with little or no supporting scientific evidence. In fact, negative online comments about true science articles have a negative effect on the reader's attitudes and understanding of the articles [57].

(a) **How do you think real safety concerns have affected public trust in vaccines?**
Public concerns regarding vaccinations increase following discussions of vaccine safety events, regardless of whether these events were ultimately associated with immunizations [58]. These fears reduce the parental confidence in the vaccination program that is needed to maintain high vaccine coverage for optimal public health protection and need to be proactively addressed in a timely manner [59].

> **Box 9.4   Personal Perspective**
> *Voices for Vaccines* is a website that advocates for vaccination and publishes stories from families who have experienced vaccine hesitancy but eventually accepted vaccines. Maranda Dynda's story starts with the moment her midwife asked her how she "felt" about vaccines [59]. Later, she writes:
>
> » "…after Ramona was born, we only took her to a doctor three times in her first year of life. Even that, we felt, was too much. We were consistently paranoid about the agenda of doctors and nurses, and deep down, we were scared that Child Protection Services would take Ramona away from us for not vaccinating due to the 'big pHARMa' conspiracy and their grip on the government… "
>
> » "While I was fine and okay with not vaccinating since I truly believed my choice was based on science, I had doubts on other things that my vaccine-denying peers believed in. Things like essential oils, chemtrails, FEMA death camps, and even AIDS denialism. I asked myself if I found no proof or reason to believe in THOSE ideas that they promote with the same fervor as their anti-vaccine views, then why am I so confident that they're right about vaccines? Why, when I had always loved and trusted science with everything in my being, did I distrust medicine and doctors? Things that I used to be so confident in I was now unsure of. I decided it was time to visit the vaccine conversation once more, with a new perspective and science mindset. It didn't take me long to go from feeling silly to feeling foolish, and finally to feeling completely stupid. I had been duped. I was flat-out lied to. The cult-like world of vaccine refusal had grabbed me by the throat and taken me for a ride."

(b) **Creative Problem-Solving:** A recent study in the journal *Science* demonstrated that false information posted to social media spread much more quickly and widely than true information, a finding that applied to science news as well as other types of stories [60]. As a future healthcare leader, what would you propose to combat scientific misinformation online?

Acknowledgments This case was adapted from earlier cases written by Cynthia Morrow, Amy Caruso Brown, and Travis Hobart.

References

1. Stadlin S, Bednarczyk RA, Omer SB. Medical exemptions to school immunization requirements in the United States – association of state policies with medical exemption rates (2004-2011). J Infect Dis. 2012;206:989–92.
2. New York State Department of Health. Immunization resources [Internet]. Rochester (NY): New York Center for School Health. [cited 2018 Mar 1]. Available from: www.schoolhealthny.com/immunizationresources.
3. New Mexico Department of Health. School immunization requirements [Internet]. Santa Fe (NM): NMDOH; 2018 Mar 1. Available from: https://nmhealth.org/about/phd/idb/imp/sreq.
4. Orenstein WA, Hinman AR. The immunization system in the United States – the role of school immunization laws. Vaccine. 1999;17:S19–24.
5. Bugenske E, Stokley S, Kennedy A, Dorell C. Middle school vaccination requirements and adolescent vaccination coverage. Pediatrics. 2012;129:1056–63.
6. Seither R, Calhoun K, Street EJ, Mellerson J, Knighton CL, Tippins A, Underwood JM. Vaccination coverage for selected vaccines, exemption rates, and provisional enrollment among children in kindergarten – United States, 2016-17 school year. MMWR Morb Mortal Wkly Rpt. 2017;66:1073–80.
7. Briss PA, Rodewald LE, Hinman AR. Reviews of evidence regarding interventions to improve vaccination coverage in children, adolescents, and adults. Am J Prev Med. 2000;18:97–140.
8. CDC. Effectiveness of a middle school vaccination law – California, 1999-2001. MMWR Morb Mortal Wkly Rep. 2001;50:660–3.
9. Averhoff F, Linton L, Peddecord KM, Edwards C, Wang W, Fishbein D. A middle school immunization law rapidly and substantially increases immunization coverage among adolescents. Am J Public Health. 2004;94:978–84.
10. Robbins KB, Brandling-Bennett D, Hinman AR. Low measles incidence: association with enforcement of school immunization laws. Am J Public Health. 1981;71:270–4.
11. Phadke BK, Bednarczyk RA, Da S, Omer SB. Association between vaccine refusal and vaccine-preventable diseases in the United States: a review of measles and pertussis. JAMA. 2016;315:1149–58.
12. Atwell JE, van Otterloo J, Zipprich J, Winter K, Harriman K, Salmon DA, Halsey NA, Omer SB. Nonmedical vaccine exemptions and pertussis in California, 2010. Pediatrics. 2013;132:624–30.
13. Omer SB, Salmon DA, Orenstein WA, deHart P, Halsey N. Vaccine refusal, mandatory immunization, and the risks of vaccine-preventable diseases. N Engl J Med. 2009;360:1981–8.
14. CDC. Vaccination coverage among children in kindergarten – United States, 2009-2010. MMWR Morb Mortal Wkly Rep. 2011;60:700–4.
15. Seither R, Calhoun K, Street EJ, Mellerson J, Knighton CL, Tippins A, Underwood JM. Vaccination coverage for selected vaccines, exemption rates, and provisional enrollment among children in kindergarten – United States, 2016-17 school year. MMWR Morb Mortal Wkly Rep. 2017;66:1073–80.
16. Wang E, Clymer J, Davis-Hayes C, Buttenheim A. Nonmedical exemptions from school immunization requirements: a systematic review. Am J Public Health. 2014;104:e62–84.
17. MacDonald NE. Vaccine hesitancy: definition, scope, and determinants. Vaccine. 2015;33:4161–4.
18. Lieu TA, Ray GT, Klein NP, Chung C, Kulldorff M. Geographic clusters in underimmunization and vaccine refusal. Pediatrics. 2015;135:280–9.
19. Smith PJ, Chu SY, Barker LE. Children who have received no vaccines: who are they and where do they live? Pediatrics. 2004;114:187–95.
20. American Academy of Pediatrics. HealthyChildren.org [Internet]. Itasca (IL): AAP. [cited 2018 May 10]. Available from: www.healthychildren.org.
21. Healy CM, Pickering LK. How to communicate with vaccine-hesitant parents. Pediatrics. 2011;127:S127–33.
22. Edwards KM, Hackell JM. Committee on infectious diseases, the committee on practice and ambulatory medicine. Pediatrics. 2016;138:e20162146. https://doi.org/10.1542/peds.2016-2146.
23. Opel DJ, Heritage J, Taylor JA, Mangione-Smith R, Showalter Salas H, DeVere V, Zhou C, Robinson JD. The architecture of provider-parent vaccine discussions at health supervision visits. Pediatrics. 2013;132:1037–46.
24. Shimabukuro TT, Nguyen M, Martin D, DeStefano F. Safety monitoring in the Vaccine Adverse Event Reporting System (VAERS). Vaccine. 2015;33:4398–405.
25. LaRussa PS, Edwards KM, Dekker CL, Klein NP, Halsey NA, Marchant C, Baxter R, Engler RJ, Kissner J, Slade BA. Understanding the role of human variation in vaccine adverse events: the clinical immunization safety assessment network. Pediatrics. 2011;127:S65–73.
26. Baggs J, Gee J, Lewis E, Fowler G, Benson P, Lieu T, et al. The vaccine safety datalink: a model for monitoring immunization safety. Pediatrics. 2011;127:S45–53.
27. CDC. Recommended childhood immunization schedule-United States, 1999. MMWR Morb Mortal Wkly Rep. 1999;48:12–6.
28. CDC. Intussusception among recipients of rotavirus vaccine – United States, 1998–1999. MMWR Morb Mortal Wkly Rep. 1999;48:577–81.
29. CDC. Withdrawal of rotavirus vaccine recommendation. MMWR Morb Mortal Wkly Rep. 1999;48:1007.
30. Offit PA. The cutter incident, 50 years later. N Engl J Med. 2005;352(14):1411–2.
31. Simoes EA. Respiratory syncytial virus infection. Lancet. 1999;354(9181):847–52.
32. Sathyanarayan Rao TS, Andrade C. The MMR vaccine and autism: sensation, refutation, retraction, and fraud. Indian J Psychiatry. 2011;53:95–6.

33. Buttenheim AM, Asch DA, Cherng ST. Provider dismissal policies and clustering of vaccine-hesitant families: an agent-based modeling approach. Hum Vaccin Immun. 2013;9:1819–24.
34. Opel DJ, Feemster KA, Omer SB, Orenstein WA, Richter M, Lantos JD. A 6-month old with vaccine-hesitant parents. Pediatrics. 2014;133:526–30.
35. Diekema DS. American Academy of Pediatrics, committee on bioethics. Responding to parental refusals of immunization of children. Pediatrics. 2005;115: e1696.
36. Halperin B, Melnychuk R, Downie J, MacDonald N. When is it permissible to dismiss a family who refuses vaccines? Legal, ethical, and public health perspectives. Paediatr Child Health. 2007;12:843–5.
37. Diekema DS. Choices should have consequences: failure to vaccinate, harm to others, and civil liability. Mich L Rev First Impressions. 2009;107:90–4.
38. Moss WJ. Measles. Lancet. 2017;390:2490–502.
39. Centers for Disease Control and Prevention. Measles [Internet]. Atlanta (GA): CDC. [cited 2018 Mar 1]. Available from: www.cdc.gov/measles/index.html.
40. Papania MJ, Wallace GS, Rota PA, Icenogle JP, Fiebelkorn AP, Armstrong GL, et al. Elimination of endemic measles, rubella, and congenital rubella syndrome from the Western hemisphere: the US experience. JAMA Pediatr. 2014;168(2):148–55.
41. Centers for Disease Control and Prevention. Human papillomavirus (HPV) [Internet]. Atlanta (GA); 2015 Sep 30 [cited 2018 Mar 1]. Available from: www.cdc.gov/hpv.
42. Diekema DS. Personal belief exemptions from school vaccination requirements. Annu Rev Public Health. 2014;35:275–92.
43. Swendiman KS. Mandatory vaccinations: precedent and current laws. Congressional Research Service, Report for Congress; 2011.
44. Goldstein ND, Suder JS, Bendistis BE. The politics of eliminating nonmedical vaccination exemptions. Pediatrics. 2017;139:e20164248.
45. Hough-Telford C, Kimberlin DW, Aban I, Hitchcock WP, Almquist J, Kratz R, et al. Vaccine delays, refusals, and patient dismissals: a survey of pediatricians. Pediatr. 2016;138:e20162127.
46. Pearson ML, Bridges CB, Harper SA, Healthcare Infection Control Practices Advisory Committee (HICPAC), Advisory Committee on Immunization Practices (ACIP). Influenza vaccination of health care personnel: recommendations of the Healthcare Infection Control Practices Advisory Committee (HICPAC) and the Advisory Committee on Immunization Practices (ACIP). MMWR Recomm Rep. 2006;55:1–16.
47. Advisory Committee on Immunization Practices, Centers for Disease Control and Prevention. Immunization of health-care personnel: recommendations of the advisory committee on immunization practices (ACIP). MMWR Recomm Rep. 2011;60:1–45.
48. Black CL, Yue X, Ball SW, Donahue SM, Izrael D, de Perio MA, et al. Influenza vaccination coverage among health-care personnel – United States, 2014-15 influenza season. MMWR. 2015;64:993–9.
49. Zimmerman RK. Ethical analysis of institutional measures to increase health care worker influenza vaccination rates. Vaccine. 2013;31(52):6172–6.
50. Lee LM. Adding justice to the clinical and public health ethics arguments for mandatory seasonal influenza immunization for healthcare workers. J Med Ethics. 2015;41:682–6.
51. Ottenberg AL, Wu JT, Poland GA, Jacobson RM, Koenig BA, Tilburt JC. Vaccinating health care workers against influenza: the ethical and legal rationale for a mandate. Am J Public Health. 2011;101:212–6.
52. Steckel CM. Mandatory influenza immunization for health care workers – an ethical discussion. AAOHN J. 2007;55:34–9.
53. Hanson KE, Koch B, Bonner K, McRee AL, Basta NE. National trends in parental HPV vaccination intentions and reasons for hesitancy, 2010-2015. Clin Infect Dis. 2018;67:1018.
54. Healy CM, Montesinos DP, Middleman AB. Parent and provider perspectives on immunization: are providers overestimating parental concerns? Vaccine. 2014;32:579–84.
55. McRee AL, Gilkey MB, Dempsey AF. HPV vaccine hesitancy: findings from a statewide survey of healthcare providers. J Pediatr Health Care. 2014;28:541–9.
56. Flaherty DK. The vaccine-autism connection: a public health crisis caused by unethical medical practices and fraudulent science. Ann Pharmacother. 2011;45:1302–4.
57. Shelby A, Emst K. Story and science: how providers and parents can utilize storytelling to combat anti-vaccine misinformation. Hum Vaccin Immunother. 2013;9:1795–801.
58. Faasse K, Porsius JT, Faasse J, Martin LR. Bad news: the influence of news coverage and Google searches on Gardasil adverse event reporting. Vaccine. 2017;35(49):6872–8.
59. Dynda M. I was duped by the anti-vaccine movement. Voices for vaccines [Internet]. 2015. Available from: http://www.voicesforvaccines.org/i-was-duped-by-the-anti-vaccine-movement/.
60. Vosoughi S, Roy D, Aral S. The spread of true and false news online. Science. 2018;359(6380):1146–51.

Further Reading on this Topic

Block SL. The pediatrician's dilemma: refusing the refusers of infant vaccines. J Law Med Ethics. 2015;43:648–53.

Lantos J. The patient-parent-pediatrician relationship: everyday ethics in the office. Pediatr Rev. 2015;36(1): 22–30.

MacDonald NE. SAGE working group on vaccine hesitancy. Vaccine hesitancy: definition, scope, and determinants. Vaccine. 2015;33:4161–4.

Wang E, Baras Y, Buttenheim AM. "Everybody just wants to do what's best for their child": understanding how pro-vaccine parents can support a culture of vaccine hesitancy. Vaccine. 2015;33:6703–9.

"Our Son's Cancer Is Gone. Why Can't We Stop Treatment?"

Thomas R. Curran Jr.

10.1 Background Questions – 186

10.2 Additional Case Information and Questions for Discussion – 186

10.3 Answers to Background Questions – 188

10.4 Responses to Discussion Questions – 193

References – 201

© Springer Nature Switzerland AG 2019
A. E. Caruso Brown et al. (eds.), *Bioethics, Public Health, and the Social Sciences for the Medical Professions*, https://doi.org/10.1007/978-3-030-03544-0_10

■ **Glossary**

Event-free survival (EFS) - Proportion of patients who are still alive and have not experienced a relapse at a defined time point after diagnosis; often contrasted with "overall survival." Time point typically varies based on the disease, although "5-year event-free survival" is commonly reported.

Germ cell tumor - Type of growth, which can be benign or malignant, that is derived from germ cell tumors. This commonly occurs in the gonads (testes, ovaries) and brain but can occur elsewhere.

Overall survival (OS) - Proportion of patients still alive at a defined time point after diagnosis; often contrasted with "event-free survival." Overall survival includes patients who have relapsed but are still living. Time point typically varies based on the disease; although "5-year overall survival" is commonly reported, studies involving a disease with a poor prognosis may report "1-year overall survival."

Learning Objectives

1. Identify ethical and social issues that are unique to certain developmental periods, such as infancy and adolescence.
2. Apply the best interests standards (BIS), the harm principle, and the right to an open future to surrogate decision-making in pediatrics.
3. Analyze the relationships among medicine, ethics, law, and culture, paying special attention to when and how biomedical cultural practices may conflict with the patient's and family's cultural practices.
4. Explain the healthcare practitioner's legal and ethical responsibilities when patients and families decline to follow medical advice.

Case Background

You are a pediatric oncologist taking care of a 14-year-old Mennonite boy, Samuel Entz, who has a pineal germinoma, a highly curable type of brain tumor. Samuel had a complete response (CR) after two cycles of chemotherapy, with no tumor seen on magnetic resonance imaging (MRI) at the time. He recently completed the recommended four cycles of chemotherapy and is scheduled to begin 6 weeks of radiation to consolidate his remission, after which he will be done with treatment. This combined treatment is associated with a 95% chance of cure. However, when you meet with Samuel and his parents, they state unequivocally that they have decided against radiation therapy.

10.1 Background Questions

1. What is the incidence of pediatric cancer? What are the most common childhood cancers?

2. How expensive is the typical treatment for childhood cancer? Where can you find this information?

3. Current treatment guidelines for central nervous system (CNS) germ cell tumors are based upon cooperative group clinical trials, such as SIOP CNS GCT 96 [1]. What type of study is this? How convincing is this evidence for the recommended treatment?

4. With respect to medical treatment of children, several approaches to medical decision-making have been proposed, including the best interests standard (BIS), the harm principle (HP), and the right to an open future (ROF). Briefly describe each of these.

5. In your state, under what circumstances can a 14-year-old make medical decisions without parents' consent?

6. What is a "mature minor"? Find an example of a court case in which a minor was allowed to refuse treatment. How is the case similar to this case? How is it different?

7. What is a mandated reporter? What is medical neglect? What is the process that healthcare practitioners should follow if they are concerned about medical neglect?

8. What does it mean to be Mennonite? Does being Mennonite explain why someone might refuse radiation therapy?

10.2 Additional Case Information and Questions for Discussion

Despite their decision about radiation, Mr. and Mrs. Entz are agreeable to your talking with and examining Samuel. Samuel states that overall, he feels very well right now. You note that he has stable permanent visual problems from the tumor that are likely, at least in part, related to his parents' delay in seeking treatment when his symptoms, including headaches and double vision, first developed.

"Our Son's Cancer Is Gone. Why Can't We Stop Treatment?"

❓ 1. What more do you want to know about this situation?

❓ 2. How do you think you should proceed and why? Apply the approach to pediatric decision-making that you think is most relevant.
   (a) Would your perspective be different if Samuel were 4 years old? What if he were 17 1/2 years old?

❓ 3. What else do you want to know about Samuel and his family?
   (a) Why might family interests be important in this case?

You learn that Samuel is 1 of 13 children and is home-schooled. His parents own a farm 80 miles from the hospital. All family members speak, read, and write in English fluently. None of the children in the family has received vaccines, except for a sibling who received a rabies vaccine series after sustaining a dog bite. In general, Samuel's parents prefer to use "natural" treatments, such as herbal supplements and vitamins, and they see a naturopathic practitioner, who referred them for the computed tomography (CT) scan that initially identified Samuel's tumor.

❓ 4. Why do you think they might not want radiation therapy?

Samuel's parents have carefully reviewed the published studies that you provided to them when Samuel was initially diagnosed. In fact, they brought highlighted and annotated copies to today's appointment. They have concluded that the risks of radiation (acutely, somnolence syndrome and radiation necrosis; longer term, mild cognitive impairment and a small but non-zero risk of secondary brain tumor) outweigh the 50% risk of relapse, given that most relapses are still curable with salvage chemotherapy and radiation.

❓ 5. Does this change your response to the situation?

❓ 6. What if the odds were different?
   (a) What if Samuel's chance of cure upfront was 50% instead of 95%?
   (b) What if Samuel's chance of cure upfront was 20%?
   (c) What if Samuel's chance of a cure after relapse was 30% instead of 80–90%?

Like most Amish, some Mennonites, because of their religious stance on government interference, choose not to apply for governmental assistance, including programs such as Medicaid or Medicare, even though they would likely be eligible for such benefits. In addition, as a general rule, they do not purchase private insurance. Instead, the community comes together to share the burden of large financial expenses, such as in the case of catastrophic illness. Samuel's family is part of one such community.

❓ 7. Does this affect your opinion on how to proceed? Should it?

❓ 8. Do you have an obligation to address the financial burden on the family?
   (a) Should you advocate for the hospital to provide charity care? Why or why not?
   (b) Should you ask if having the treatment discounted would change the family's decision? Why or why not?

You ask Samuel's family whether financial assistance would change their decision. They reply that it would not. They also indicate they believe that a higher power is ultimately in control of Samuel's fate and that regardless of their decision, whatever happens is God's will. The family further states that they feel that they have already compromised and have taken a moderate position with respect to Samuel's medical care because they accepted both Western medical therapy (chemotherapy) and herbal treatments and prayer.

❓ 9. How would you respond to their statement?

You discuss this case with your colleagues. The majority of pediatric oncologists at the teaching hospital are in favor of referral to Child Protective Services. However, you, as the primary oncologist, disagree and request an ethics consultation. The ethics consultant believes that the parents' decision is not unreasonable and does not meet the threshold set by the harm principle. You also consult a neuro-oncologist at a major children's hospital, who agrees that, while he would not recommend forgoing radiation, he would not seek legal action to compel this treatment.

❓ 10. **Now how would you proceed?**

An Idaho newspaper recently reported that the state's Child Fatality Review Team had identified 10 preventable pediatric deaths between 2011 and 2013 that were attributed to parents' decision not to seek medical attention for their children, based on their religious beliefs [2].

❓ 11. **Self-reflection: Do your beliefs influence your personal medical decisions or advice for your family members? If so, how so?**

❓ 12. **Do you think that parents' religious beliefs make a stronger argument for respecting their decision compared to parents making the same decision on nonreligious grounds? Why or why not?**
   (a) Do you support state laws that exempt parents from medical neglect if they are motivated by religious beliefs? Why or why not?
   (b) Self-reflection: How would you feel about caring for a child whose parents refused a recommended treatment based on their religious beliefs? Would it be difficult to show compassion for these parents?

10.3 Answers to Background Questions

✅ 1. **What is the incidence of pediatric cancer? What are the most common childhood cancers?**
The overall incidence of pediatric cancer in patients among persons aged 0–19 years during 2001–2009 in the USA was 171.01 per million persons. Boys had a slightly higher incidence than girls; adolescents aged 15–19 had a higher rate than children aged 0–14 years; and white children had a higher rate than African American children [3].
   The most common cancers for children aged 0–14 years are acute lymphoblastic leukemia (26%), brain/central nervous system tumors (21%), neuroblastoma (7%), and non-Hodgkin lymphoma (6%). The most common cancers for adolescents aged 15–19 are Hodgkin lymphoma (15%), thyroid carcinoma (11%), brain/central nervous system tumors (10%), and acute lymphoblastic leukemia (8%) [4].

✅ 2. **How expensive is the typical treatment for childhood cancer? Where did you find the information?**
Based on data from 2009, the average pediatric cancer hospitalization cost in the USA was $40,400, while the average cost for any other type of pediatric hospitalization was $8100 [5]. In a population-based childhood cancer cohort in British Columbia, overall pre-diagnosis mean costs per patient (in Canadian dollars) were $4633; mean costs per patient of the first year of therapy were $97,780; and mean costs per patient of the final year of life (for patients who died) were $284,201 [6].
   This is just the tip of the iceberg with respect to the cost of treating childhood cancer. Children are dependent on their families to shuttle them to their inpatient and outpatient visits, which often interferes with parents' work schedules. It is not unusual for these work disruptions to lead to financial difficulties, and indeed, research has shown that 6 months after their children's cancer diagnoses, 25% of families have lost more than 40% of household income and 30% face household material hardship despite utilizing available support services [7, 8]. These financial burdens are even worse for rural families. Facilities that offer cancer follow-up are typically located in urban areas. Rural patients may require longer travel distance to get care and require lodging outside the home, potentially contributing to greater financial burden [9].

✅ 3. **Current treatment guidelines for pineal germinoma are based upon cooperative group clinical trials, such as SIOP CNS GCT 96 [1]. What type of study is this? How convincing is this evidence for the recommended treatment?**
This study was a prospective, multinational, nonrandomized trial. This study was limited in that it was

non-randomized and the choice of treatment strategy was made in accordance with a national decision; thus the comparison made is not totally independent from external factors. The choice of a combined treatment or radiotherapy alone in localized germinoma might have been affected by local decisions on patients' status at diagnosis [1].

While the gold standard for clinical trial is a randomized control trial, non-randomized controlled trials can still be convincing. They may still be used to detect associations between an intervention and an outcome, although they cannot rule out the possibility that the association was caused by a third factor linked both to intervention and outcome. Random allocation ensures no systematic differences between intervention groups in factors, known and unknown, that may affect outcome. In some circumstances, a randomized, controlled trial may not be feasible because of difficulties with randomization or recruitment [10].

4. **Several approaches to medical decision-making have been proposed, including the best interests standard (BIS), the harm principle, and the right to an open future. Briefly describe each of these.**
The treatment of children generally starts with deference toward parental preferences, assuming, in part, that parents are best situated to make optimal decisions for their children, as well as most invested in the outcome. However, parental rights are not absolute. Medical decision-making in minors has historically been centered on the best interests standard (BIS), which directs the surrogate to maximize benefits and minimize harms to the minor. This approach has been criticized for being vague and inconsistently applied, for being excessively demanding, and for failing to respect the family unit and the interests of other family members, interests that may be compromised by unmitigated support for the "best interests" of a single member [11]. The BIS is typically applied when the parents do not agree with each other or when a third party questions the parents' wishes. A classic example is that of parents who are Jehovah's Witnesses and who refuse to consent to blood transfusions for their child based on religious grounds. In such cases, parents indicate that they are willing to allow their religious beliefs to supersede the child's best interests. In the vast majority of cases, the courts have ordered transfusions over the parents' objections, thus protecting the child's best interests [12].

The harm principle is a second approach to surrogate medical decision-making in pediatrics. The intent of the harm principle is not to identify a single course of action that is in the minor's best interest or is the healthcare practitioner's preferred approach but rather to identify a threshold of harm below which treatment is required, regardless of parental wishes. In other words if, for example, failure to treat poses a significant risk of a poor outcome (harm) for a child, outside intervention is indicated to protect that child [13]. Like the best interests standard, the harm principle is criticized for not clearly delineating when parental refusal of recommended treatment is permissible.

The third approach, the right to an open future, centers on the concept that pediatric surrogates should not make choices that constrain a child's range of future options. It is an effort to protect the child's future autonomy. This has been applied to the practice of testing children for adult-onset genetic disorders. In order to preserve a child's open future, parents and healthcare practitioners should not test for conditions that do not have potential health impact during childhood (either in terms of symptoms appearing in childhood or interventions being performed in childhood to prevent or ameliorate later effects). Instead, children, when they reach maturity, should be permitted to decide whether they want to seek such testing [14]. This is discussed in detail in ▶ Chapter 16. A significant critique of this approach is that, outside the healthcare setting, parents are routinely allowed to make choices for their children that constrain their future options. For example, a child may be a talented pianist, yet parents may choose not to pay for music lessons.

5. **In your state, under what circumstances can a 14-year-old make medical decisions without parents' consent?**
Adolescents capable of giving informed consent are legally authorized in most states to make their own medical decisions for health care related to mental health concerns, substance use, pregnancy, contraception, and sexually transmitted diseases [15]. Professional healthcare organizations such as the American Academy of Pediatrics (AAP) have provided guidance to facilitate policy-making that supports the delivery of confidential health services for adolescents, including laws intended to increase the likelihood that adolescents will seek and receive the healthcare services they need.

Some states allow minors to consent to treatment themselves (rather than providing assent with parental permission) for conditions such as psychiatric illness, sexually transmitted infections (STI), contraception, pregnancy, sexual assault, and substance use. In such cases, confidentiality is also owed to the minor patients, and parents cannot be informed of their children's medical information—confidentiality follows consent. This is distinct from the concept of a "mature" minor who can make a healthcare decision outside of these categories [15]. Some services, such as treatment of STIs and provision of contraception, are permitted in many states, whereas other services, such as abortion, require parental notification or permission (or a waiver from a judge) in a majority of states [16].

Clinicians should familiarize themselves with the relevant legislation in their state and, if faced with a situation in which a minor clearly needs services but will not consider informing their parents, should consider seeking guidance from an adolescent medicine specialist and ethics consultant. It should be noted that some states with such laws have parallel legislation that absolves parents of financial responsibility for medical care provided without their consent [17]. Some adolescent medicine clinics are able to bill for routine services using codes that do not reveal the specific treatment provided on the statements sent to parents or guardians, while others use a sliding scale in order to allow adolescents to pay out of pocket.

Such policies and practices have been supported by evidence showing that without confidentiality, adolescent patients do not seek necessary treatment and are less likely to use contraception and consequently more likely to become pregnant [18]. While concerns have been raised that usurping parental authority has the effect of removing the parents from the decision-making process just when they are most needed, in fact, most adolescents from supportive homes do involve their parents in decision-making. The presumption here is that parents possess what the adolescent lacks in judgment, maturity, and experience. Unfortunately, in some situations, this is not the case, and in others, adolescents may be at risk for physical and emotional abuse or neglect, including being kicked out of the family home. This topic is explored further in ▶ Chapter 21.

6. **What is a "mature minor"? Find an example of a court case in which a minor child was allowed to refuse treatment. How is the case similar or different to this case?**
"Mature minors" are adolescents under the age of 18 (in the USA) who have been recognized to demonstrate the ability to make rational, responsible healthcare decisions for themselves. This is primarily an ethically derived right and does not necessarily translate into a legal right for mature adolescents to consent to treatment in general medical settings, independent of their parents or guardians [17]. Fewer than 20% of states have a formal mature minor exception to the typical requirement that consent be provided by a parent or guardian on behalf of a child. In some of the remaining

states, the concept is supported by case law, while in others, such as New York, there is neither; yet many hospitals in those states have supported the decision-making of older adolescents who sought to refuse recommended treatment.

Adolescent decision-making is dependent on several factors, including cognitive ability, maturity of judgment, and moral authority, which may not all proceed to maturation along the same timeline. Dissent by the adolescent should carry considerable weight when the proposed intervention is not essential and/or can be deferred without substantial risk. If the likely benefits of treatment in conditions with a good prognosis outweigh the burdens, parents should choose a treatment plan over the objections or dissent of the minor. In general, adolescents should not be able to refuse lifesaving treatment even when parents agree with them [19].

An example of a mature minor being allowed to refuse treatment is the case of Abraham Cheerix in 2006. He and his parents refused additional chemotherapy for Hodgkin lymphoma because of adverse effects he had experienced earlier in his therapy. Instead, the family chose to pursue prayer and herbal remedies, thus leading to legal action, including possible loss of custody by the parents. In the end, a compromise was reached. The parents retained custody, and Abraham underwent herbal treatments along with radiation therapy (but no chemotherapy) [20]. In both Samuel's case and the Cheerix case, the parents and the adolescent were in agreement about refusing the medical team's recommended therapy.

✓ 7. **What is a mandated reporter? What is medical neglect? What is the process that healthcare practitioners should follow if they are concerned about medical neglect?**
All U.S. states and territories require identified professionals (including healthcare practitioners as well as teachers, social workers, and child care workers among others) to report suspected child abuse or neglect to Child Protective Services (CPS). It is important to note that mandatory reporting laws do *not* require certainty and failure to make a report to CPS for suspected child abuse can result in civil and/or criminal penalties for the healthcare practitioner. The parents' privacy is overridden in such situations in order to protect vulnerable children from imminent risk of serious harm. All states also protect healthcare practitioners for good faith reporting through some type of immunity from civil or criminal penalties [21].

Medical neglect occurs when parents or caregivers are not meeting a child's medical needs. The neglect most commonly results from either the caregiver's failure to heed obvious signs of serious illness or failure to follow a medical professional's advice once treatment has been sought. It is distinct in most cases from treatment refusal, in which parents decline recommended treatment for various reasons, such as conflict with religious or cultural practices, concern about adverse effects, preference for other treatment options (including complementary and alternative medicine), or simply distrust in the healthcare practitioner making the recommendation. Concepts related to medical neglect include treatment abandonment and medical child abuse. Treatment abandonment is a related term that describes situations in which families who have initially accepted treatment subsequently stop seeking care, often because of adverse effects or burden on the family. The latter is particularly problematic in low-resourced countries [22]. Medical child abuse (previously called Munchausen syndrome by proxy or "factitious disorder imposed on another") occurs when a caregiver, often the mother, seeks and obtains diagnostic and therapeutic interventions for fabricated or feigned illness in a child [23]. The diagnosis can be controversial because in some prominent cases, the situation appears to have been a genuine diagnostic dilemma, compounded by disagreements between physicians regarding the best course of treatment [24].

Once child abuse or neglect, including medical neglect, has been identified, the child's well-being should be the healthcare practitioner's first concern. When the child is medically stable, a meticulous history should be obtained. This typically includes detailed interviews with the parents or other caregivers regarding the circumstances surrounding the child's injuries or illness. Although it is not the healthcare practitioners' role to identify the perpetrator, they are obligated to obtain and document a thorough medical history, commence an appropriate medical workup, and refer the child to specialists who have expertise in completing the medical workup and investigation needed for suspected abuse.

In cases of suspected medical neglect, the first task is to identify the underlying problem that led to the neglect, as it may be unintentional; for instance, language or educational barriers or limited resources may have prevented the parents from recognizing the severity of the illness or providing necessary treatment. In general, the least restrictive, most collaborative approach should be utilized. For example, if the primary problem stems from a language barrier, the healthcare team should help the family access services in the community to facilitate translation. However, in more serious cases in which the neglect is putting the child's health and survival in jeopardy, a referral to CPS is required to ensure that the necessary medical care is received [25].

8. **What does it mean to be Mennonite? Does being Mennonite explain why someone might refuse radiation therapy?**
Mennonites are Christian Anabaptists. There are two major groups of Mennonites: Plain Mennonites and assimilated Mennonites. Plain Mennonites share many of the same religious beliefs and cultural ideologies with the Amish and other Anabaptist sects. The Plain Mennonites are not a monolithic group. For example, with respect to transportation, some use a horse and buggy for transportation, while others drive all-black cars. Assimilated Mennonites are typically indistinguishable from other members of mainstream American society.

The Mennonites commonly believe that commercial health insurance plans undermine the religious duty of community accountability. Their sense of community is strengthened by the belief that most modern technology brings a worldliness that detracts from their lifestyle. To compensate for their lack of commercial insurance, the Mennonites turn to their own community. They do not participate in or receive benefits from Social Security or Medicare. When handling medical costs that exceed an individual's ability to pay out of pocket, voluntary contributions from the congregation, called alms, may be used. Alms are consistent with the tradition of sharing individual burdens in the community [26].

Mennonite beliefs and practices do not offer a straightforward explanation for why someone might refuse radiation therapy. In fact, it is worth noting that even in situations in which healthcare practitioners may assume that a patient's beliefs or cultural practices clearly prohibit a certain treatment (such as refusal of blood transfusions by Jehovah's Witnesses or avoidance of contraception by Roman Catholics), they should still always offer all available treatment options and ask questions about refusal, rather than make assumptions. It is rare that a healthcare practitioner and patient or family will truly share a cultural background, since Western medicine is itself a cultural system, and a better understanding of the family's culture can help guide the discussion. One concern that might be considered for a Mennonite family—though not unique to Mennonites—is the cost of care, since many do not have private health insurance and do not accept public funds as discussed above. Another consideration may be that many members of the Mennonite community endorse the belief that life on Earth is a prelude to heavenly rewards and they rejoice in the promise of eternal life.

"Our Son's Cancer Is Gone. Why Can't We Stop Treatment?"

Thus, they do not view death on Earth as the end of their existence, which might also influence medical decisions.

10.4 Responses to Discussion Questions

Despite their decision about radiation, Mr. and Mrs. Entz are agreeable to your talking with and examining Samuel. Samuel states that overall, he feels very well right now. You note that he has stable permanent visual problems from the tumor that are likely at least in part related to his parents' delay in seeking treatment when his symptoms, including headaches and double vision, first developed.

✓ 1. **What more do you want to know about this situation?**
There are two key lines of inquiry to pursue. First, further information is needed regarding the medical diagnosis, treatment, and prognosis for Samuel's disease. Second, further information is needed regarding Samuel's background—his personal beliefs, values, goals, and interests—and those of his family, community, religion, culture, and so forth. The second of these two is explored in Discussion Question 3 below.

The 5-year event-free survival (EFS) for this type of brain cancer is 95% after standard of care therapy (radiation alone) or with the newer option of chemotherapy plus lower total doses of radiation. This means fewer than 5% of patients who receive one of these treatment options relapse (i.e., have an "event"). The risk of relapse after chemotherapy alone is 50%; however, the chance of cure after relapse is still 80% to 90% with additional chemotherapy *and* higher doses of radiation. This additional therapy, following relapse, is often called salvage therapy; unlike some other cancers, for which the only good chance of cure is with upfront therapy, pineal germinoma has a high rate of salvage [1].

✓ 2. **How do you think you should proceed and why? Apply the approach to pediatric decision-making that you think is most relevant.**

Possible options include, but are not limited to, the following:
1. Accept the parents' refusal without further intervention.
2. Explore the reasons for the refusal further.
3. Speak with Samuel alone and attempt to evaluate his maturity and decision-making capacity.
4. Request an ethics consultation or refer the case to the hospital ethics committee.
5. Obtain a second opinion from another pediatric oncologist.
6. Transfer the patient's care to another healthcare practitioner.
7. Attempt to find a compromise in the approach to treatment.
8. Tell the family that you will need to report them to Child Protective Services if they refuse.
9. Contact hospital counsel in order to pursue obtaining a court order for treatment over their objections.

In this case, Samuel's best interests are not immediately clear. If he proceeds with radiation, he faces a small risk of potentially significant adverse effects but has an excellent chance of being cured. If he does not proceed with radiation, he still has a 50% chance of avoiding all further adverse effects, as well as a very good chance of cure in the event of relapse. Using the harm principle, it might be argued that the situation is above the threshold—Samuel is likely to have a good outcome with either approach, although the distribution and timing of risks are different.

Practically, more than one of the above options can be combined, and doing so may, in fact, be the best approach for a particular disagreement regarding treatment. Some of the above are not ethically acceptable provided the limited information available at this point in the case. For instance, accepting the parents' refusal *may* be appropriate, but not without exploring their reasons, speaking with the patient and understanding his perspective, and perhaps attempting to compromise. If the patient's health is not in immediate jeopardy, seeking legal recourse immediately—before engaging in further discourse and exploring compromises—is also not the best approach. An ethics consultation and second opinion from another expert in the field may be very helpful and, indeed, necessary for optimal resolution of the conflict, but again, they should follow efforts by the primary oncologist to truly understand the situation.

(a) **Would your perspective be differ-**

> **Box 10.1   Teaching Tip**
> Encourage learners to brainstorm options and then prioritize them. An important part of clinical reasoning is recognizing the "best next step" in a given situation.

ent if Samuel were 4 years old? What if he were 17 1/2 years old?
It is important to think about the limits of parental decision-making authority in the context of their child's ability to express informed opinions. In general, the opinions of older children and adolescents are given more weight, while healthcare practitioners are quicker to intervene on behalf of children who are too young to express an opinion.

With respect to treatment refusal, it is worth noting that among cases similar to Samuel's, some of which have received significant media attention, the legal and medical outcomes vary dramatically and not predictably. For instance, in 2015, Cassandra Fortin Callender was a 17 1/2-year-old girl diagnosed with Hodgkin lymphoma in the state of Connecticut. She and her mother wanted to refuse chemotherapy in favor of complementary and alternative medicine; however, the court ruled that she was not mature (that she did not understand the implications of her decision). Cassandra was forcibly treated over her objections.

In contrast, Starchild Abraham Cherrix, another adolescent with Hodgkin lymphoma and similar objections, living in Virginia in 2006, was allowed to refuse further chemotherapy in favor of a modified radiation schedule. In other cases, different courts have allowed the parents of younger children with curable diseases to decline treatment and ordered that older teens with harder-to-treat cancers receive treatment. Unfortunately, the legal system is not designed to effectively and efficiently resolve these types of conflicts [27]. On the other hand, for cases in which an older teen desires to pursue treatment, virtually every healthcare practitioner pursues legal action to authorize that decision in the face of parental refusal.

3. **What else do you want to know about Samuel and his family?**
The details about Samuel's home life are an essential part of developing a well-reasoned, ethical, and culturally appropriate medical plan. The oncologist needs to know more about Samuel's community, home, and family life, including his parents' previous approaches to health care. Questions to be considered might include [28]:

> **Box 10.2   Teaching Tip**
> If learners struggle to generate ideas, consider providing them with Arthur Kleinman's "Eight Questions," which can be used to help healthcare practitioners understand how their patients view their illnesses and what they expect of health care [28].

1. Who lives at home? Does Samuel have brothers and sisters? How does he define his family? Are grandparents or other family members important parts of his support system?
2. What kind of schooling and education has Samuel had? Have his parents had? What language(s) do they speak at home? Can they read and write? How do they make a living? What

"Our Son's Cancer Is Gone. Why Can't We Stop Treatment?"

are Samuel's goals in life, regarding further education, employment, and family?
3. Do they regard themselves as spiritual or religious? What traditions or practices are important to them? Does Samuel share his parents' beliefs?
4. How do Samuel and his family think about physical health? Do they have specific beliefs about the cause of illness? What types of health care do they use? Where do they go for routine care or minor injuries? Do they take any vitamin, herbs, or other supplements? What prior experiences—positive or negative—have they had with Western medicine? What role do they expect the oncologist to play in their son's care?

For example, some families might expect a healthcare practitioner to give a clear recommendation for a particular course of action, while others expect a menu of options to be presented. Some families may want to read studies, do independent research, or obtain second opinions. They may expect the oncologist to be involved in their considerations, or they may prefer to deliberate privately. The oncologist should be careful to avoid assumptions about which style this family prefers.

5. How does the family make decisions?

Although the healthcare practitioner should respect and give consideration to the individual voices—Samuel's, his mother's, and his father's—it can help to understand how the family typically approaches decision-making. Does Samuel expect to be included in the decision-making process? Do his parents make decisions by consensus, does his father expect to decide unilaterally, or do both parents defer to grandparents? Are there religious or community elders who will be consulted?

(a) **Why might family interests be important in this case?**

Consideration of family interests is important in pediatrics, because children and adolescents do not exist in isolation—they grow and develop as part of family unit, and decisions that seem to be in the best interests of one child, the patient, but negatively affect other family members or the family unit as a whole may ultimately have negative consequences for the patient [29]. This is part of the rationale for parental authority: first, that parents "are more likely than others to be able to weigh the competing interests of other family members alongside the patient's interests" and "that there are clear individual and societal benefits to allowing parents to raise their children in accordance with their values, beliefs, and customs" [27].

In this case, Samuel's parents may be concerned about the emotional or financial impact on their other children and on the family if they have to expend resources they do not have on treatment.

You learn that Samuel is 1 of 13 children and is home-schooled. His parents own a farm 80 miles from the hospital. All family members speak, read, and write in English fluently. None of the children in the family has received vaccines, except for a sibling who received a rabies vaccine series after sustaining a dog bite. In general, Samuel's parents prefer to use "natural" treatments, such as herbal supplements and vitamins, and they see a naturopathic practitioner, who referred them for the computed tomography (CT) scan that initially identified Samuel's tumor.

✓ 4. **Why do you think they might not want radiation therapy?**

At this point in the case, the motivation of Samuel's parents remains unclear and requires further clarification, although the oncologist is probably beginning to develop some hypotheses. The parents' rationale for refusal of radiation might include the risks of therapy, lack of understanding of disease severity, costs of therapy, mistrust of Western medicine, or mistrust of the individual healthcare practitioner's expertise and

> **Box 10.3   Teaching tip**
> Learners may benefit when facilitators are clinicians who can offer analogous cases from their own practice in order to help learners understand that the key lessons are generalizable across patient populations, diseases, and practice settings.
>   A comparable case involved a young infant with a fever in whom bacterial meningitis needed to be ruled out. The child's parents, who spoke English as a second language, initially declined to provide consent for the lumbar puncture (LP). Because the child had been born at home, the treating team assumed that the refusal was based on a desire to minimize medical interventions; they elected to proceed with empiric antibiotic therapy. Some days later, when the child required long-term venous access and was going to be sedated, the parents asked whether the LP could be performed at the same time. The team learned that the parents' initial refusal was rooted in concern about their child experiencing pain. If the parents had been asked upfront about their reasons, the treating team might have been able to address them and proceed with the procedure, sparing the child unnecessary antibiotics.

recommendation. Mistrust might also stem from prior experiences with the U.S. healthcare system and with Western medical providers. It is critical that the oncologist continue to ask questions and maintain an open mind, rather than jump to conclusions based on the limited information above.

Samuel's parents have carefully reviewed the published studies that you provided to them when Samuel was initially diagnosed. In fact, they brought highlighted and annotated copies to today's appointment. They have concluded that the risks of radiation (acutely, somnolence syndrome and radiation necrosis; longer term, mild cognitive impairment and a small but non-zero risk of secondary brain tumor) outweigh the 50% risk of relapse, given that most relapses are still curable with salvage chemotherapy and radiation.

✓ 5. **Does this change your response to the situation?**
    It is important to reflect on the fact that Samuel's chance of relapse is 50/50, but he will most likely still be cured. This also means he has a 50% chance of avoiding radiation entirely. Many adult cancer patients might view his situation similarly, and while their oncologists might disagree, they would also regard the risk-benefit assessment as rational or, simply, "not unreasonable."

    Risk assessments are fundamentally unique and personal. Questions about Samuel's background, such as those in Discussion Question 3 below, can help the oncologist understand the context in which Samuel and his parents view such risks. For one patient and family, mild cognitive impairment might seem irrelevant in the face of the patient being alive; for another, it might be devastating and unacceptable. Similar discussions have taken place in medical therapies that have infertility as a side effect. In fact, large pediatric clinical trials for Hodgkin disease have been designed with an aim of maintaining high cure rates while decreasing high rates of male infertility, suggesting that many families and patients consider long-term quality of life worthy of consideration even when survival is also at stake [30].

✓ 6. **What if the odds were different?**

> **Box 10.4   Teaching Tip**
> The following questions are designed to explore understanding and explaining differences in risk and how this might affect parental decision-making authority.

It is not unusual for people to be confused by or struggle with probability; patients and families find it equally challenging, if not more so. Most people are not confronted with numbers and statistics in this way in their daily lives. It is also important to understand that a person's perception of risk does not depend on the precise numbers as much as the personal and cultural expectations of what risks are acceptable and what benefits are valuable.

Many oncologists also take care to emphasize the nature of statistics in this way: When it is said that the "chance of cure is 80%," for example, it is meant that 80% of all patients with a particular disease are cured. For the individual patient, the number has very little tangible meaning—any individual patient is either cured or not cured, a binary reality. Further, probabilities are affected by assorted risk factors (such as age, staging, etc.), as well as treatment era. For some rare diseases, treatments have become more effective, but published data may pool outcomes for patients across decades. Finally, as medicine advances, highly personalized prognostication (based on genetics and more) has become more feasible but is still far from perfect.

(a) **What if Samuel's chance of cure upfront was 50% instead of 95%?**
In this scenario, his chance of cure is only 50/50 even with radiation. Some healthcare practitioners, ethicists, and lay people would argue that every child deserves a shot at a cure, regardless of the family's beliefs and even if it is only a 50% chance. However, others would favor respecting the family's wishes, arguing that they should not be forced to accept a 50% chance of dying while suffering the adverse effects of therapy. This is a situation in which greater weight might be placed on the wishes of a "mature minor."

(b) **What if Samuel's chance of cure upfront was 20%?**
In this scenario, his chances of cure are poor. When the treatment is not very effective and risk of mortality is high, families are generally granted more leeway to refuse treatment. This is sometimes described as lying within the zone of parental discretion (ZPD) [31].

(c) **What if Samuel's chance of a cure after relapse was 30% instead of 80–90%?**
In this scenario, the chance of relapse has not changed: He still has a 50% chance of avoiding radiation entirely by being cured with upfront chemotherapy alone, but he now has a 35% chance of ultimately dying from his disease (versus 5–10%). This changes the stakes considerably—if he does relapse, the odds no longer favor cure. That shifts the risk/benefit ratio in favor of upfront radiation therapy and increases the potential for significant harm if he does not receive radiation, suggesting that the situation may now be below the threshold for outside intervention based on the harm principle.

Like most Amish, some Mennonites, because of their religious stance on government interference, choose not to apply for governmental assistance, including programs such as Medicaid or Medicare, even though they would likely be eligible for such benefits. In addition, as a general rule, they do not purchase private insurance. Instead, the community comes together to share the burden of large financial expenses, such as in the case of catastrophic illness. Samuel's family is part of one such community.

7. **Does this affect your opinion on how to proceed? Should it?**
Amish and Mennonite communities are not fundamentally opposed to utilizing complex or technologically sophisticated medical interventions such as dialysis, cancer treatment, and transplantation. However, the emphasis on the community, rather than the individual, means that care that is too costly may be rejected because the expense will too heavily burden the community. Furthermore, because of their deep faith in God, they may reject extraordinary measures to save a life because such measures may be viewed as an attempt to usurp the will of God [32].

In this case, it is important to understand these potential roots of refusal, as it is quite different from refusal that is rooted in a religious objection to a specific aspect of treatment. Examples of the latter might include a patient who is a Jehovah's Witness and refuses blood transfusion, believing that the acceptance of such an intervention will lead to ostracism from the community or condemnation of the patient's soul, or a Jewish or Muslim patient who rejects a porcine (pig-derived) cardiac valve replacement. All of these situations carry the potential for unique solutions and compromises, but finding such compromises requires a thorough understanding of the patient's and family's concerns.

8. **Do you have an obligation to address the financial burden on the family?**
In Samuel's case, the medical team members would be remiss if they did not factor in the cost of treatment as potentially affecting the decision to forgo radiation.

As discussed previously, health care can be costly to the Mennonite population. Without an insurance company to negotiate group rates, self-insured individuals are typically charged full price by hospitals. Exploring and presenting alternative sources of charitable healthcare financing could potentially address cost concerns.

As an example of noninsurance-based sources of funding, all public general hospitals in New York State take part in a plan to distribute state money to the hospitals from the "indigent care pool." The hospitals must offer a financial plan to patients and families who cannot afford care (those under 300% of the federal poverty line) on a sliding scale based on their income. This almost always means that the hospital is paid less than usual, which is why the state supplements the payment. This could be an option for Samuel's family depending on their beliefs.

It is worth noting that a recent review of the literature supported a "strong recommendation for the assessment of financial hardship as a component of comprehensive psychosocial care in pediatric oncology" and this was confirmed in guidelines for the same [33, 34]. Further, there is some evidence that unmet financial need is associated with greater risk for relapse in acute lymphoblastic leukemia (the most common type of childhood cancer and thus the most easily studied cohort)—a consideration of particular importance for a family who is already reluctant to accept treatment [35].

(a) **Should you advocate for the hospital to provide charity care? Why or why not?**
Seeking creative ways to finance expensive medical care for self-insured populations should certainly be encouraged. If healthcare practitioners limit their role to only treating the biomedical aspects of care, they run the risk of limiting themselves to only managing disease rather than optimizing health. Enhancing the health of patients requires thinking about health care not just as service delivered to individual patients but also as an effort to effect change at the practice, population, and system levels [36]. In Samuel's case, the physician can serve as an advocate by demonstrating cultural humility and recognizing that the cost of therapy may be a factor in choosing a treatment plan for Samuel's Mennonite community.

(b) **Should you ask if having the treatment discounted would change the family's decision? Why or why not?**
Addressing the cost of care is important because some families may be reluctant to ask for financial assistance, and yet the potential burden on their other children and their communities may lead them to decline treatment. Excluding this as a factor may open up new avenues for resolution of the conflict and allows the healthcare practitioners caring for Samuel to focus on the core values and beliefs at stake.

You ask Samuel's family whether financial assistance would change their decision. They reply that it would not. They also indicate they believe that a higher power is ultimately in control of Samuel's fate and that regardless of their decision, whatever happens is God's will. The family further states that they feel that they have already compromised and have taken a moderate position with respect to Samuel's medical care because they accepted both

"Our Son's Cancer Is Gone. Why Can't We Stop Treatment?"

Western medical therapy (chemotherapy) as well as herbal treatments and prayer.

✅ 9. **How would you respond to their statement?**
The family makes a fair point. It is important for the medical team to be mindful of Mennonite culture. They will seek out Westernized medical care and technology when necessary. Mennonites are not prohibited by church law from taking medication or seeking care from a doctor; however, they will often use prayer and naturopathic therapies in addition to conventional Western therapy [37].

It is critical that healthcare practitioners recognize the requirement for cultural awareness and sensitivity in order to understand, appreciate, and work with individuals from disparate cultures. With respect to the Mennonite culture, it is important to recognize that they may prefer traditional healers to Western medical practitioners. While caregivers may bring a family member to a Western physician, they often pick and choose which therapies to employ. A humble healthcare practitioner will recognize that both the traditional healer and the physician come to the medical encounter with their own deeply rooted beliefs and values and that the goal is a good outcome for the patient, not for one side to "win." This sets the table for open and honest communication and maximizes the possibility of developing a shared care plan based on compromise and cultural sensitivity [38]. One might argue that compromise itself is an ethical behavior [39, 40].

You discuss this case with your colleagues. The majority of pediatric oncologists at the teaching hospital are in favor of referral to Child Protective Services. However, you, as the primary oncologist, disagree and request an ethics consultation. The ethics consultant believes that the parents' decision is not unreasonable and does not meet the threshold set by the harm principle. You also consult a neurooncologist at a major children's hospital, who agrees that, while he would not recommend forgoing radiation, he would not seek legal action to compel this treatment.

✅ 10. **Now how would you proceed?**

> **Box 10.5   Teaching Tip**
> With a show of hands, poll the students on which group they support, the primary oncologist and ethics consultant or the rest of the pediatric oncology team.

Physicians and other healthcare practitioners frequently disagree with each other. In Samuel's case, the differing opinions of the healthcare practitioners present particularly stark alternatives. One faction feels that he should be treated over his parents' objection, requiring the involvement of Child Protective Services and the legal system, while the other feels the parents' authority and the patient's evolving autonomy should be honored. Parents are generally better situated than others to understand the unique needs of their children and to make appropriate, caring decisions regarding their children's health care. This is not an absolute legal right, however, because the state also has a societal interest in protecting the child from harm and can challenge parental authority in situations in which a minor is put at significant risk of serious harm or neglect [20].

An Idaho newspaper recently reported that the state's Child Fatality Review Team had identified 10 preventable pediatric deaths between 2011 and 2013 that were attributed to parents' decision not to seek medical attention for their children, based on their religious beliefs [2].

✅ 11. **Self-reflection: Do your beliefs influence your personal medical decisions or advice for your family members? How so?**

> **Box 10.6   Teaching Tip**
> The goal of this question is to encourage reflection on one's personal, familial, cultural, and religious beliefs. Each facilitator and each learner brings their own beliefs to the practice of medicine. Identifying and acknowledging those beliefs is the first step to providing effective care to patients who do not share the same beliefs.

12. **Do you think that parents' religious beliefs make a stronger argument for respecting their decision compared to parents making the same decision on nonreligious grounds? Why or why not?**
    In general, all reservations—on the part of the parents, the child, or both—regarding whether to accept a particular treatment deserve careful consideration, regardless of whether they are motivated by religious beliefs, cultural practices, or other concerns. While individuals over the age of 18 are permitted to decline medical treatment when it conflicts with their religious beliefs, this constitutional guarantee of freedom of religion does not sanction the harming of another person, particularly a child, in the practice of one's religion [41]. Many people have argued that it is unjust to allow some children to die because of their parents' beliefs—beliefs that the children are too young to hold independently.
    In practice, religious beliefs—and some cultural ones—may be given more weight by healthcare practitioners because they are situated in broader contexts than simply the individual family unit and, to some extent, can be assessed in that context [42]. However, this does not make them more valid than other types of beliefs or concerns.
    (a) **Do you support state laws that exempt parents from medical neglect if they are motivated by religious beliefs? Why or why not?**
        It may be useful to consider the background on such laws. Several Supreme Court cases have addressed legal aspects of this issue, though not always in a medical context, including:
        - *Prince v. Massachusetts* (1944 Child labor laws and Jehovah's Witnesses): Court held that the government does have the authority to protect children and restrict parents' authority. "Parents may be free to become martyrs themselves. But it does not follow they are free, in identical circumstances, to make martyrs of their children before they have reached the age of full and legal discretion when they can make that choice for themselves."
        - *Wisconsin v. Yoder* (1972): Court upheld right of Old Order Amish to end children's schooling before age 16. "Court held that individual's interests in the free exercise of religion under the First Amendment outweighed the State's interests in compelling school attendance beyond the eighth grade."
        - *McKown v. Lundman* (1996): Court refused to hear appeal of a decision by the Minnesota Court of Appeals regarding the death of an 11-year-old boy whose mother and stepfather tried to use prayer to heal his diabetes. Court awarded $1.5 million to the child's father. Stephen Carter, a professor of law, wrote about this decision: "It is perfectly OK to believe in the power of prayer, so long as one does not believe in it so sincerely that one actually expects it to work—a peculiar fate indeed for our 'most inalienable' right."

        The disadvantages of state laws that exempt parents from medical neglect if they are motivated by religious beliefs are fairly straightforward. The most obvious disadvantage is that forgoing treatment for religious reasons when such treatment is likely to prevent substantial harm or suffering or death of a child fails to protect the future autonomy of the child. A classic example would be a child with appendicitis. The surgical treatment of appendicitis is considered to be a low-risk/high-reward intervention. Failure to seek medical care in this case for religious reasons could lead to catastrophic consequences for the child. In this type of medical neglect, the focus is on the caregiver's motivations or justifications rather than the child's needs [25]. The AAP considers failure to seek medical care in such cases to be child neglect, regardless of the motivation. Parents and others who deny a child necessary medical care on religious ground should not be exempt from civil or criminal action. Constitutional guarantees of freedom of religion do not permit children to be harmed through religious practices [43].

However, using the court system as the ultimate arbiter in religiously based disputes between families and clinicians presents its own host of potential problems. For an older child, from a practical standpoint, it is extremely difficult to force treatment without their assent, especially if the parents support their refusal. With the loss of trust that accompanies court-mandated treatment, parents will be less likely to seek help in the future for that particular child and for other children in the family and may be less willing to accept a compromise position that would lead to a good practical outcome for the sick child. From a pragmatic standpoint, court-mandated treatment may be useful for a single-treatment medical regimen such as surgery for appendicitis. On the other hand, cancer treatment typically takes place over months or years. Court-mandated treatment becomes much more problematic when the therapy requires ongoing treatment over time as circumstances change [39].

(b) **Self-reflection: How would you feel about caring for a child whose parents refused a recommended treatment based on their religious beliefs? Would it be difficult to show compassion for these parents?**

> **Box 10.7 Case Conclusion**
>
> The entire team eventually decided to support the family's decision to forgo radiation after an ethics consultation and consideration of the second opinion of the external neuro-oncologist. However, the primary oncologist asked them to meet with the radiation oncologist one more time before they left the clinic on the last day of chemotherapy.
>
> The day after that final discussion with the pediatric radiation oncologist, Samuel's father asked to speak with the primary oncologist again. He reported that they had carefully considered all the available information, prayed together, and decided to accept radiation therapy. A key factor in the decision seemed to have been that the radiation oncologist gave more explicit information about the differences in dose and treatment area between upfront and salvage radiation therapy. The family had not previously understood that relapse therapy would include much more extensive radiation. This information altered their assessment of the relative risks and benefits of each approach.
>
> Arrangements were also made to provide the radiation therapy free of charge. Samuel's schedule of follow-up appointments, tests, and imaging was thoroughly discussed with the family, and they agreed upon a modified schedule, intended to better balance the costs and low likelihood of recurrence with the benefits of detecting a treatable recurrence early.

Acknowledgments This case was adapted from an earlier case written by Amy Caruso Brown and published, in part, in the journal *Pediatrics* in 2017.

References

1. Calaminus G, Kortmann R, Worch J, Nicholson JC, Alapetite C, Garrè ML, et al. SIOP CNS GCT 96: final report of outcome of a prospective, multinational nonrandomized trial for children and adults with intracranial germinoma, comparing craniospinal irradiation alone with chemotherapy followed by focal primary site irradiation for patients with localized disease. Neuro Oncol. 2013;15(6):788–96.
2. Berry H. Idaho legislature's faith healing panel offers no recommendations: many want the open door for faith-healing permanently closed. Boise Weekly [Internet]. 2016 Aug 10 [cited 2018 Mar 4]. Available from: https://www.boiseweekly.com/boise/idaho-legislatures-faith-healing-panel-offers-no-recommendations/Content?oid=3866334.

3. Siegel DA, King J, Tai E, Buchanan N, Ajani UA, Li J. Cancer incidence rates and trends among children and adolescents in the United States, 2001–2009. Pediatrics. 2014;134(4):e945–55.
4. Ward E, DeSantis C, Robbins A, Kohler B, Jemal A. Childhood and adolescent cancer statistics, 2014. CA Cancer J Clin. 2014;64(2):83–103.
5. Price RA, Stanges E, Elixhauser A. Pediatric cancer hospitalizations, 2009. Statistical brief #132. Agency for Healthcare Research and Quality; 2012 May.
6. McBride ML, Duncan R, Bremner K, De Oliveira C, Liu N, Nathan PC, et al. Total and cancer-attributable phase-based costs for childhood cancer care: a population-based study in British Columbia. Canada J Clin Oncol. 2017;35(5 Suppl):11.
7. Bona K, London WB, Guo D, Frank DA, Wolfe J. Trajectory of material hardship and income poverty in families of children undergoing chemotherapy: a prospective cohort study. Pediatr Blood Cancer. 2016;63(1):105–11.
8. Bilodeau M, Ma C, Al-Sayegh H, Wolfe J, Bona K. Household material hardship in families of children post-chemotherapy. Pediatr Blood Cancer. 2018;65(1):1–4.
9. Warner EL, Kirchhoff AC, Nam GE, Fluchel M. Financial burden of pediatric cancer for patients and their families. J Oncol Pract. 2014;11(1):12–8.
10. Sibbald B, Roland M. Understanding controlled trials: why are randomised controlled trials important? BMJ Br Med J. 1998;316(7126):201.
11. Salter EK. Deciding for a child: a comprehensive analysis of the best interest standard. Theor Med Bioeth. 2012;33(3):179–98.
12. Carbone J. Legal applications of the "best interest of the child" standard: judicial rationalization or a measure of institutional competence? Pediatrics. 2014;134:S111–20.
13. Diekema D. Parental refusals of medical treatment: the harm principle as threshold for state intervention. Theor Med Bioeth. 2004;25:243.
14. Bredenoord AL, de Vries MC, van Delden H. The right to an open future concerning genetic information. Am J Bioeth. 2014;14(30):21–3.
15. English A, Ford CA. More evidence supports the need to protect confidentiality in adolescent health care. J Adolesc Health. 2007;40(3):199–200.
16. Guttmacher Institute. An overview of minor's consent law [Internet]. New York: The Institute. [cited 2018 Mar 4]. Available from: https://www.guttmacher.org/state-policy/explore/overview-minors-consent-law.
17. Coleman DL, Rosoff P. The legal authority of mature minors to consent to general medical treatment. Pediatrics. 2013;131(4):786–93.
18. Copen CE, Dittus PJ, Leichliter JS. Confidentiality concerns and sexual and reproductive health care among adolescents and young adults aged 15–25. NCHS data brief, no. 266. Atlanta (GA): CDC; 2016 Dec [cited 2018 Mar 4]. Available from: https://www.cdc.gov/nchs/data/databriefs/db266.pdf.
19. Katz AL, Webb SA. Informed consent in decision-making in pediatric practice. Pediatrics. 2016;138(2):136–41.
20. Mercurio M. An adolescent's refusal of medical treatment: implications of the Abraham Cheerix case. Pediatrics. 2007;120(6):1357–8.
21. Christian CW. The evaluation of suspected child abuse. Pediatrics. 2015;135(5):e1337–54.
22. Weaver MS, Arora RS, Howaed SC, Salaverria CE, Liu YL, Ribeiro RC, et al. A practical approach to reporting treatment abandonment in pediatric chronic conditions. Pediatr Blood Cancer. 2015;62:565–70.
23. Burton MC, Warren MB, Lapid MI, Bostwick JM. Munchausen syndrome by adult proxy: a review of the literature. J Hosp Med. 2015;10(1):32–5.
24. Gardner K, Ruest S, Cummings B. Diagnostic uncertainty and ethical dilemmas in medically complex pediatric patients and psychiatric boarders. Hospital Pediatrics. 2016;6(11):689–92.
25. Jenny C. Recognizing and responding to medical neglect. Pediatrics. 2007;120(6):1385–9.
26. Rohrer K, Dundes L. Sharing the load: Amish health-care financing. Healthcare. 2016;4(4):92.
27. Caruso Brown AE, Slutzky AR. Refusal of treatment of childhood cancer: a systematic review. Pediatrics. 2017;140(6):e20171951.
28. Kleinman A, Benson P. Anthropology in the clinic: the problem of cultural competency and how to fix it. PLoS Med. 2006;3(10):e294.
29. Groll D. Four models of family interests. Pediatrics. 2014;134(Supplement 2):S81–6.
30. Fallat ME, Hutter J. Preservation of fertility in pediatric and adolescent patients with cancer. Pediatrics. 2008;121(5):e146–1469.
31. Gillam L. The zone of parental discretion: an ethical tool for dealing with disagreement between parents and doctors about medical treatment for a child. Clinical Ethics. 2016;11(1):1–8.
32. Graham LL, Cates JA. Health care and sequestered cultures: a perspective from the old Amish order. J Multicultural Nursing Health. 2006;12(3):60–6.
33. Pelletier W, Bona K. Assessment of financial burden as a standard of care in pediatric oncology. Pediatr Blood Cancer. 2015;62(S5):S619–31.
34. Wiener L, Kazak AE, Noll RB, Patenaude AF, Kupst MJ. Standards for the psychosocial care of children with cancer and their families: an introduction to the special issue. Pediatr Blood Cancer. 2015;62(S5):S419–24.
35. Bona K, Blonquist TM, Neuberg DS, Silverman LB, Wolfe J. Impact of socioeconomic status on timing of relapse and overall survival for children treated on Dana-Farber Cancer Institute ALL consortium protocols (2000–2010). Pediatr Blood Cancer. 2016;63(6):1012–8.
36. Hubinette MM, Regehr G, Cristancho S. Lessons from rocket science: reframing the concept of the physician advocate. Acad Med. 2016;91(10):1344–7.
37. Weyer SM, Hustey VR, Rathbun L, Armstrong VL, Anna SR, Ronyak J, et al. A look into the Amish culture: what should we learn? J Transcult Nurs. 2003;14(2):139–45.
38. Prielipp RC, Wahr JA. The PLAIN truth: caring for the Amish: what every anesthesiologist should know. Anesth Analg. 2017;24(5):1387–8.

39. Gray B, Brunger F. (Mis)understandings and uses of 'culture' in bioethics deliberations over parental refusal of treatment: children with cancer. Clinical Ethics. 2017;6:1477750917738109.
40. Weinstock D. On the possibility of principled moral compromise. Crit Rev Int Soc Polit Philos. 2013;16:537–56.
41. American Academy of Pediatrics Committee on Bioethics. Religious exemptions from child abuse statutes. Pediatrics. 1988;81(1):169–71.
42. Sulmasy DP. Spirituality, religion, and clinical care. Chest. 2009;135(6):1634–42.
43. American Academy of Pediatrics Committee on Bioethics. Religious objections to medical care. Pediatrics. 1997;99(2):279–81.

Further Reading on this Topic

Brown AC. At the intersection of faith, culture, and family dynamics: a complex case of refusal of treatment for childhood Cancer. J Clin Ethics. 2017;28(3):228–35.

Diekema DS, Mercurio MR, Adam MB, editors. Clinical ethics in pediatrics: a case-based textbook. Cambridge, UK: Cambridge University Press; 2011.

Fadiman A. The spirit catches you and you fall down: a Hmong child, her American doctors and the collision of two cultures. New York: Farrar, Straus and Giroux; 1997.

McDougall R, Delany C, Gillam L, editors. When doctors and parents disagree: ethics, paediatrics and the zone of parental discretion. Annandale: Federation Press; 2016.

Miller G, editor. Pediatric bioethics. Cambridge, UK: Cambridge University Press; 2010.

Trauma, Violence, and Mental Health

Contents

Chapter 11 "He Has a Gun and Wants to Kill Himself" – 207

Chapter 12 "Bleeding Too Much" (In the Words of a Refugee) – 231

"He Has a Gun and Wants to Kill Himself"

Christopher R. Botash

11.1   Background Questions – 208

11.2   Additional Case Information and Questions for Discussion – 208

11.3   Answers to Background Questions – 210

11.4   Additional Case Information and Responses to Questions for Discussion – 215

References – 228

Learning Objectives

1. Describe barriers in mental health care in the United States.
2. Articulate the ethical and legal basis for the duty of confidentiality.
3. Evaluate and choose among potential courses of action when the physician's obligation to maintain patient confidentiality conflicts with the duty to protect or warn others of potential harm.
4. Recognize situations in which a patient's consent can be overridden and defend a course of action in such situations.
5. Apply the socio-ecological model to a complex situation, analyzing the extent to which societal factors can affect the outcome.

Case Background

You are a psychiatrist who has been seeing Christian DuPont, a 20-year-old man who has screened positive for depression. As part of your assessment, you ask him about gun possession. He reports that he owns a gun and tells you about how he safely stores the gun. He does not express any suicidal ideation. You prescribe fluoxetine and refer him to see a counselor.

11.1 Background Questions

1. What are the leading causes of death in the United States for people aged 15–24 years? What are the leading causes of unintentional injury death in this category? Where are firearms on this list?
2. How common is homicide? What are the most commonly used weapons? For homicides involving firearms, how often are handguns reported compared to rifles/assault rifles?
3. What are the long-term trends in deaths associated with firearms (homicide, suicide, and unintentional deaths)?
4. How are data on morbidity and mortality associated with firearm-related violence collected? What are some limitations of these data?
5. Under what circumstances is it legally permissible to breach a patient's confidentiality? Under what circumstances is it legally permissible to treat an adult patient without their consent?
6. What proportion of young adults who experience a major depressive episode does not receive mental health services in the U.S.? What proportion of adults diagnosed with mental illness does not receive mental health services in the U.S.? Consider why this may be.
7. Are physicians required to report patients who are at-risk for serious harm to self (or others) to their local or state government?
8. One argument against firearm-related legislation is that having a gun in the home makes the homeowner safe from a possible intruder. What evidence is available to support or refute this claim? How does the risk of personal or family injury change by having a gun in the home? How might past efforts to improve motor vehicle safety inform future research on firearm injury prevention?

11.2 Additional Case Information and Questions for Discussion

In 2004, the U.S. Food and Drug Administration (FDA) began requiring a "black box" warning on all antidepressants, indicating that initiation of use might increase rates of suicidal ideation and behavior among young people. Through social media, Christian has heard about this warning and asks your opinion about the warning. His parents, who often accompany him to his visits, also express concern.

1. **How would you balance the potential benefits of antidepressant use with the risks?**
   (a) **What would you tell Christian?**
   (b) **What would you tell his parents?**

In a follow-up phone call with Christian a month later, you learn that he has not seen a counselor. He states he gave up trying to see one after calling two providers who did not take his insurance. The "gold standard" for Christian would likely have been a combination of psychotherapy and pharmacotherapy, with close monitoring. Shortages of mental health providers, particularly those willing to accept certain types of insurance, including Medicaid, make the former more difficult to obtain. You are worried now because Christian has missed his last two appointments and is not returning your phone calls.

? 2. What was your obligation to Christian when he failed to find a therapist and did not follow up with you?

Christian's parents, Mr. and Mrs. DuPont, call you today requesting to meet with you as soon as possible. When Mr. DuPont walked into the garage earlier today, he found Christian loading a handgun. Christian told his father that he wanted to kill himself. Mr. DuPont was able to convince him to come to your office.

You arrange to meet with Christian and his parents. You spend some time with him alone as well as with his parents present. You assess that he is actively suicidal. You discuss your assessment and a recommendation for immediate hospitalization with Christian. He becomes agitated and refuses to be admitted. His parents are able to calm him down, but he still refuses admission.

? 3. Given that Christian is actively suicidal, what are your professional obligations in this situation?
   (a) What would you do next?

? 4. A death from a gunshot wound (or any other death from traumatic injury) is usually the result of "breakdowns" on multiple levels. Identify events in this case that might have ultimately resulted in Christian taking his own life using a gun.
   (a) Where in this "chain of events" are opportunities for action or prevention?
   (b) What role can physicians take in these opportunities?

You are able to get Christian involuntarily admitted to the psychiatric unit. He does well with his inpatient course. His parents again reach out to you before Christian is discharged. They are concerned about Christian's long-term safety and ask you for advice about what they can do within their home to decrease his risk of self-harm. You assure them that you will continue to work with Christian and that you found a counselor who accepts his insurance. You also discuss firearm safety with his parents and learn that even before becoming a legal adult, Christian had access to guns in the home. Christian's father was formerly in the military. Both men are avid hunters who also enjoy target shooting at a gun range. You advise Mr. and Mrs. DuPont to change the way firearms are stored in the home so that your patient has no access unless supervised.

In multiple states, there have been attempts to restrict physicians from discussing gun ownership with patients. For example, the Firearms Owners' Privacy Act passed by the Florida legislature in 2011 restricted physicians from asking patients about gun ownership—unless there was concern for the patient's safety or the safety of others [1]. The Florida legislature passed this law after hearing multiple complaints about healthcare providers asking invasive questions on gun ownership and requesting parents to remove guns from their homes. In one case, a pediatrician asked a mother to find another physician after the mother refused to answer questions about firearms in the home [2].

? 5. Self-reflection: How do you feel about the pediatrician asking the family to find another physician because the mother refused to discuss firearm ownership?

While key components of the law were struck down as unconstitutional in 2017, such legislation continues to be the subject of much debate [2, 3].

? 6. Are laws that limit what physicians can discuss with patients ethical?

? 7. How are gun injury and gun violence portrayed in the media? Do you think the major causes are covered adequately?

Over the past several months, you have worked closely with Christian's counselor. She calls you today after a follow-up visit with her. She is concerned that Christian has been making some vague but concerning comments, such as saying that he "understands how the killers in Las Vegas and Virginia Tech must have felt."

8. What is your obligation in this situation?
   (a) What if Christian made a more specific threat (e.g., toward an individual person)?

9. What are the ethical implications of mandated reporting laws designed to remove firearms from those who are at-risk to harm themselves or others?
   (a) Is it ethical to take away someone's rights because of their underlying medical condition, including depression or other mental illness?

10. What are some of the challenges of evaluating the impact of policy change? How would you conduct a study to assess the impact of such changes?

11. Creative Problem-Solving: How would you construct a law or policy to decrease deaths associated with firearms?

11.3 Answers to Background Questions

1. **What are the leading causes of death in the United States for people aged 15–24 years? What are the leading causes of unintentional injury death in this category? Where are guns on this list?**
   As detailed in the National Vital Statistics System, in 2015, the six leading causes of death for individuals aged 15–24 were:
   1. Accidents (unintentional injuries)
   2. Intentional self-harm (suicide)
   3. Assault (homicide)
   4. Malignant neoplasms
   5. Diseases of the heart
   6. Congenital malformations, deformations, and chromosomal abnormalities

   Other leading causes included influenza and pneumonia, chronic lower respiratory diseases, and diabetes mellitus; 30,494 deaths from all causes were reported in this age group [4]. The "unintentional injuries" label encompasses a wide range of ICD-10 codes. The most commonly cited accidents in younger age groups—particularly teenagers—include transportation (i.e., motor vehicle accidents), drowning, poisoning, fires, and sports-related injuries. Alcohol use is frequently a contributing factor in some of these injuries [5].

   Though the leading causes of death may change their order slightly from year-to-year, these top six causes have remained generally consistent since 1999. Stratification of the data by gender shows that homicide consistently ranks second for males (77,781 deaths recorded from 1999 to 2016) versus third for females (12,247 deaths). As these statistics allude, absolute numbers for male homicide/suicide/unintentional injury are much greater than those for females. Race also impacts the order of this list: Accidents are more than double the number of suicides/homicides in white people ages 15–24 (from 1999 to 2016: 209,419 vs. 101,157, respectively). Conversely, for black people in the same age group and time period, homicides (52,960) exceed the number of accidents and suicides combined (37,895) [6].

   Firearms may be involved in any of these top three causes. For ages 15–24, the U.S. Centers for Disease Control (CDC) reports more than 106,350 deaths involving firearms from 2001 to 2016, with more than 7500 in 2016 alone. Of these, 63% were homicides, 32% were suicides, and 2% were deemed unintentional. Once again, stratification by race and gender reveals significant differences: 95,557 of these recorded firearm-related deaths were male. There were 54,480 recorded firearm-related deaths among white young people during this time period, and an estimated 53% were suicides. Conversely, among black young people, there were 48,437 firearm-related deaths, but an estimated 88% were homicides [6].

> **Box 11.1 Teaching Tip**
> Many of the statistics in this section were obtained using the CDC's WISQARS database (Web-based Injury Statistics Query and Reporting System). Available at ▶ https://www.cdc.gov/injury/wisqars/index.html.
> The fatal injury component of this system includes a unique data visualization tool that allows for sorting of fatalities by intent and mechanism, as well as filtering of results by state, race, ethnicity, sex, and age.

2. **How common is homicide? What are the most commonly used weapons? For homicides involving guns, how often are handguns reported compared to rifles/assault rifles?**
   In a 2011 article on homicide, the homicide rate in America was estimated at 4.7 per 100,000 people. The murder rate for males was higher (7.4) compared to females (2.0). Handguns were identified as the most commonly used weapon in homicides, for both sexes. Among male homicides from 1992 to 2011, handguns were used an average of 57% of the time, compared to 16% use of other firearms (including rifles, shotguns, and automatic weapons). This tendency persists among female homicide victims, with an average 35% handgun use and 13% other firearms [7].
   The U.S. Federal Bureau of Investigation's (FBI) expanded homicide data showed that almost 65% of homicides by firearm were attributed to handguns, whereas less than 5% were attributable to rifles or shotguns. The type of firearm used was not specified in approximately 30% of reported homicides [8].

3. **What are the long-term trends in deaths associated with guns (homicide, suicide, and unintentional deaths)?**
   In 1992, the homicide rate in America was 9.3 per 100,000 people, about twice the rate found in 2011. Though the rate of homicides involving a firearm mirrored this downward trend (declining 49%), the percentage of homicide victims killed by a firearm (67%) remained about the same [7]. As detailed above, trends must also be considered in the context of race, age, and ethnicity: For example, the overall firearm-related homicide rate in 2016 was calculated to be 4.6 victims (age-adjusted using 2000 as a standard year) per 100,000 individuals. Among white people, this rate was 2.3, while among black people, the rate was 17.7. Rates in both groups have begun to increase again in recent years [6]. Note that different organizations (e.g., Bureau of Justice Statistics, CDC) make calculations using data gathered from different sources (see below); therefore, reported figures may not always match, although the trends are generally comparable.
   From 1999 to 2014, the age-adjusted suicide rate in the United States increased by 24% (10.5 to 13.0 per 100,000 people). The annual percent increase also trended upward. The proportion of suicides involving firearms decreased during this same time period. Among females, firearms were the most common means to commit suicide (36.9%) in 1999 but decreased to 31.0% in 2014. Similarly among males, firearm use in suicide decreased from 61.7% to 55.4%. Of note, intentional poisoning overtook firearms as the leading method of suicide among females in 2014 (34.1%), while firearm use remained the most common method among males [9].
   Regarding unintentional or accidental deaths associated with guns, available data is less reliable as "accidental death" by firearm does not have a universally accepted definition. Limitations of data gathering methods must also be considered (refer to Background Question 4). Querying the WISQARS database (Web-based Injury Statistics Query and Reporting System), provided through the CDC Website, unintentional gun deaths are trending down: 824 deaths were recorded in 1999, while just 495 were identified in 2016 [6]. Unintentional firearm injuries have been trending down since the early 1900s, possibly secondary to changes in firearm safety or cause-of-death coding practices [10].

4. **How are data on morbidity and mortality associated with gun violence collected? What are some limitations of these data?**
   Data on morbidity and mortality associated with gun violence are collected by multiple government agencies. There are two national sources of homicide data: the FBI's Supplementary Homicide Reports (SHR) and the CDC's National Vital Statistics System (NVSS). For the SHR, participating law enforcement agencies voluntarily submit annual summaries on homicide cases, which generally include information about the event (e.g., weapon used), victim, and offender. The SHR does not include cases handled by federal law enforcement, and homicides are defined based on police investigation, rather than a medical examiner, court, or coroner determination. The NVSS provides more general demographic information on homicide victims, as its data are based on death certificates filed with state governments. Cause of death is coded using the ICD-10 system. Unlike the SHR, the physician or other medical professional responsible for completing the death certificate makes a determination of homicide as the cause of death [7]. The NVSS is also the primary source of data for suicides via fatal gun injury. Nonfatal firearm injuries are catalogued in the National Electronic Injury Surveillance System (NEISS), operated by the U.S. Consumer Product Safety Commission (CPSC). This system extrapolates its data from a sample of U.S. hospital emergency departments (EDs), which strive to report cause of injury, race/ethnicity, sex, and disposition (e.g., hospitalized, transferred, or discharged).

   Discussion of these various collection methods and agencies alludes to some of the inherent limitations of the collected data. Though the SHR data contains important detail on the weapons used in homicides, it is potentially limited by a selection effect: Only murders known to law enforcement are captured in this system, including homicide victims who are not U.S. residents [7]. Similarly, the NEISS is dependent on participating hospitals submitting relevant data. The potential for non-response bias must be adjusted for in any statistics derived from these data [10]. Though the NVSS captures its data from a more extensive record system, it remains vulnerable to measurement errors. Suicide, homicide, and unintentional fatal injury from firearms may initially appear as distinct categories, but discrepancies in how deaths are classified can cause inaccurate counts. This phenomenon is particularly relevant when examining accidental firearm-related deaths. Previous research has shown both type 1 and type 2 errors. Young victims in particular are more likely to be misclassified as homicides, rather than accidents. Certain coding practices (in ICD-10) have also been suspected to contribute to "over-reporting" of accidental deaths when information on how the victim died is not available at the time of death certificate completion [11].

5. **Under what circumstances is it legally permissible to breach a patient's confidentiality? Under what circumstances is it legally permissible to treat an adult patient without their consent?**
   The answers to these questions continue to be debated and revised and may change based on clinical setting (e.g., hospital versus prison). However, there are certain generally recognized exceptions to both the duty of confidentiality and informed consent. Philip Merideth, MD, JD, uses a "Five Cs" mnemonic to identify exceptions to confidentiality: Consent, Court Order, Continued Treatment, Comply with the Law, and Communicate a Threat. Medical professionals may release confidential information with consent of their patients (or patients' legal guardians) or on court order. Relevant information may also be released between providers caring for the same patient (e.g., during transition of care). In some situations and regions, clinicians are required to report confidential information to law enforcement or government agencies. For example, all U.S. states have

statutes requiring medical professionals to report suspected child maltreatment to the appropriate authority/agency. "Communicate a Threat" refers to a physician's duty to protect others from violence by a patient—also known as the "Tarasoff exception" [12].

There are also three general circumstances where exceptions to direct informed consent by the patient are commonly cited: (1) emergency situations where a patient cannot communicate a choice or their life is imminently threatened, (2) situations where a patient is found to lack decision-making capacity, and (3) situations in which a patient has legally waived their right to information disclosure (e.g., the patient does not want to know the risks of a particular therapy). Often the most contentious circumstances are those involving evaluation of decision-making capacity. There are four main abilities essential to making any decision:

Patients must (1) communicate a consistent choice, (2) understand the relevant information regarding that choice, (3) appreciate the situation and consequences of their choice, and (4) rationally process information regarding other treatment options. If a physician feels that a patient lacks any of these four abilities, the patient is considered to lack decision-making capacity and cannot give informed consent [13]. In non-emergency situations, a guardian is generally chosen to make decisions.

✅ 6. **What proportion of young adults who experience a major depressive episode does not receive mental health services in the United States? What proportion of adults diagnosed with mental illness does not receive mental health services in the United States? Consider why this may be.**

These questions are inherently difficult to answer, as individuals who do not receive treatment are difficult to capture in any surveillance system. Even among individuals who do participate, classification of their mental illness (according to the DSM) is reliant on their ability to accurately recall and report symptoms. Keeping these limitations in mind, the Substance Abuse and Mental Health Services Administration (SAMHSA) issues reports on key mental health indicators in the United States—utilizing data gathered from the National Survey on Drug Use and Health (NSDUH). In a 2016 report, SAMSHA identified that among the 3.7 million young adults (ages 18–25 years) who had a major depressive episode that year, just 1.6 million (44.1%) received mental health treatment. Also in 2016, there were an estimated 44.7 million adults (ages 18 and older) who met criteria for "any mental illness," and 19.2 million (43.1%) received mental health services [14].

There are myriad reasons that may contribute to this discrepancy between diagnosis and treatment. Perhaps the most latent barrier is a persistent stigma associated with mental health. The National Alliance on Mental Illness (NAMI) attempts to shed light on how people living with mental health conditions are often alienated, perceived as dangerous, less likely to be hired, and more likely to be criminalized [15]. Therefore, people living with mental illness do not feel comfortable sharing their struggle with others. Stigma can be internalized and lead to minimization of symptoms. Previous research has shown a "desire to handle the problem on one's own" as a commonly cited reason for not seeking or abandoning treatment [16]. As awareness of mental illness has increased in recent years, so too has awareness of practical barriers to obtaining mental health care. Despite passage of the Mental Health Parity and Addictions Equity Act in 2008 (an attempt to require equal health insurance coverage for both mental and physical health benefits), a 2016 NAMI survey found many health plan provider networks to be much smaller for mental health compared to primary care [17]. When patients cannot find a provider in their network, out-of-pocket costs for mental health care substantially increase.

7. **Are physicians required to report patients who are at-risk for serious harm to self (or others) to their local or state government?**
   Nearly all states do have laws either permitting or requiring mental health professionals to disclose information about patients who pose a potential risk for serious harm to self or others [18]. In New York State, for example, Mental Hygiene Law 9.46 (otherwise known as the NY Secure Ammunition and Firearms Enforcement Act [NYSAFE]) was passed in March 2013. The law requires mental health professionals—including physicians, psychologists, registered nurses, and licensed social workers—to report to their local director of community services (DCS) when in their "reasonable professional judgment" one of their patients is "likely to engage in conduct that would result in serious harm to self or others" [19]. The NYSAFE law goes a step further than other states, allowing law enforcement to potentially remove firearms owned by reported persons. Many of these "duty to warn" laws are derived from the landmark California Supreme Court Case *Tarasoff v. Regents of the University of California* in 1976.

8. **One argument against firearm-related legislation is that having a gun in the home makes the homeowner safe from a possible intruder. What evidence is available to support or refute this claim? How does the risk of personal or family injury change by having a gun in the home? How might past efforts to improve motor vehicle safety inform future research on firearm injury prevention?**
   There is little reliable evidence to provide a clear answer regarding firearm ownership improving homeowner safety. Part of the reason for this dearth of data involves a provision (referred to as the Dickey Amendment) in a 1996 spending bill passed by the U.S. Congress that prohibited the CDC from using federal funds "to advocate or promote gun control." Though not explicitly banning research on firearm-related violence, the provision resulted in general avoidance of the divisive subject. Furthermore, Congress revised the CDC's budget for 1997 so that funds previously used for studying firearm-related injury were reallocated to other areas of research [20]. Funding remained scarce over the next 20 years, despite President Obama signing an executive order in 2013 directing the CDC to resume research on firearm-related violence.

   The Dickey Amendment arose after a 1993 *New England Journal of Medicine* article found that keeping a gun in the home is "independently associated" with an increased risk of homicide—frequently at the hands of a family member or intimate acquaintance. The study did not find evidence of a protective benefit from gun ownership, even in cases where resistance to home invasion was attempted. Controversially, the article stated, "People should be strongly discouraged from keeping guns in their homes" [21]. After this publication received significant media coverage, the National Rifle Association (NRA) responded by lobbying for elimination of the center that funded the study: the CDC's National Center for Injury Prevention [20].

   Previous research has not only suggested a link between firearm ownership and risk of homicide but also risk of suicide. One review of case-control studies found that gun ownership increases suicide risk 2 to 10 times that of homes without guns (dependent on sample population and method of firearm storage). Compelling ecological data has revealed that states with higher rates of household gun ownership also have higher rates of firearm-related suicide [22]. A common criticism of this type of evidence notes that association does not necessarily imply causation. Case-control studies are particularly susceptible to confounding variables.

   Increased suicide risk with gun ownership does not just apply to the gun owner but also extends to his or her family. As discussed in Background Question 1, suicide is one of the leading causes of death for American youth. Based on an array of past research, the American Academy of Pediatrics (AAP) has concluded that the presence of firearms in the home further increases the risk of suicide among adolescents, which guides their recommendations on physician counseling of parents (refer to Discussion Question 4) [23]. Perhaps not unexpectedly, previous research has also revealed that safe storage (e.g., locked, unloaded) of firearms is associated with at least a 60% reduction in firearm-related suicide risk [24].

Unintentional transportation-related deaths far exceed the number of unintentional firearm-related fatalities. In the 15–24 age group, from 1999 to 2016, the CDC reports 158,947 deaths from car accidents. Comparatively, there were 2,908 recorded fatalities related to accidental firearm discharge. This number rises to 116,955 when accounting for firearm-related suicides and homicides [6]. Though still less than motor vehicle accidents, this significant number prompts comparison between motor vehicle-related and firearm-related injury prevention efforts. The National Highway Traffic Safety Administration (NHTSA), first established in 1970, has spent years attempting to "save lives, prevent injuries and reduce economic costs due to road traffic crashes, through education, research, safety standards and enforcement activity" [25]. The NHTSA's efforts have influenced the transportation industry in countless ways, including improving structural integrity of vehicles, vehicle safety systems, and roadway design. These improvements are based on a massive catalogue of research incorporating everything from examining biomechanics during traumatic events (i.e., crashes) to evaluating the efficacy of recent active braking and pedestrian detection technologies. The Fatality Analysis Reporting System (FARS) reveals a generally steady decline in fatal car crashes, from 42,130 in 1988 to 34,439 in 2016 [26].

Though firearms represent a markedly different safety challenge regarding their design, sale, distribution, and intent for use, there are lessons to be learned from motor vehicle injury prevention. Significantly, the NHTSA does not just limit their funded research to vehicle and road design; behavioral and human factor researches are also top priorities. Studies have been conducted on who wears seatbelts, where and when people speed, alcohol-impaired driving, and the extent of cell phone use while driving—just to name a few. In the same way, research on firearms might examine individuals' reasons for firearm ownership, methods of firearm storage, and knowledge of firearm safe-handling practices. The Crash Injury Research (CIREN) of the NHTSA utilizes multidisciplinary teams of physicians, crash investigators, and mechanical engineers to determine injury causation [27]. The circumstances leading to firearm-related injury are often similarly uncertain and require thorough investigation. Of course, reliable and effective research requires extensive funding. The NHTSA has utilized billions of dollars over the years, with their 2018 budget request allocating approximately $900 million [25]. As already mentioned, firearm-related research will require allotment of significant additional funding in order to positively effect safer firearm use.

11.4 Additional Case Information and Responses to Questions for Discussion

In 2004, the FDA began requiring a "black box" warning on all antidepressants, indicating that initiation of use might increase rates of suicidal ideation and behavior among young people. Through social media, Christian has heard about this warning and asks your opinion about the warning. His parents, who often accompany him to his visits, also express concern.

✓ 1. **How would you balance the potential benefits of antidepressant use with the risks?**
Every medical decision involves weighing potential risks against desired ben-

efits. Counseling patients and family about potential adverse reactions of medications becomes challenging when evidence regarding these reactions is inconsistent or unclear. Conversations become even more complex when patients are already wary of trying pharmacotherapy. In 2004, based on extensive meta-analysis of randomized controlled trials, the FDA issued a black box warning for all antidepressants stating that there is an increased risk of suicidal thoughts or behavior in children and adolescents who use these medications. In 2006, an advisory committee to the FDA recommended extending the warning to include all young adults up to age 25 [28]. The initial warning was based on the FDA's finding that the rate of suicidality appeared to double among patients assigned to receive an antidepressant compared to a placebo.

Though appearing straightforward, this conclusion is clouded in controversy. For example, there is significant debate about methodological issues plaguing the meta-analyses. No completed suicides occurred among the children in the included trials, and there have been questions regarding the validity of the assessment of suicidality in these trials [28, 29]. Furthermore, though the rate of suicidality "doubled" with antidepressants, it remained a rare 4%. Suicide in general is a rare event (i.e., small sample size), which means that it can be hard to detect an effect in randomized controlled trials with appropriate statistical power. The problem is further complicated by the fact that depression itself increases the risk for suicide. Therefore, ascertaining causality with antidepressant use is extremely difficult.

After the issuance of the black box warning, antidepressant use trended down not only among adolescents but also adults—for whom the advisory was never intended to apply. Though association does not necessarily imply causation, it is reasonable to suggest that given the intense media coverage of the FDA advisory, physicians were more reluctant to prescribe antidepressants, and patients less likely to take them [29]. Years later, research continues to present conflicting results on the efficacy of antidepressant use. In their 2016 recommendation statement, the USPSTF cited multiple previous systematic reviews that appeared to establish benefits of treating adults with antidepressants [30]. Yet researchers have also called attention to the complexity of analyzing antidepressant effects. Patients with major depressive disorder are prone to non-specific treatment effects and often have a high rate of response to placebo. Over the last several decades, the estimated effect size of antidepressants has continued to decrease [31].

(a) **What would you tell Christian?**
(b) **What would you tell his parents?**

Both of these questions allude to the concept of "informed consent." Christian is an adult; therefore, he is responsible for providing informed consent in this case. In cases involving minors (e.g., children, adolescents), or where an adult is found to lack capacity, parents and/or family may become responsible to provide informed consent. Exceptions to informed consent were reviewed in Background Question 5. Informed consent does not simply mean that a patient has said "yes" to treatment. The "informed" component refers to the process of giving consent. The American College of Physicians' *Ethics Manual* (sixth edition) summarizes this process in several statements:

> » The principle and practice of informed consent rely on patients to ask questions when they are uncertain about the information they receive; to think carefully about their choices; and to be forthright with their physicians about their values, concerns, and reservations about a particular recommendation. Once patients and physicians decide on a course of action, patients should make

every reasonable effort to carry out the aspects of care under their control or inform their physicians promptly if it is not possible to do so. The physician must ensure that the patient or the surrogate is adequately informed about the nature of the patient's medical condition and the objectives of, alternatives to, possible outcomes of, and risks of a proposed treatment. [32]

As mentioned, one component of consent is to inform regarding alternatives to the proposed treatment. For many patients, psychotherapy is often tried as an initial treatment for mild depression. Even if there is no perceived benefit by the patient, psychotherapy can help to further qualify a patient's depression and whether antidepressant medication is appropriate [28]. The Treatment for Adolescents with Depression Study (TADS) found that a combination of medication (fluoxetine) and psychotherapy (cognitive behavioral therapy) was superior to either modality offered alone [33].

When obtaining informed consent for starting an antidepressant, as with many other medical decisions, it is challenging to be honest while neither frightening families unnecessarily nor downplaying rare but serious risks. Risks (e.g., side effects, adverse reactions) should never be dismissed, no matter how rare or how widely used a medication may be. Along with discussion of risks, physicians should emphasize how these risks will be managed. When starting an SSRI in a child, for example, both the doctor and the parent will need to carefully monitor for potential side effects (e.g., suicidal ideation) and create a plan for dealing with these adverse events (e.g., take the child to an ED for evaluation), should they arise.

The risks of going without treatment must also be part of informed consent. In the case of antidepressants, the morbidity and mortality risks of untreated depression may greatly exceed the small risks associated with treatment—particularly when the risks of medication can be managed by careful monitoring [29]. After presenting patients (or surrogate decision-makers) with all of this information, a key final step is to check for their understanding. Perhaps the easiest way to perform this check is to ask the patient to explain their understanding in their own words. Patients must feel comfortable with their physician in order to convey their concerns and ask relevant questions. Only then can informed consent be provided.

In a follow-up phone call with Christian a month later, you learn that he has not seen a counselor. He states he gave up trying to see one after calling two providers who did not take his insurance. The "gold standard" for Christian would likely have been a combination of psychotherapy and pharmacotherapy, with close monitoring. Shortages of mental health providers, particularly those willing to accept certain types of insurance, including Medicaid, make the former more difficult to obtain. You are worried now because Christian has missed his last two appointments and is not returning your phone calls.

2. **What was your obligation to Christian when he failed to find a therapist and did not follow up with you?**
   This is a complicated question with a multitude of reasonable answers that depend on the circumstances of Christian's case. In general, the primary care provider (PCP) is responsible for coordinating his or her patient's overall care, which includes communicating with the patient and liaising with consultants [32]. Given this enormous responsibility, physicians are encouraged to seek competent consultation when they (or their patients) need assistance in determining (or providing) appropriate management. Referring Christian to a psychotherapist was intended to provide an additional treatment modality beyond what his physician could offer. When Christian was unable to establish care with a counselor due to insurance limitations, his PCP had an ongoing responsibility to continue offering potential alternative therapy options. Even if Christian's PCP desired to transfer Christian's care to a psychiatrist (or other PCP), he or she would remain responsible for Christian's care until the formal transfer is complete (i.e., after Christian has begun to see his new provider).

> **Box 11.2 Teaching Tip**
> If background questions were assigned, encourage students with Background Question 6 to comment on the difficulties of accessing mental health care in the United States. How much time is reasonable for a PCP to devote to trying to find a therapist who accepts Christian's insurance? There is no universal answer to this question. Consider ways to potentially lessen the burden on PCPs. Some practices have social workers and case managers assigned to patients, who assist in the process of finding covered providers.

In Christian's case, his PCP decided to pursue a pharmacotherapy route and prescribed fluoxetine. However, Christian then missed his follow-up appointment, and despite repeated attempts of his PCP to follow up, Christian did not return any phone calls. As an adult, Christian is presumed competent—a legal determination—unless declared otherwise in court [32]. It is important to distinguish competency from capacity—a clinical determination (refer to Background Question 5). Competent patients who do not lack capacity are generally free to make their own decisions—even if that decision is to terminate follow-up with their physician. In cases of mental illness, patients can meet criteria for assisted outpatient treatment (AOT)—essentially a court order mandating compliance with an approved outpatient treatment plan. Different states have their own laws in place to facilitate AOT. In New York State, for example, Kendra's Law (MHL Section 9.60) allows physicians to petition a supreme or county court if there is concern that their patient is unable to reside safely in the community without supervision. Patients must meet a number of criteria, including but not limited to being at least 18 years of age, suffering from a diagnosed mental illness, being unlikely to survive without supervision, being nonadherent to treatment resulting in hospitalization at least twice in the last 3 years, and posing a threat to self or others within the last 4 years [34]. Patients who refuse to comply with their AOT order (under Kendra's Law) may be detained and transported to a hospital, where they can be held for up to 72 hours while awaiting psychiatric assessment for potential involuntary hospitalization.

AOT orders apply to a limited subset of mental health patients. Based on the presented background information, it is unlikely that Christian would have met criteria for an AOT order—at least prior to his current suicide attempt. In Christian's case, his PCP is expected to take reasonable steps to ensure follow-up but cannot compel him to do so. "Reasonable" does not have a precise definition, but calling Christian to check in after missed appointments and offering multiple follow-up times are essential practices. If a doctor has any concern about an individual's health or safety, calling the police and asking for a "welfare check" is another potential option. Generally, no patient care information is required to perform a welfare check. Friends and family can take this step as well. Police will then visit a patient's residence to verify his or her wellbeing.

Christian's parents, Mr. and Mrs. DuPont, call you today requesting to meet with you as soon as possible. When Mr. DuPont walked into the garage earlier today, he found Christian loading a handgun. Christian told his father that he wanted to kill himself. Mr. DuPont was able to convince him to come to your office.

You arrange to meet with Christian and his parents. You spend some time with him alone as well as with his parents present. You assess that he is actively suicidal. You discuss your assessment and a recommendation for immediate hospitalization with Christian. He becomes agitated and refuses to be admitted. His parents are able to calm him down, but he still refuses admission.

3. **Given that Christian is actively suicidal, what are your professional obligations in this situation?**
   When a patient is at imminent risk of "harm-to-self" (or others), the physician is

> **Box 11.3 Teaching Tip**
> Background Question 5 introduces physician responsibilities regarding maintaining confidentiality and treatment without consent. Consider discussing broad "exceptions to the rule" prior to addressing this particular case.

obligated to intervene, including taking appropriate steps to ensure the patient's immediate safety (or the safety of others). Key components of this particular case include that Christian appears in "imminent" (i.e., emergent) danger of harming himself, and there is also some question of a major depressive disorder, as suggested by his previous positive depressive screen. Furthermore, it is unclear if he has been compliant with his prescribed antidepressant. These elements simplify the physician's role at the present moment: Christian is in need of further evaluation and potential treatment.

The commonly cited principle of beneficence suggests that physicians have a duty to act in the best interests of their patients [32]. Christian is communicating a consistent choice (refusing hospitalization and attempting to end his life) and has been informed of the relevant information regarding his options (benefits of hospitalization vs. risks of returning home). Based on the presented case information, there is currently no evidence that Christian appreciates the consequences of his choice or is rationally processing information to arrive at his decision. Put simply, the average 20-year-old male does not typically want to die.

In this case, there is significant concern that the patient may lack capacity to make a decision regarding inpatient psychiatric admission due to an underlying (and untreated) mental illness. This argument is used in many guidelines and laws regarding involuntary hospitalization, but it also raises the idea of whether suicidality can exist without mental illness or be considered rational in certain circumstances. For example, the situation seems to change significantly if Christian was recently diagnosed with terminal bone cancer and is currently experiencing significant pain. Or perhaps he was diagnosed with severe, recurrent major depressive disorder years ago and continues to suffer despite a variety of medication trials. These characterizations allude to a broader underlying conflict between the principles of beneficence and respect for patient autonomy. Despite this conflict, there is significant legal precedent that physicians have a responsibility to intervene (or potentially face malpractice claims related to negligence) when a patient has been identified as suicidal [35]. In many outpatient scenarios, including Christian's, the physician lacks enough information or resources to ensure the patient's immediate safety at home. Therefore, sending the patient for emergency assessment is often the safest course of action.

(a) **What would you do next?**
An acutely suicidal patient should not be allowed to leave the office until an appropriate treatment plan is in place—in this case, emergency assessment in the ED for potential psychiatric admission. Staff should remain with the patient while transport arrangements are made. Even in cases where the suicidal patient is seeking treatment, the safest course of action is for ambulance or law enforcement to assist with transport [36]. Suicidal patients should never be discharged from the office alone unless they are threatening staff. In this case, local law enforcement may be contacted to issue a "pickup order" to bring the patient to the ED for emergency evaluation. Once in the ED, comprehensive workup includes evaluation for potential comorbid conditions (e.g., substance intoxication, delirium). Throughout evaluation, patients are often continuously observed, and steps are taken to ensure a safe environment (e.g., removal of sharp objects, heavy equipment, ropes, or strings).

Once Christian is "medically cleared," he would most likely be admitted to an inpatient psychiatric unit. States have varying statutes in place to facilitate psychiatric admission, but in general, patients are either voluntary or involuntarily hospitalized. As an example, in New York State, there are three methods to involuntarily commit an individual to psychiatric treatment: medical certification, certification by a director of community services (DCS), or emergency admission [37].

1. Medical certification (MHL Section 9.27) involves two physicians certifying that the patient requires treatment in a psychiatric facility. An application for admission is made by a third individual who is familiar with the patient (or a government official).
2. DCS certification (MHL Section 9.37) must be completed by a director of community services or their designee (physicians can apply for this designation) and is used when a person is believed to have a mental illness that will likely result in serious harm to self or others—necessitating immediate inpatient care.
3. Emergency admission (MHL Section 9.39) similarly involves an evaluating physician claiming that a patient has a mental illness likely to result in serious harm to self or others—requiring immediate treatment.

These three forms of involuntary commitment differ in how soon after admission a patient must be re-evaluated by a staff psychiatrist. They also differ in the length of time they may be used to hold a patient involuntarily. The common theme among these forms is that they are used when a patient is deemed to be at significant risk of serious harm to self or others.

Most mental hygiene laws (MHLs) contain provisions designed to guide clinicians in determining whether use of the law is appropriate. For example, the application for 9.37 defines "likely to result in serious harm" as:

» A substantial risk of physical harm to the person as manifested by threats of or attempts at suicide or serious bodily harm or other conduct demonstrating that the person is dangerous to himself or herself ('other conduct' shall include the person's refusal or inability to meet his or her essential need for food, shelter, clothing, or healthcare, provided that such refusal or inability is likely to result in serious harm if there is not immediate hospitalization) OR a substantial risk of physical harm to other persons as manifested by homicidal or other violent behavior by which others are placed in reasonable fear of serious harm [38].

By this definition, Christian meets criteria for involuntary admission to a psychiatric unit and would likely be admitted after evaluation in the ED. Of note, patients who sign voluntary admission have a right to request discharge, but if their caregivers have acute safety concerns, a voluntary status may be converted to involuntary. Patients may also object to involuntary admission and, in New York State, have a right to legal representation through the Mental Hygiene Legal Service [37].

4. **A death from a gunshot wound (or any other death from traumatic injury) is usually the result of "breakdowns" on multiple levels. Identify events in this case that might have ultimately resulted in Christian taking his own life using a gun.**

"He Has a Gun and Wants to Kill Himself"

> **Box 11.4 Teaching Tip**
> Encourage students to order events chronologically and begin the timeline prior to the circumstances depicted in the case background. Presume that Christian's father did not walk in on him loading the handgun.

The sequence of events might look like the following:
1. Christian is in good physical health and is comfortable with his state of wellbeing.
2. Christian becomes depressed.
3. The depression worsens and Christian develops passive thoughts of dying.
4. These suicidal thoughts evolve into thoughts of killing himself, although Christian does not yet have a specific plan.
5. These active suicidal thoughts continue to develop; Christian develops a plan to use a gun to end his life.
6. Christian decides he is going to act on this plan.
7. Christian has access to a gun and ammunition in his home.
8. Christian uses the gun in an attempt to end his life.

A "Haddon matrix," conceived by William Haddon, can be used to help organize the host, agent, and environmental factors (both risk and protective) that exist during a sequence of events resulting in injury or death [39]. A sample matrix is outlined in ▪ Table 11.1, with case events separated into "pre-event," "event," and "post-event" phases. The factors identified do not represent an exhaustive list. In this matrix, the event or "injury" is defined as Christian developing and acting on his suicidal thoughts. References to the chain of events above are noted in the phase column.

(a) **Where in this "chain of events" are opportunities for action/prevention?**
All of the factors listed in ▪ Table 11.1 represent potential points of intervention. The broadly outlined "phases" roughly correlate to opportunities for primary prevention, secondary prevention, and tertiary prevention, respec-

▪ **Table 11.1** Sample Haddon matrix for Christian's case

Phase	Host (Christian)	Agent (gun)	Environment (community)
Pre-suicidal ideation (events 1 and 2)	Unaware of the symptoms of depression Depression impairs ability to seek help Perceived stigma of receiving mental health treatment	Ease of access (ability to purchase) Properly maintained and stored (to prevent accidental discharge)	Actual stigma surrounding mental health care Access to primary care services (and subsequent referral to mental health services) Knowledgeable primary care provider (PCP) and effective screening/treatment
Suicidal ideation and attempt (events 3–8)	Knowledge of guns and previous use History of suicidal ideation or behavior	Safe storage (locked in safe, ammunition kept separately) Gun safety switch or trigger lock	Community and PCP recognition of safety concerns Community culture around firearm use Provider awareness of appropriate mental health resources Supportive/invested family or friends Religious or cultural opposition to suicide
After nonfatal suicide attempt	First aid skills Access to emergency services	Accuracy/reliability of the weapon	Emergency response time/access to emergency medical care Availability of inpatient mental health facility

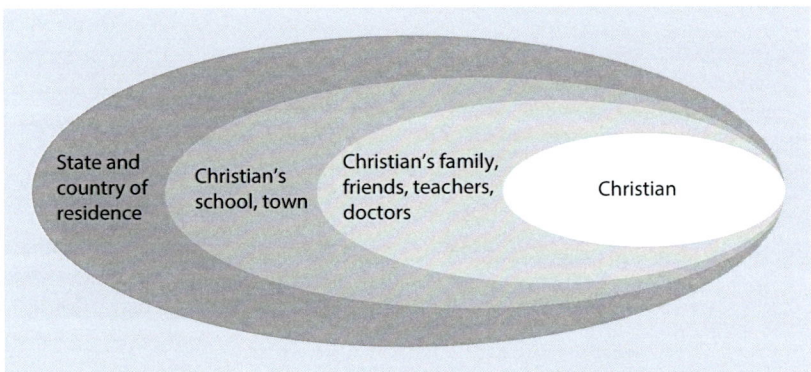

☐ **Fig. 11.1** A depiction of the social-ecological model levels as they apply to the risk-factor categories of Christian's case

tively. Haddon also proposed generic "countermeasures" that may be applied as preventative strategies to reduce risk from numerous types of adverse events/injuries [39]. For example, he recommended preventing release of the agent, which translates to restricting sales of firearms in this case. He also suggested separating the agent from the host: In this case, Christian's family could have stored the family handgun in a fingerprint safe, preventing him from having any access to the weapon.

Another framework for analyzing preventive strategies is the social-ecological model. The CDC identifies four overlapping categories of factors that influence a person's risk for violence: individual, relationship, community, and societal [40]. In the context of suicide, the individual level includes biological and personal history factors that increase risk for self-harm. The relationship level explores aspects of an individual's social network (i.e., friends, family) that may influence the individual's behavior. The community level examines the setting in which these social relationships occur (i.e., school, neighborhood). The societal level involves the broad societal factors that might contribute to an environment where suicide is a considerable option. The social-ecological model, as applied to the components of this case, is depicted in ☐ Fig. 11.1.

Of note, different versions of the social-ecological model may utilize additional levels. Each of the levels overlaps to represent the interplay between all of the different factors. Utilizing this model, opportunities for prevention begin to emerge—many of which are alluded to by the Haddon matrix: Christian did not receive adequate treatment for his depression, potentially because he does not recognize his depressive symptoms or associates a negative stigma with seeking mental health treatment. As detailed in Background Question 6, stigma regarding mental illness is a very real psychological barrier to young adults seeking treatment. Perhaps Christian's high school further perpetuated this stigma with limited education on the prevalence of depression in youth (increasing Christian's social isolation). There may have been a lack of resources or training for teachers and guidance counselors to screen their students for depression. If this is not Christian's first suicide attempt, and he was able to gain access to the family's handgun, it is possible that his family also does not fully appreciate the severity of his illness. Christian was also unable to see a therapist due to limitations in his insurance coverage—representing a failure of societal health policies.

(b) **What role can physicians take in these opportunities?**
The U.S. Preventive Services Task Force (USPSTF) recommends screening for major depressive disorder in adolescents and adults, aged 12 and older, with a grade B recommendation. "Screening should be implemented with adequate systems in place to ensure accurate diag-

nosis, effective treatment, and appropriate follow-up" [30, 41]. Grade B indicates "There is a high certainty that the net benefit is moderate or there is moderate certainty that the net benefit is moderate to substantial" [42]. Ensuring accurate diagnosis, effective treatment, and appropriate follow-up can be a challenge for primary care providers. The USPSTF identified several evidence-based screening tests that are used in primary care settings, including the Patient Health Questionnaire (PHQ, various forms, such as the PHQ-A for adolescents) and the Beck Depression Inventory (BDI). The Columbia-Suicide Severity Rating Scale (C-SSRS) is an evidence-based screening tool that allows providers to stratify suicide risk for their patients and provides appropriate "next steps," such as emergent evaluation vs. an ambulatory behavioral health consult. Part of the C-SSRS involves identifying risk and protective factors similar to those noted in Table 11.1. For example, risk factors can include a history of mood disorder, social isolation, and non-compliance with treatment. Protective factors might involve religious beliefs, responsibility to children, and supportive family [43].

Previous research has shown that primary care physicians are treating a substantial proportion of patients with mental illness [44]. Classically, complex or treatment-resistant patients are then referred for more specialized psychiatric care. From 2008 to 2011, approximately one-third of children receiving mental health care appeared to only visit their PCP. Pediatricians were also found to prescribe medications to a higher percentage of children, compared to psychiatrists [45].

In terms of modifying gun availability, physicians have limited power to directly intervene—though in some cases are required to report gun ownership to law enforcement when there is concern for risk of serious harm to self or others. Nevertheless, physicians can take an active role in counseling patients and their families about firearm ownership and safety. The American Academy of Pediatrics has long recommended that pediatricians inform parents about the dangers of guns around the home. Providers are urged to include questions about guns into their history taking and are advised to counsel parents regarding removing or safely storing weapons [23].

Most major medical organizations view firearm deaths as a public health concern and have proposed multiple policy-level interventions, including universal background checks of gun purchasers, repeal of physician "gag laws" designed to limit counseling on firearm safety, restricting the sale of assault-style weapons and large-capacity magazines, and increasing support for research on reducing firearm-related injuries and deaths [46]. Any physician interested in advocacy can become further involved in these efforts to reduce firearm availability and/or increase firearm safety.

You are able to get Christian involuntarily admitted to the psychiatric unit. He does well with his inpatient course. His parents again reach out to you before Christian is discharged. They are concerned about Christian's long-term safety and ask you for advice about what they can do within their home to decrease his risk of self-harm. You assure them that you will continue to work with Christian and that you found a counselor who accepts his insurance. You also discuss firearm safety with his parents and learn that even before becoming a legal adult, Christian had access to guns in the home. Christian's father was formerly in the military. Both men are avid hunters who also enjoy target shooting at a gun range. You advise Mr. and Mrs. DuPont to change the way firearms are stored in the home so that your patient has no access unless supervised.

In multiple states, there have been attempts to restrict physicians from discussing gun ownership with patients. For example, the Firearms Owners' Privacy Act passed by the Florida legislature in 2011 restricted physicians from asking patients about gun ownership—unless there was concern for the patient's safety or the safety of others [1]. The Florida legislature passed this law after hearing multiple complaints about healthcare providers asking invasive questions on gun ownership and requesting parents to remove guns from their homes. In one case, a pediatrician asked a mother to find another physician after the mother refused to answer questions about firearms in the home [2].

- 5. **Self-reflection: How do you feel about the pediatrician asking the family to find another physician because the mother refused to discuss firearm ownership?**

While key components of the law were struck down as unconstitutional in 2017, such legislation continues to be the subject of much debate [2, 3].

- 6. **Are laws that limit what physicians can discuss with patients ethical?**
  Most physicians feel that these types of laws are not ethical, as they interfere with the patient-physician relationship and inappropriately restrict physicians' professional autonomy. In 2015, the American Academy of Family Physicians, American Academy of Pediatrics, American College of Emergency Physicians, American Congress of Obstetricians and Gynecologists, American College of Physicians, American College of Surgeons, and American Psychiatric Association issued a joint statement opposing such "gag laws." Physicians are widely expected to discuss behaviors that can adversely affect health (e.g., smoking tobacco, failing to wear a seatbelt). They are usually then required to document these conversations in the medical record. Similarly, these organizations argued that "physicians must be allowed to speak freely to their patients in a nonjudgmental manner about firearms, provide patients with factual information about firearms relevant to their health and the health of those around them, fully answer their patients' questions and advise them on the course of behaviors that promote health and safety without fear of liability or penalty" [46].
  Some advocates have expressed concerns that counseling on firearm safety is an unwanted intrusion into patients' private lives, with consequences for the patient-physician relationship. This position is based on a common misconception: Previous research has shown that most patients are open to "nonjudgmental education," particularly if they belong to a group at high risk for firearm-related violence. High-risk groups include people who voice suicidal or homicidal ideation, possess risk factors for future violence (e.g., history of violence, drug use, or severe mental illness), or are members of at-risk demographic groups (e.g., older white males) [47].
  Private gun ownership is an inextricable component of American culture. Therefore, counseling on firearm safety requires a certain degree of cultural competence on the part of the physician. There are multiple subpopulations of gun owners whose perspectives on firearm-counseling likely vary based on their reasons for owning firearms [48]. Of note, parents and patients who feel uncomfortable discussing gun ownership with their physician are free to refuse counseling—just as they can refuse counseling on any behavior or medical practice (e.g., vaccination or tobacco cessation).

- 7. **How are gun injury and gun violence portrayed in the media? Do you think the major causes are covered adequately?**
  In recent years, firearm-related violence has become a recurrent subject of media reporting and political discourse. Tragically, discussion of this subject has repeatedly coincided with public mass shootings. Though there is no universally accepted definition of a "public mass shooting," the U.S. Congressional Research Service (CRS) has suggested that such events involve four or more deaths—not including the shooter(s)—where gunmen select victims somewhat indiscriminately in relatively public places (e.g., schools, workplaces, and restaurants). This definition excludes incidents where perpetrators kill for criminal profit or terrorist ideologies. Under this definition, the CRS identifies 78 public mass shootings that occurred in the United States from 1983

to March 2013. These shootings claimed an estimated 547 lives and led to an additional 476 injuries [49]. Though devastating, public mass shootings account for a small portion of fatal firearm injuries in the United States. In 2015 alone, for example, there were an estimated 36,252 deaths associated with firearm use [6]. Furthermore, most firearm-related deaths are suicides (60.5% on average per year from 2002 to 2012), not homicides [50].

The media can have a profound impact on how Americans perceive firearm-related violence. A 2013 Kaiser Family Foundation telephone poll found more than 40% of sampled Americans were at least "somewhat worried" about being a victim of gun violence [51]. In an analysis of print and broadcast media produced from 1997 to 2012, McGinty, Webster, Jarlenski, and Barry examined the timing of news coverage regarding gun violence and how these stories framed discussion of "serious mental illness." Of the 364 sampled news stories, 51% were published in 2007, 2011, and 2012—coinciding with four public mass shootings. McGinty and colleagues identified five evidence-based facts about mental illness, including that people with serious mental illness are often stigmatized and most people with serious mental illness are nonviolent. Less than 10% of published news stories were found to mention these key points [52].

Over the past several months, you have worked closely with Christian's counselor. She calls you today after a follow-up visit with her. She is concerned that Christian has been making some vague but concerning comments, such as saying that he "understands how the killers in Las Vegas and Virginia Tech must have felt."

8. **What is your obligation in this situation?**
   It is important to recognize that Christian appears to be actively participating in outpatient mental health follow-up. Though Christian's comments may be disconcerting to his therapist, they do not identify a particular target or establish a specific threat to himself or others. Psychotherapy frequently involves patients sharing uncomfortable personal thoughts. This level of communication and rapport is often essential for a therapist to help their patient address internal conflicts. Based on the presented information, there is no obligation to intervene by notifying law enforcement or pursuing involuntary admission. Furthermore, if there were no safety concerns (i.e., the physician is not communicating a threat), discussing the patient's comments outside of his treatment setting would constitute a violation of the patient's right to confidentiality. Refer to Background Question 5 for information on exceptions to confidentiality.

   (a) **What if Christian made a more specific threat (e.g., toward an individual person)?**
   As previously considered in Background Question 7 as well as Discussion Question 3, specific and credible threats toward self or others should be reported to the appropriate authorities. This situation is a generally recognized exception to maintaining patient confidentiality. If acute safety concerns are present (i.e., the patient is considered to be at imminent risk for harm to self or others), psychiatric admission—voluntary or involuntary—should be pursued for stabilization. Of note, these recommendations are based on "duty to protect" or "Tarasoff laws," of which there is considerable interstate variation. Despite this variation, these laws arise from a single moral imperative: Potential victims have the right to avoid preventable death or injury. Physicians and mental healthcare providers are often in a position to protect potential victims (e.g., by warning identified victims and alerting law enforcement) and have an ethical obligation to do so while minimizing the extent to which patient confidentiality is compromised [53].

✅ 9. **What are the ethical implications of mandated reporting laws designed to remove firearms from those who are at-risk to harm themselves or others?**
Mandated reporting laws were discussed in Background Question 7. It is helpful to explore this question in the context of the presented case: Assume Christian lives in New York, where his physician has an obligation to report him under the NY SAFE Act, given that Christian tried to end his own life using a handgun. If Christian has a gun permit, it will likely be revoked, and law enforcement may seize any guns that he owns. The physician was required to override patient confidentiality in order to disclose patient information in compliance with state law. This compromise of privacy is generally considered acceptable because the physician is attempting to act in the best interests of the patient (i.e., prevent future harm to self or others). However it is important to recognize that at this point, Christian's Second Amendment right to keep and bear arms was compromised.

(a) **Is it ethical to take away someone's rights because of their underlying medical condition, including depression or other mental illness?**
In cases like Christian's, where there is clear evidence that an individual represents a significant (and imminent) threat to himself or others, this serious consequence is more readily justified. The issue becomes complicated when deciding where to set the threshold for mandated reporting. Put another way, when assessing a patient's risk for harm, at what "level of risk" is a physician justified (or required) to discharge their "duty to protect"? If Christian had no history of suicide attempts, and made a passing comment about wanting to die, should he have his guns taken away? Previous research has shown that isolated severe mental illness does not predict future violence, though history of violence, legal trouble, and substance use are all positively associated factors [54]. Still, firearm-related suicide attempt injuries are reportedly more common in states with less strict gun laws [55]. These questions and findings allude to the enormous challenge of creating ethical yet effective and protective gun control policies.

> **Box 11.5 Teaching Tip**
> Though this chapter focused on gun ownership, consider other instances where rights or privileges might be withheld in the setting of medical illness. For example, when patients are diagnosed with epilepsy, physicians must consider restricting their driving privileges—which can have life-changing consequences.

Answering this question involves weighing the principle of beneficence against that of patient autonomy. The principle of non-maleficence (or more colloquially to "do no harm") is also relevant when discussing the rights of other individuals who may be harmed by a patient. In short, there is no single answer to this question. Every patient and presentation is unique. As with all medical decision-making, physicians must strive to promote their patients' well-being while encouraging their active engagement in care. Physicians are also crucial advocates for their patients and play an essential role in ensuring their safety as well as the safety of the community they serve.

✅ 10. **What are some of the challenges of evaluating the impact of policy change? How would you conduct a study to assess the impact of such changes?**
Though restrictions on firearm-related research (discussed in Background Question 8) have slowed the accumulation

of evidence, available data suggests that access to firearms increases risk for violence, both homicide and suicide [56]. Some of the deadliest single-day mass shootings in American history have occurred within the last 5 years, including 49 killed in Orlando, FL (6/12/16), 58 killed in Las Vegas, NV (10/1/17), 26 killed in Sutherland Springs, TX (11/5/17), and 17 killed in Parkland, FL (2/14/18) [57]. As these horrific events have continued to recur, a rallying call for change has grown stronger. Proposed policies have included mandating gun purchase permits, banning individuals with a violent crime history from purchasing guns, allowing repossession of firearms from serious domestic violence offenders, and temporarily banning alcohol abusers from firearms [58]. Other legislative proposals have included banning military-style assault weapons and restricting firearms in public places. Many American citizens have expressed that such measures may be used to infringe upon the Second Amendment right to "keep and bear arms." There is also widespread skepticism that gun control laws positively impact Americans' safety.

Assessing the impact of firearm policy change is inherently difficult. For example, introduction of a new medication often involves a series of randomized controlled trials. Subjects are randomly assigned to receive the new medication, and specific outcomes are compared with control groups (who have received either a placebo or the currently accepted standard of care). An experiment in which guns are distributed to a community, and homicide rates are compared to those in a community where all guns have been removed, is not only impossible, but also unethical. For these, and other reasons, past research has frequently utilized observational studies in which data for a specific population are compared to a historic control (same community before and after policy implementation) or to a contemporary control (similar community at same point in time that did not implement the policy). Observational studies are prone to confounding. Researchers must identify specific outcome measures, such as homicide or suicide rates, and allot an adequate observational time period in order to perceive an effect. Appropriate statistical tests must be used to analyze the obtained data. As an example, in 2001, Wintemute, Wright, Drake, and Beaumont published a retrospective, population-based cohort study evaluating the effects of a 1991 California law that restricts the sale of handguns to individuals with a history of violent convictions. They specifically examined arrest records for 3 years following actual handgun purchase (1989–1990, approved before the law) or attempted purchase (1991, denied after the law was passed). Results revealed that misdemeanants who were able to purchase a gun before the law went into effect were nearly 30% more likely to be arrested for a firearm-related and/or violent crime compared with misdemeanants who were denied purchase [59].

A 2017 systematic review of peer-reviewed literature, published in the *Journal of the American Medical Association*, examined 34 studies (dating back to 1975) to evaluate the association between certain firearm laws and firearm-related homicides. All 34 of the studies utilized an ecological design. In aggregate, laws strengthening background checks and requiring permits to purchase firearms were found to be associated with reduced firearm homicide rates [60]. The researchers also noted that the quality of reviewed studies was variable: Multiple studies did not conduct multivariate analyses, and for some, the robustness of the results to changes in variables was not assessed. More research, with adequate funding, is essential to further ascertain the effectiveness of various firearm policies.

11. **Creative Problem-Solving: How would you construct a law or policy to decrease deaths associated with firearms?**

Acknowledgments The author and editors wish to thank Margaret Formica, for her review of this chapter. This case was adapted from an earlier case written by Cynthia Morrow, Travis Hobart, and Amy Caruso Brown.

References

1. Privacy of firearm owners. CS/CS/HB 155 No 2011-112 The Florida Senate Session 2011. [cited 2018 Feb 5]. Available from: https://www.flsenate.gov/Session/Bill/2011/0155.
2. *Wollschlaeger v Governor of the State of Florida*, No. 12-14009, The United States Court of Appeals (11th Cir. 2017). [cited 2018 Feb 5]. Available from: https://www.documentcloud.org/documents/3467440-Wollschlauger-v-Fla-11thCircuitDecision.html.
3. Parmet WE, Smith JA, Miller M. Physicians, firearms, and free speech – overturning Florida's firearm-safety gag rule. N Engl J Med. 2017;376(20):1901–3.
4. Heron M. Deaths: leading causes for 2015. National vital statistics reports. Hyattsville: National Center for Health Statistics. 2017;66(5):1
5. Sleet DA, Ballesteros MF, Borse NN. A review of unintentional injuries in adolescents. Annu Rev Public Health. 2010;31:195–212. 4 p following 212.
6. Centers for Disease Control and Prevention. Web-based Injury Statistics Query and Reporting System (WISQARS™). National Center for Injury Prevention and Control [Internet]. Atlanta (GA): CDC. [cited 2018 Apr 29]. Available from: https://www.cdc.gov/injury/wisqars/index.html.
7. Smith EL, Cooper AD. Homicide in the U.S. known to law enforcement, 2011. Washington, DC: Bureau of Justice Statistics; 2013; NCJ 243035.
8. Federal Bureau of Investigation. 2016 Crime in the United States [Internet]. Clarksburg (WV): Uniform Crime Reporting (UCR) Program. [cited 2018 May 13]. Available from: https://ucr.fbi.gov/crime-in-the-u.s/2016/crime-in-the-u.s.-2016/tables/expanded-homicide-data-table-4.xls.
9. Curtin SC, Warner M, Hedegaard H. Increase in suicide in the United States, 1999–2014. Hyattsville: National Center for Health Statistics; 2016. NCHS data brief, no 241.
10. Fowler KA, Dahlberg LL, Haileyesus T, Annest JL. Firearm injuries in the United States. Prev Med. 2015;79:5–14.
11. Barber C, Hemenway D. Too many or too few unintentional firearm deaths in official U.S. mortality data? Accid Anal Prev. 2011;43(3):724–31.
12. Merideth P. The five c's of confidentiality and how to deal with them. Psychiatry (Edgmont). 2007;4(2):28–9.
13. Wei M. Capacity evaluation and informed consent. In: Rosenquist JN, Nykiel S, Chang T, Sanders K, editors. The Massachusetts General Hospital/McLean Hospital residency handbook of psychiatry. Philadelphia (PA): Lippincott Williams & Wilkins; 2010. p. 47.
14. Ahrnsbrak R, Bose J, Hedden SL, Lipari RN, Park-Lee E. Key substance use and mental health indicators in the United States: results from the 2016 National Survey on Drug Use and Health. Rockville: Center for Behavioral Health Statistics and Quality, Substance Abuse and Mental Health Services Administration; 2017;HHS Publication No. SMA 17-5044, NSDUH Series H-52.
15. StigmaFree Me [Internet]. Arlington (VA): National Alliance on Mental Illness (NAMI) [cited 2017 Dec 3]. Available from: https://www.nami.org/Get-Involved/Take-the-stigmafree-Pledge/StigmaFree-Me.
16. Mojtabai R, Olfson M, Sampson NA, Jin R, Druss B, Wang PS, et al. Barriers to mental health treatment: results from the National Comorbidity Survey Replication. Psychol Med. 2011;41(8):1751–61.
17. Diehl S, Honberg R, Kimball A, Douglas D. The doctor is out: continuing disparities in access to mental and physical health care. Arlington: National Alliance on Mental Illness (NAMI); 2017.
18. Mental health professionals' duty to warn. National Conference of State Legislation; 2015 [cited 2017 Dec 3]. Available from: http://www.ncsl.org/research/health/mental-health-professionals-duty-to-warn.aspx.
19. NYSAFE Act: mental health frequently asked questions. NY gov [Internet]. [cited 2017 Dec 3]. Available from: https://safeact.ny.gov/mental-health-faq.
20. Jamieson C. Gun violence research: history of the federal funding freeze. Psychol Sci Agenda. 2013;27(2). Available from: https://www.apa.org/science/about/psa/2013/02/gun-violence.aspx.
21. Kellermann AL, Rivara FP, Rushforth NB, Banton JG, Reay DT, Francisco JT, et al. Gun ownership as a risk factor for homicide in the home. N Engl J Med. 1993;329(15):1084–91.
22. Miller M, Hemenway D. Guns and suicide in the United States. N Engl J Med. 2008;359(10):989–91.
23. Dowd MD, Sege RD, Council on Injury, Violence, and Poison Prevention Executive Committee, American Academy of Pediatrics. Firearm-related injuries affecting the pediatric population. Pediatrics. 2012;130(5):e1416–23.
24. Shenassa ED, Rogers ML, Spalding KL, Roberts MB. Safer storage of firearms at home and risk of suicide: a study of protective factors in a nationally representative sample. J Epidemiol Community Health. 2004;58(10):841–8.
25. United States Department of Transportation. National Highway Traffic Safety Administration. Budget information [Internet]. Washington, D.C.; 2018 [cited 2018 Apr 29]. Available from: https://www.nhtsa.gov/about-nhtsa/nhtsa-budget-information.
26. United States Department of Transportation. National Highway Traffic Safety Administration. Traffic safety facts annual report tables [Internet]. Washington, D.C.; 2016 [cited 2018 Apr 29]. Available from: https://cdan.nhtsa.gov/tsftables/tsfar.htm#.
27. United States Department of Transportation. National Highway Traffic Safety Administration. Crash injury research [Internet]. Washington, D.C. [cited 2018 Apr

29]. Available from: https://www.nhtsa.gov/research-data/crash-injury-research.
28. National Institute of Mental Health. Antidepressant medications for children and adolescents: information for parents and caregivers [Internet]. Bethesda (MD): NIMH. [cited 2017 Dec 17]. Available from: https://www.nimh.nih.gov/health/topics/child-and-adolescent-mental-health/antidepressant-medications-for-children-and-adolescents-information-for-parents-and-caregivers.shtml.
29. Friedman RA. Antidepressants' black-box warning: 10 years later. N Engl J Med. 2014;371(18):1666–8.
30. Siu AL, US Preventive Services Task Force (USPSTF), Bibbins-Domingo K, Grossman DC, Baumann LC, Davidson KW, et al. Screening for depression in adults: US Preventive Services Task Force recommendation statement. JAMA. 2016;315(4):380–7.
31. Khan A, Brown WA. Antidepressants versus placebo in major depression: an overview. World Psychiatry. 2015;14(3):294–300.
32. Snyder L, American College of Physicians Ethics, Professionalism, and Human Rights Committee. American College of Physicians ethics manual: sixth edition. Ann Intern Med. 2012;156(1 Pt 2):73–104.
33. March J, Silva S, Petrycki S, Curry J, Wells K, Fairbank J, et al. Fluoxetine, cognitive-behavioral therapy, and their combination for adolescents with depression: treatment for adolescents with depression study (TADS) randomized controlled trial. JAMA. 2004;292(7):807–20.
34. Brennan KJ, New York State Office of Mental Health. Appendix A, Kendra's Law overview and statute. 2009 June 30 [cited 2018 May 14]. Available from: https://www.omh.ny.gov/omhweb/resources/publications/aot_program_evaluation/appendix_a.html.
35. Ho AO. Suicide: rationality and responsibility for life. Can J Psychiatr. 2014;59(3):141–7.
36. Carrigan CG, Lynch DJ. Managing suicide attempts: guidelines for the primary care physician. Prim Care Companion J Clin Psychiatry. 2003;5(4):169–74.
37. Rights of inpatients in New York State Office of Mental Health psychiatric centers. Albany (NY): The Joint Commission. [cited 2017 Dec 14]. Available from: https://www.omh.ny.gov/omhweb/patientrights/inpatient_rts.htm.
38. State of New York Office of Mental Health. Application for involuntary admission on certificate of a director of community services or designee -Section 9.37 Mental Hygiene Law. Form OMH 475.
39. McDowell I. Preventing injuries: the Haddon matrix [Internet]. Ottawa: University of Ottawa. [cited 2017 Dec 15]. Available from: http://www.med.uottawa.ca/sim/data/Injury_Prevention_Haddon_e.htm.
40. Centers for Disease Control and Prevention. The social-ecological model: a framework for prevention [Internet]. Atlanta (GA): CDC; 2015 [cited 2017 Dec 15]. Available from: https://www.cdc.gov/violenceprevention/overview/social-ecologicalmodel.html.
41. Siu AL, US Preventive Services Task Force (USPSTF). Screening for depression in children and adolescents: US Preventive Services Task Force recommendation statement. Pediatrics. 2016;137(3):e20154467.
42. U.S. Preventive Services Task Force. Grade definitions [Internet]. Rockville (MD): USPSTF; 2017 [cited 2017 Dec 15]. Available from: https://www.uspreventiveservicestaskforce.org/Page/Name/grade-definitions.
43. The Columbia Lighthouse Project. C-SSRS for communities and healthcare [Internet]. New York: Columbia University; 2016 [cited 2017 Dec 15]. Available from: http://cssrs.columbia.edu/the-columbia-scale-c-ssrs/cssrs-for-communities-and-healthcare/#filter=.general-use.english.
44. Abed Faghri NM, Boisvert CM, Faghri S. Understanding the expanding role of primary care physicians (PCPs) to primary psychiatric care physicians (PPCPs): enhancing the assessment and treatment of psychiatric conditions. Ment Health Fam Med. 2010;7(1):17–25.
45. Anderson LE, Chen ML, Perrin JM, Van Cleave J. Outpatient visits and medication prescribing for US children with mental health conditions. Pediatrics. 2015;136(5):e1178–85.
46. Weinberger SE, Hoyt DB, Lawrence HC III, Levin S, Henley DE, Alden ER, et al. Firearm-related injury and death in the United States: a call to action from 8 health professional organizations and the American Bar Association. Ann Intern Med. 2015;162(7):513–6.
47. Wintemute GJ, Betz ME, Ranney ML. Yes, you can: physicians, patients, and firearms. Ann Intern Med. 2016;165(3):205–13.
48. Betz ME, Wintemute GJ. Physician counseling on firearm safety: a new kind of cultural competence. JAMA. 2015;314(5):449–50.
49. Bjelopera JP, Bagalman E, Caldwell SW, Finklea KM, McCallion G. Public mass shootings in the United States: selected implications for federal public health and safety policy. Congressional Research Service. 2013;R43004.
50. Wintemute GJ. The epidemiology of firearm violence in the twenty-first century United States. Annu Rev Public Health. 2015;36:5–19.
51. Kaiser Family Foundation. One in five Americans know a victim of gun violence; worry reaches even more broadly. Kaiser health tracking poll: February 2013 [Internet]. San Francisco (CA): KFF; 2013 [cited 2018 May 3]. Available from: https://www.kff.org/disparities-policy/poll-finding/kaiser-health-tracking-poll-february-2013/.
52. McGinty EE, Webster DW, Jarlenski M, Barry CL. News media framing of serious mental illness and gun violence in the United States, 1997–2012. Am J Public Health. 2014;104(3):406–13.
53. Johnson R, Persad G, Sisti D. The Tarasoff rule: the implications of interstate variation and gaps in professional training. J Am Acad Psychiatry Law. 2014;42(4):469–77.
54. Elbogen EB, Johnson SC. The intricate link between violence and mental disorder: results from the National Epidemiologic Survey on Alcohol and Related Conditions. Arch Gen Psychiatry. 2009;66(2):152–61.

55. Alban RF, Nuno M, Ko A, Barmparas G, Lewis AV, Margulies DR. Weaker gun state laws are associated with higher rates of suicide secondary to firearms. J Surg Res. 2018;221:135–42.
56. Moyer M. More guns do not stop more crimes, evidence shows. Sci Am. 2017;317(4):54–63. Available from: https://www.scientificamerican.com/article/more-guns-do-not-stop-more-crimes-evidence-shows/.
57. CNN Library. Deadliest mass shootings in modern US history fast facts. CNN [Internet]; 2018 Dec 15 [cited 2018 May 5]. Available from: https://www.cnn.com/2013/09/16/us/20-deadliest-mass-shootings-in-u-s-history-fast-facts/index.html.
58. Moyer M. 4 laws that could stem the rising threat of mass shootings. Sci Am; 2017 Nov 10. Available from: https://www.scientificamerican.com/article/4-laws-that-could-stem-the-rising-threat-of-mass-shootings/.
59. Wintemute GJ, Wright MA, Drake CM, Beaumont JJ. Subsequent criminal activity among violent misdemeanants who seek to purchase handguns: risk factors and effectiveness of denying handgun purchase. JAMA. 2001;285(8):1019–26.
60. Lee LK, Fleegler EW, Farrell C, Avakame E, Srinivasan S, Hemenway D, et al. Firearm laws and firearm homicides: a systematic review. JAMA Intern Med. 2017;177(1):106–19.

Further Reading on This Topic

Branas CC, Richmond TS, Culhane DP, Ten Have TR, Wiebe DJ. Investigating the link between gun possession and gun assault. Am J Public Health. 2009;99(11):2034–40.

Dembosky A. Depressed teen's struggle to find mental health care in rural California. NPR [Internet]. 2016 Jun 23. Available from: https://www.npr.org/sections/health-shots/2016/06/23/481764541/depressed-teen-s-struggle-to-find-mental-health-care-in-rural-california.

Ganzini L, Volicer L, Nelson WA, Fox E, Derse AR. Ten myths about decision-making capacity. J Am Med Dir Assoc. 2005;6(3 Suppl):S100–4.

Koven S. Should mental health be a primary-care doctor's job? The New Yorker [Internet]. 2013 Oct 21. Available from: https://www.newyorker.com/tech/elements/should-mental-health-be-a-primary-care-doctors-job.

Luo M, McIntire M. Children and guns: the hidden toll. The New York Times [print]. 2013;29(56274):A1–20.

Watkins AM, Lizotte AJ. Does household gun access increase the risk of attempted suicide? Evidence from a national sample of adolescents. Youth Soc. 2013;45(3):324.

"Bleeding Too Much" (In the Words of a Refugee)

Andrea V. Shaw

12.1 Background Questions – 232

12.2 Additional Case Information and Questions for Discussion – 232

12.3 Answers to Background Questions – 235

12.4 Responses to Discussion Questions – 238

References – 246

Glossary

Female genital mutilation - The World Health Organization (WHO) defines female genital mutilation (FGM) as comprising "all procedures that involve partial or total removal of the external female genitalia, or other injury to the female genital organs for non-medical reasons" [1]. Also called female genital cutting or, historically, female circumcision.

Human trafficking - The United Nations Office on Drugs and Crime defines human trafficking as "the recruitment, transportation, transfer, harboring or receipt of persons, by means of the threat or use of force or other forms of coercion, of abduction, of fraud, of deception, of the abuse of power or of a position of vulnerability or of the giving or receiving of payments or benefits to achieve the consent of a person having control over another person, for the purpose of exploitation" [2].

Rape myth - False beliefs about sexual violence, perpetrators, and survivors that often serve to excuse sexual aggression and create hostility toward victims and can bias criminal prosecution. Examples in the United States include that women often lie about rape, that women "ask for it" or increase their chances of being assaulted by how they act or dress or where they go, and that "real victims" behave in predictable ways (such as fighting back physically and seeking medical attention immediately) [3].

Learning Objectives

1. Describe some unique health issues facing refugees.
2. Understand the high level of traumatic experience from the initial persecution refugees faced in their home country, the risks of the journey to a second country of asylum, the uncertainty of life awaiting resettlement, and the process of assimilation into a new life.
3. Develop strategies to overcome barriers to communication.

Case Background

You are a primary care practitioner (PCP) and your first appointment of the day is with Ms. Mariam Gebreselassie, a 22-year-old woman from Eritrea. Ms. Gebreselassie lived in Egypt for 2 years before she was resettled in the United States 1 month ago. She presents to establish care and address her periods, which have been irregular for the past few months. Ms. Gebreselassie is accompanied by an advocate from the resettlement agency who tells you that her primary language is Tigrinya—a language that neither you nor the advocate speak.

12.1 Background Questions

1. Find and compare the definitions of the following: refugee, internally displaced person, stateless person, asylum seeker, immigrant, and migrant.

2. The United Nations High Commissioner for Refugees (UNHCR) provides global estimates for refugees, internally displaced persons, and asylum seekers. Summarize the most recent numbers. Describe one of the humanitarian emergencies identified by UNHCR as producing refugees.

3. How many refugees does the United States accept every year? How have these trends changed over the years? Where in the United States do most refugees resettle?

4. How do we support newly resettled refugees in the United States?

5. What are the most common health conditions seen in refugees at the time of resettlement in the United States? What health conditions are refugees at risk for post-resettlement in the United States?

6. How prevalent in gender-based violence around the world? What are some of the challenges to collecting this type of data? How do rates in the United States compare with other countries?

7. What is generally known about the relationship between traumatic events in childhood and adult health? Describe the methods of the Adverse Childhood Experiences Study (ACES) and report some of their findings.

8. What is "trauma-informed care"?

12.2 Additional Case Information and Questions for Discussion

As you wait for an interpreter to connect on the phone line, you observe a pleasant young woman sitting comfortably across from you. She smiles

"Bleeding Too Much" (In the Words of a Refugee)

and makes good eye contact when you greet her. You note that she seems to have difficulty with her right leg. She uses her arms to push off when she gets up from her chair to greet you, and she pivots on her left foot as she gets up and down. However, she does not appear to be in any significant discomfort. She waits calmly for the phone interpreter.

1. **What are some of the nonverbal cues that you might look for during this part of the encounter?**

Once the interpreter comes on the line and after initial introductions have taken place, you proceed to inquire about where she is from, where she has been, and the circumstances that brought her to see you today. Although Ms. Gebreselassie is young, she is quite mature for her age. She attended secondary school in Eritrea and is literate in her family's primary language, Tigrinya.

Ms. Gebreselassie shares with you that she fled Eritrea 3 years ago. She explains that her father was forced to serve in the military for many years, and he feared her mandatory enlistment when she turned 18. He understood that the path to getting an education in Europe was not without risk to his daughter, and her fleeing the service years was not without risk to his family, but he felt that it was the only choice to secure his daughter's future. With Europe as her final goal, she fled Eritrea on foot, crossed the border into Sudan, and began the long trek northward to Egypt.

At this point in her story, her head turns down, her hands start to tremble, and her voice becomes soft and a tear falls down her cheek.

2. **How might you acknowledge a distressing moment for a patient?**

3. **How do you practice trauma-informed care in this situation?**

After giving her a moment in case she would like to offer more and hearing nothing further, you decide to move on. At this point, you feel it is important to shift your questions to her medical concerns. Ms. Gebreselassie states that her periods are irregular with "bleeding too much" over the past few months. She denies sexual activity. Upon further questioning about past surgical history, she tells you that she had "a vaginal procedure" done in a health facility shortly after she arrived at the United Nations (UN) refugee registration site in Egypt. She reports that there was a complication with the procedure that required her to be transferred to a nearby hospital for treatment. She says that she woke up there, was told she needed a blood transfusion, and was monitored for 2 days afterward. For 1 year after that procedure, she did not menstruate.

You try to ask her more about the procedure but receive only vague answers. Despite asking different ways, the only reply you get from the interpreter is "a procedure," and it is not clear to you what the procedure was for.

4. **What else do you want to know? How might you create a safer space for your patient to answer your questions?**

5. **Communication is essential with any patient. How do you make a patient feel comfortable when you are not able to communicate in their primary language?**

You pause, ask her how she may feel more comfortable discussing this with you. She states that she is uncomfortable with the male telephone interpreter and would like an in-person female interpreter. She knows a woman in the community she feels more comfortable with and can arrange to come to a follow-up appointment with her. Mariam returns the following morning with a woman from the community, who works as a case manager during the day. As part of her training as a case manager, she took classes in medical interpretation.

Mariam now feels comfortable sharing more of her history. She reports that she started menses at the age of 15 and had normal monthly periods until 2 years ago. When she arrived in Egypt, she found out she was pregnant; the UN refugee clinic offered her termination and options for contraception. During the termination procedure, she experienced a hemorrhage, which led to her transfer to a nearby hospital. She believes that she had something implanted after that procedure to prevent pregnancy as she was given an option for implantation in her arm or uterus and did not want something visible in her arm.

Six months ago, she started to have irregular cycles with variable days of bleeding. She denies any associated symptoms, including abdominal pain. You feel that you have made good progress in your understanding of her situation and feel that you should ask about her other medical history, including her limp. When you do, she lowers her eyes to the floor and becomes quiet. You do not press the issue.

❓ 6. Having learned more about Mariam's experience, what kinds of challenges do you think *you* face as her primary care provider?

You ask and receive permission to do a pelvic exam to further assess the abnormal bleeding. You note that her clitoris is absent and there is scarring over her anterior genitalia, which you recognize as female genital cutting (or female genital mutilation [FGM]), a common cultural practice in Eritrea. You are able to visualize the string to an intrauterine device (IUD). You believe that the IUD may be the cause of the irregular menses and schedule an ultrasound to assess its placement. You plan to see her back in 2 weeks for close follow-up.

Before she leaves, you want to ensure that she has appropriate supports in place. She states that she will receive 1 more month of case management support from the resettlement agency. She lives in a low-income housing apartment with another single young woman who speaks her language. She recently took a minimum-wage job cleaning a nearby hotel, so she will lose her public assistance this month. She receives Supplemental Nutrition Assistance Program (SNAP) benefits and is able to walk to a market nearby her home.

❓ 7. What does cultural humility mean to you in the context of this case?

❓ 8. What kinds of challenges do you think Ms. Gebreselassie faces daily in her new life in the United States?

You continue to follow up with M. Gebreselassie regularly. You want to better understand the origin of her limp to see if you can help her improve her gait and function. She continues to deny that she has any pain in her leg. Eventually, you feel you have established a good relationship, and you decide to ask again. This time, Mariam states simply, "They beat me." She emotionally recounts the experience she had navigating through Sudan. The people she paid to assist in arranging her travel turned out to be human smugglers.

Through tears, she describes repeated beatings, rapes, burns, and many hours being tied down, which resulted in her right leg having significant contractures of the muscle and her skin being scarred from the burns. She was only able to escape from the smugglers in Egypt because she was too weak to move quickly, and the smugglers viewed her as more of a liability than an asset. She reports being seen by physicians in Egypt but was told that surgery for her leg would not help and might make it worse. She did not feel that the local doctors took her seriously: "They acted as if my problems were my fault." She also states "I have learned to live with it. If I think about the pain, I feel pain all over my body. If I just keep moving, it does not hold me down."

❓ 9. As a primary care healthcare practitioner, how do you assess Mariam's mental health, in light of the traumatic experiences she has survived?

❓ 10. Self-reflection: How do you think Mariam's experience is different than experiences of young women who report sexual assault in social settings in the United States? How is it the same?

Some months later, while you are out of town, Mariam presents with new abdominal pain. The physician covering your practice obtains blood work and a urinalysis, which are normal, and urine pregnancy test, which is negative. He feels she looks stable and this pain may be psychosomatic given her traumatic past. He is already over an hour behind and asks her to come back in a week for follow-up. Mariam agrees to call if anything changes before then, but later that night, she goes to the emergency department, where she is diagnosed with appendicitis.

❓ 11. What do you think happened?
   (a) Women are more likely than men to have "medically unexplained symptoms," a term to describe situations in which people have physical symptoms for prolonged periods of time for which no

"Bleeding Too Much" (In the Words of a Refugee)

medical explanation is identified [4, 5]. In the past, the term "somatization" was frequently used to describe similar situations. What do you think are contributing factors for this?

12. What is a healthcare practitioner's role in advocating for vulnerable patients? Are you ethically obligated to "do more" for patients with fewer resources and greater needs in order to allow them to reach the same level of health as those with greater resources and fewer needs?
    (a) Do our current care models support patients with fewer resources and greater needs? Do our societal norms? In what ways? How can healthcare practitioners spur change?

12.3 Answers to Background Questions

1. **Find and compare the definitions of the following: refugee, internally displaced person, stateless person, asylum seeker, immigrant, and migrant.**
   Refugee: The United Nations (UN) defines refugee as "someone who has been forced to flee his or her country because of persecution, war or violence. A refugee has a well-founded fear of persecution for reasons of race, religion, nationality, political opinion or membership in a particular social group" [6].
   Internally displaced person: The UN defines an internally displaced person, or IDP, as someone who has been forced to flee their home but never cross an international border. Unlike refugees, IDPs are not protected by international law or eligible to receive many types of aid because they are legally under the protection of their own government [6].
   Stateless person: The UN defines a stateless person as someone who is not a citizen of any country [6].
   Asylum seeker: An asylum seeker is a person who flees their own country and seeks sanctuary in another country. An asylum seeker must demonstrate that his or her fear of persecution in his or her home country is well-founded [6].
   Immigrant: An immigrant is an individual who moves to another country to take up permanent legal residence.
   Migrant: The UN states that "while there is no formal legal definition of an international migrant, most experts agree that an international migrant is someone who changes his or her country of usual residence, irrespective of the reason for migration or legal status" [7]. Migrant workers are individuals who change country of usual residence for work reasons (often seasonal work).

2. **The United Nations High Commissioner for Refugees (UNHCR) provides global estimates for refugees, internally displaced persons, and asylum seekers. Summarize the most recent numbers. Describe one of the humanitarian emergencies identified by UNHCR as producing refugees.**
   In 2017, the UNHCR estimated the number of refugees to be 25.4 million, internally displaced persons to be 40 million, stateless persons to be 10 million, and asylum seekers to be 3.1 million. More than 50% of refugees come from three countries: Syria, Afghanistan, and South Sudan, and more than 50% are under 18 years of age. Only 102,800 were resettled in 2017 [8]. Three key forces that drive people to become refugees are persecution, conflict, and generalized violence; however, there is growing concern that climate change will be an increasing force in the coming decades.
   One example of a humanitarian emergency involves the Rohingya, a stateless Muslim minority population that have suffered extreme violence in Myanmar. This violence is described by Human Rights Watch as a result of ethnic cleansing by the Burmese government. Rohingya refugees are disproportionately women, children, and elderly. After fleeing Myanmar, they have settled in

refugee camps in Bangladesh [9]. Unfortunately, these already vulnerable populations continue to be at risk as the camps have inadequate safe water and shelter, little access to basic sanitation and health care, and are at risk of being destroyed by flooding due to climate change [10].

3. **How many refugees does the United States accept every year? How have these trends changed over the years? Where in the United States do most refugees settle?**
The number of refugees that the United States accepts fluctuates significantly every year and is primarily dependent on the global situation coupled with U.S. priorities. Fewer than 30,000 refugees were admitted to the U.S. in 2002 and 2003, following the 9/11 attacks in 2001. The peak influx post-9/11 of 85,000 resettled refugees occurred in 2016; the number was down to 33,400 in 2017 [11]. In recent years, California, Texas, and New York have resettled the most refugees [12]. According to UNHCR data, many European countries, Canada, and Australia accept more resettled refugees per capita than the United States currently does. However, it should be noted that most refugees remain displaced without a designated status, while being supported by resource-limited countries that neighbor conflict zones around the world [13].

4. **How do we support newly resettled refugees in the United States?**
The International Organization for Migration (IOM), an affiliate of the United Nations, has approximately 170 member states and a mission of "promoting dignified, orderly, and safe migration for the benefit of all" [14]. The IOM provides travel assistance for refugees who are being resettled through the United Nations High Commissioner for Refugees and provides policy guidance for its member states to optimally support refugees. However, the travel assistance (e.g., cost of a plane ticket) is in fact an interest-free loan and must be repaid within 1 year.

There is federal support through the U.S. Office of Refugee Resettlement (ORR) to provide support during the first 3 months of resettlement.

There are also local programs and organizations that provide grants and support services to refugees who no longer qualify for ORR support.

5. **What are the most common health conditions seen in refugees at the time of resettlement in the United States? What health conditions are refugees at risk for post-resettlement in the United States?**
In general, the risk of health problems in refugees is usually similar to that of the general population in their country of origin; however, newly arrived refugees are at greater risk for health problems that may be associated with crowded living conditions in the refugee camps (e.g., gastrointestinal illness or respiratory illnesses), as well as those associated with the root causes of their displacement and migration (e.g., physical, sexual and/or psychological trauma, malnutrition). Because access to health care is disrupted during the process of fleeing their home countries, refugees with noncommunicable diseases (e.g., diabetes, heart disease, and periodontal disease) are at risk for poor health outcomes. Mental health problems, especially major depression, are also more common in refugees than the general population, both immediately after resettlement and during the post-resettlement process [17–20].

Ideally, the support services during the resettlement and post-resettlement period allow refugees access to medical care to treat any acute illnesses, as well as to re-establish consistent care to address chronic diseases. However, a recent study of barriers to health care during the post-resettlement period in a refugee population in San Diego, CA, demonstrated that significant barriers persist through the post-settlement period. Consequently, many refugees did not regularly access healthcare services. Identified barriers included language and communication, transportation, and cultural differences in health care. In particular, some refugees did not utilize preventive health care, as the benefits of preventive care were unfamiliar and not explained to them [21].

Refugees who resettle in the United States undergo pre-departure health screenings (and treatment if indicated) 6 months prior to coming to the United States. This includes a general history and physical exam, as well as screening for tuberculosis, hepatitis, parasites, and venereal disease. In addition, within 90 days of arrival in the United States, refugees undergo a comprehensive health screening that includes a complete medical history, physical examination, and pertinent testing [22].

6. **How prevalent is gender-based violence around the world? What are some of the challenges to collecting this type of data? How do rates in the United States compare with other countries?**
The United Nations defines sexual or gender-based violence (GBV) as any act that is perpetrated against a person's will and is based on gender norms and unequal power relations, including threats and coercion [23]. Examples include intimate partner violence (IPV), rape, female genital cutting (also called female genital mutation or FGM), trafficking, and child marriage. The rates of GBV vary greatly depending on the specific measure used. For example, it is estimated that approximately one-third of women across the world have experienced intimate partner violence, with a reported range from approximately 10% to more than 60% of women impacted (the latter is reported in the Democratic Republic of Congo) [24, 25]. For countries in Africa in which FGM is practiced, reported rates of any type of FGM range from 1% in Uganda to more than 95% in Somalia [26].

In the United States, almost 45% of women have reported "contact sexual violence," defined as "a combined measure that includes rape, being made to penetrate someone else, sexual coercion, and/or unwanted sexual contact," based on the 2015 National Intimate Partner and Sexual Violence Survey [27].

There are many challenges in collecting this type of data. As previously noted, there is significant variation in definitions of gender-based violence and of the categories within GBV. There is also variation in how data are measured and collected (e.g., self-reporting, which has major limitations, as survivors may themselves use different definitions and may be reluctant to disclose their experiences) and in the robustness of data collection system in different countries.

7. **What is generally known about the relationship between traumatic events in childhood and adult health? Describe the methods of the Adverse Childhood Experiences Study (ACES) and report some of their findings.**
Over the past two decades, evidence has accrued to support a strong association between traumatic events in childhood and poor health outcomes (as measured by both risk factors and disease occurrence) in adults [28, 29].

In 1998, Felitti and colleagues published a study exploring the relationship between adverse childhood experiences and leading causes of death in adults. The authors surveyed more than 13,000 adults and asked about exposure to seven categories of adverse childhood experiences: psychological, physical, or sexual abuse; violence against one's mother; or living with household members who were substance abusers, mentally ill, or suicidal, and/or incarcerated. The authors then assessed the adult participants' risk behaviors and health status. They found that there was an association between the cumulative number of adverse childhood experiences and poor health outcomes. Specifically, the authors noted that "(p)ersons who had experienced four or more categories of childhood exposure, compared to those who had experienced none, had 4–12-fold increased health risks for alcoholism, drug abuse, depression, and suicide attempt; a 2- to 4-fold increase in smoking, poor self-rated health, ≥50 sexual intercourse partners, and sexually transmitted disease; and 1.4–1.6-fold increase in physical inactivity and severe obesity" [30]. See Fig. 12.1 [31].

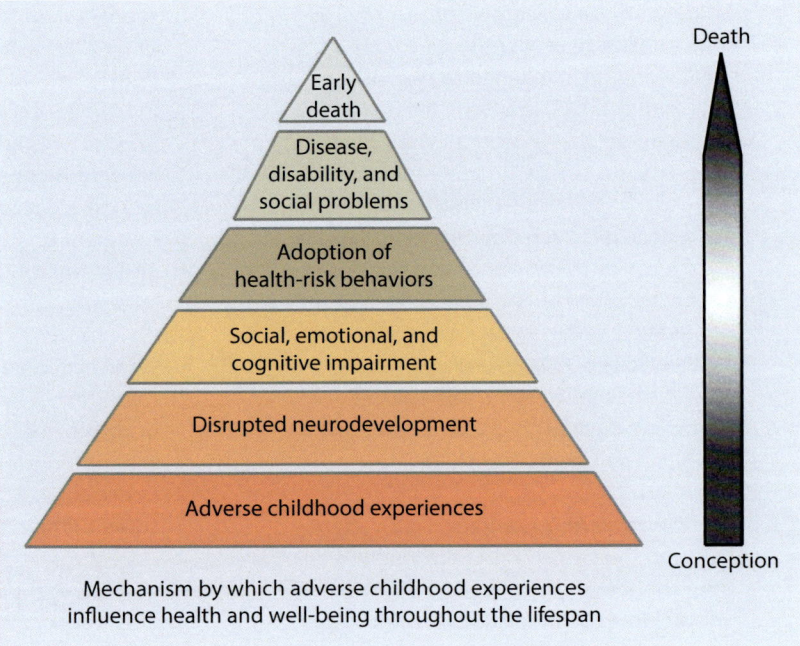

Fig. 12.1 The ACE pyramid. (Modified from [31])

8. **What is "trauma-informed care"?**
   The Substance Abuse and Mental Health Services Administration (SAMHSA) provides a framework for a trauma-informed approach to healthcare. According to SAMHSA, a health care institution or system provides such care when it [32]:
   1. Realizes the widespread impact of trauma and understands potential paths for recovery
   2. Recognizes the signs and symptoms of trauma in clients, families, staff, and others involved with the system
   3. Responds by fully integrating knowledge about trauma into policies, procedures, and practices
   4. Seeks to actively resist re-traumatization

   In order to deliver such care, it is important that health care providers demonstrate responsiveness to patients' needs, including the need to be informed and respected, understand the association between trauma and the physical and psychologic impact of that trauma, and work together in a team to address the range of supports from which patients and their families can benefit [32].

12.4 Responses to Discussion Questions

As you wait for an interpreter to connect on the phone line, you observe a pleasant young woman sitting comfortably across from you. She smiles and makes good eye contact when you greet her. You note that she seems to have difficulty with her right leg. She uses her arms to push off when she gets up from her chair to greet you, and she pivots on her left foot as she gets up and down. However, she does not appear to be in any significant discomfort. She waits calmly for the phone interpreter.

1. **What are some of the nonverbal cues that you might look for during this part of the encounter?**
   It is important that healthcare practitioners remember to engage, connect, and be present with patients. This can be more challenging with utilization of electronic medical records (EMR), which may draw practitioners' eyes away from patients and toward computer screens for much of the day in a busy primary care practice. Eye contact, hand gestures, body posture, and other signs of nonverbal communication are essential to ensure that translation in either direction is being effective. For

patients, these cues convey the practitioners' commitment to being fully present for the encounter. For practitioners, it allows them to gain valuable information that they might otherwise miss. Such gestures become even more meaningful when practitioners and patients do not share the same language and culture.

Once the interpreter comes on the line and after initial introductions have taken place, you proceed to inquire about where she is from, where she has been, and the circumstances that brought her to see you today. Although Ms. Gebreselassie is young, she is quite mature for her age. She attended secondary school in Eritrea and is literate in her family's primary language, Tigrinya.

Ms. Gebreselassie shares with you that she fled Eritrea 3 years ago. She explains that her father was forced to serve in the military for many years, and he feared her mandatory enlistment when she turned 18. He understood that the path to getting an education in Europe was not without risk to his daughter, and her fleeing the service years was not without risk to his family, but he felt that it was the only choice to secure his daughter's future. With Europe as her final goal, she fled Eritrea on foot, crossed the border into Sudan, and began the long trek northward to Egypt.

At this point in her story, her head turns down, her hands start to tremble, and her voice becomes soft and a tear falls down her cheek.

✓ 2. **How might you acknowledge a distressing moment for a patient?**
It is important to acknowledge distress (either by gesture, such as an outstretched hand to touch her shoulder, or by sympathetic words like "I'm sorry") and then to pause, allowing the patient time and space. Depending on the rapport that the healthcare practitioner has with the individual patient, she may or may not divulge further information to explain her emotions at this moment in time. It is important to acknowledge a patient's stressful moment and wait for them to move the conversation forward. Many healthcare practitioners feel uncomfortable with silence and quickly rush into the next line of questioning, but this approach should be avoided.

Refugees, by definition, have survived some form of trauma. Many refugees may have layers of trauma acquired over time, from initially fleeing persecution or desperate situations in their home country to living an exposed life with uncertain future amidst the transitional period to post-resettlement stressors that this vulnerable population commonly faces. It is important to avoid re-traumatizing patients around topics that are painful for them while keeping an awareness of how these experiences affect their current and future health.

✓ 3. **How do you practice trauma-informed care in this situation?**
We want to avoid retraumatization. Give a moment for silence to happen in case she would like to offer more, if not, there is no pressing need to learn more about this experience at this moment and you can shift your questions to the medical concerns she brought to your attention. Healthcare practitioners should take the time necessary to build rapport and establish a relationship with the patient first. Depending on the individual, traumatic experiences may be revealed fairly early in the therapeutic relationship, years later, or perhaps never. Pieces of traumatic stories may naturally surface during clinical interactions, or symptoms of stress (e.g., anxiety, insomnia, change in affect) may lead the healthcare practitioner to inquire further about a traumatic experience. Clinical tools, such as the validated Refugee Health Screener-15, have been designed to screen for symptoms of anxiety, depression, and post-traumatic stress disorder (PTSD) through a series of questions at the time of the initial interaction [33]. Translations are publicly available in several major languages, including Tigrinya.

After giving her a moment in case she would like to offer more and hearing nothing further, you decide to move on. At this point, you feel it is important to shift your questions to her medical concerns. Ms. Gebreselassie states she has periods that are irregular with "bleeding too much" over the past few months. She denies sexual activity. Upon further questioning about past surgical history, she tells you that she had "a vaginal procedure" done in a health facility shortly after she arrived at the United Nations refugee registration site in Egypt. She reports that there was a complication with the procedure that required her to be transferred to a nearby hospital for treatment. She says that she woke up there, was told she needed a blood transfusion, and was monitored for 2 days afterward. For 1 year after that procedure, she did not menstruate.

You try to ask her more about the procedure but receive only vague answers. Despite asking different ways, the only reply you get from the interpreter is "a procedure," and it is not clear to you what the procedure was for.

✓ 4. **What else do you want to know? How might you create a safer space for your patient to answer your questions?**
Many healthcare practitioners would have questions at this point about the nature of the procedure, as well as Ms. Gebreselassie's medical history before and after. Some might wonder whether the patient was disclosing everything. Unfortunately, it is common for patients to not fully understand the nature of and rationale for medical interventions received. Furthermore, Ms. Gebreselassie was traveling alone and may not have had anyone to help her understand what was happening. Although she denies sexual activity, this can be a difficult concept to translate, and in many cultures, it is not acceptable to discuss it outside of the context of marriage and children. Ms. Gebreselassie might also have been exposed to sexual violence that she does not consider "sexual activity."

Although the PCP has explained the confidential nature of the interaction and involvement with an interpreter, Ms. Gebreselassie still may not feel comfortable revealing information. Such discomfort might stem from the nature of the questions (and their answers), from lack of trust in the healthcare practitioner, or from the presence of the interpreter. Medical and health professions training encourage healthcare practitioners to take a complete history; however, it is important to assess the situation and determine whether such a thorough approach is appropriate. Healthcare practitioners can use several techniques to create a safer space and address sensitive areas, such as:

- Asking permission to address the question
- Being transparent about why you are asking the question
- Normalizing the situation and reassuring the patient
- Providing the patient with the opportunity to decline to answer the question

If a healthcare practitioner is unable to get further information after utilizing these techniques, then it is best to move on to the next part of the history. There will be other opportunities during this encounter and during future encounters in which further inquiry will be appropriate. Healthcare practitioners should not place thorough documentation or "checking all the boxes" above the well-being of the patient in front of them—nor should they avoid asking important questions because of their *own* discomfort.

✓ 5. **Communication is essential with any patient. How do you make a patient feel comfortable when you are not able to communicate in their primary language?**
Excellent interpretation is essential when a healthcare practitioner and patient are not fluent in a common language. This may involve an in-person, telephone, or video monitor interpreter. Telephone and video monitor services are available for all major languages, but challenges may still arise due to different dialects, discomfort with interpreters of the opposite sex from the patient, or fears that confidentiality will not be protected by the third party. It is always appropriate to confirm a patient's preference for language and comfort with the mode of translation available; gender may not be a choice but can be considered if possible.

In an ideal situation, the healthcare practitioner may have a moment to connect with the interpreter to share the

"Bleeding Too Much" (In the Words of a Refugee)

context of the visit they are joining to assist with, in order to explain the goals for the encounter and to identify any potential barriers to successful interpretation (such as social barriers that might result from a shared community of origin). Once the healthcare provider, patient, and interpreter are all introduced, it is important to review the roles of each participant and discuss confidentiality. Healthcare practitioners should address the patient rather than the interpreter and should reevaluate the patient's needs whenever they feel there is difficulty connecting with the patient. Finally, when the interview is concluding, it is important to assess the patient's understanding of the encounter and ensure any questions are answered [34].

Although healthcare practitioners may find it convenient to use a patient's family member (even a young child) or friend as an interpreter, a professionally trained interpreter should be used whenever possible. (In urgent situations, this may not be possible.) It is certainly permissible for patients with limited English language proficiency to choose to have older children or friends listen to discussions, as they may turn to them at home for reminders about the contents of a healthcare discussion.

Any agency receiving federal funds from medicaid or medicare has a legal obligation to provide appropriate language assistance to those who are not proficient in English. The Centers for Medicare and Medicaid Services (CMS) specifically direct that "interpreters, translators and other aids needed to comply with this policy shall be provided without cost to the person being served" [35].

You pause, ask her how she may feel more comfortable discussing this with you. She states that she is uncomfortable with the male telephone interpreter and would like an in-person female interpreter. She knows a woman in the community she feels more comfortable with and can arrange to come to a follow-up appointment with her. Mariam returns the following morning with a woman from the community, who works as a case manager during the day.

As part of her training as a case manager, she took classes in medical interpretation.

Mariam now feels comfortable sharing more of her history. She reports that she started menses at the age of 15 and had normal monthly periods until 2 years ago. When she arrived in Egypt, she found out she was pregnant; the UN refugee clinic offered her termination and options for contraception. During the termination procedure, she experienced a hemorrhage, which led to her transfer to a nearby hospital. She believes that she had something implanted after that procedure to prevent pregnancy as she was given an option for implantation in her arm or uterus and did not want something visible in her arm.

Six months ago, she started to have irregular cycles with variable days of bleeding. She denies any associated symptoms, including abdominal pain. You feel that you have made good progress in your understanding of her situation and feel that you should ask about her other medical history, including her limp. When you do, she lowers her eyes to the floor and becomes quiet. You do not press the issue.

✓ 6. **Having learned more about Mariam's experience, what kinds of challenges do you think *you* face as her primary care practitioner?**

> **Box 12.1   Teaching Tip**
> Encourage learners to reflect on their own cultural background as a starting point. What kind of life experiences do they bring to this encounter? Do they feel judgment toward the patient's community? Have they ever experienced judgment from their own families or communities? This is an opportunity to cultivate cultural humility.

Cultural differences may make it more difficult for healthcare practitioners to empathize. Imagining themselves in her shoes becomes harder when life experiences are very different (or when people believe them to be very different). Practitioners may show and interpret affective responses (e.g., sadness, grief, anger, compassion, etc.) differently than patients, resulting in miscommunication. Differences in communication styles exist even with good interpretation.

Conversations like these can be stressful for healthcare practitioners themselves. Some have personally experienced traumatic events in their own pasts; others have listened to many, many stories of trauma or treated patients directly after trauma; and still others may have very limited experience with trauma. Refugees, by definition, have survived traumatic events. It is important for practitioners to learn to recognize their own emotional responses to patients' stories of suffering and to find ways to practice self-care and avoid compassion fatigue [36].

You ask and receive permission to do a pelvic exam to further assess the abnormal bleeding. You note that her clitoris is absent and there is scarring over her anterior genitalia, which you recognize as female genital cutting (or female genital mutilation [FGM]), a common cultural practice in Eritrea. You are able to visualize the string to an intrauterine device (IUD). You believe that the IUD may be the cause of the irregular menses and schedule an ultrasound to assess its placement. You plan to see her back in 2 weeks for close follow-up.

Before she leaves, you want to ensure that she has appropriate supports in place. She states that she will receive 1 more month of case management support from the resettlement agency. She lives in a low-income housing apartment with another single young woman who speaks her language. She recently took a minimum-wage job cleaning a nearby hotel, so she will lose her public assistance this month. She receives Supplemental Nutrition Assistance Program (SNAP) benefits and is able to walk to a market nearby her home.

✅ 7. **What does cultural humility mean to you in the context of this case?**
In this case, cultural humility requires being open to learning about Mariam's culture (e.g., language, values, beliefs, practices, etc.), from Mariam herself. Healthcare practitioners who aspire to cultural humility accept that they cannot know or understand everything about other cultures and actively seek to learn from their patients. This stance can enhance clinicians' ability to connect with their patients and build trusting relationships. As one author states, "It takes away the need to know everything about a certain culture and encourages us to approach every patient encounter acknowledging that we will humble ourselves, learn what is important to the patient, and leave having learned something from the interaction" [37]. Essential components of cultural humility include a commitment to learning about others, to engaging in self-reflection, and to considering power differentials in the relationship [38].

For example, in this case, the PCP may be aware that female genital cutting is a common practice in North Africa (including in Mariam's home country, Eritrea), but approach matters more than knowledge. Directly asking her (e.g., "Have you ever undergone 'female circumcision' or 'female genital mutilation'?") may demonstrate awareness of the issue but may not help Mariam to feel comfortable talking about it. Instead, it may be helpful to say something like, "I have heard about a tradition of cutting to prepare girls for marriage in Eritrea. Are you familiar with this? What are your thoughts about it?"

It is important to acknowledge that some clinicians may find it challenging to listen in an empathetic and nonjudgmental way. They may be inclined to condemn Mariam's mother for performing genital cutting. However, it is important to validate her experiences without criticizing her family, community, faith, or culture. Reflective language is very useful in these types of conversations.

✅ 8. **What kinds of challenges do you think Ms. Gebreselassie faces daily in her new life in the United States?**
Like most refugees, Mariam is likely to face a myriad of challenges, including but not limited to:
— Adapting to a new culture
— Learning a new language
— Understanding and accessing public transportation or finding other means of transportation
— Adapting to and being able to maintain the housing that was made available upon resettlement

"Bleeding Too Much" (In the Words of a Refugee)

- Adjusting a traditional diet to that of the new country
- Finding and sustaining employment
- Adapting to a new education system for herself (or, in other cases, for one's children)
- Accessing publicly available services
- Potentially adjusting to loss of social status
- Adjusting to loss of community/social support
- Facing discrimination based upon her ethnicity, religion, or immigration status

---

Box 12.2   Personal Perspective

Foua Yang, a Hmong refugee whose daughter Lia's story is told in Anne Fadiman's 1997 book *The Spirit Catches You and You Fall Down: A Hmong Child, Her American Doctors, and the Collision of Two Cultures,* describes her experience thus:

> When you think about Laos and about not having enough food and those dirty and torn clothes, you don't want to think. Here it is a great country. You are comfortable. You have something to eat. But you don't speak the language. You depend on other people for welfare. If they don't give you money you can't eat, and you would die of hunger. What I miss in Laos is that free spirit, doing what you want to do. You own your own fields, your own rice, your own plants, your own fruit trees. I miss that feeling of freeness. I miss having something that really belongs to me. (p. 105) [39]

You continue to follow up with Ms. Gebreselassie regularly. You still want to better understand the origin of her limp to see if you can help her improve her gait and function. She continues to deny that she has any pain in her leg. Eventually, you feel you have established a good relationship, and you decide to ask again. This time, Mariam states simply, "They beat me." She emotionally recounts the experience she had navigating through Sudan. The people she paid to assist in arranging her travel turned out to be human smugglers.

Through tears, she describes repeated beatings, rapes, burns, and many hours being tied down, which resulted in her right leg having significant contractures of the muscle and her skin being scarred from the burns. She was only able to escape from the smugglers in Egypt because she was too weak to move quickly, and the smugglers viewed her as more of a liability than an asset. She reports being seen by physicians in Egypt but was told that surgery for her leg would not help and might make it worse. She did not feel that the local doctors took her seriously: "They acted as if my problems were my fault." She also states "I have learned to live with it. If I think about the pain, I feel pain all over my body. If I just keep moving, it does not hold me down."

✓ 9. **As a primary care healthcare practitioner, how do you assess Mariam's mental health, in light of the traumatic experiences she has survived?**

PCPs are the first gatekeepers to mental health assessment and manage a great deal of mental health care because of the high demand for care and limited availability of mental health services. This is true for both the general population and the refugee population. The same approach is used for both diagnosing and treating mental health for both populations; however, PCPs should be aware of the increased risk for mental health problems among refugees. In order to optimize health care for this vulnerable population, it is imperative that PCPs have a contextual understanding for mental health within the patient's culture of origin and an appreciation for the barriers they may face (e.g., communication challenges, cultural differences, etc.). (Discussion Question 3 provides information on screening tools to assist with the diagnosis of mental illness in refugee populations.)

Patients with known history of psychiatric illness or concerns that are clearly beyond a PCP's scope of practice will need referral to a psychiatrist or psychologist. Ensuring that the patient is safe and has appropriate supports in her community is the first step. Close follow-up with a consistent primary care provider can also help her to work through social, cultural, and other resettlement-related issues that may be contributing to symptoms of anxiety, depression, and post-traumatic stress.

Recounting a traumatic experience to many individuals is not typically therapeutic for trauma survivors, but once they have entrusted their story, they often prefer to return to someone who understands their situation and can help them move forward without rehashing the details of that history. Among recently resettled refugees, it is not uncommon for such symptoms to persist for 6–12 months, as they relate to an adjustment reaction. Healthcare practitioners should avoid prematurely labeling patients with a formal diagnosis of mental illness, as this may lead to a breakdown in rapport and to shame or stigmatization within their communities [21, 34].

10. **Self-reflection: How do you think Mariam's experience is different than experiences of women who report sexual assault in social settings in the United States? How is it the same?**

> **Box 12.3 Teaching Tip**
> Encourage learners to reflect on rape myths and whether they perceive some survivors as more "deserving" of justice than others. Mariam faced discrimination of this same type in Egypt. Some learners may admit that they are more willing to accept Mariam's experiences as traumatic than the experiences of more privileged American women and that they are more likely to hold American women culpable for behavior that is perceived to "encourage" sexual assault. These attitudes are difficult to discuss but very important to explore.

Some months later, while you are out of town, Mariam presents with new abdominal pain. The physician covering your practice obtains blood work and a urinalysis, which are normal, and urine pregnancy test, which is negative. He feels she looks stable and this pain may be psychosomatic given her traumatic past. He is already over an hour behind and asks her to come back in a week for follow-up with you. Mariam agrees to call if anything changes before then, but later that night, she goes to the emergency department, where she is diagnosed with appendicitis.

11. **What do you think happened?**
    It is difficult to truly assess the adequacy of the work-up with the benefit of hindsight. Was he able to get a detailed history? Was interpretation adequate? Was he able to ascertain how this presentation was different from her prior concerns about too much bleeding? Knowing how long it took the PCP to develop a trusting relationship with Mariam, it is understandable that he was not able to do so in the space of a single encounter. What led him to believe that the symptoms were associated with psychosocial issues? Did biases or assumptions about patients like Mariam contribute? Regardless of whether the diagnosis was "missed" or delayed, this situation highlights the benefits of continuity of care, as Mariam's PCP may have been more likely to recognize that this was not consistent with Mariam's previous concerns. (For a more extensive exploration of how cognitive biases can result in decisional errors and missed diagnoses, see ▶ Chapter 17.)
    (a) **Women are more likely than men to have "medically unexplained symptoms," a term to describe situations in which people have physical symptoms for prolonged periods of time for which no medical explanation is identified** [6, 7]. **In the past, the term "somatization" was frequently used to describe similar situations. What do you think are contributing factors for this?**
    The term "medically unexplained symptoms" (MUS), historically referred to as "somatization," is used to describe situations in which a biomedical diagnosis has not been identified to explain physical symptoms that have persisted for an extended period of time. It is important to note that the term itself does not mean that no biomedical explanation exists nor that the symptoms are "not real" or "all in the patient's head."
    In a recent report of MUS in primary care, women, adolescents, and "non-native speakers" had the

highest reported rates of MUS; previous studies also identified female gender as a risk factor for *being identified* as having MUS [6, 7]. This may be due to gender bias, in which physicians and other healthcare practitioners dismiss women's physical complaints because they feel they are "emotional" responses or physiologic differences. Pain perception, for instance, is associated with hormonal differences, sociocultural context, history of trauma, and other factors [6, 7].

12. **What is a healthcare practitioner's role in advocating for vulnerable patients? Are you ethically obligated to "do more" for patients with fewer resources and greater needs in order to allow them to reach the same level of health as those with greater resources and fewer needs?**
Healthcare practitioners indeed have an ethical obligation to "do more" for patients with fewer resources. Patients with fewer resources may not have money to pay for medications or procedures. They may not have access to transportation, which can result in their missing appointments. They may live in houses that increase their risk of respiratory disease. They may not have access to healthful food or clean, safe water. For all these reasons and more, patients with fewer resources are at greater risk for not being healthy. The concept of distributive justice obliges practitioners to treat their patients equitably—that is, to ensure that all have access to beneficial healthcare resources that are available to any. Without justice, it is impossible to enact the other principles of ethics. Practitioners cannot act beneficently if the best available medical care is not accessible to a particular patient because of language or cultural barriers, for instance. They may inadvertently cause harm for the same reason. Autonomy cannot exist if patients' voluntary choices are constrained by immigration status, socioeconomic resources, or subtler forms of discrimination.

(a) **Do our current care models support patients with fewer resources and greater needs? Do our societal norms? In what ways? How can healthcare practitioners spur change?**
At this time, while there are pockets of support in the U.S. healthcare system for vulnerable populations, there are not widespread, significant financial incentives or public policies in place to support healthcare practitioners in the pursuit of health equity. (Constraints within the U.S. healthcare system are discussed further in ▶ Chapter 8.) There is, however, growing attention to the importance of addressing social determinants of health to improve population health. The Centers for Medicaid and Medicare Services have funded the "Accountable Health Communities Model" pilot program in 31 communities across the country. Health systems that are funded through this program develop, implement, and evaluate screening and referral for social determinants of health, using patient navigators and other resources [40]. The program is based upon:

» …emerging evidence that addressing health-related social needs through enhanced clinical-community linkages can improve health outcomes and reduce costs. Unmet health-related social needs, such as food insecurity and inadequate or unstable housing, may increase the risk of developing chronic conditions, reduce an individual's ability to manage these conditions, increase healthcare costs, and lead to avoidable healthcare utilization [40].

Though programs like this are encouraging, policies related to immigration, refugee resettlement, and universal healthcare access remain politically controversial in the United States, with Americans voicing divided opinions on these issues. This politicization creates obstacles for healthcare practitioners who wish to promote health equity.

The ethical principle of justice requires that healthcare practitioners provide all patients with optimal health care. Sometimes fulfilling that principle requires advocacy for communities or populations, beyond simply advocating for an individual patient's needs. Physicians and other healthcare practitioners have a uniquely intimate perspective on the lives and experiences of their patients. Particularly for physicians, the role and title often confers inherent respect or prestige in their communities, which their patients usually do not have. They also have resources and connections—professional, financial, educational—that most patients lack. Where healthcare practitioners' position has granted them insight into a public issue, they should speak up—this is part of the social contract between physicians and society [41, 42].

> **Box 12.4 Teaching Tip**
> Ask learners to brainstorm specific ways that they can advocate for patients.

References

1. World Health Organization. Female genital mutilation [Internet]. Geneva: WHO; 2018 Jan 31 [cited 2018 Jul 5]. Available from: http://www.who.int/news-room/fact-sheets/detail/female-genital-mutilation.
2. United Nations Office on Drugs and Crime. Human trafficking [Internet]. Vienna: UNODC [cited 2018 5 Jul]. Available from: https://www.unodc.org/unodc/en/human-trafficking/what-is-human-trafficking.html.
3. Ubuntu! Myths about sexual assault- rape myths [Internet]. Durham (NC). [cited 2018 Jul 5]. Available from: https://iambecauseweare.wordpress.com/myths-about-sexual-assault-rape-myths/.
4. Mobini S. Psychology of medically unexplained symptoms: a practical review. Cogent Psychol. 2015;2:1.
5. Page L, Wessely S. Medically unexplained symptoms: exacerbating factors in the doctor-patient encounter. J R Soc Med. 2003;96(5):223–7.
6. The UN Refugee Agency. What is a refugee? [Internet]. Geneva: UNHCR [cited 2018 Jul 6]. Available from: https://www.unrefugees.org/refugee-facts/what-is-a-refugee/.
7. United Nations Migrants and Refugees. Definitions [Internet]. UN. [cited 2018 Jul 6]. Available from: https://refugeesmigrants.un.org/definitions.
8. The UN Refugee Agency. Figures at a glance [Internet]. Geneva: UNHCR. [cited 2018 Jul 6]. Available from: http://www.unhcr.org/en-us/figures-at-a-glance.html.
9. Human Rights Watch. Rohingya crisis [Internet]. New York: Human Rights Watch. [cited 2018 Jul 10]. Available from: https://www.hrw.org/tag/rohingya-crisis.
10. The UN Refugee Agency. Rohingya emergency [Internet]. Geneva: UNHCR. [cited 2018 Jul 10]. Available from: http://www.unhcr.org/en-us/rohingya-emergency.html.
11. US Department of State Refugee Processing Center. Admissions and arrivals. Historical arrivals broken down by region (1975 – present) [Internet]. Arlington (VA): RPC. [cited 2018 Jul 12]. Available from: http://www.wrapsnet.org/admissions-and-arrivals/.
12. Krogstad JM, Radford J. Key facts about refugees to the U.S. Pew Research Center, Facttank [Internet]. Washington, D.C.: Pew Research Center; 2017 Jan 30 [cited 2018 Jul 12]. Available from: http://www.pewresearch.org/fact-tank/2017/01/30/key-facts-about-refugees-to-the-u-s/.
13. United Nations High Commissioner for Refugees. Statistical yearbook 2016. Table 26: indicators of host country capacity and contributions. Geneva: UNHCR; 2016 [cited 2018 July 30]. Available from: http://www.unhcr.org/en-us/statistics/country/5a8ee0387/unhcr-statistical-yearbook-2016-16th-edition.html.
14. International Organization of Migration. About IOM [Internet]. Le Grand-Saconnex: IOM. [cited 2018 Jul 12]. Available from: https://www.iom.int/about-iom.
15. US Department of Health and Human Services, Administration for Children and Families. Office of Refugee Resettlement. Washington, D.C.: ACF. [cited 12 Jul 2018]. Available from: https://www.acf.hhs.gov/office-of-refugee-resettlement.
16. The UN Refugee Agency. US resettlement agencies [Internet]. Geneva: UNHCR. [cited 2018 Jul 6]. Available from: http://www.unhcr.org/en-us/us-resettlement-agencies.html.
17. World Health Organization. Migration and health: key issues [Internet]. Geneva: WHO. [cited 2018 Jul 12]. Available from: http://www.euro.who.int/en/health-topics/health-determinants/migration-and-health/migrant-health-in-the-european-region/migration-and-health-key-issues.
18. McNeely CA, Morland L. The health of the newest Americans: how US public health systems can support Syrian refugees. Am J Public Health. 2016;106(1):13–5.
19. US Department of Health and Human Services, Administration for Children and Families. Office of Refugee

Resettlement. Refugee health [Internet]. Washington, D.C.: ACF. [cited 2018 Jul 12]. Available from: https://www.acf.hhs.gov/orr/programs/refugee-health.
20. Giacco D, Priebe S. WHO Europe Policy brief on migration and health: mental health care for refugees. Geneva: World Health Organization. [cited 2018 Jul 15] Available from: http://www.euro.who.int/__data/assets/pdf_file/0006/293271/Policy-Brief-Migration-Health-Mental-Health-Care-Refugees.pdf.
21. Morris MD, Popper ST, Rodwell TC, Brodine SK, Brouwer KC. Healthcare barriers of refugees post-resettlement. J Community Health. 2009;34(6):529–38.
22. Centers for Disease Control and Prevention. Refugee health guidance [Internet]. Atlanta (GA): CDC. [cited 2018 Jul 15]. Available from: https://www.cdc.gov/immigrantrefugeehealth/guidelines/refugee-guidelines.html.
23. The UN Refugee Agency. Sexual and gender-based violence [Internet]. Geneva: UNHCR. [cited 2018 Jul 12]. Available from: http://www.unhcr.org/en-us/sexual-and-gender-based-violence.html.
24. The US Agency for International Development. Preventing and responding to gender-based violence. [Internet]. Washington, D.C.: USAID. [cited 2018 Jul 12]. Available from: https://www.usaid.gov/gbv.
25. United Nations Women. Facts and figures: ending violence against women [Internet]. New York: UN Women [cited 2018 Jul 15]. Available from: http://www.unwomen.org/en/what-we-do/ending-violence-against-women/facts-and-figures.
26. United Nations Population Fund. FGM dashboard [Internet]. New York: UNFPA. [cited 2018 Jul 17]. Available from: https://www.unfpa.org/data/dashboard/fgm.
27. Smith SG, Zhang X, Basile KC, Merrick MT, Wang J, Kresnow, et al. National intimate partner and sexual violence survey: data brief [Internet]. Atlanta (GA): Centers for Disease Control and Prevention; 2014 Nov [cited 2018 Jul 17]. Available from: https://www.cdc.gov/violenceprevention/pdf/2015-data-brief.pdf.
28. Centers for Disease Control and Prevention. ACEs definitions. Data and statistics. [Internet]. Atlanta (GA): CDC; 2016 Jun [cited 2018 Jul 5]. Available from: https://www.cdc.gov/violenceprevention/acestudy/about.html.
29. Felitti V. Adverse childhood experiences and adult health. Acad Pediatr. 2009;9:131–2.
30. Felitti VJ, Anda RF, Nordenberg D, Williamson DF, Spitz AM, Edwards V, Koss MP, Marks JS. Relationship of childhood abuse and household dysfunction to many of the leading causes of death in adults: the adverse childhood experiences (ACE) study. Am J Prev Med. 1998;14:245–58.
31. Centers for Disease Control and Prevention. Adverse childhood experiences (ACEs) [Internet]. Atlanta (GA): CDC; 2016 [cited 2018 Aug 22]. Available from: https://www.cdc.gov/violenceprevention/acestudy/index.html.
32. Substance Abuse and Mental Health Services Administration. Trauma informed approach and trauma specific interventions [Internet]. Rockville (MD): SAMHSA. [cited 2018 Jul 17]. Available from: https://www.samhsa.gov/nctic/trauma-interventions.
33. Hollifield M, Verbillis-Kolp S, Farmer B, Toolson EC, Woldehaimanot T, Yamazaki J, Holland A, St. Clair J, SooHoo J. The Refugee Health Screener-15 (RHS-15): development and validation of an instrument for anxiety, depression, and PTSD in refugees. Gen Hosp Psychiatry. 2013;35(2):202–9.
34. Kirmayer L, Narasiah L, Munoz M, Rashid M, Ryder A, Guzder J, Hassan G, Rousseau C, Pottie K. Common mental health problems in immigrants and refugees: general approach in primary care. CMAJ. 2011;183(12):E959–67.
35. Centers for Medicare & Medicaid Services. Strategic language access plan (lap) to improve access to CMS federally conducted activities by persons with limited English proficiency (lep). Baltimore (MD): CMS; 2014 Feb 28 [cited 2018 Jul 15]. Available from: https://www.cms.gov/About-CMS/Agency-Information/OEOCRInfo/Downloads/StrategicLanguageAccessPlan.pdf.
36. Figley CR. Compassion fatigue: coping with secondary traumatic stress disorder in those who treat the traumatized. New York: Routledge; 2013.
37. Ansari A. Battling biases with the 5 Rs of cultural humility. The Hospitalist [Internet]. Philadelphia (PA); 2017 [cited 2018 Jul 12]. Available from: https://www.the-hospitalist.org/hospitalist/article/136529/leadership-training/battling-biases-5-rs-cultural-humility.
38. Tervalon M, Murray-Garcia J. Cultural humility versus cultural competence: a critical distinction in defining physician training outcomes in multicultural education. J Health Care Poor Underserved. 1998;9(2):117–25.
39. Fadiman A. The spirit catches you and you fall down: a Hmong child, her American doctors and the collision165413 of two cultures. New York: Farrar, Straus and Giroux; 1997.
40. Centers for Medicare & Medicaid Services. Accountable Health Communities Model [Internet]. Baltimore (MD): CMS. [cited 2018 Jul 12]. Available from: https://innovation.cms.gov/initiatives/ahcm/.
41. Cruess RL, Cruess SR. Expectations and obligations: professionalism and medicine's social contract with society. Perspect Biol Med. 2008;51(4):579–98.
42. Cruess SR, Cruess RL. Professionalism and medicine's social contract with society. Virtual Mentor. 2004;6(4):185–88.

Further Reading on This Topic

Beah I. A long way gone: memoirs of a boy soldier. New York: Sarah Crichton Books; 2007.

Korn F, Eichhorst S. Born in the big rains: a memoir of Somalia and survival. New York: The Feminist Press at CUNY; 2008.

Mire S. The girl with three legs: a memoir. Chicago (IL): Lawrence Hill Books; 2011.

Rawlence B. City of thorns: nine lives in the world's largest refugee camp. New York: Picador; 2016.

Wamariya C, Weil E. The girl who smiled beads: a story of war and what comes after. New York: Crown Publishing Group; 2018.

Choices and Care at the End of Life

Contents

Chapter 13   "My Father Would Not Want to Live Like This" – 251

Chapter 14   "We're Not Ready to Give Up" – 269

"My Father Would Not Want to Live Like This"

Sharon A. Brangman

13.1 Background Questions – 252

13.2 Additional Case Information and Questions for Discussion – 252

13.3 Answers to Background Questions – 254

13.4 Additional Case Information and Responses to Questions for Discussion – 258

References – 266

© Springer Nature Switzerland AG 2019
A. E. Caruso Brown et al. (eds.), *Bioethics, Public Health, and the Social Sciences for the Medical Professions*, https://doi.org/10.1007/978-3-030-03544-0_13

- **Glossary**

Artificial nutrition and hydration - May include any method of administering nutrition or hydration to a patient who cannot take food or fluids independently or with assistance from a caregiver, such as placement of a nasogastric tube for administration of enteral formula or intravenous fluids.

Limitation of intervention - Refers to decisions or orders to restrict the types of interventions (both diagnostic and therapeutic) administered to a particular patient. Usually made when certain interventions cannot achieve the patient's goals of care. These commonly include (but are not exclusive to) Do Not Resuscitate and Do Not Intubate orders.

Learning Objectives

1. Apply ethical principles and legal rules to analyze and resolve end-of-life issues.
2. Identify strategies to increase ease with initiating and facilitating end-of-life discussions with patients.
3. Recognize how personal, family, cultural, and religious beliefs can impact healthcare decisions.

> **Case Background**
>
> You are a physician who is taking care of an 85-year-old man, Antonio Alvarez. Mr. Alvarez was admitted to the hospital with bacterial endocarditis and pneumonia almost 2 weeks ago. He developed acute kidney failure as a complication of antibiotic therapy, and dialysis was initiated. Unfortunately, soon after starting dialysis, Mr. Alvarez experienced cardiac arrest. He was resuscitated and is now breathing with continuous positive airway pressure (CPAP) support; however, he suffered significant cerebral hypoxia. Mr. Alvarez, who has a history of Alzheimer disease, has not returned to his baseline cognitive state. At this time, he is unable to take adequate nutrition or hydration by mouth.

13.1 Background Questions

1. From a physiologic perspective, what does death from withdrawal of artificial nutrition or hydration typically look like? How long does it take?

2. What does your state's law say about withdrawal of artificial nutrition and hydration? In the case of a healthcare agent? In the case of a surrogate decision-maker?

3. There is no ethical or legal difference between withholding (not initiating) and withdrawing life-sustaining treatments. Is there evidence of psychological differences in attitudes toward withholding (not initiating) compared with withdrawing life-sustaining treatments?

4. Many states have medical or physician "orders for life-sustaining treatment" (MOLST/POLST) forms. What does your state have? Under what circumstances should a patient have a MOLST/POLST?

5. Physicians and many other healthcare practitioners are expected to discuss end-of-life issues with their patients and their families. Does Medicare cover end-of-life counseling and care?

6. Define and distinguish between the following conditions: coma, vegetative state, persistent vegetative state, permanent vegetative state, minimally conscious state, and brain death.

7. What is the Uniform Determination of Death Act (UDDA)? How did it come to be, and what are some of the criticisms of it?

8. Several religious traditions have specific positions on end-of-life decision-making (such as declaration of brain death in Orthodox Judaism or withdrawal of artificial nutrition and hydration in Catholicism). Choose one and research the basis of the position.

13.2 Additional Case Information and Questions for Discussion

When Mr. Alvarez was initially admitted, you reviewed the electronic health records for advance directives but were unable to find any. The social worker on the healthcare team has worked diligently to try to identify any information about the patient's wishes. Together you learned that the patient's daughter, Justine Levi-Alvarez, was previously appointed as his healthcare proxy. The signed document was drawn up 20 years earlier, after the

"My Father Would Not Want to Live Like This"

death of her mother, Mr. Alvarez's first wife. Ms. Levi-Alvarez is often at her father's side when you round. Today she has requested a Do Not Resuscitate (DNR) order for any further arrest. She has also stated that she is in favor of withdrawing nutrition via nasogastric tube and is currently considering whether hydration is appropriate.

1. Do you agree with Justine's decisions on her father's behalf? Why or why not?

2. Self-reflection: How would you approach this situation if this were your father or mother?

Mr. Alvarez's second wife, Sophia Alvarez, whom he married 15 years ago, is often at her husband's bedside as well. She tries to visit him every day but is not always able to do so because of her own health issues and her dependence on family and friends for transportation. You learn that Mr. Alvarez has two other children, a son who calls in every few days to check on his father and a second daughter who has been estranged from the family for several years.

3. What are your obligations to his wife?

4. What are your obligations to his other children?

Before you can take any action on Ms. Levi-Alvarez's request, the nurse calls you to the patient's bedside to speak with Mrs. Alvarez, who vehemently disagrees with her stepdaughter regarding withdrawal of nutrition. She tells you that Mr. Alvarez's wishes have evolved since the healthcare proxy was first written. She agrees that when she first met her husband, he was adamant that he would not want to live in this sort of condition; however, 2 years ago, he was diagnosed with Alzheimer disease, and he became much more easygoing. She says that early on, after his diagnosis but before his cognitive deficits became severe, they had numerous discussions about their values. She describes her husband's appreciation for simply being alive for its own sake and his acceptance of his decreasing ability to accomplish certain tasks.

5. Does this change how you would manage his care? If so how, and how would you proceed?

6. Under what circumstances, if any, can a healthcare proxy be overruled?

7. How do you weigh the patient's recent statements against his past wishes?
   (a) Does the diagnosis of Alzheimer disease mean that you must disregard the patient's more recent statements?
   (b) Creative problem-solving: How would you construct a law or policy to take into account the above?
   (c) Would you regard Mr. Alvarez's stated wishes differently if he had a lifelong cognitive impairment, such as Down syndrome?

Another patient on your service is Eleanor Rosenstein, a 79-year-old woman who, like Mr. Alvarez, has suffered significant hypoxia after a cardiac arrest and is unable to communicate or take adequate nutrition or hydration by mouth. She is not married and has no known children or friends.

8. How would you approach decisions for Ms. Rosenstein's medical care in this situation?

A new diagnosis, such as Alzheimer disease, or a change in prognosis of an underlying medical condition, is often a good time for healthcare practitioners to revisit their patients' advance directives, but Mrs. Alvarez tells you that their regular healthcare practitioner did not do this. In fact, she had no idea that Mr. Alvarez's daughter's decisions, as the healthcare proxy, would supersede her decisions, as his wife.

9. Why do you think physicians have difficulty discussing patients' wishes with them before their condition becomes terminal?

10. How might physicians be encouraged to have such conversations?
    (a) What kinds of hospital policies can drive this?
    (b) Should such discussions be incentivized monetarily? Why or why not?

An ethics consultation is called to help determine the best course of action in light of the family's conflicts. The ethics consultants carefully review all of the available information, including the original healthcare proxy. They realize that the healthcare proxy does not explicitly address artificial nutrition and hydration. When they inquire how Mr. Alvarez felt about this specific issue, Ms. Levi-Alvarez indicates that she is not sure whether her father ever expressed his wishes at this level of detail.

11. **Given this information and your understanding of your state's law, is withholding artificial nutrition and hydration a legal option?**

When the discussion turns to the DNR order, Mrs. Alvarez tearfully shares that her mother recently died after a long illness. She and her extended family pressed for aggressive care despite the poor prognosis because they felt they owed it to their mother to "fight for her." They all felt that doing so was an act of love and devotion. Furthermore, she says, "You can't trust hospitals. Some of the doctors and nurses care but some don't. Telling them we don't want everything done for my husband is like telling him to hurry up and die. That one doctor only talked to us for 5 minutes this morning. I bet we won't see a doctor at all once Justine has her way."

12. **How would you respond to this?**

13. **Do you think it is important to ask about faith and culture in your discussions with the members of the Alvarez family? Why or why not?**
    (a) What would you want to know about Mrs. Alvarez's background and beliefs?

14. **Do you think that understanding Mr. Alvarez's faith and culture is important to your interactions with Mr. Alvarez himself? Why or why not?**
    (a) How might you learn more about Mr. Alvarez's background and beliefs since he cannot tell you himself?

13.3 Answers to Background Questions

1. **From a physiologic perspective, what does death from withdrawal of artificial nutrition or hydration typically look like? How long does it take?**
   Artificial nutrition and hydration (ANH) are considered medical interventions for individuals who are not able to take adequate food and fluids by mouth. The discussion of artificial nutrition and hydration is common in end-of-life decision-making (both with respect to initiating the intervention or to withdrawing it) because many people naturally have a decrease in appetite and a reduction in the need for food and fluids as the end of life approaches. There are many myths about what patients may experience after withdrawal of such medical interventions, including that they experience pain associated with starvation or thirst. There are many studies that disprove these myths [1]. Such deaths are generally described as being peaceful. Some patients report feelings of euphoria after a few days without food or water. Many patients appear to be very comfortable and transition from being alert to being in a coma and then to death. This process can take approximately 1–2 weeks but may occur more quickly depending on the patient's degree of disease burden.

   From a physiologic perspective, a lack of fluids leads to an imbalance of electrolytes such as sodium and potassium. Muscle and fat stores are mobilized by the body and used to make energy. Significant complications and discomfort can occur if ANH is initiated [2]. As an example, feeding a person who is at the end of life may result in physical discomfort such as abdominal cramps, diarrhea, nausea, and vomiting since gastrointestinal function is declining.

2. **What does your state's law say about withdrawal of artificial nutrition and hydration? In the case of a healthcare**

agent? In the case of a surrogate decision-maker?

States have different laws and regulations about the withdrawal of ANH. In New York, for example, adults who have decision-making capacity have the right to make any medical decisions, including the withdrawal of ANH. In the case of an adult who does not have decision-making capacity, a healthcare agent (as designated by a healthcare proxy) typically has the authority to make decisions on the patients' behalf either based on knowledge of the patient's wishes or in accordance with the patient's best interests. Some states have specific rules for withholding or withdrawing artificially provided fluids and nutrition. For example, in New York the patient's wishes regarding artificial nutrition and hydration must be "reasonably known" or can be reasonably ascertained. If there is no reason to believe the patient would want to forego a feeding tube under the circumstances, then the proxy has no authority to make this decision [3]. Surrogate decision-makers who are not appointed as healthcare proxy can most often make decisions to forego life-sustaining treatment, including artificial nutrition and hydration, but are sometimes limited in this regard by state law. For adults without decision-making capacity who are in a hospital or nursing home, a surrogate decision-maker can make decisions for adults who do not have decision-making capacity and who do not have a designated healthcare agent or legal guardian. In most states, if not all, there is a hierarchical list of surrogates: spouse or domestic partner, followed by adult child, parent, sibling, and friend, in that order [3].

3. **The consensus view is that there is no legal or ethical difference between withholding (not initiating) and withdrawing a life-sustaining treatment (LST). Is there evidence of psychological differences in attitudes toward withholding (not initiating) compared with withdrawing life-sustaining treatments?**

While experts commonly agree that there is neither a legal or ethical distinction between withholding and withdrawing LST, there are several studies that demonstrate that healthcare practitioners find it more emotionally difficult to withdraw treatment than to withhold it.

In a recent study, more than 90% of physicians who participated in a survey about LST agree that they should always comply with orders to withdraw LST; however, more than 60% reported that they felt that withdrawing LST was more "psychologically difficulty" than withholding it. The authors identified that religious characteristics, specialty, and levels of experience caring for dying patients influenced their opinions. Physicians' religiosity was associated with more distress about withdrawing LST, while end-of-life specialists appeared less likely to distinguish between withdrawing and withholding LTS [4].

4. **Many states have medical or physician "orders for life-sustaining treatment" (MOLST/POLST) forms. What does your state have? Under what circumstances should a patient have a MOLST/POLST?**

Advance directives are written instructions that indicate an individual's instructions for future healthcare decisions that ideally result from advance care planning. While there is state-to-state variation in the terminology and some of the details of advance directives, usually there are two key pieces of information included: what the individual would want in the event of a future hypothetical situation (e.g., ventilatory support after significant brain injury) and who should make medical decisions if the individual loses decision-making capacity. The intent of such directives is to increase the likelihood that an individual's wish (autonomy) is respected in medical decision-making. All adults should have advance directives. Ideally, this discussion should occur before a medical crisis occurs and should involve the patient, his or her healthcare proxy, as well as healthcare providers to ensure that the wishes of the patient are followed. It is important to understand that, while these directives are useful tools to facilitate the discussion about end-of-life care, they are not intended to replace ongoing, open communication between healthcare practitioners and patients about difficult decisions.

In contrast, the MOLST (medical orders for life-sustaining treatment) and POLST (physician orders for life-sustaining treatment) are more detailed documents, allowing a person to determine with greater specificity the type of medical care they want to receive. It may include instructions on chemotherapy, antibiotics, and other medical interventions in addition to cardiopulmonary resuscitation, intubation, feeding tubes, and other life-sustaining treatments. One important distinction between MOLST/POLST and advance directives is that MOLST/POLST documents are typically completed with either the patient or the patient's proxy or surrogate, contemporaneously in consideration of the patient's current condition, whereas advance directives may be written many years before onset of serious illness when the patient is in good health. Ideally, healthcare practitioners should discuss MOLSTs/POLSTs with patients who are seriously ill, medically vulnerable, or nearing the end of their life [5].

5. **Physicians and many other healthcare practitioners are expected to discuss end-of-life issues with their patients and their families. Does Medicare cover end-of-life counseling and care?**
While early versions of the Affordable Care Act contained provisions for healthcare practitioners to discuss end-of-life care with their patients and for programs to provide public education about advanced care planning, these provisions were eliminated in the final bill [6]. As of January 2016, physicians have been allowed to bill Medicare for advance care planning including discussion about patients' wishes about care they desire at the end of life. According to the Kaiser Family Foundation, "Medicare now covers advance care planning provided in medical offices and facility settings, including hospitals. As with most other physician services, beneficiaries are subject to cost sharing for advance care planning provided by their physician or health professional. If Medicare beneficiaries desire advance care planning during their annual wellness visit, physicians and other health professionals may provide it during the visit and bill Medicare separately for it. However, beneficiaries will not have any cost sharing liability for advance care planning provided in conjunction with their annual wellness visits" [6].

6. **Define and distinguish between the following conditions: coma, vegetative state, persistent vegetative state, permanent vegetative, minimally conscious state, and brain death.**
A **coma** is a "state of unresponsiveness" in which patients cannot be awakened (aroused) and have no awareness of self or surroundings, lasting for at least an hour in order to exclude transient causes of loss of consciousness. Most patients who survive begin to awaken within 2–4 weeks [7].

A **vegetative state** is a condition in which patients are awake but unaware of themselves and their environment. It is qualified as "persistent" if it lasts at least 1 month and as "permanent" if it lasts more than 3 months after non-traumatic injury (such as drug overdose) or 12 months after traumatic brain injury [7].

A **minimally conscious state** describes circumstances in which patients do show some limited or occasional awareness of themselves or their environments. Proposed criteria for assessing this include reproducible or sustained behaviors, such as following simple commands, responding affirmatively or negatively with words or gestures, making intelligible speech, or other purposeful behavior [7].

Brain death is the irreversible loss of all functions of the entire brain, including the brain stem. Among the criteria for determination of brain death are "unresponsiveness and lack of receptivity, the absence of movement and breathing, the absence of brain stem reflexes, and coma whose cause has been identified." Conditions that might confound the diagnosis, such as shock or hypothermia, must be identified and treated before a determination of death can be made [8].

7. **What is the Uniform Determination of Death Act (UDDA)? How did it come to be, and what are some of the criticisms of it?**

   The Uniform Determination of Death Act (UDDA) is a piece of model legislation (nonbinding statutory text intended to serve as a guide for state legislation) that was approved in 1981 by the National Conference of Commissioners on Uniform State Laws, in cooperation with the American Medical Association (AMA), the American Bar Association (ABA), and the President's Commission for the Study of Ethical Problems in Medicine and Biomedical and Behavioral Research. It provides two circumstances in which a person may be declared dead: (1) irreversible cessation of circulatory and respiratory functions and (2) irreversible cessation of all functions of the entire brain, including the brain stem (whole-brain death) [9].

   The UDDA originated in the work of an ad hoc committee at Harvard University in the late 1960s. As technology advanced, an increasing number of patients thought to be permanently comatose were maintained on ventilators. At the same time, there was growing success in the field of organ transplantation. This group argued for neurological criteria for death in order to address two concerns: the burden of families and hospitals of providing "futile" care to these patients and ethical concerns surrounding the procurement of organs from such patients [10]. The UDDA, or some version of it, has been adopted as law in all 50 states and the District of Columbia. The UDDA reflects the medical, legal, and societal consensus that the irreversible cessation of all brain functions, including the brain stem, constitutes the death of the individual, and that consensus still exists today. At the bedside, the determination of brain death is crucial in deciding when a patient can be disconnected from life-sustaining technology. As a general matter, there is no obligation to continue to treat a dead body. Determination of brain death (also known as neurological death) is also critical to the procurement of organs. The continued beating of the heart with ventilator and other support increases the chances of procuring viable organs. Under the widely accepted "dead donor rule," death must be determined before organs can be removed from the donor.

   Some of the biological underpinnings of the concept of "brain death" have been questioned in recent years, particularly the assertion that integrative functioning of the organism as a whole is dependent upon the brain. Cases in which patients were diagnosed as meeting the criteria for brain death but went on to maintain integrative functioning (temperature regulation, circulation, etc.) with only mechanical ventilation and nursing care have been described [10].

8. **Several religious traditions have specific positions on end-of-life decision-making (such as declaration of brain death in Orthodox Judaism or withdrawal of artificial nutrition and hydration in Catholicism). Choose one and research the basis of the position.**

> **Box 13.1 Teaching Tip**
>
> Although it is potentially more challenging for instructors, asking learners to select which faith they wish to research in response to this question avoids the risks of singling out specific religious beliefs for scrutiny and permits them to decide for themselves whether to discuss their own beliefs. A sample response is included below.

In the above example, there is a division within the Orthodox Jewish community as to whether "whole-brain death" means death of the individual. For some Orthodox Jews, so long as the heart still beats, even if artificially sustained, the patient still lives. Others accept the prevailing consensus that death occurs when there is irreversible cessation of brain function even if the heart is beating. Jewish law, known as *halacha*, prohibits one life being taken to save the life of another person, under any circumstances. Halachic laws require that death be clearly established, otherwise removal of an organ for transplant or discontinuing life-support is considered to be equivalent to murder. In situations where the family states that their loved one does not accept brain death as death, it is important for the healthcare team to understand their patient's perspective and seek guidance from the appropriate spiritual leader for their patient as well as an ethics consultation team, if available [11–14]. A handful of states recognize by law a right of conscientious objection to the determination of death based on whole-brain death criteria [15, 16].

With respect to the second example, the Catholic tradition has long held that human life is sacred and that humans have the responsibility to preserve it if they have the means, including ANH, to do so [17]. Furthermore, the Vatican issued a position statement on this issue: "The administration of food and water even by artificial means is, in principle, an ordinary and proportionate means of preserving life. It is therefore obligatory to the extent to which, and for as long as, it is shown to accomplish its proper finality, which is the hydration and nourishment of the patient. In this way suffering and death by starvation and dehydration are prevented" [18]. Leading Catholic bioethics scholars disagree with the Vatican's position. Central to their position is the principle of proportionality—that counsels' consideration of the benefits and burdens of ANH as the proper approach to determining whether to forego ANH is ethically justified. The patient's own wishes play a central role in this analysis as well [19].

13.4 Additional Case Information and Responses to Questions for Discussion

When Mr. Alvarez was initially admitted, you reviewed the electronic health records for advance directives but were unable to find any. The social worker on the healthcare team has been working diligently to try to identify any information about the patient's wishes. Together you learn that the patient's daughter, Justine Levi-Alvarez, was previously appointed as his healthcare proxy. The signed document was drawn up 20 years earlier, after the death of her mother, Mr. Alvarez's first wife. Ms. Levi-Alvarez is often at her father's side when you round. Today she has requested a Do Not Resuscitate (DNR) order for any further arrest. She has also stated that she is in favor of withdrawing nutrition via nasogastric tube and is currently considering whether hydration is appropriate.

1. **Do you agree with Justine's decisions on her father's behalf? Why or why not?**
   There are a few issues that should be addressed before making a decision: First, the healthcare proxy form is 20 years old, so the healthcare team (the physician with support from the social worker and perhaps later the ethics consultation team) may want to confirm that there is not a more recent document. The healthcare practitioner will also want to know whether the healthcare proxy includes any specific directives or limitations on the decisions that the proxy can make. Two key concepts are that:
   - Healthcare proxies are expected to make decisions first, in accordance with the patient's known wishes (advance directives); second, based upon what the patient would have wanted if he were able to communicate ("substituted judgment"); and third, the patient's best interests (if the other two are not possible).

- As stated earlier, states sometimes have specific rules governing withdrawal or withholding of artificial nutrition and hydration. It is important to be familiar with your local law.

In addition, the healthcare practitioner should ask Ms. Levi-Alvarez what is motivating her decision and whether she had prior conversations with her father about this situation, Do Not Resuscitate orders, and/or withdrawal of artificial nutrition and hydration.

2. **Self-reflection: How would you approach this situation if this were your father or mother?**

> **Box 13.2   Teaching Tip**
> This question is intended to encourage learners to reflect on the challenges of predicting one's own future actions in hypothetical circumstances since such prediction is at best dubious. "I would never…" and "I would always…" statements are intrinsically dangerous. Furthermore, temptations to guide patients with one's own opinions on such matters may lead into places that are not psychologically or ethically appropriate. Encourage learners to consider these issues and to realize that recognizing one's personal reactions to issues is the first step in becoming a more compassionate physician or healthcare provider.

Mr. Alvarez's second wife, Sophia Alvarez, whom he married 15 years ago, is often at her husband's bedside as well. She tries to visit him every day but is not always able to do so because of her own health issues and her dependence on family and friends for transportation. You learn that Mr. Alvarez has two other children, a son who calls in every few days to check on his father and a second daughter who has been estranged from the family for several years.

3. **What are your obligations to his wife?**

> **Box 13.3   Teaching Tip**
> It is important for facilitators to be familiar with the law regarding healthcare proxies in their state. The below answers reflect a typical approach to healthcare proxies.

In the absence of a healthcare proxy, the patient's spouse, if available and capable of making decisions, would be the patient's surrogate decision-maker. However, when a healthcare proxy has been appointed, that person supersedes other potential surrogate decision-makers.

Regardless of the law, healthcare practitioners have a professional responsibility to maintain good communication with the patient's wife as well as his daughter. The healthcare practitioner should ideally have discussion of Mr. Alvarez's current condition, prognosis, and options with both women. While it is not part of this case, this would be a much more difficult scenario for communication if the patient's daughter, as his proxy, requested that information be withheld from his wife or other family members.

4. **What are your obligations to his other children?**

Mr. Alvarez's son may have knowledge of his father's wishes and may also have insight into whether he intentionally maintained Ms. Levi-Alvarez as healthcare proxy or whether he simply did not think about it. His son is also deserving of honest and compassionate information about his father's condition.

The estranged daughter is unlikely to know her father's wishes, and the physician does not have an obligation to call every relative to notify them of decisions being made; however, it might be appropriate for a member of the team to encourage Ms. Levi-Alvarez or Mrs. Alvarez to consider whether this daughter would like to know about her father's condition.

Before you can take any action on Ms. Levi-Alvarez's request, the nurse calls you to the patient's bedside to speak with Mrs. Alvarez, who vehemently disagrees with her stepdaughter regarding withdrawal of nutrition. She tells you that Mr. Alvarez's wishes have evolved since the healthcare proxy was first written. She agrees that when she first met her husband, he was adamant that he would not want to live in this sort of condition; however, 2 years ago, he was diagnosed with

Alzheimer disease, and he became much more easygoing. She says that early on, after his diagnosis but before his cognitive deficits became severe, they had numerous discussions about their values. She describes her husband's appreciation for simply being alive for its own sake and his acceptance of his decreasing ability to accomplish certain tasks.

5. **Does this change how you would manage his care? If so, how and how would you proceed?**
   It may make the team ask more questions of all family members, regarding both Mr. Alvarez's stated wishes and his relationships with his daughter, wife, and son. Involvement of others—such as the social worker or the ethics consultation team—might be useful, and a family meeting will be important to clarify all the issues and positions. Family members—including Ms. Levi-Alvarez as healthcare proxy—may need to be educated regarding the standards for surrogate decision-making. The fact that Mr. Alvarez appointed her as his healthcare proxy does not necessarily mean that anyone discussed what that role entails with her. It will also be important to help Mr. Alvarez's family distinguish their own feelings or values from what they think he would have wanted.

6. **Under what circumstances, if any, can a healthcare proxy be overruled?**
   Both law and practice can vary from state to state on this question. As a general rule, a proxy who is not able, available, or willing to serve as proxy needs not be recognized as the proxy, and the healthcare team may look for others, first to the alternate proxy if one has been designated and then to family members in order of priority under state law, to make decisions on the patient's behalf. If there is a proxy who is, or wants to be, involved in the patient's care, he or she may not simply be overruled when others disagree. Rather the proxy may be formally overruled or removed from his or her authority as proxy only on specific grounds. Law and ethics typically support overriding the proxy if he or she is disregarding the patient's wishes, is acting with ill motive (perhaps seeking to keep the patient alive to continue collection of social security benefits), or has become unable, unwilling, or unavailable to meet the responsibilities of the proxy appointment. Depending on the reason(s) for questioning the proxy's authority, it may be necessary to contact hospital counsel and/or to seek court involvement in order to remove a proxy or override his or her decisions. Like other advance directives, proxy forms are legal documents that typically require the formalities of signing and witnessing in order to be legally valid. On occasion these formalities are not met, calling into question the validity of the proxy appointment. In general, before concluding that a proxy document is not legally valid, hospital policy and/or hospital counsel should be consulted.

7. **How do you weigh the patient's recent statements against his past wishes?**
   Some persons with Alzheimer disease or a related dementia may change their preferences for care over time. The full course of dementia can span 10–15 years, and disease progression can vary significantly between individuals. It is not unusual to find that once patients are in the midst of a chronic or life-threatening disease, they might find that their quality of life is not as bad as they had predicted at the time of initial diagnosis. If patients have capacity to make medical decisions, then their proxy document does not apply. This may be an appropriate time to ask the patient to revisit, and perhaps re-execute, their advance directive. Ideally, these discussions should occur with the patient and healthcare proxy periodically throughout the course of the disease, for as long as the patient maintains capacity to make healthcare decisions.

(a) **Does the diagnosis of Alzheimer disease mean that you must disregard the patient's more recent statements?**

This depends on whether Mr. Alvarez is deemed to have the capacity to make his own healthcare decisions. The diagnosis of Alzheimer disease does not itself make Mr. Alvarez decisionally incapable. Rather, this is determined based on a careful capacity assessment. Recall that capacity is decision-specific. Mr. Alvarez may be able to make his own healthcare decisions for now. Even if he cannot make all of his own decisions, he may still be able to reaffirm or change his choice of his daughter as healthcare proxy.

Many patients experience mild personality changes early in the course of dementia syndromes, and some have described a shifting of their priorities similar to what Sophia describes. With the progressive loss of memory and capacity for integrated thought that marks more advanced stages of dementia, families, physicians, and others who have known the patient for some time may see such a deterioration of psychological continuity from the past to the present that they begin to think of the patient as being a different person than he once was. If at some point Mr. Alvarez is no longer the same person he once was, should his former self control decisions for his present self? Should his previously expressed wishes, interests, and values, including his proxy directive, take precedence over his current experiences and interests as a seriously ill patient? The prevailing and widely accepted view in both law and ethics is that the patient's advance directive, along with other evidence of his previously expressed wishes and values, is to be respected. Writing an advance directive is an act of prospective autonomy, the very point of which is to ensure that one's wishes, intentions, and values for future care—the exercise of autonomy now—will control decisions in the future when capacity is lost due to the ravages of illness, disease, or disability [20, 21]. Nonetheless, the psychosocial view that the patient with advanced dementia has become a different person is a real concern, especially for families, and might pull loved ones to focus on decision-making based on the patient's current best interests more so than on the patient's prior wishes and directives [22, 23]. Some have made the stronger ethical argument that if the patient is no longer the same person he once was (is he a new person?), then current best interests, not previously expressed wishes, should be the controlling standard for decisions [24]. Some have also argued against assuming that a person has a single self or identity that exists continuously from birth until death [25].

> **Box 13.4 Teaching Tip**
> Explore this topic in depth with interested learners. Would they want their end-of-life decisions made on the basis of their beliefs, values, and choices when they were fully "well" or when they had begun to understand what it was like to be ill? Is it possible to determine which persona is more truly reflective of Antonio? These are difficult questions about the nature of the self and of living—there is not a single correct answer.

Again, Ms. Levi-Alvarez has decisional authority, but she should certainly be aware of all available information about her father's wishes and must also take into account his current interests, including the benefits and burdens of continued interventions and the avoidance of pain and suffering. It is the fiduciary responsibility of the proxy (and family) to consider both the patient's previously expressed wishes and his current best interests, giving priority to respect for the patient's advance directive and prior autonomy. It might be helpful to have additional insights from Mr. Alvarez's primary care provider and his other children or other close friends or relatives, in order to explore further his continuity of self with the past and to further inform understanding of his previously expressed wishes and how they apply to Mr. Alvarez's current medical condition [26].

> **Box 13.5 Personal Perspective**
>
> A *New York Times Magazine* essay explored the story of Sandy Bem, a Cornell University professor diagnosed with early-onset Alzheimer disease who eventually chose to end her own life [26]. The reporter describes events after the birth of Dr. Bem's first grandchild:
>
> » [Dr. Bem] told [her daughter] Emily that her 'new brain' might actually make her better suited to being a grandmother than her focused, hyperanalytical 'old brain.' She seemed to have found a way of being that she liked, content to sing silly songs and make nonsense sounds for hours on end.
>
> » Emily liked her mother this way, too. It had sometimes been difficult to be Sandy's daughter. As a child, Emily wanted to wear her hair long and take ballet lessons; Sandy, ever vigilant about gender stereotypes, nudged her to cut her hair and play soccer instead. But now Sandy didn't seem to care about such things. Emily thought that her mother was taking pleasure in life in a way that the old Sandy could not have anticipated—and she found herself hoping that the joy her mother took in Felix might make her reconsider her intention to end her life quite so soon.
>
> » The others in Sandy's inner circle saw her relationship with Felix and wondered what it would mean for her original plan. The old Sandy, who valued her rationality and her agency, had been clear that she would be unwilling to keep living when she could no longer articulate coherent thoughts. But this newer Sandy didn't seem unhappy living her life in this compromised way. Ultimately, who should make the decision to die, the old Sandy or the new one?

(b) **Creative problem-solving:** How would you construct a law or policy to take into account the above?

(c) Would you regard Mr. Alvarez's stated wishes differently if he had a lifelong cognitive impairment, such as Down syndrome?
Legally, there are subtle differences: a determination of competence would likely have been made at the time the patient turned 18, and he would likely be under guardianship of his parents or someone else if he could not make his own decisions.

Another patient on your service is Eleanor Rosenstein, a 79-year-old woman who, like Mr. Alvarez, has suffered significant hypoxia after a cardiac arrest and is unable to communicate or take adequate nutrition or hydration by mouth. She is not married and has no known children or friends.

8. **How would you approach decisions for Ms. Rosenstein's medical care in this situation?**
Making medical treatment decisions for unbefriended older adults poses special challenges. These are typically individuals who lack the capacity to make a healthcare decision, have no advance directive, and have no family, friends, or legally authorized surrogate to help make medical decisions. Many healthcare organizations have policies that address the needs of this group of older adults to make sure that decisions are not made in an ad hoc manner. Currently, there are significant state-to-state differences in

the approach for medical decision-making for unbefriended older patients.

A judge can appoint a guardian, but this is often a costly and time-consuming process. In the absence of specific hospital policies or in the midst of an urgent medical situation, decisions might be made by team consensus that can include two or more physicians. This approach has the goal of avoiding decisions made by a single individual who may have a conflict of interest. An ethics consultation team, if available, could also be authorized to assist with medical decision-making [27].

A new diagnosis, such as Alzheimer disease, or a change in prognosis, is often a good time for healthcare practitioners to revisit their patients' advance directives, but Mrs. Alvarez tells you that their regular healthcare practitioner did not do this. In fact, she had no idea that Mr. Alvarez's daughter's decisions, as the healthcare proxy, would supersede her decisions, as his wife.

9. **Why do you think physicians have difficulty discussing patients' wishes with them before their condition becomes terminal?**
Although patients and families want their healthcare practitioners to discuss end-of-life care with them, there are many reasons why a physician might be reluctant to have these conversations. Healthcare practitioners may feel that they have "failed" the patient and are uncomfortable addressing this fear or may feel unequipped or uncomfortable in how to approach the discussion. These discussions often take time, and it can be difficult to review the issues thoroughly in a busy hospital or office setting [28]. In addition, many diseases have an unpredictable course, and it may be difficult to predict when an individual will reach the final stages of a disease. Furthermore, healthcare practitioners may be uncomfortable making decisions in the context of a disease process that might have a variety of outcomes. Sometimes healthcare practitioners do not want to be the bearers of bad news or make a patient feel as though the current situation is hopeless. Some healthcare practitioners feel it is their duty to offer every option and medical intervention to patients even if that care may be futile.

Despite these factors, it is important that physicians understand the importance of their role in introducing the discussion of end-of-life care in a compassionate manner with the ultimate goal of ensuring that their patients' wishes are understood and respected in medical decision-making.

10. **How might physicians be encouraged to have such conversations?**
First, these skills should be taught during medical school and residency training programs, so that they become an expected part of a physician's clinical responsibilities. Some large medical systems offer continuing education programs that are helpful for practicing physicians who may not have learned these skills during their training. In addition, some electronic medical records give physicians reminders that indicate when an advance care directive is needed. Other members of the healthcare team, such as nurses and social workers, can be helpful in getting the conversation started with patients and families. There are also websites and educational materials for patients' and families' use. These may be helpful in answering basic questions so that physician time is used more effectively.

(a) **What kinds of hospital policies can drive this?**
In 2014, La Crosse, Wisconsin, made national news when it was found that 96% of its citizens had an advance directive when they died [29, 30]. This accomplishment has been attributed to the community "engaging in a disruptive communicative innovation" [29]. Throughout the community, physicians and nurses inspired by Bernard Hammes, a medical ethicist with Gundersen Health Services, talked with their patients about advance care plans.

In *Dying in America Improving Quality and Honoring Individual Preferences Near the End of Life*, the Institute of Medicine's 2014 report on end-of-life care "a person-centered, family-oriented approach that honors individual preferences and promotes quality of life through the end of life" is recommended. The report provides specific recommendations for health systems, healthcare practitioners, and payers [31].

Training healthcare practitioners to provide advance care planning as well as the use of a standardized form or template and the provision of incentives to do so (such as quality measures) can drive the conversation. Health systems should ensure that their electronic health records are structured to facilitate promotion and improve access to advance directives. Some hospitals require that, upon admission, all patients be asked to identify a healthcare proxy or to state their status with regard to Do Not Resuscitate (DNR) and Do Not Intubate (DNI). Nursing homes or long-term care facilities may also employ such policies.

(b) **Should such discussions be incentivized monetarily? Why or why not?**
Because of time constraints placed on healthcare practitioners, it is important that contributions to meaningful care, such as advance care planning, have a billable code to ensure sustainability of providing that care. Since January 2016, Medicare has provided reimbursement for physicians, nurse practitioners, and physician assistants who discuss end-of-life care with their patients. The advance care planning codes (there are two distinct codes that can be applied) can be used as often as needed to discuss advance directives and to complete MOLST or other advance directive forms [32]. Despite this, many health systems have been slow to bill Medicare for these discussions [33].

It is important to note that while monetary incentives are helpful, as discussed above, a comprehensive, systematic approach to adopting advance care planning is needed to move the conversation.

An ethics consultation is called to help determine the best course of action in light of the family's conflicts. The ethics consultants carefully review all of the available information, including the original healthcare proxy. They realize that the healthcare proxy does not explicitly address artificial nutrition and hydration. When they inquire how Mr. Alvarez felt about this specific issue, Ms. Levi-Alvarez indicates that she is not sure whether her father ever expressed his wishes at this level of detail.

✅ 11. **Given this information and your understanding of your state's law, is withholding artificial nutrition and hydration a legal option?**
As discussed in the Answers to Background Questions 2, in some states, the law requires a higher standard for withdrawal of ANH: in New York State, in order to withdraw ANH, the proxy must "reasonably know" the wishes of the patient. That is, withholding or withdrawal of ANH is not legally allowed unless it is known that the patient desired it.

When the discussion turns to the DNR order, Mrs. Alvarez tearfully shares that her mother recently died after a long illness. She and her extended family pressed for aggressive care despite the poor prognosis because they felt they owed it to their mother to "fight for her." They all felt that doing so was an act of love and devotion. Furthermore, she says, "You can't trust hospitals. Some of the doctors and nurses care but some don't. Telling them we don't want everything done for my husband is like telling him to hurry up and die. That one doctor only talked to us for 5 minutes this morning. I bet we won't see a doctor at all once Justine has her way."

✅ **12. How would you respond to this?**
Physicians and other healthcare practitioners commonly use language such as "Do you want everything done?" Phrases such as "do everything" or "withdrawal of care" can inadvertently lead patients to believe that refusal or discontinuation of certain life-sustaining interventions means that the patient will be abandoned. These phrases can also suggest that these are "all-or-nothing" decisions, when in fact there is often a range of options, including some of which are only available if the goal is palliative care. For instance, an elderly patient with a high risk of aspiration might be allowed to eat by mouth, for pleasure, if the goal is not to prolong life as long as possible but to provide comfort.

In this situation, it is important to acknowledge Mrs. Alvarez's feelings and fears and to reassure her that her husband will not be abandoned. This is a common fear among families considering "limitations of intervention" orders and palliative care and may be more prevalent in some communities that have experienced discrimination and bias from the healthcare system (historically or currently). However, the healthcare practitioner or ethics consultant will need to take care that Mrs. Alvarez does not feel her fears were dismissed or trivialized. A family or care conference may be helpful in achieving this goal.

✅ **13. Do you think it is important to ask about faith and culture in your discussions with the members of the Alvarez family? What would you want to know about Mrs. Alvarez's background and beliefs?**
Her religious or cultural views will certainly impact her perspective. For instance, although Mrs. Alvarez's religious beliefs are not explicitly stated, the Catholic Church has an official position opposing withdrawal of ANH (as discussed in the Answers to Background Questions 7) on the grounds that it is ordinary care, not medical care. They may also help the physician to better understand the disagreement between Mr. Alvarez's wife and daughter. Their respective positions may be a direct product of their own values; however, it may also be that their different backgrounds influenced what they discussed with Mr. Alvarez and how they interpreted his statements.

Additionally, many people hold ideas about what their faith requires them to choose in medical situations that are stricter than those held by leaders in their faith community or described in their sacred texts: believing, for example, and, as noted above, that withholding an intervention is permissible but that withdrawing it once started is not. While people are always free to choose how to interpret their faith, in such situations, it can be helpful to encourage the patient or family to discuss their situation with a faith or religious leader who can better guide them. It is, however, inappropriate and unprofessional for a physician to tell a patient or a patient's family member that she does not understand her own religion.

✅ **14. Do you think that understanding Mr. Alvarez's faith and culture is important to your interactions with Mr. Alvarez himself? Why or why not?**
Especially in the absence of any specific advance directives from Mr. Alvarez (e.g., a DNR/DNI order), the family will have to make decisions based on what they imagine he would have wanted. Asking the family to reflect on their loved one's beliefs may help his daughter make better substituted judgments. It also might help to bridge the divide between Ms. Levi-Alvarez and Mrs. Alvarez. For instance, if Mr. Alvarez did not share his wife's faith, it might be helpful for her to acknowledge that and to discuss how they built a life together in spite of their different religious backgrounds. The same might apply to his daughter—she might be encouraged to discuss what she respected and loved about her father, including his beliefs, even if they diverge from her own.

Other patients may have specific beliefs about what should happen during the end of their lives and after their death. Understanding of these beliefs and practices can help the team make appropriate arrangements and accommodations. Specific examples include:
- Arranging for a Catholic priest to administer last rites
- Withdrawing care at a specific time to allow family to be present and for burial within 24 hours per Jewish tradition

Acknowledgment The author and editors wish to thank Robert Olick, for his review of this chapter. This case was inspired by earlier cases written by Robert Olick and revised by Amy Caruso Brown.

References

1. Arenella C. Artificial nutrition and hydration at end of life: beneficial or harmful? [Internet]. American Hospice Foundation. [cited 2018 Apr 29]. Available from: https://americanhospice.org/caregiving/artificial-nutrition-and-hydration-at-the-end-of-life-beneficial-or-harmful/.
2. Coyle N, Todaro-Franceschi V. Chapter 6 Artificial nutrition and hydration. In: End-of-life ethics a case study approach. Washington, DC: Hospice Foundation of America; 2012.
3. New York State Department of Health. Deciding about health care: a guide for patients and families [Internet]. Albany: NYSDOH; 2018 [cited 2018 Apr 29]. Available from: https://www.health.ny.gov/publications/1503.pdf.
4. Chung GS, Yoon JD, Rasinski KA, Curlin FA. US Physicians' opinions about distinctions between withdrawing and withholding life-sustaining treatment. J Relig Health. 2016;55(5):1596–606.
5. National POLST Paradigm. POLST and advanced directives [chart]. Washington, D.C. [cited 2018 Apr 25]. Available from: http://polst.org/advance-care-planning/polst-and-advance-directives/.
6. Kaiser Family Foundation. 10 FAQs: Medicare's role in end-of-life care [Internet]. San Francisco: KFF; 2016 [cited 2018 Apr 30]. Available from: https://www.kff.org/medicare/fact-sheet/10-faqs-medicares-role-in-end-of-life-care/.
7. Laureys S, Owen AM, Schiff ND. Brain function in coma, vegetative state, and related disorders. Lancet Neurol. 2004;3(9):537–46.
8. Wijdicks EF. The diagnosis of brain death. N Engl J Med. 2001;344(16):1215–21.
9. What is the Uniform Declaration of Death Act (UDDA)? FindLaw [Internet]. Eagan (MN): Thomson Reuters. Available from: https://healthcare.findlaw.com/patient-rights/what-is-the-uniform-declaration-of-death-act-or-udda.html.
10. Shah SK, Truog RD, Miller FG. Death and legal fictions. J Med Ethics. 2011;37:719. https://doi.org/10.1136/jme.2011.045385.
11. Olick RS, Braun EA, Potash J. Accommodating religious and moral objections to neurological death. J Clin Ethics. 2009;20(2):183–91.
12. Vaad Halacha of the Rabbinical Council of America. Halachic issues in the determination of death and in organ transplantation, including an evaluation of the neurological 'Brain Death' Standard. Sivan 5770–2010 Jun. Available from: http://www.rabbis.org/pdfs/Halachi_%20Issues_the_Determination.pdf.
13. Breitowitz Y. The brain death controversy in Jewish law. Jewish Law Articles [Internet]. [cited 2018 May 1]. Available from: https://www.jlaw.com/Articles/brain.html.
14. Tendler M. Halakhic death means brain death [Internet]. Jewish Rev. 1990 Jan/Kislev, 5750;3(3). [cited 2018 May 1]. Available from: http://thejewishreview.org/articles/?id=114.
15. Pope T. Brain death rejected: expanding legal duties to accommodate religious objections. In: Lynch HF, Cohen IG, Sepper E, editors. Law, religion, and health in the United States. New York: Cambridge University Press; 2017. p. 293–305.
16. Olick RS. Brain death, religious freedom, and public policy: New Jersey's landmark legislative initiative. Kennedy Inst Ethics J. 1991;1(4):275–88.
17. O'Rourke K. Artificial nutrition and hydration and the catholic tradition. Health Progress. J Cathol Health Assoc U. S.; 2007 May-Jun [cited 2018 May 1]. Available from: https://www.chausa.org/publications/health-progress/article/may-june-2007/artificial-nutrition-and-hydration-and-the-catholic-tradition.
18. Levade W, Amato Angelo. Congregation for the doctrine of the faith. Responses to certain questions of the united states conference of catholic bishops concerning artificial nutrition and hydration [Internet]. The Vatican; 2007 [cited 2018 May 1]. Available from: http://www.vatican.va/roman_curia/congregations/cfaith/documents/rc_con_cfaith_doc_20070801_risposte-usa_en.html.
19. Mackler AL. Introduction to Jewish and Catholic bioethics: a comparative analysis. Washington, DC: Georgetown University Press; 2003.
20. Cantor NL. Advance directives and the pursuit of death with dignity. Bloomington: Indiana University Press; 1993.
21. Olick RS. Taking advance directives seriously: prospective autonomy and decisions near the end of life. Washington, DC: Georgetown University Press; 2001.
22. Jaworska A. Respecting the margins of agency: Alzheimer's patients and the capacity to value. Philos Public Aff. 1999;28(2):105–38.
23. Dworkin R. Life's dominion: an argument about abortion, euthanasia, and individual freedom. New York, NY: Vintage; 2011.

24. Dresser RS, Robertson JA. Quality of life and nontreatment decisions for incompetent patients: a critique of the orthodox approach. Law Med Health Care. 1989;17(3):234–44.
25. Foster C, Herring J. Identity, personhood and the law. Cham: Springer; 2017.
26. Henig RM. The last day of her life. The New York Times [Internet]. 2015 May 14. Available from: https://www.nytimes.com/2015/05/17/magazine/the-last-day-of-her-life.html.
27. Farrell TW, Widera E, Rosenberg L, Rubin CD, et al. AGS position statement: making medical treatment decisions for unbefriended older adults. J Am Geriatr Soc. 2017;65:14–5.
28. The John A. Hartford Foundation. Improving advance care planning: research results from the "Conversation Starters" focus groups and "Conversation Stopper" physician survey [Internet]. New York: JAHF; 2016 [cited 2018 May 6]. Available from: https://www.johnahartford.org/newsroom/view/advance-care-planning-poll.
29. Hatkoff C, Kula I, Levine Z. How to die in America: welcome to La Crosse, Wisconsin. Forbes [Internet]; 2014 Sep 33 [cited 2018 May 6]. Available from: https://www.forbes.com/sites/offwhitepapers/2014/09/23/how-to-die-in-america-welcome-to-la-crosse/#31aa6dcde8c6.
30. Joffe Walt C. The town where everyone talks about death [web streaming audio]. Washington, D.C.: National Public Radio; 2014 Mar 5 [cited 2018 May 6]. Available from: https://www.npr.org/sections/money/2014/03/05/286126451/living-wills-are-the-talk-of-the-town-in-la-crosse-wis.
31. Institute of Medicine. Dying in America: improving quality and honoring individual preferences near the end of life. Washington, D.C.: National Academies; Sep 2014.
32. Centers for Medicare & Medicaid Services. Frequently asked questions about billing the physician: fee schedule for advance care planning services [Internet]. Baltimore (MD): CMS; 2016 Jul 14 [cited 2018 May 6]. Available from: https://www.cms.gov/Medicare/Medicare-Fee-for-Service-Payment/PhysicianFeeSched/Downloads/FAQ-Advance-Care-Planning.pdf.
33. Hayhurst C. Why providers are slow to adopt new Medicare codes. AthenaInsight [Internet]. 2017 Nov 9 [cited 2018 May 6]. Available from: https://www.athenahealth.com/insight/why-providers-are-slow-adopt-new-medicare-codes.

Further Reading on This Topic

Doorway thoughts: cross-cultural care for older adults. Brangman S, Periyakoil V, editors. Volumes 1–3. New York: American Geriatrics Society; 2014. Available from: www.americangeriatrics.org.

"We're Not Ready to Give Up"

Anthony Michael Zepeda

14.1　Background Questions – 270

14.2　Additional Case Information and Questions for Discussion – 270

14.3　Answers to Background Questions – 272

14.4　Additional Case Information and Responses to Questions for Discussion – 275

References – 283

Learning Objectives

1. Apply ethical and legal principles to analyze and resolve end-of-life issues.
2. Analyze the implications of healthcare costs at the end of life for the United States (U.S.) healthcare system.
3. Describe conversational tools to facilitate being able to discuss death and dying openly with patients and respect their beliefs, opinions, and choices when they differ from one's own.
4. Explain the benefits of palliative care and when to offer it to patients.
5. Recognize and address your own biases as healthcare practitioners involved in end-of-life conversations.

> **Case Background**
>
> You are an oncologist caring for Kate Jaraczeski, a 43-year-old woman with relapsed metastatic breast cancer. She was initially diagnosed with breast cancer while pregnant with her third child and was treated during the second and third trimesters of her pregnancy. She went into remission but relapsed a year later. At that time, she was enrolled in a clinical trial and appeared to be doing well for several months. She presented to her primary care practitioner yesterday with symptoms of pneumonia. Imaging revealed new lesions in her lungs, prompting her visit with you today.

14.1 Background Questions

1. In the U.S., how much is spent on health care, on average per patient, in the last year of life? What percentage of total healthcare costs are incurred in the last year of life? In the last month?

2. What is palliative care? What is hospice care? How are they different?

3. Describe trends in utilization of hospice care in the U.S.

4. With respect to a hospice referral, how do healthcare practitioners make the determination that a person meets the criterion of having a prognosis of 6 months or less?

5. What is the National Institute for Health and Care Excellence (NICE) in the UK? How does NICE decide whether the UK government will pay for certain drugs?

6. What process, if any, does the U.S. government have to determine whether a costly medical therapy should be made available to individuals on public insurance? What about private insurance companies?

7. What is the Consolidated Omnibus Budget Reconciliation Act (COBRA)? What is the cost? What does it cover? How long can families be on COBRA?

8. What is the definition of "medical futility"? In clinical scenarios in which there is a perception of "medical futility," which principles of ethics are likely to be in conflict and why?

14.2 Additional Case Information and Questions for Discussion

You speak with the patient and her wife, Leila, about their options, including a new drug that has been shown to extend survival by 2–3 months. They agree that they want to try it. They have an 8-year-old, a 3-year-old, and a 15-month-old, and they feel that it is worth fighting for every day together. Kate has excellent private insurance through her wife's employer, so she is able to get the drug and pay relatively little out of pocket.

1. **What do you think is motivating the family's decision? Consider the perspectives of Kate, her wife, and their children.**

2. **Self-reflection: What would you choose if you were in this position?**

Initially, Kate had positive response to the new drug and felt well for the following 3 months. During that time, she was able to enjoy her time with her children and spend time outdoors. However, she recently began to feel short of breath, and scans show new lesions in her lungs. She meets with you again. You feel you should discuss palliative care and end-of-life preferences; however, before you have the opportunity to discuss it, Kate's wife asks what treatment options are left. You feel that they want to continue fighting, and you suggest another drug regimen even though you do not think it will work. Kate chooses to pursue it. When you are about to leave, you ask if they have any other questions that you have not answered. Kate and Leila ask how they should tell their 8-year-old and 3-year-old about the cancer.

> 3. If a healthcare practitioner feels that a medical intervention is medically futile, does she have an obligation to provide it if the family requests it?

> 4. What should the mothers tell their children? How would you counsel them to proceed?

Surprisingly, Kate's tumors appear to stabilize, though not regress, and Kate and her family report being very optimistic. She hopes to become eligible for a Phase 1 clinical trial of a new drug. You are less optimistic, but when you suggest at least a consultation with the palliative care team, Kate shakes her head and says, "You mean giving up? No. Not yet." You tell her it is not giving up and that it would be worth going just to hear another perspective.

> 5. How should you talk to Kate at this point about her disease and prognosis? How would you approach discussing difficult conversations?
>   (a) Would you encourage her to accept palliative care? Hospice care? Why or why not?
>   (b) What would be the benefits of offering palliative care at this time?
>   (c) Why do you think palliative care was not offered earlier than this?

Kate and Leila meet with the palliative care team. The palliative care physician assesses the understanding that Kate has about her disease and introduces the idea of setting goals and creating advance directives, a legal written statement that describes some aspects of medical care what a person would want in certain situations, although Kate asks to discuss this at another time. The care team spends time discussing the social situation of the family and discusses how they are emotionally and spiritually coping during the treatment. Leila reveals that she has been stressed after the recent layoffs in her company. She and Kate have been stressed about this since Leila lost her job earlier last month, and they have been paying for COBRA. This has been expensive, leading to increased debt. They are contemplating seeking out discounted care in the safety net system versus buying insurance through healthcare exchanges.

> 6. If Kate lived in a country like the UK, do you think her initial treatment options would have been different?

> 7. If Kate had to purchase individual insurance through the healthcare marketplace and was only able to afford a plan with a low monthly premium but a high deductible (meaning that the initial cost of treatment falls on the patient before the insurance company begins to provide reimbursement), how do you think her treatment options would change?

> 8. What are some of the advantages to having an advance directive?

> 9. How would you approach discussing do not resuscitate (DNR) status?

Kate enrolls in a local safety net cancer clinic and meets her new oncologist. For the first time, Kate finds out that the drug she has been receiving is very expensive (about $12,000 per month of treatment) and that not all patients are on it. Kate and Leila also learn that the drug is not covered by many insurance companies. In recent years, several newspapers and medical journals have published editorials from respected oncologists criticizing the pricing structure for drugs in the U.S., including a letter signed by more than 100 oncologists in 2016 [1]. Their proposals include allowing Medicare to directly negotiate pricing with pharmaceutical companies (currently prohibited by a 2003 law) and restricting companies' ability to delay the release of generic versions.

> 10. Is 3 months of life worth $36,000? To you personally? To society?

- 11. Is this pricing structure just? Why or why not?

- 12. What do you think about these proposals? Are they likely to be effective?

- 13. What is your responsibility, as a healthcare practitioner, to address this problem?

More recently, a form of immunotherapy for certain relapsed and refractory leukemias and lymphomas has been approved by the U.S. Food and Drug Administration (FDA) and priced at $500,000. It has been suggested that pricing might be prorated based upon efficacy; that is, patients who do not respond to the therapy would not pay for it [2].

- 14. What is the value of a 10% chance of cure? A 1% chance of cure?
  - (a) **Creative problem-solving:** Who should decide—patients, healthcare practitioners, hospitals, insurance companies, or governments?

- 15. What do you think about only paying for a high-cost pharmaceutical if it actually works? What are some of the practical challenges and ethical issues of this approach?

- 16. **Creative problem-solving:** How would you construct a law or policy to address drug pricing? You want to encourage innovation but avoid exploitation of patients. Keep in mind that many innovative therapies are developed through partnerships with public and private universities and supported by federal grants, all of which, in turn, are at least partly funded by taxpayers.

Several weeks later, Kate presents to the emergency department with hip pain and is found to have a pathologic fracture. An orthopedic surgeon comes to the emergency room to discuss operating. Kate is taken to the operating room for what was described as a "relatively simple procedure." Postoperatively, she struggles with being taken off the ventilator and is admitted to the intensive care unit (ICU). Within a few days, Kate develops sepsis and has to be emergently resuscitated. In the ICU, during a brief period of lucidity, Kate and Leila finally discuss her wishes, and the family decides to withdraw support. She dies a few hours later.

- 17. **Self-reflection:** Did Kate have a good death? Why or why not?

- 18. What factors make a good death? Who decides?

14.3 Answers to Background Questions

1. **In the U.S., how much is spent on health care, on average per patient per year? What percentage of total healthcare costs are incurred in the last year of life?**
   In 2016, the U.S. spent $3.3 trillion on health care or $10,348 per capita. Of this, 32% was hospital care, 3% was home health care, and 5% was nursing facilities [3]. The sickest 5% of the population consume 50% of healthcare cost [4]. There are several studies about healthcare cost attributable to care during the last year with estimates ranging from approximately 8% to 13% of total healthcare costs [5]. More recent studies have identified that while there are high-cost utilizers in their last year of life, they account for only 2 million (11%) of the 18.2 million high-cost population. The majority of high-cost utilizers of services are those with chronic conditions and/or functional limitations who live for several years [6].

2. **What is palliative care? What is hospice care? How are they different?**
   Palliative care is defined by the World Health Organization (WHO) as "an approach that improves the quality of life of patients and their families facing the problem associated with life-threatening illness, through the prevention and relief of suffering by means of early identification and impeccable assessment and treatment of pain and other problems, physical, psychosocial and spiritual" [7]. Palliative care is typically delivered by an interdisciplinary

team using a multidimensional approach. While it is common for people receiving palliative care services to be receiving end-of-life care, it is not limited to people with a "terminal" disease. Patients who receive palliative care may continue to receive cure-directed therapy. Palliative care has been shown to have numerous benefits including, for example, a decrease in symptom burden without an increase in family distress for patients with in-home palliative care [8]. It should be considered for any patient diagnosed with a life-threatening condition as early as possible. Both adult and pediatric palliative care are now board-certified subspecialties, requiring fellowship training (beyond residency) and a written examination, to improve quality of life in patients at the end of life.

Hospice care is palliative care specifically defined as care for those who are expected to live fewer than 6 months. With respect to coverage of costs associated with hospice care, coverage typically requires (by Medicaid, Medicare, and most insurance companies) that the patient forgoes all curative treatment and most additional life-prolonging treatment. It can include some palliative interventions with limited benefit (such as blood transfusions) with the intent of alleviating discomfort. Several states have Medicaid waivers in place for pediatric concurrent care, meaning that pediatric patients can receive hospice care while continuing to receive cure-directed treatment. Hospice care can be received at home, a nursing home, or an inpatient hospice facility—although in pediatrics, receiving hospice care in the hospital is not uncommon. Hospice programs may accept both children and adults, but many smaller programs may be uncomfortable with or inexperienced in the care of pediatric patients. The modern hospice movement is attributed to Dame Cicely Saunders who advocated for treating total pain, which she defined as pain that had elements of physical, emotional, social, and spiritual suffering.

3. **Describe and explain trends in utilization of hospice care in the U.S.**
   Hospice utilization has grown dramatically in the past 20 years. In 2000, approximately 500,000 people in the U.S. received hospice care; in 2016, 1,430,000 patients received hospice care [9]. Approximately eight out of ten Americans who died in 2014 were Medicare recipients. Of those, almost 50% received hospice care, compared to 21% in 2000. With respect to demographic variations, higher rates of utilization of hospice services were seen among women compared to men, white beneficiaries compared to beneficiaries of other races, and persons over the age of 85 years [10]. The increasing trend in hospice utilization is in part attributable to awareness of hospice and to a positive impression of hospice care. A Kaiser Family Foundation study showed that more than 7 out of 10 Americans have heard of hospice care, and of those, about 85% have a positive opinion about the benefits of hospice care [11].

4. **With respect to a hospice referral, how do healthcare practitioners make the determination that a person meets the criterion of having a prognosis of 6 months or less?**
   More than half of patients who are enrolled in hospice die within 3 weeks of enrollment, with more than 35% dying within 1 week [12]. Healthcare practitioners use their understanding of the natural history of disease to determine whether a person meets the criteria for having a prognosis of 6 months or less based on the most current clinical data. For many years, healthcare professionals have also applied the Palliative Performance Scale (PPS), a tool that measures a person's functional status (based on level of ambulation, activity level and evidence of disease, self-care, oral intake, and level of consciousness) to predict survival rates [12, 13]. The lower the score, the lower the survival rates. For example, a PPS score of 40 had an approximately 50% survival rate at 14 days, whereas a PPS score of 70 had a survival rate of

approximately 50% at 60 days [12]. Healthcare practitioners may apply the PPS and other variables—such as type and location of disease, age, gender, and where the patient is receiving care—to determine a 6-month prognosis.

5. **What is the National Institute for Health and Care Excellence (NICE) in England? How does NICE decide whether the government will pay for certain drugs?**
The National Health Service (NHS) in the UK, based on the idea that good health care should be available, was founded in 1948 to provide healthcare services to all UK residents. It currently provides services for more than 64 million people. The NHS is funded through taxation [14]. NICE is a non-department public entity with a vision to drive and enable excellence across the health and social care system in the UK, although its legal jurisdiction is England. It was initially established as the National Institute of Clinical Excellence in 1999 to assure access and quality of NHS treatments and care but has since evolved both in structure and scope. Independent committees establish guideline [15]. One of NICE's roles is determining whether government will pay for certain drugs based on quality-adjusted life-years attributable to the drug, stakeholder input, and cost-effectiveness. Based on responses to delayed access to newer drugs, in 2016 NICE established that new treatments would be reviewed within 90 days of being licensed [16].

6. **What process, if any, does the U.S. government have to determine whether a costly medical therapy should be made available to individuals on public insurance? What about private insurance companies?**
Unlike NICE and NHS, the U.S. does not set the cost of medications at the national level and instead allows pharmaceutical companies to do this on an individual basis. Reimbursement rates are set by private insurers and by the U.S. Centers for Medicare and Medicaid Services (CMS) for these medications. Private insurers conduct their own review of the medications and determine what they will cover. The Food and Drug Administration reviews whether medications can be released to the public but does not review the cost-effectiveness of these medications.
CMS' primary criterion for determining whether any medical therapy should be made available to its beneficiaries is "whether the item or service is reasonable and necessary for the diagnosis or treatment of illness or injury or to improve the functioning of a malformed body member for the affected Medicare beneficiary population" [17]. If an external entity requests that CMS review a new medical therapy, a "formal evidence review is then undertaken to determine whether or not an unbiased interpretation of the available evidence base supports or refutes the requested coverage in whole or in part. A proposed decision is normally issued for public comment within 6 months of opening the [national coverage determination or NCD] review" [17].

7. **What is the Consolidated Omnibus Budget Reconciliation Act (COBRA)? What is the cost? What does it cover? How long can families be on COBRA?**
COBRA requires that health insurance coverage must be offered to employees and their dependents after a qualifying event, such as a job transition, reduction in hours, or job loss. The employee is given 60 days to sign up for COBRA. COBRA can last 18 to 36 months, depending on the qualifying event. While COBRA cannot cost more than 102% of a similar plan, without the employer subsidizing the insurance, the premiums can be much larger than what the employees were previously paying. COBRA is required to be the identical coverage for the employee [18].

8. **What is the definition of "medical futility"? In clinical scenarios in which there is a perception of "medical futility," which principles of ethics are likely to be in conflict and why?**
There is much controversy about the term "medical futility" both in terms of its definition and in terms of its application

and utility in the practice of medicine. In a 2011 article, Lawrence Schneiderman addressed many proposed definitions of medical futility, including his own, that "medical futility is the unacceptable likelihood of achieving an effect that the *patient* has the capacity to appreciate as a *benefit*" [19]. Other proposals incorporate concepts such as patient goals, physiologic effect on body, and prolonging life into the definition [20]. Medical futility is frequently considered as having quantitative and qualitative components. Quantitative futility is applicable when an intervention or treatment has less than 1% chance of being effective. Qualitative futility is applicable when an intervention or treatment will not lead to a change in quality of life [20]. There is no general consensus around a single definition of futility to be applied or invoked uniformly across cases. The idea of physiological futility comes closest to a consensus view.

In bioethics, principlism is an approach that utilizes four major principles—autonomy, beneficence, nonmaleficence, and justice—and encourages the practitioner to weigh them accordingly. In this case, assuming that the patient or the patient's family is requesting an intervention that the clinician has deemed "futile," patient autonomy is likely going to be in direct conflict with the other three principles [21]. In this case, the clinician has presumably assessed that there is no benefit to the patient (beneficence is not applicable); all interventions carry some risk (thus nonmaleficence would be violated); and healthcare resources are limited (by cost or availability) and therefore should be utilized for those for whom there is benefit (justice).

14.4 Additional Case Information and Responses to Questions for Discussion

You speak with the patient and her wife, Leila, about their options, including a new drug that has been shown to extend survival by 2 to 3 months. They agree that they want to try it. They have an 8-year-old, a 3-year-old, and a 15-month-old, and they feel that it is worth fighting for every day together. Kate has excellent private insurance through her wife's employer, so she is able to get the drug and pay relatively little out of pocket.

✓ 1. **What do you think is motivating the family's decision? Consider the perspectives of Kate, her wife, and their children?**
In this case, it seems quite clear that prolonged time with family is likely motivating the decision to pursue care that extends life. Since the cost of the medication is covered and the benefit is extended survival, the initial decision is quickly rationalized in this scenario. However, in such situations, it is important to ensure that the discussion also includes information about side effects of the new medication and what that time may look like because extended survival does not necessarily equate to quality time.

✓ 2. **Self-reflection: What would you choose if you were in this position?**

> **Box 14.1 Teaching Tip**
> As with all the self-reflection questions, this is an opportunity for learners to acknowledge that in many, if not all, clinical situations, they will have a personal preference or a sense of what they would do in the same situation. Learners may also want to consider how they would answer the question, "What would you do if it were you [or your child], Doctor?" and when, if ever, it is appropriate to give a direct answer. The latter question is also explored in ▶ Chapters 16 and 19.

Initially, Kate had positive response to the new drug and felt well for the following 3 months. During that time, she was able to enjoy her time with her children and spend time outdoors. However, she recently began to feel short of breath, and scans show new lesions in her lungs. She meets with you again. You feel you should discuss palliative care and end-of-life preferences; however, before you have the opportunity to discuss it, Kate's wife asks what treatment options are left. You feel that they want to continue fighting, and you suggest another drug regimen even though you do not think it will work. Kate chooses to pursue it. When you are about to leave, you ask if they have any other questions that you have not answered. Kate and Leila ask how they should tell their 8-year-old and 3-year-old about the cancer.

3. **If a healthcare practitioner feels that a medical intervention is medically futile, does she have an obligation to provide it if the family requests it?**
One of the most challenging situations for healthcare providers arises when they disagree with patients or patients' families about medical decision-making, particularly with end-of-life care. While the U.S. healthcare delivery system places high value on respect for patient and family autonomy, upon considered analysis of the relevant principles, the physician will sometimes reach the conclusion that saying no to the request for medically futile treatment is justified. But is this position supported by the law? A number of states have laws pertinent to the medical futility dilemma. These laws, some modeled after the Uniform Health-Care Decisions Act, variously declare that there is no duty to provide treatment that is "medically inappropriate," "medically ineffective," or "contrary to accepted medical standards" [22]. However, these laws have rarely been tested, and some commentators believe they do not offer strong protections when the physician and hospital seek to override a family's decision to continue life-sustaining interventions. To the extent these laws rely on the judgment that the treatment in question offers "no significant benefit," different views of what counts as significant (meaningful) are often at the heart of the disagreement between patient, family, and physician [23].

Obligation to provide life-sustaining treatment, such as performing cardiopulmonary resuscitation (CPR), can be complicated. Some states have laws that allow physicians to refuse to provide care that cannot achieve the intended goal, as long as they follow the prescribed decision-making process. However, in other states, physicians may not be able to withhold a resuscitation attempt if a patient or family has requested it, even if the physician feels CPR will not be successful. As discussed in the Answer to Background Question 7, part of the challenge is defining "futility" and its corollary "success." For CPR, if a patient's cardiovascular status is stabilized for a short time but the patient dies within hours or days, is that success? In some cases, the goal of CPR may be seen not as saving the patient's life, but rather as giving the family the comfort of knowing everything was done [24].

With respect to "medically ineffective healthcare," there are some notable areas of consensus in the medical profession. For example:
- Chemotherapy does not have to be given for severely debilitated patients at high risk of mortality from chemotherapy and with limited benefit due to the extent of disease.
- Surgeons are known to refuse to perform a procedure that presents with an unacceptably high risk of mortality.

Disagreements about "futile" interventions can arise either with respect to withdrawal of the intervention or withholding it. When the physician deems an intervention that is sometimes appropriate for the treatment of this condition to be "inappropriate" or "futile" in this case, is there a duty to discuss this as an option with the patient and family? Advocates of patient and family autonomy will argue that there is; others contend that in some cases the "option" should not be presented unless the physician is prepared to provide the intervention at the request of the patient and family. These types of discussions tend to engender strong emotional responses, much like the term "rationing," as they seem to suggest that physicians, hospital administrators, or even politicians will be able to withhold care from sick people. Physicians should clearly explain why an intervention should not be offered.

"We're Not Ready to Give Up"

4. **What should the mothers tell their children? How would you counsel them to proceed?**

   The age and maturity of the children determine what should be addressed in such discussions. For the youngest children, under age 2, generally nothing needs to be shared. They do not need to have "closure." Children ages 2–5 years typically need to have some common questions answered. These include the following: "Is this contagious?" and "Was it something they did or said that caused this?" They may also want to know who will care for them in the way that Kate did, after she is gone. It is important to dispel magical thinking. It is also important to start working on any transitions that will be required. School-age children may need answers to these same questions, but they should also be asked if they want to know what is happening with respect to the care being provided. For adolescents, the focus should be on providing appropriate information, being honest, and respecting privacy both of the family and the teen. Often adolescents benefit from talking with a member of the healthcare team who will keep things confidential, as long as the adolescent is not a danger to themselves or others [25, 26].

Surprisingly, Kate's tumors appear to stabilize, though not regress, and Kate and her family report being very optimistic. She hopes to become eligible for a Phase 1 clinical trial of a new drug. You are less optimistic, but when you suggest at least a consultation with the palliative care team, Kate shakes her head and says, "You mean giving up? No. Not yet." You tell her it is not giving up and that it would be worth going just to discuss her cancer.

5. **How should you talk to Kate at this point about her disease and prognosis? How would you approach discussing difficult conversations?**

   There are multiple models that help with end-of-life conversations. One of the more well accepted is "SPIKES" [27]:
   - *S*etting up the interview
   - Assessing the patient's *P*erception
   - Obtaining the patient's *I*nvitation to discuss the news
   - Giving *K*nowledge and information to the patient
   - Addressing the patient's *E*motions with an *E*mpathetic response
   - *S*trategy and *S*ummary

   Another model is "COMFORT." COMFORT helps practitioners become more cognizant of some major barriers in delivering bad news. It stands for communication, orientation, mindfulness, family, ongoing, reiterative, and team [28, 29].

> **Box 14.2 Teaching Tip**
>
> If this conversation is difficult for learners, consider introducing outside resources to stimulate discussion. Learners can read and reflect on a poem or short story at this point in the case, such as Raymond Carver's poem, "What the Doctor Said," which reflects on this sort of interaction [29]. Alternatively, if computers or tablets are readily available, they can explore Five Wishes, a plain-language "living will" available online that helps patients to begin to have these conversations and to consider their options (► https://www.agingwithdignity.org/).
> The questions address:
> - "The person I want to make care decisions for me when I can't"
> - "The kind of medical treatment I want or don't want"
> - "How comfortable I want to be"
> - "How I want people to treat me"
> - "What I want my loved ones to know"

Although the case does not describe the previous discussions in detail, it is implied that Kate's physician might have done better to be forthcoming about her likelihood of death. However, it is also possible that she was fully informed and chose an aggressive approach to treatment despite the odds. Certainly, at this point, questions about what she and her family value most and what they want out of her final days are important. Most patients have fears about what dying and death will be like, and an honest conversation can alleviate some of those fears and allow patients to make concrete plans for the end-of-life period. For instance, the physician might ask Kate whether she would tolerate some discomfort in order to be more awake to interact with her wife and children. If so, how much? These will be familiar questions for most oncologists; however, consultation with a palliative care specialist may be helpful. Another question that may be helpful to ask is "How do you want this story to end?" or "What kind of an ending do you want to give this story?" [30].

(a) **Would you encourage her to accept palliative care? Hospice care? Why or why not?**
With its intent to improve quality of life, palliative care should be encouraged for all persons facing very serious or life-threatening diseases such as cancer. It is important to ensure that the patient and the patient's family understand what is meant by such care. For example, patients may not understand that curative care can still be administered with palliative care. It is important for the healthcare practitioners to be clear about the risks and benefits of all treatment options, introducing the topic of palliative care as a multidisciplinary team holistic approach to care. (See Background Question 8 for more information.) Hospice should be encouraged when the patient's life expectancy is less than 6 months and the patient is ready to forgo curative care.

(b) **What would be the benefits of offering palliative care at this time?**
Palliative care is beneficial since it is more holistic and will give her and her family access to a multidisciplinary team. The focus would shift to addressing the physical, emotional, social, and spiritual suffering of Kate and her family.

(c) **Why do you think palliative care was not offered earlier than this?**
Although all the details are not known, it is reasonable to believe that the physician may have felt uncomfortable bringing it up because of Kate's clear desire to continue to aggressively treat her disease. For many reasons, both patients and healthcare practitioners often hesitate to discuss death directly. Physicians and other healthcare practitioners may feel like they have "failed" in their role when they cannot cure their patients. They also may worry about "taking away hope." Many scholars have noted a "cultural imperative to maintain hope" in Western medicine and that optimism can be performative [31]. Many patients believe in the power of optimism to improve their health and may feel that discussing death is inherently pessimistic. However, a good clinician's job is to be able introduce the discussion of death and dying compassionately. Healthcare practitioners will often emphasize redirecting hope, toward hope for a certain kind of death or hope of achieving other goals in the time remaining and can explain that being prepared for dying does not mean "giving up." They can also reassure patients that palliative care is part of the continuum of all medical care: Shifting to a primarily palliative orientation does not mean that the healthcare practitioner will stop providing caring or abandon the patient [32].

Box 14.3   Patient Perspective

The neurosurgeon Paul Kalanithi was diagnosed with metastatic lung cancer at the age 36 and wrote extensively about his illness and dying. His memoir, *When Breath Becomes Air*, was published in 2016, after his death. The following is an excerpt from his essay, "How Long Have I Got Left?":

» The path forward would seem obvious, if only I knew how many months or years I had left. Tell me three months, I'd just spend time with family. Tell me one year, I'd have a plan (write that book). Give me 10 years, I'd get back to treating diseases. The pedestrian truth that you live one day at a time didn't help: What was I supposed to do with that day? My oncologist would say only: 'I can't tell you a time. You've got to find what matters most to you.'

I began to realize that coming face to face with my own mortality, in a sense, had changed both nothing and everything. Before my cancer was diagnosed, I knew that someday I would die, but I didn't know when. After the diagnosis, I knew that someday I would die, but I didn't know when. But now I knew it acutely. The problem wasn't really a scientific one. The fact of death is unsettling. Yet there is no other way to live [32].

Kate and Leila meet with the palliative care team. The palliative care physician assesses the understanding that Kate has about her disease and introduces the idea of setting goals and creating advance directives, a legal written statement that describes some aspects of medical care what a person would want in certain situations, although Kate asks to discuss this at another time. The care team spends time discussing the social situation of the family and discusses how they are emotionally and spiritually coping during the treatment. Leila reveals that she has been stressed after the recent layoffs in her company. She and Kate have been stressed about this since Leila lost her job earlier last month, and they have been paying for COBRA. This has been expensive, leading to increased debt. They are contemplating seeking out discounted care in the safety net system versus buying insurance through healthcare exchanges.

6. **If Kate lived in a country like the UK, do you think her initial treatment options would have been different?**
   The National Institute for Health and Care Excellence (NICE) in the UK makes determinations about technologies and medications based on clinical and economic considerations. It is possible that Kate's new drug would not be provided by the UK's National Health Service (NHS). (See Answer to Background Question 5 for more detail.)

   It is important to note, however, that many drugs are priced much lower for markets in countries like the UK, in which the government negotiates directly with pharmaceutical companies. In the U.S., a 2003 law prohibits the Centers for Medicaid and Medicare Services from negotiating with drug companies.

7. **If Kate had to purchase individual insurance through the healthcare marketplace and was only able to afford a plan with a low monthly premium but a high deductible (meaning that the initial cost of treatment falls on the patient before the insurance company begins to provide reimbursement), how do you think her treatment options would change?**
   Many of the low-premium, high-deductible plans (typically "Bronze plans" under the Affordable Care Act [ACA]) would not cover these drugs or would have co-pays that are out of the reach of most patients. For example, some of these plans charge a $10,000 co-pay for a 1-month supply of dasatinib, a tyrosine kinase inhibitor (TKI) that is effective first-line therapy for chronic myeloid leukemia (CML); patients with CML can live relatively normal lives but require dasatinib or another TKI indefinitely. Some pharmaceutical companies do offer financial assistance programs for patients. While these programs can benefit individual patients, they may be harmful to society as a whole, as they allow pharmaceutical companies to justify their higher prices and may reduce competition by eliminating the market for cheaper generic forms. The supplemental article "There is No Such Thing As 'Free' Vaccines" discusses this in more detail, in a global health context.

   Bronze plans typically have low monthly premiums but high deductibles, meaning that the initial cost of treatment falls on the patient before the insurance company begins to provide reimbursement, even for those expenses that are clearly covered. Many conservative and libertarian politicians and many in the current administration favor increasing the use of high-deductible plans. Supporters of these plans argue that they encourage people to use health care more wisely and to search out the best prices for services; however, most research indicates that people with these plans tend to "roll the dice" (trying to guess whether medical care is necessary for a given condition and delaying preventive care) and that most consumers are not sufficiently knowledgeable about medicine to make rational, informed choices (National Public Radio [NPR] recently interviewed a physician who tested having a high-deductible plan on himself and his family) [33].

8. **What are some of the advantages to having an advance directive?**
   The most important advantage about having an advance directive is that it provides an opportunity for individuals to express their wishes before a medical crisis occurs. Ideally, an advance directive is the outcome of a comprehensive discussion about medical decision-making with individuals, their loved ones, and their healthcare practitioner. Advance directives are portable, legal documents that increase the likelihood that a person's expressed wishes are respected.

9. **How would you approach discussing DNR status?**
   As discussed in Response to Discussion Question 5, there are multiple models for having difficult conversations about bad news. In ideal situations, advance directives would be explored gradually, over several visits, not at a time that is otherwise deeply emotionally stressful, and by healthcare practitioners who already have an established relationship with the patient. It is important to be open, honest, and remember that most people are not aware of what resuscitation means, its results, and what exactly is involved. Often, an individual's perception of resuscitation is informed by what is portrayed on television or film. Describing the various components of resuscitation in detail and continually assessing for comprehension is an important step in discussing DNR. It is also important to explain that DNR status is not an "all-or-nothing" decision: Patients can choose to have some interventions and refrain from others. Some patients may want, for example, a trial of intubation in circumstances where they may only need mechanical ventilation for a brief period of time. Others may want to avoid cardioversion. It is important, however, to discuss the "why" behind the decisions to ensure that the patient's goals are being met and that the healthcare practitioner has dispelled any falsehoods. However, the patient's rationale needs only to support the decision—the healthcare practitioner does not have to validate the science or logic behind it.

Kate enrolls in a local safety net cancer clinic and meets her new oncologist. For the first time, Kate finds out that the drug she has been receiving is very expensive (about $12,000 per month of treatment) and that not all patients are on it. Kate and Leila also learn that the drug is not covered by many insurance companies. In recent years, several newspapers and medical journals have published editorials from respected oncologists criticizing the pricing structure for drugs in the U.S., including a letter signed by more than 100 oncologists in 2016 [1]. Their proposals include allowing Medicare to directly negotiate pricing with pharmaceutical companies (currently prohibited by a 2003 law) and restricting companies' ability to delay the release of generic versions.

10. **Is 3 months of life worth $36,000? To you personally? To society?**
    There is, of course, no "right" answer to this question. Different people are likely to place a different value on life. Some may feel strongly that the quality of life during that month is a much more important measure than the quantity (30 days).
    With respect to how "society" measures this, there are two commonly used analyses for economic decision-making in healthcare: cost-effectiveness analysis (CEA) and cost-benefit analysis (CBA). In medicine, CEA is a fairly straightforward approach to determining cost of an intervention relative to a specific health outcome, such as a quality-adjusted life-year (QALY). CEA typically asks the question: Which intervention should I use given this budget? If used appropriately, this type of analysis can be used to determine medical spending to optimize health outcomes in a population. In general, it is considered a unidimensional approach: dollars in versus life-years gained out. In contrast, a cost-benefit analysis is a more complex (and expensive) analysis allowing for "aggregation of health and non-health benefits because it values benefits, as well as costs in money terms" [34]. For example, with the new medication, does it increase or decrease other healthcare costs related to Kate's care and increase or decrease childcare needs or Kate and her wife's ability to work? A CBA would attempt to take all of these measures into account.

Both types of analyses can be, and are, used to drive policy decisions about what payors (government, private insurers) are willing to pay for.

✓ 11. **Is this pricing structure just? Why or why not?**
This system is not just because it does not ensure that all patients have access to the same treatment, regardless of ability to pay. A 2015 study found no relationship between price of anticancer drugs and (1) novelty or (2) time-to-event endpoints (how long the drug delayed relapse or death), and the authors argued that "current pricing models are not rational but simply reflect what the market will bear" [35].

Although pharmaceutical companies do not openly disclose this information, in many cases, marketing expenditures exceed research and development (R&D) expenditures. Note that a proposed California bill would require these disclosures. A BBC report also reported profit margins and research and development versus marketing spending for several of the largest pharmaceutical companies [36].

✓ 12. **What do you think about these proposals? Are they likely to be effective?**
Because of its size and scope, Medicare often sets the agenda for what private insurance covers and how much treatments and interventions cost, so it is possible that this proposal would lower prices across the board. However, Medicare overall tends to pay less for therapies than private insurance, so those with private insurance might still have higher costs than Medicare. Unfortunately, they are not politically feasible in the current climate. The complexity of the healthcare system makes it difficult to effectively educate and mobilize voters to generate change.

✓ 13. **What is your responsibility, as a healthcare practitioner, to address this problem?**
All healthcare practitioners have some level of responsibility to advocate for their patients, but to what extent is greatly determined by their level of interest in any given area. There are multiple avenues for advocacy. It may be as simple as pushing an insurer to cover the cost of a high-priced intervention for an individual patient or as involved as working with elected officials to write and pass legislation to address a root cause of an issue, such as the ones described above. A common approach for advocacy is to join large professional organizations that advocate and even lobbying for policy changes to improve health. In addition, healthcare practitioners can also call, write, or visit their elected officials to advocate for change.

More recently, a form of immunotherapy for certain relapsed and refractory leukemias and lymphomas has been approved by the Food and Drug Administration (FDA) and priced at $500,000. It has been suggested that pricing might be prorated based upon efficacy; that is, patients who do not respond to the therapy would not pay for it [2].

✓ 14. **What is the value of a 10% chance of cure? A 1% chance of cure?**
The individual assessment of value of cure is subjective. However, at the societal level, when resources are not infinite, choices may need to be made based on subjective values. A month of life or a 1% chance may appear very different to different stakeholders. For a patient or loved one, these chances are likely to be both terrifying and devastating but for some, any chance, no matter how small, may represent a source of hope. For healthcare practitioners, all of the above may be true, but a subjective sense of futility and even personal failure may also be present ("If I care for 50 patients in this situation, it's likely that I will watch all of them die").

For an administrator or policymaker, this assessment is likely to be less personal and is weighed against other ways in which resources might be allocated. The FDA assesses safety and efficacy of medical therapies during the approval process but does not address their cost.

(a) **Creative problem-solving: Who should decide—patients, healthcare practitioners, hospitals, insurance companies, or governments?**

✅ 15. **What do you think about only paying for a high-cost pharmaceutical if it actually works? What are some of the practical challenges and ethical issues of this approach?**
From a practical perspective, this proposal raised several questions. First, even if cure is, in an abstract sense, priceless, how will average patients be able to afford such expensive treatment? While many of the better employer-based plans will no doubt cover it, what about patients with less comprehensive coverage? What would happen to patients who switched insurance plans midway through treatment or before their response was determined? Which insurance company would cover the bill?

One argument in favor of such a pricing schema was that it would encourage insurance companies to cover novel therapies for novel indications, in cases where the insurance companies might otherwise balk at paying for something unlikely to work. On the other hand, would such practices simply encourage more pharmaceutical companies to set outrageous prices?

To put this number in context, the average lifetime earnings of a male in the U.S. with a high school degree is $1.5 million, associate's degree is $1.7 million, bachelor's degree is $2.1 million, and master's degree is $2.5 million. A female with a high school degree is $870,000, associate's degree is $1 million, bachelor's degree is $1.3 million, and master's degree is $1.6 million. Doctoral degrees are $3.4 million, and professional degrees are $4.4 million [37]. Is asking someone to pay 30% to 60% of their entire lifetime earnings just?

It is important to note that the therapy in question (chimeric antigen receptor [CAR] T cell therapy), like many others, was developed through partnerships between researchers at private universities, supported by federal grants and benefitting from tax-exempt status, in partnership with pharmaceutical companies. Although it has been argued that researchers and pharmaceutical companies have the right to make some profit from their ideas and hard work, when such work is at least partly funded by taxpayers, it is particularly difficult to defend restricting access to treatment only to those who can pay.

✅ 16. **Creative problem-solving: How would you construct a law or policy to address drug pricing? Keep in mind, as noted above, that many innovative therapies are developed through partnerships with public and private universities and supported by federal grants, all of which, in turn, are at least partly funded by taxpayers.**

> **Box 14.4   Teaching Tip**
> This question is an opportunity for learners to brainstorm solutions to real dilemmas in healthcare practice and policy. Ask learners to propose different possible decision makers, and consider how they would work in practice.

Several weeks later, Kate presents to the emergency department with hip pain and is found to have a pathologic fracture. An orthopedic surgeon comes to the emergency room to discuss operating. Kate is taken to the operating room for what was described as a "relatively simple procedure." Postoperatively, she struggles with being taken off the ventilator and is admitted to the intensive care unit. Within a few days, Kate develops sepsis and has to be emergently resuscitated. In the ICU, her wife finally discusses her wishes during a brief period of lucidity, and the family decides to withdraw support. She dies a few hours later.

✅ 17. **Self-reflection: Did Kate have a good death? Why or why not?** [38].

"We're Not Ready to Give Up"

> **Box 14.5 Physician Perspective**
>
> Writer and surgeon Atul Gawande has argued that healthcare practitioners focus too much on curative therapies with little chance of success. He has written:
>
> » The simple view is that medicine exists to fight death and disease, and that is, of course, its most basic task. Death is the enemy. But the enemy has superior forces. Eventually, it wins. And, in a war that you cannot win, you don't want a general who fights to the point of total annihilation. You don't want Custer. You want Robert E. Lee, someone who knew how to fight for territory when he could and how to surrender when he couldn't, someone who understood that the damage is greatest if all you do is fight to the bitter end. More often, these days, medicine seems to supply neither Custers nor Lees. We are increasingly the generals who march the soldiers onward, saying all the while, 'You let me know when you want to stop.' All-out treatment, we tell the terminally ill, is a train you can get off at any time—just say when. But for most patients and their families this is asking too much. They remain driven by doubt and fear and desperation; some are deluded by a fantasy of what medical science can achieve. But our responsibility, in medicine, is to deal with human beings as they are. People die only once. They have no experience to draw upon. They need doctors and nurses who are willing to have the hard discussions and say what they have seen, who will help people prepare for what is to come—and to escape a warehoused oblivion that few really want [38].

✅ 18. **What factors make a good death? Who decides?**

This varies for each patient and is very individual. A good death generally would include meeting goals of care. Goals of care need to be discussed with patients since each patient will have different answers. There does seem to be common themes in many patients. These include involvement and closure with family, maintaining dignity and respect, pain control, level of awareness of surrounding, ability to communicate, and so forth. The patient should be empowered to determine what a death looks like, ideally with complete support from their loved ones and the healthcare team.

As noted in the Gawande excerpt above, healthcare practitioners have to balance their expertise, which often includes a more intimate knowledge of death, with respect for patients' autonomy, which includes respect for their values, beliefs, and personal preferences. It is a delicate balance.

> **Box 14.6 Teaching Tip**
>
> By the end of this session and after these final two questions, learners should have a sense of respect for patients' own choices alongside a beginning understanding of how they help patients make the best choices for their individual situations. They should not automatically conclude that Kate's death was "bad," but they should consider the possibility that it could have been better.

Acknowledgments This case was adapted from an earlier case written by Amy Caruso Brown and was inspired in part by Atul Gawande's essay, "What Should Medicine Do When It Can't Save You?," published in the *New Yorker* magazine in August 2010.

References

1. Whalen J. Doctors object to high cancer-drug prices. Wall Street Journal [Internet]. 2015 Jul 23 [cited 2018 May 20]. Available from: https://www.wsj.com/articles/doctors-object-to-high-cancer-drug-prices-1437624060.

2. Novartis receives first ever FDA approval for a CAR-T cell therapy, Kymriah(TM) (CTL019), for children and young adults with B-cell ALL that is refractory or has relapsed at least twice [Internet]. Novartis: Basel (Switzerland). [cited 2018 May 20]. Available from: https://www.novartis.com/news/media-releases/novartis-receives-first-ever-fda-approval-car-t-cell-therapy-kymriahtm-ctl019.
3. Centers for Medicare and Medicaid Services. National health expenditures 2016 highlights [Internet]. Baltimore (MD): CMS. [cited 2018 May 21]. Available from: https://www.cms.gov/research-statistics-data-and-systems/statistics-trends-and-reports/nationalhealthexpenddata/downloads/highlights.pdf.
4. Riley GF, James D, Lubitz JD. Long-term trends in medicare payments in the last year of life. Health Serv Res. 2010;45(2):565–76.
5. French EB, McCauley J, Aragon M, Bakx P, Chalkley M, Chen SH, et al. End-of-life medical spending in last twelve months of life is lower than previously reported. Health Aff. 2017;36(7):1211–7.
6. Aldridge MD, Keeley AS. The myth regarding the high cost of end-of-life care. Am J Public Health. 2015;105(12):2411–5.
7. World Health Organization. WHO definition of palliative care [Internet]. Geneva: WHO. [cited 2018 May 31]. Available from: http://www.who.int/cancer/palliative/definition/en/.
8. Gomes B, Calanzani N, Curiale V, McCrone P, Higginson IJ. Effectiveness and cost-effectiveness of home-based palliative care services for adults with advanced illness and their caregivers. Cochrane Database Syst Rev. 2013;6(6):CD007760. Available from: http://www.cochrane.org/CD007760/SYMPT_effectiveness-and-cost-effectiveness-home-based-palliative-care-services-adults-advanced-illness-and.
9. National Hospice and Palliative Care Organization. Facts and figures of hospice care in America [Internet]. Alexandria (VA): NHPCO; 2017 [cited 2018 May 31]. Available from: https://www.nhpco.org/hospice-statistics-research-press-room/facts-hospice-and-palliative-care.
10. Kaiser Family Foundation. 10 FAQs: Medicare's role in end-of-life care [Internet]. San Francisco (CA): KFF; 2016 Sep 26 [cited 2018 May 31]. Available from: https://www.kff.org/medicare/fact-sheet/10-faqs-medicares-role-in-end-of-life-care/.
11. Hamel L, Wu B, Brodie M. Views and experiences with end-of-life medical care in the U.S. [Internet]. San Francisco (CA): Kaiser Family Foundation; 2017 [2018 Jun 1]. Available from: https://www.kff.org/report-section/views-and-experiences-with-end-of-life-medical-care-in-the-us-findings/.
12. Lau F, Downing M, Lesperance M, Karlson N, Kuziemsky C, Yang J. Using the palliative performance scale to provide meaningful survival estimates. J Pain Symptom Manag. 2009;38(1):134–44.
13. Harris PS, Stalam T, Ache KA, Harrold JE, Craig T, Teno J, et al. Can hospices predict which patients will die within six months? J Palliat Med. 2014;17(8):894–8.
14. National Health Service. About the NHS [Internet]. London (UK): NHS. [cited 2018 Jun 2]. Available from: https://www.nhs.uk/NHSEngland/thenhs/Pages/thenhshome.aspx.
15. The National Institute for Health and Care Excellence (NICE). Who we are [Internet]. London (UK): NICE. [cited 2018 Jun 2]. Available from: https://www.nice.org.uk/about.
16. The National Institute for Health and Care Excellence (NICE). Changes to NICE drug appraisals: what you need to know [Internet]. London UK: NICE. [cited 2018 Jun 2]. Available from: https://www.nice.org.uk/news/feature/changes-to-nice-drug-appraisals-what-you-need-to-know.
17. Department of Health and Human Services, Centers for Medicare & Medicaid Services. Medicare program: revised process for making national coverage determinations. Notices. Federal Register. 2013 August 7; 78(152). [cited 2018 Jun 2]. Available from: https://www.cms.gov/Medicare/Coverage/Determination-Process/Downloads/FR08072013.pdf.
18. U.S. Department of Labor. FAQs on COBRA continuation health coverage [Internet]. Baltimore (MD): Centers for Medicare and Medicaid Services. [cited 2018 Jun 2]. Available from: https://www.dol.gov/sites/default/files/ebsa/about-ebsa/our-activities/resource-center/faqs/cobra-continuation-health-coverage-consumer.pdf.
19. Schneiderman L. Defining medical futility and improving medical care. J Bioeth Inq. 2011;8(2):123–31.
20. Oxford American handbook of hospice and palliative medicine and supportive care. Yennurajalingam S, Bruera E, editors. 2nd ed. New York: Oxford American Handbooks in Medicine; 2016. Chapter 24: 302.
21. Beauchamp T, Childress JF. Principles of biomedical ethics. 7th ed. New York: Oxford University Press; 2012.
22. National Conference of Commissioners on Uniform State Laws. The Uniformed Health-care Decisions Act [Internet]. 1993 [cited 2018 Jun 5]. Available from: http://www.uniformlaws.org/shared/docs/health%20care%20decisions/uhcda_final_93.pdf.
23. Pope T. Medical futility statutes: no Safe Harbor to unilaterally refuse life-sustaining treatment. Tennessee L Rev. 2007;71(Fall):1–81.
24. Lantos JD. Bethann's death. J Emerg Med. 1995; 13(6):843–5.
25. Zhukovsky DS, Robert R. Chapter 27. Pediatric palliative care. In: Yennurajalingam S, Bruera E, editors. Oxford American handbook of hospice and palliative medicine and supportive care. 2nd ed. New York: Oxford University Press; 2016. p. 357–8.
26. McCue K. How to help children through a parent's serious illness. New York: St. Martin's Press; 1994.
27. Baile WF, Buckman R, Lenzi R, Glober G, Beale EA, Kudelka AP. SPIKES—a six-step protocol for delivering bad news: application to the patient with cancer. Oncologist. 2000;5(4):302.
28. Villagran M, Goldsmith J, Wittenberg-Lyles E, Baldwin P. Creating COMFORT: a communication-based model for breaking bad news. Commun Educ. 2010;59(3):220–34. https://doi.org/10.1080/03634521003624031.
29. Carver R. What the doctor said. The Washington Post [Internet]. 1991 Apr 16 [cited 2018 Jun 1]. Available from:

29. https://www.washingtonpost.com/archive/lifestyle/wellness/1991/04/16/what-the-doctor-said-2f7f3a97-12d9-4eb3-afef-b76b7a2b87f0/?noredirect=on&utm_term=.581e23bb4894.
30. Montello M. Narrative ethics. Hastings Cent Rep. 2014;44:S2–6. https://doi.org/10.1002/hast.260.
31. Del Vecchio Good MJ, Good BJ, Schaffer C, Lind SE. American oncology and the discourse on hope. Cult Med Psychiatry. 1990;14(1):59–79.
32. Kalanithi P. How long have I got left? The New York Times [Internet]; 2014 Jan 24. Available from: https://www.nytimes.com/2014/01/25/opinion/sunday/how-long-have-i-got-left.html.
33. Gorenstein D. Do high-deductible plans make the health care system better? Marketplace [Internet]; 2017 Jan 18 [cited 2018 Jun 6]. Available from: https://www.marketplace.org/2017/01/18/health-care/cost-vs-care.
34. Russell L. The science of making better decisions about health [Internet]. Rockville (MD): Agency for Healthcare Research and Quality. [cited 2018 Jun 7]. Available from: https://www.ahrq.gov/professionals/education/curriculum-tools/population-health/russell.html.
35. Mailankody S, Prasad V. Five years of Cancer drug approvals: innovation, efficacy, and costs. JAMA Oncol. 2015;1(4):539–40. https://doi.org/10.1001/jamaoncol.2015.0373.
36. Anderson R. Pharmaceutical industry gets high on fat profits. BBC [Internet]; 2014. Nov 6 [cited 2018 Jun 7]. Available from: https://www.bbc.com/news/business-28212223.
37. Tamborini CR, Kim C, Sakamoto A. Education and lifetime earnings in the United States. Demography. 2015;52:1383. https://doi.org/10.1007/s13524-015-0407-0.
38. Gawande A. What should medicine do when it can't save you? The New Yorker [Internet]. 2010 Aug 2. Available from: https://www.newyorker.com/magazine/2010/08/02/letting-go-2.

Further Reading on This Topic

Cone J. There is no such thing as "free" vaccines: why we rejected Pfizer's donation offer of pneumonia vaccines. New York: Doctors Without Borders; 2016. Available from: https://www.doctorswithoutborders.org/article/there-no-such-thing-%E2%80%9Cfree%E2%80%9D-vaccines-why-we-rejected-pfizer%E2%80%99s-donation-offer-pneumonia.

Kalanithi P. When breath becomes air. New York: Random House; 2016.

Riggs N. The bright hour: a memoir of living and dying. New York: Simon & Schuster; 2017.

Quality of Life in the Context of Disability

Contents

Chapter 15 "Please Look Beyond My Disability" – 289

Chapter 16 "It Runs in the Family" – 315

"Please Look Beyond My Disability"

Jeremy French-Lawyer and Margaret A. Turk

15.1 Background Questions – 290

15.2 Additional Case Information and Questions for Discussion – 290

15.3 Answers to Background Questions – 293

15.4 Responses to Discussion Questions – 297

References – 312

Learning Objectives

1. Describe the complexity of the healthcare and healthcare finance systems in relation to people with disability and their interactions with medical professionals.
2. Recognize the importance of treating people with disability as individuals who are defined not only by their functional limitation and/or diagnosis but by many other things that also influence their health and well-being.
3. Identify skills needed to provide high-quality and respectful care to all patients, including patients with recognized or unrecognized disability.
4. Analyze healthcare practitioners' legal obligations to accommodate patients under the Americans with Disabilities Act (ADA) and how those legal obligations relate to ethical ones.

> **Case Background**
>
> *You are an internist taking care of Dr. Grace Anker, a 33-year-old woman with multiple sclerosis (MS). Dr. Anker first noticed something was wrong 5 years ago, when she developed numbness and visual field disturbances. These symptoms were especially alarming because she is a pulmonologist with an active practice including performing bronchoscopies. The symptoms resolved and had not returned until 3 years later, at which time she had an episode of fleeting vision changes. She was diagnosed with relapsing-remitting MS following this second episode.*

15.1 Background Questions

1. What is the prevalence of disability in the United States? What definition did your source use? Find at least one other example of a definition for disability.

2. People with disability experience both health differences and health disparities. Identify examples of disparities in both health outcomes and in healthcare services received among persons with disability.

3. What is the Americans with Disabilities Act (ADA)? What is "reasonable accommodation"? What does the ADA require that healthcare practitioners do to accommodate their patients?

4. What is "universal design"? Find an example of universal design in the healthcare setting.

5. What is the "socio-ecological model"? What are some examples of the socio-ecological model related to disability?

6. Who are the different members of the care team for a person with disability? What are each of their roles? Be sure to consider all team members, such as physical therapists, occupational therapists, and speech and language pathologists.

7. What is multiple sclerosis? What is relapsing-remitting multiple sclerosis? What range of severity of functional limitation might there be for a person with a diagnosis of relapsing-remitting MS?

8. Most schools for health professionals have "technical standards" as part of the application process. What are technical standards? Research a school's technical standards.

15.2 Additional Case Information and Questions for Discussion

Dr. Anker was diagnosed with relapsing-remitting MS following the second episode of fleeting vision changes. She started immunomodulating treatment to temporarily ameliorate her symptoms and has had no relapses since. At this time, 2 years after initiation of treatment, she has no functional limitations.

1. How do you define ability?

2. How do you define disability?

3. As a primary care provider, how would you know what ability your patient has, or what changes in ability your patient may be experiencing?
    (a) How might this situation be different if Dr. Anker were experiencing an ongoing disability that was more easily perceived by those

around her, such as tetraplegia from a spinal cord injury and use of a power wheelchair full time?

4. How do you provide effective care for people with disability? Is this different than care for all patients?
   (a) What is your responsibility as a healthcare practitioner if your patient has a disability that is not within your area of expertise?
   (b) How might you approach a treatment plan or referral for a patient with disability?

5. Self-reflection: When you envision your future career, do you imagine yourself caring for patients with disability?

6. Do all healthcare practitioners interact with people with disability as a part of their practice?

You continue to follow Dr. Anker for 15 years, and you have formed a good relationship with her. It has now been 20 years since Dr. Anker was diagnosed with MS. She is now 53. In the past year, Dr. Anker experienced a significant relapse in her MS. After this most recent episode, she is experiencing continued vision impairment in one eye, numbness in her hands, and now uses a left leg brace and a cane to walk.

Dr. Anker is adjusting to the change in her ability. She sees both an occupational therapist (OT) and a physical therapist (PT). She has found visits with her OT particularly helpful and notes that the OT has helped her find many accommodations for her vision issues and completing tasks of daily living.

7. What practical accommodations might Dr. Anker need to consider related to her disability? How might you support her?

8. How might you support Dr. Anker through her emotional process as she adjusts to her disability?
   (a) Your values may be different from your patients' values. If this is the case, what can you do to understand and respect the values of your patient?

9. Self-reflection: Think of someone you have met or with whom you have interacted who has a disability. What was your experience of the interaction?

You are surprised to discover that Dr. Anker has not seen her gynecologist in the past year. When you ask why, she explains that she found the office was difficult to navigate with her limited vision and using her brace and cane. She felt precarious trying to get onto the fixed-height exam table, and she had difficulty keeping her balance on the scale. Dr. Anker also says that she avoids making appointments with her gynecologist because she stopped asking about sexual activity.

10. How does accessibility of the office affect a person's ability to access quality care?

11. Considering disability in a societal context, how can the socio-ecological model help us to understand disability and accessibility in Dr. Anker's case?

12. Why do you think Dr. Anker has experienced a change in her gynecological care? What assumptions were made?
    (a) What are the ramifications of focusing on Dr. Anker's disability, rather than viewing her as a whole person?
    (b) Patients with disability are less likely to get recommended screening tests, such as mammograms and Pap tests. Why do you think that is?
    (c) How can healthcare practitioners address this disparity in primary care?

Another patient that you follow, Anna Drever, is a 55-year-old receptionist at a car dealership. She also has a diagnosis of relapsing-remitting MS. Ms. Drever completed high school, has previous experience as a receptionist, and has been working at the dealership for the past 15 years. Ms. Drever has been using a cane and braces intermittently for several years and more recently has transitioned to using a wheelchair for mobility over long distances.

13. How does Ms. Drever's education and socioeconomic status affect her management of her disability?
    (a) How does Ms. Drever's work environment affect her ability?
    (b) How would Ms. Drever's work environment affect her ability if she worked in a position that did not allow her to spend a significant amount of her time seated?

In the past 2 years, Ms. Drever has used a brace and a cane to walk and has used a wheelchair for mobility over long distances. With her most recent relapse, she has found that she cannot stand for long periods and increasingly has had difficulty propelling her wheelchair. Human Resources at the dealership has approved her use of a wheelchair to enable her to perform the essential functions of her position, but this has become more difficult due to difficulty propelling her chair. Unfortunately, her insurance company has denied her claim for a powerchair. The reason they have rejected the claim is that Ms. Drever does not currently have a vehicle that will allow her to transport the power wheelchair between locations (such as home and work). Ms. Drever is very worried about being able to continue working without this accommodation.

14. Do you agree or disagree with the insurance company? Why or why not?
    (a) Should insurance companies be required to cover items like this?
    (b) How does insurance coverage influence the ability of a person with disability? Is this influence the same as a person without disability?

Dr. Anker is divorced. She lives alone, but has a 22-year-old son who attends college and lives with her in the summer. Dr. Anker also has a 28-year-old daughter who is married, has one young child, and is pregnant with her second. Dr. Anker's daughter has talked about moving closer to home, but Dr. Anker does not want her to feel obligated to do so. Dr. Anker's parents moved to Sweden, their country of origin, after they retired; they are now in their 80s and find it difficult to travel.

Ms. Drever lives with her husband, who is 54 years old and has chronic back pain from his job. He helps her with many daily living tasks, but she does not want to feel like a burden. Ms. Drever realizes that her husband will not be able to continue to be a primary caregiver for her as her MS progresses, due to his own health problems. Her parents are in their late 80s and live in Florida, while her husband's parents are deceased. They have no other close family members.

15. What options do Dr. Anker and Ms. Drever have for daily care?
    (a) How do each of these women's situations influence the options available to them?

16. What responsibility do Dr. Anker's and Ms. Drever's families have to care for them?
    (a) How might Ms. Drever's husband's medical conditions influence her decision regarding daily care?
    (b) How might Dr. Anker's concern about being a burden influence her decisions?

With respect to her medical practice, Dr. Anker is no longer participating in procedures but is still seeing patients. She is feeling some pressure from her colleagues to decrease her role further and is frustrated with feeling like they are constantly looking over her shoulder.

17. Are there minimum physical requirements for being a physician?

18. Who has the responsibility to monitor physician ability?
    (a) Self-reflection: Do you trust yourself to be objective in assessing your own ability?
    (b) Self-reflection: Do you trust your peers to be objective in assessing their own ability? Would you feel differently if your peers were responsible for assessing your ability?
    (c) What are some potential misuses of this responsibility? What biases might come into play?

19. Creative problem-solving: How should ability be assessed in cases related to aging or medical events?

20. Self-reflection: Did you approach aspects of this case differently because

"Please Look Beyond My Disability"

the first patient was a physician? Do you think physicians treat physician-patients differently from non-physician patients? How so?

15.3 Answers to Background Questions

1. **What is the prevalence of disability in the United States? What definition did your source use? Find at least one other example of a definition for disability.**

   Prevalence of disability varies considerably depending on the definition. For example, in 2016 the Centers for Disease Control and Prevention (CDC) has reported that one in four adults (more than 61 million people in the United States) have a disability that impacts major life activities and has noted significant state-level variation (as low as one in six adults or 16.4% of the adult population in Minnesota and as high as one in three or 31.5% in Alabama in 2013). These statistics were based on the Morbidity and Mortality Weekly Report (MMWR) from those years, which summarizes present Behavioral Risk Factors Surveillance System (BRFSS) data. The BRFSS survey asks respondents to report specific functional types of disability such as mobility (defined as "serious difficulty walking or climbing stairs"), cognitive ("serious difficulty concentrating, remembering, or making decisions"), vision ("serious difficulty seeing"), self-care ("difficulty dressing or bathing"), and independent living ("difficulty doing errands alone") [1, 2].

   The Americans with Disabilities Act (ADA) defines disability as "a physical or mental impairment that substantially limits one or more major life activities, a person who has a history or record of such an impairment, or a person who is perceived by others as having such an impairment. The ADA does not specifically name all of the impairments that are covered" [3]. Alternatively, the *Merriam-Webster English Dictionary* defines disability as "a physical, mental, cognitive, or developmental condition that impairs, interferes with, or limits a person's ability to engage in certain tasks or actions or participate in typical daily activities and interactions" [4].

2. **People with disability experience both health differences and health disparities. Identify examples of disparities in both health outcomes and in healthcare services received among persons with disability.**

   Health disparities are differences in health care or health outcomes that are an "expression of discrimination" [5]. Health differences, meanwhile, are differences in health care or outcomes that are caused by problems in the healthcare system that unequally impact people in a certain group. For example, poor coordination in healthcare services may disproportionately impact people with disability, who are often frequent users of health services and engage with multiple health professionals [5]. Understanding that each of these terms is relevant can help health professionals identify inequality in healthcare service delivery or health outcomes and also identify the source of that inequality in order to better address the underlying cause.

   The article *Persons With Disabilities as an Unrecognized Health Disparity Population* reports differences in health outcomes for people with disability. Findings include 27.0% of people with disability reporting "needing to see a doctor but did not because of cost" [6] compared to 12.1% of people without disability; 70.7% of women with disability report having a current mammogram, compared to 76.6% of women without disability. Similarly, 78.3% of women with disability report a current Pap test, compared to 82.3% [6].

   Other health data that show disparities for people with disability include adults who engage in no leisure-time physical activity (people with disability 54.2%, people without disability 32.2%). Also, 44.6% of adults with disability are obese, compared to 34.2% of people without disability [6].

The article provides population statistics that demonstrate differences and disparities for people with disability. The report also includes weighted population estimates of adults with disability and adults without disability. Information about social determinants of health—often considered root causes of health disparities—such as income level, employment status, transportation, Internet access, and education is also provided [6]. For more information, visit ▶ https://www.cdc.gov/ncbddd/disabilityandhealth/features/unrecognizedpopulation.html.

3. **What is the Americans with Disabilities Act (ADA)? What is "reasonable accommodation"? What does the ADA require healthcare practitioners do to accommodate their patients?**
"The Americans with Disabilities Act of 1990 (ADA) [amended in 2008] is a federal civil rights law that prohibits discrimination against individuals with disabilities in everyday activities, including medical services. Section 504 of the Rehabilitation Act of 1973 (Section 504) is a civil rights law that prohibits discrimination against individuals with disabilities on the basis of their disability in programs or activities that receive federal financial assistance, including health programs and services. These statutes require medical care providers to make their services available in an accessible manner" [7].

"Both Title II and Title III of the ADA and Section 504 require that medical care providers offer individuals with disabilities:
   - Full and equal access to their healthcare services and facilities.
   - Reasonable modifications to policies, practices, and procedures when necessary to make healthcare services fully available to individuals with disabilities, unless the modifications would fundamentally alter the nature of the services (i.e., alter the essential nature of the services)" [7].

For information regarding specific situational questions, such as accessible medical equipment, assistance, staff training, and accessibility of facilities, several resources may be useful.
   - "Access To Medical Care For Individuals With Mobility Disabilities" provides extensive information about appropriate accommodation for persons with mobility disabilities [7].
   - "Accessibility of Medical Diagnostic Equipment—Implications for People with Disability" provides information about accessible medical equipment [8].
   - "Practical Recommendations for Enhancing the Care of Patients with Disability" provides continuing education on access to medical care for people with disability in general. These modules provide information and recommendations related to accessibility, clinical management, and communication techniques [9].

4. **What is "universal design"? Find an example of universal design in the healthcare setting.**
Universal design is "the process of creating products that are accessible to people with a wide range of abilities, disabilities, and other characteristics. Universally designed products accommodate individual preferences and abilities; communicate necessary information effectively (regardless of ambient conditions or the user's sensory abilities); and can be approached, reached, manipulated, and used regardless of the individual's body size, posture, or mobility" [10].

An example of universal design in health care is a height-adjustable exam table. It allows people with mobility limitations, difficulties with balance, or other limitations to get onto the table or to do so more safely. This is universal design because the table can be at a variety of heights to meet the different needs of many people. Some other examples include large-print signage for people with vision impairment and clear well-placed signage for those with intellectual disability. Other examples include flashing lights on fire alarm systems to inform people with hearing impairment that evacuation is necessary.

☐ Fig. 15.1 The socio-ecological model

✅ 5. **What is the "socio-ecological model"? What are some examples of the socio-ecological model related to disability?**

The socio-ecological model is a model that puts health and wellness into the context of not only the individual but also the interpersonal, organizational, community, and policy that impact that individual (☐ Fig. 15.1) [11]. Some examples related to disability are listed as follows:
  – **Individual:** Personal values and preferences, specific medical diagnosis and treatment, habits
  – **Interpersonal:** Stigma and assumptions, helping and caregiving, supportive relationships and social network
  – **Organizational:** Accessibility, inclusion/exclusion, accommodation
  – **Community:** Accessibility, media representation, inclusion/exclusion, disparities in access to health care
  – **Policy:** Inclusion/exclusion, required accessibility, historical abuse

✅ 6. **Who are the different members of the care team for a person with disability? What are each of their roles? Be sure to consider all team members, such as physical therapists, occupational therapists, and speech and language pathologists.**
  – Primary care physicians, physician assistants, and nurse practitioners assess and address concerns related to health and wellness of the patient, including weight management, chronic condition management, and preventative health screening in most cases.
  – **Specialty physicians** treat specific concerns, including gynecological care for women, preventative health screening, and treatment of health issues outside of the expertise of the primary care physician [12].
  – **Disability specialty physicians** manage the medical and health conditions of the patient/consumer within the rehabilitation process, providing diagnosis, treatment, or management of disability-specific issues (e.g., need and direction for focused therapies; identification and prescription of appropriate brace or wheelchair; recognition of a medical condition commonly associated with the disability [spasticity management in spinal cord injury] or one superimposed on the pre-existing condition [myelopathy in cerebral palsy]) [12].
  – **Social workers** "may provide case management or coordination for persons with complex medical conditions and needs; help patients navigate the paths between different levels of care; refer patient to legal, financial, housing, or employment services; assist patients with access to entitlement benefits, transportation,

assistance, or community-based services; identify, assess, refer, or offer treatment for such problems as depression, anxiety, or substance abuse; or provide education or support programming for health or related social problems" [12].
- **Physical therapists** "assess movement dysfunction and use treatment interventions such as exercise, functional training, manual therapy, assistive and adaptive devices and equipment, and physical agents, including electrotherapy, massage and manual traction" [12].
- **Occupational therapists** "typically work with patients/consumers through functional activities in order to increase their ability to participate in activities of daily living (ADLs) and instrumental activities of daily living (IADLs) in school and work environments using a variety of techniques" [12].
- **Speech and language pathologists (SLP)** "assess, treat, and help to prevent disorders related to speech, language, cognition, voice, communication, swallowing and fluency" [12].

✓ 7. **What is multiple sclerosis? What is relapsing-remitting multiple sclerosis? What range of severity of functional limitation might there be for a person with a diagnosis of relapsing-remitting MS?**
Multiple sclerosis is an autoimmune disease that is characterized by focal demyelinated plaques within the central nervous system. Myelin degeneration can affect multiple areas of neurological function including vision, movement, touch, cognition, and emotion. In the United States, the estimated prevalence is 100–150 per 100,000 population, for a total of 300,000–400,000 persons with MS. The mean age of onset is between 28 and 31 years old. MS affects more women (70%) than men (30%). There are geographic differences in MS prevalence, previously thought to be caused by racial differences, and then explained with latitudinal differences. Although there is data to support a higher frequency of MS in Europe (including Asian Russia), southern Canada, Northern United States, New Zealand, and southeast Australia, factors such as survival time, diagnostic accuracy, and ascertainment probability may influence reported rates. All prevalence estimates are impacted by inconsistent registration, tracking, and reporting of MS. There are three main types of MS: relapsing-remitting, secondary progressive, and primary progressive [13].

The most common type of MS is relapsing-remitting MS (or RRMS), representing approximately 85–90% of cases. RRMS is characterized by acute immune attacks followed by stable periods. During these stable periods, patients often maintain a high level of functional ability and, in the early stages of MS, may not experience any symptoms between relapses. As the patient experiences more relapses, often over the course of several years, symptoms tend to increase and functional ability may decrease. Thus, over the course of an individual's life, there is a broad range of severity of functional limitation. Treatment options for RRMS include disease-modifying drugs and immunosuppressive treatment. Symptom management is an important aspect of treatment and can include physical, occupational, and speech and language therapy [13].

✓ 8. **Most schools for health professionals have "technical standards" as part of the application process. What are technical standards? Research a school's technical standards.**
- **Technical standards** are "the nonacademic criteria essential for the student to participate in the program; they include the attitudes, experiences and physical requirements the student must possess to learn and perform the essential requirements of the program. Technical standards are usually divided into five key areas: Perception/Observation, Motor/Tactile, Cognition, Communication and Professionalism" [3].

- **Essential requirements** are "the expected or desired outcomes acquired through participation in the program and include the skills, knowledge and judgments that all students must demonstrate to graduate, with or without reasonable accommodation" [3].

Medical schools set technical standards based on individual programs and the requirements of those programs. Recently, medical schools have begun to focus on what students must be able to do, and focus has moved away from how students go about meeting those requirements. This move, from "organic" technical standards to "functional" technical standards, allows people with disability to be included as health profession students with the use of adaptive equipment or other modifications [14, 15].

15.4 Responses to Discussion Questions

Dr. Anker was diagnosed with relapsing-remitting MS following the second episode of fleeting vision changes only. She started immunomodulating treatment to temporarily ameliorate her symptoms and has had no relapses since. At this time, 2 years after initiation of treatment, she has no functional limitations.

✓ 1. **How do you define ability?**
   People often have different understandings of ability. Ability includes the range of intelligence, athleticism, and other abilities. People with disability may view ability differently than expected by people who do not have a disability. All people—not only those with a disability—perceive differences in ability and work to improve or adjust in order to be independent and successful.

> **Box 15.1   Teaching Tip**
> Learners should be encouraged to offer their personal definitions first. The facilitator can then lead a discussion exploring the difficulties of defining this type of concept. Learners should recognize that ability may be defined differently for different people and that they should not impose their definitions on others (i.e., their patients).

✓ 2. **How do you define disability?**

> **Box 15.2   Teaching Tip**
> Some definitions were already provided in the Answer to Background Question 1, and learners may have discovered their own. Encourage learners to synthesize available information and experience to develop their own definitions.

   Some definitions include:
   - "A disability is defined as a limitation in performing certain roles and tasks that society expects an individual to perform. Disability is the expression of the gap between a person's limitations with social and physical environmental factors. Many disabling conditions are thus preventable or reversible with proper and adequate rehabilitation, including environmental modification" [16].
   - "[Disability] is defined by the ADA as a physical or mental impairment that substantially limits one or more major life activities, a person who has a history or record of such an impairment, or a person who is perceived by others as having such an impairment. The ADA does not specifically name all of the impairments that are covered" [3].

   Like definitions of ability, definitions of disability can be personal and related to individual experiences. Healthcare practitioners should recognize that:

- A patient's concept of disability may not align with their healthcare practitioner's.
- Disability is a broad definition and includes intellectual, developmental, sensory, physical, and learning disability.
- Disability can affect anyone at any time in their life.
- Disability can be temporary or permanent.
- Disability is *not a diagnosis* but rather a difference in function.
- Ability is not the converse of disability; rather, the definition of disability encompasses a variety of differences in ability. Disability is a mismatch of capacity with environment—disability is defined by environment, context, and societal expectations. A person with a mobility impairment may not identify with a disability if the environment has been modified to allow for independence.
- Environment limits ability for everyone, not just people with disability. Loss of power in one's home limits one's ability to maintain routine activities (assuming one is accustomed to having power), regardless of whether the individual has a "disability."

3. **As a primary care provider, how would you know what ability your patient has, or what changes in ability your patient may be experiencing?**
A short appointment in a controlled exam room environment is unlikely to provide you with an accurate impression of ability. A healthcare practitioner should directly ask a patient about their ability or any changes in their ability. All people have different abilities, and these may need to be addressed with the help of a healthcare practitioner. Questions should focus on recent changes in daily tasks such as walking, climbing stairs, self-care, getting in and out of the bathtub, and driving. Discussing cognitive changes and mental health may require sensitivity. Some specific examples are listed as follows:
- Is your walking any more difficult?
- Are you having problems with swallowing?
- Do you drive yourself to errands or appointments?
- Are you needing any help getting ready for work?
- Have you noticed that your signature has changed?
- Have you noticed any changes in your reading?
- Have you had trouble remembering appointments?
- Have you had any changes in interests or in your hobbies?
- Have you recently seen less of your family or friends?

(a) **How might this situation be different if Dr. Anker were experiencing an ongoing disability that was more easily perceived by those around her, such as tetraplegia from a spinal cord injury and use of a power wheelchair full time?**
If Dr. Anker had a disability that was more apparent, she might have been more likely to:
- Be asked about changes in her ability by her healthcare practitioner
- Receive unsolicited comments from people in her life, including strangers, about her disability
- Have her colleagues focus on and openly voice concerns about her changes in ability

Dr. Anker may not need to modify her practice for some time. However, if her functional impairments become more apparent, she will be subject to more of the judgments and opinions of others regarding her

> **Box 15.3 Teaching Tip**
> Be sure that learners consider not only people with disability but also individuals who are aging or who experience changes due to weight gain or due to chronic conditions—all of these can influence ability.

ability to perform essential functions of her position. The perceptions of others may not be accurate regarding Dr. Anker's ability. Conversely, if Dr. Anker's ability changes in a way that is not apparent, such as a persistent change in her vision, she may not be able to provide clinical care to her patients without modification and may put patients at risk if accommodations are not made.

4. **How do you provide effective care for people with disability? Is this different than care for all patients?**
Ultimately, caring for patients with disability requires that a healthcare practitioner treat each person with respect. This is true not only for patients with disability but also for all patients. Healthcare practitioners should work to enter each patient interaction with the full understanding that the patient is an individual with their own values and needs. If the healthcare provider does everything they can to respect the patient, the relationship is more likely to be considerate and genuine for both patient and provider [17].

> **Box 15.4 Patient Perspective**
> In the article *"Paying the Price to Get There": Motherhood and the Dynamics of Pregnancy Deliberations among Women with Disabilities*, women with disability offered some of their insight in focus groups. One woman stated the following:
>
> » If doctors could just say "This is the problem… these are three roads that we could go down to treat whatever you have… What would best fit your life? What would you want? And these are the side-effects that come with each one," you know? Not the doctor deciding, "Ok, I choose this one. This one is the best one for you even though I don't know anything about you outside of this room" [17].

(a) **What is your responsibility as a healthcare practitioner if your patient has a disability that is not within your area of expertise?**

The following are three strategies healthcare practitioners might use to meet the needs of a patient with disability when they are unfamiliar with a specific disability diagnosis.

- *Acknowledge that it is not the healthcare practitioner's area of expertise*. Once a healthcare practitioner acknowledges a deficit in their understanding, he or she can work to expand and update their knowledge by referring to an evidence-based, point-of-care resource for clinical decision support. One example of such a resource is *UpToDate*, which is authored, edited, and peer reviewed by physicians and can be used to access information at the point of care [18]. Resources like *UpToDate* have content that can help clinicians maintain their professional knowledge and find information that may be outside their area of expertise.
- *Recognize that the patient may know the most about their lived experience and disability*. Asking questions of the patient is a good way to learn what they need, what their health priorities are, and how a healthcare provider can best help them.
- *Consult experts*. Healthcare practitioners who specialize in working with people with disability, such as physical medicine and rehabilitation physicians, developmental pediatricians, and specialty neurologists, are an important resource, both in terms of consultation and referral. Although each healthcare practitioner should be sure to do everything they can for their patients, they should also recognize when consulting an expert is the best option.

(b) **How might you approach a treatment plan or referral for a patient with disability?**

Treating patients with disability or referring them to specialty physicians requires some discussion with the patient about the goals for the treatment or referral. The patient's insurance, transportation needs, and other concerns should be considered. For example, depending on the patient's insurance plan, pre-authorization may be needed for a referral. In addition, the patient may need to arrange travel through a transportation system, which may require an out-of-pocket expense (i.e., not covered by insurance). Regardless of whether it is the patient, the patient's family, or the healthcare practitioner who is making the arrangements, these may be difficult to organize.

Acknowledging these and other concerns can help the patient access appropriate and necessary care. Having clear conversations with the patient can also help foster a positive relationship with the patient. Without such conversations, the patient may feel that the healthcare practitioner does not have an interest and is unwilling to expend the effort or time to provide care.

5. **Self-reflection: When you envision your future career, do you imagine yourself caring for patients with disability?** [1].

> **Box 15.5   Teaching Tip**
> If learners are quiet or simply reply "yes," ask them to estimate the proportion of patients they expect to see with disabilities. If they expect some physicians to see no patients with disability, or describe only a small proportion of their patients, ask why, and encourage them to reflect on the prevalence of disability in the United States reported in the Answer to Background Question 1.
>
> For reference, the CDC has reported that one in four adults (more than 61 million people in the United States) has a disability and has noted significant state-level variation (as low as one in six adults or 16.4% of the adult population in Minnesota and as high as one in three or 31.5% in Alabama) [1, 2].

6. **Do all healthcare practitioners interact with people with disability as a part of their practice?**
    All healthcare practitioners encounter people with disability and should work to provide these patients with high-quality, respectful care and to feel confident in doing so:
    - Aging populations, people with chronic health conditions, and other disability populations are all patients with disability whom healthcare practitioners may see in their practice.
    - People with disability are seen by healthcare practitioners in all specialties everyday as they, like others, need surgery, cancer care, vision care, dental care, and other healthcare services. Furthermore, primary care physicians (PCPs) see aging patients with hearing loss and mobility issues, and surgeons see patients with short-term disability who are recovering from surgery. Other examples exist for almost every specialty.

You continue to follow Dr. Anker for 15 years, and you have formed a good relationship with her. It has now been 20 years since Dr. Anker was diagnosed with MS. She is now 53. In the past year, Dr. Anker experienced a significant relapse in her MS. After this most recent episode, she is experiencing continued vision impairment in one eye, numbness in her hands, and now uses a left leg brace and a cane to walk.

Dr. Anker is adjusting to the change in her ability. She sees both an occupational therapist (OT) and a physical therapist (PT). She has found visits with her OT particularly helpful and notes that the OT has helped her find many accommodations for her vision issues and completing tasks of daily living.

7. **What practical accommodations might Dr. Anker need to consider related to her disability? How might you support her?**
    Dr. Anker will need to make many practical accommodations to her disability. She may need assistance with activities of daily living (ADLs) that are required for personal care—such as bathing, dressing, and eating—or she may need to use

adaptive equipment that allows her to perform these tasks independently. Dr. Anker uses a cane and orthotics (braces) and may need a wheelchair for long-distance mobility. Dr. Anker may need assistance or adaptive equipment to perform "more complex tasks required for independent living in the immediate environment," called instrumental activities of daily living (IADLs). These include activities such as community mobility, meal preparation, and financial management. Dr. Anker's ability and need for assistance may change (increase or decrease) as her MS progresses [19].

Dr. Anker currently works with physical and occupational therapists. She may need a referral to a speech and language therapist in the future. These clinicians can help Dr. Anker work through practical adaptations to maximize her independence. Asking Dr. Anker at appointments about her ability will help to identify ways that Dr. Anker's ability could be affected with adaptive equipment or modifications to her environment. To support a patient with disability, a healthcare practitioner should ask about ability, changes in ability, or concerns that a patient is currently having and, based on the responses, provide appropriate support and recommendations.

✓ 8. **How might you support Dr. Anker through her emotional process as she adjusts to her disability?**

One of the most important ways you might support Dr. Anker through her emotional adjustment is to ask her about the process, listen to her needs, and, when appropriate, help her meet them. Some people may want more emotional support, others want less. Each person will have a different emotional process as they come to terms with their disability. Additionally, this process may change over the life course or as there are changes in that person's ability or disability.

(a) **Your values may be different from your patients' values. If this is the case, what can you do to understand and respect the values of your patient?**

Respecting a patient's values is key to being a supportive clinician and providing quality care. This may be challenging when a healthcare practitioner's values differ significantly from those of their patients. Imposition of your values on your patient can result in loss of trust and decreased openness on the part of the patient, as they may choose not to share information about their life and their decisions.

✓ 9. **Self-reflection: Think of someone you have met or with whom you have interacted who has a disability. What was your experience of the interaction?**

Box 15.6   Teaching Tip

Create an atmosphere in which learners feel comfortable disclosing fears, as well as discussing experiences with disability in their personal or professional life, including family members, friends, peers, and colleagues who have a disability. Take care to recognize that one or more learners may have an unrecognized disability, and encourage learners to use sensitive language.

The key to this question is having students realize that people with disability are not just some abstract, stereotypical group. They are people with whom the students interact as an integrated part of their lives. Many students will have parents, grandparents, friends, or extended family members who have some kind of functional limitation that requires accommodation, understanding, and support. Once students realize they are comfortable interacting with people with disability with whom they are familiar, and see as being like themselves, it will likely be obvious that most of the discomfort people feel when interacting with a person with disability comes from a lack of familiarity.

Any time a person is faced with something outside of what they know, they tend to feel some discomfort. One approach learners can take to increase their comfort is to focus on similarities. This minimizes the focus on differences.

You are surprised to discover that Dr. Anker has not seen her gynecologist in the past year. When you ask why, she explains that she found the office was difficult to navigate with her limited vision and using her brace and cane. She felt precarious trying to get onto the fixed-height exam table, and she had difficulty keeping her balance on the scale. Dr. Anker also says that she avoids making appointments with her gynecologist because she stopped asking about sexual activity.

✓ 10. **How does accessibility of the office affect a person's ability to access quality care?**
   As noted in Background Question 3, physicians and healthcare practitioners are responsible for ensuring the accessibility of their facilities and providing reasonable accommodation as laid out by the ADA. If a clinical space is not accessible, the patient may not be able to access care, may have limited access, or may experience discomfort [9]. Some examples often discussed by people with disability include:
   - Appropriate parking space and curb cuts at office buildings
   - Space between chairs in the waiting room sufficient for adaptive equipment
   - Height of the reception counter that allows a person using a wheelchair to see over the counter
   - Office staff loudly discussing health information through glass windows
   - Health information not accessible to hearing (verbally provided)- or vision (written information)-impaired individuals
   - No accessible scales
   - Fixed-height tables without accommodations to help with balance, such as rails
   - Staff assisting patients without asking—pushing a wheelchair
   - Clinicians moving a wheelchair out of reach
   - Clinicians not informing vision-impaired patients verbally of the physical space or individuals in the room
   - Addressing a caregiver rather than speaking directly to a patient
   - Other behaviors that make the patient feel they are being treated differently or are not welcome in the office [9]

   A thoughtful approach and utilization of universal design can help ensure access. Asking patients with disability if they had any problems with accessibility in the space can help identify areas for improvement. Healthcare practitioners not only have legal obligations under the ADA but they also have ethical obligations to advocate for universal design in their workspaces.

✓ 11. **Considering disability in societal context, how can the socio-ecological model help us to understand disability and accessibility in Dr. Anker's case?**
   The socio-ecological model emphasizes health at multiple levels (◯ Fig. 15.1). Some examples specific to this case at each level are listed below (For a more general overview of the socio-ecological model, see Background Question 5 above.):
   - **Individual:** Dr. Anker values her work as a pulmonologist and her independence. She has a diagnosis of relapsing-remitting MS and has had immunomodulating treatment to ameliorate her symptoms. She goes on regular walks in her neighborhood and enjoys spending time with her first grandchild, eating out with her friends, and attending local productions of plays and musicals.
   - **Interpersonal:** Dr. Anker's coworkers are concerned that she will not be able to perform her work as a physician, and her friends and neighbors now treat her as though she is delicate. Her daughter often offers to help with errands, even though she lives far away. Dr. Anker's neighbors sometimes help with yardwork, and her friends are always happy to listen and provide social support for Dr. Anker.
   - **Organizational:** Dr. Anker has some trouble moving around her office, especially using the restroom, as the closest accessible restroom is far away from her office.

- **Community:** MS is portrayed in television programs and popular media as an insurmountable or debilitating disability. There is an MS support group that Dr. Anker has begun to attend that has helped her to begin exploring community resources, such as home care options for her future. Dr. Anker's gynecologist's office is not functionally accessible for her, and she feels stigmatized by her healthcare practitioners there.
- **Policy:** The ADA mandates that there are parking spaces, curb cuts, and other accommodations in the United States. There is advocacy by disability groups at local, state, and federal government levels and advocacy and research initiatives related to MS in particular. There is a history of institutionalization, underrepresentation, and abuse of people with disability in the United States.

12. **Why do you think Dr. Anker has experienced a change in her gynecological care? What assumptions were made?**
    Dr. Anker's healthcare practitioners are making assumptions about her ability, goals, and life. By not asking her about her sexual activity, her gynecologist has provided her with sub-par care and assumed that sexual health and well-being is not an important part of her life. It may not be, but that needs to be asked, not assumed. Assumptions about her disability and how it might limit her sexual activity or other aspects of her daily life are inappropriate and inaccurate.
    (a) **What are the ramifications of focusing on Dr. Anker's disability, rather than viewing her as a whole person?**
    First, they have led to Dr. Anker no longer accessing care. Second, an incomplete medical history can have an impact on treatment. For example, if Dr. Anker had invasive cervical cancer, neglecting to ask about her recent sexual history might lead to missing information about postcoital bleeding, one of the symptoms at presentation of invasive cervical cancer [20]. Furthermore, there would be no information about the human papillomavirus (HPV) status of her recent sexual partners, as there may not be *any* current record of her sexual activity or partners. Finally, Dr. Anker would not have had a recent Pap test. An oversight in sexual history taking could have serious ramifications for the patient, who might have been diagnosed and treated earlier.

> **Box 15.7 Teaching Tip**
> Ask learners to hypothesize about other potential consequences of such assumptions and consider situations in which such assumptions might lead to unnecessary testing, misdiagnosis, or mistreatment. Physician-educators have long told medical students that "90% of diagnosis is history"— healthcare practitioners need to avoid making assumptions in order to ask the right questions and obtain a thorough history.
> Another potential (and not uncommon) problem is for healthcare practitioners to assume that symptoms are related to a patient's disability, which may lead them to overlook health issues that are unrelated to the disability diagnosis.

   (b) **Patients with disability are less likely to get recommended screening tests, such as mammograms and Pap tests. Why do you think that is?**
    Health disparities and differences for people with disability are an area of ongoing research. There are many factors that may contribute to disparities and differences for people with disability. Some examples of these factors, based on current research and advocacy work, are:
    - Social determinants of health including income, employment, education, transportation, and other factors that impact access to health services

- Assumptions and prejudice in the healthcare system, including assumptions about values, independence, ability, health goals, wellness, quality of life, career goals, and inherent worth of people with disability
- Accessibility and comfort for people with disability in the clinical space
- Increased health needs, frequency of health services use, and engagement with multiple healthcare professionals by people with disability

Health professionals should participate in training in how to create accessible, respectful environments for people with disability. Patient advocacy is also necessary to address health disparities.

(c) **How can healthcare practitioners address this disparity in primary care?**
Healthcare practitioners can address this disparity first by changing the way they interact with people with disability—such as recognizing their own assumptions and ensuring that they are respectful toward people with disability. Then, they can consider accessibility when making decisions about clinical spaces and equipment, such as large exam rooms, accessible restrooms, and height-adjustable tables. They can also act as advocates for people with disability in and out of the clinical space.

Another patient that you follow, Anna Drever, is a 55-year-old receptionist at a car dealership. She also has a diagnosis of relapsing-remitting MS. Ms. Drever completed high school, has previous experience as a receptionist, and has been working at the dealership for the past 15 years. Ms. Drever has been using a cane and braces intermittently for several years and more recently has transitioned to using a wheelchair for mobility over long distances.

13. **How does Ms. Drever's education and socioeconomic status affect her management of her disability?**
Unlike Dr. Anker, Ms. Drever has less flexibility in her work to accommodate her disability. Consider how Ms. Drever may not be able to transition to a job that gives her more flexibility or accommodation. Because of her lower education level compared to Dr. Anker, Ms. Drever is also more likely to have lower health literacy, or ability to understand health information, compared to Dr. Anker [21]. Health literacy is "the degree to which individuals have the capacity to obtain, process, and understand basic health information and services needed to make appropriate health decisions" [22]. Health literacy can be assessed in a clinical setting using tools such as the Short Assessment of Health Literacy—Spanish and English (SAHL-S&E), which uses flash cards to assess comprehension of some medical terms. For more information and to access the SAHL-S&E, visit ► https://www.ahrq.gov/professionals/quality-patient-safety/quality-resources/tools/literacy/index.html.

(a) **How does Ms. Drever's work environment affect her ability?**
Ms. Drever may need to stand in order to perform some of the essential functions of her position. She may not be able to adjust her environment to accommodate changes in her ability, particularly as she is in a position with much less autonomy than Dr. Anker's position.

(b) **How would Ms. Drever's work environment affect her ability if she worked in a retail position that did not allow her to spend a significant amount of her time seated?**
If Mrs. Drever worked in a highly mobile retail position, or another environment that required her to be standing and active throughout the day, she may not be able to keep her job. Some accommodations may be possible, but they would be challenging, and her employer may not

be supportive in terms of accommodating her needs. She may feel uncomfortable at work and leave her job. If Ms. Drever was unable to find another place of employment that was able or willing to accommodate her, she might then need to apply for SSDI (Social Security Disability) despite being able and willing to work.

> **Box 15.8 Teaching Tip**
> Remind learners that disability is defined by a mismatch between the environment and the ability of the individual. If the environment is truly accessible, then the individual may not experience a disability. Learners should consider the ways in which this is the case for Ms. Drever.

In the past 2 years, Ms. Drever has used a brace and a cane to walk and has used a wheelchair for mobility over long distances. With her most recent relapse, she has found that she cannot stand for long periods and increasingly has had difficulty propelling her wheelchair. Human Resources at the dealership has approved her use of a wheelchair to enable her to perform the essential functions of her position, but this has become more difficult due to difficulty propelling her chair. Unfortunately, her insurance company has denied her claim for a powerchair. The reason they have rejected the claim is that Ms. Drever does not currently have a vehicle that will allow her to transport the power wheelchair between locations (such as home and work). Ms. Drever is very worried about being able to continue working without this accommodation.

14. **Do you agree or disagree with the insurance company? Why or why not?**
    There are several perspectives to consider in this situation. First, there is Ms. Drever: She will not be able to afford to buy a power wheelchair (which, according to a 2010 survey of individuals with amyotrophic lateral sclerosis [ALS], cost $26,404 on average with a range of $19,376 to $34,311 [23]), but she needs one to accommodate her recent changes in ability. Second, there is the insurance company: The company has a clear policy and a financial interest in arguing that Ms. Drever does not meet the conditions to have a power wheelchair. Ms. Drever's healthcare practitioner might be able to help by exploring other options that fulfill the requirements of the insurance company and work within Ms. Drever's means and by taking the time to appeal the decision on her behalf.

    > **Box 15.9 Teaching Tip**
    > Learners often ask about the appeal process for insurance. The appeal process varies by state or plan and can involve either internal or external appeals. More information can be found at
    > ▶ https://www.healthcare.gov/appeal-insurance-company-decision/appeals/.
    >
    > Physicians may be asked to write a letter in support of an appeal. This is an important part of patient advocacy.

    (a) **Should insurance companies be required to cover items like this?**
    Not only may Ms. Drever be unable to work, but she is undergoing a high level of stress through the process. The ability to acquire a power wheelchair, not her MS, is preventing her from continuing to work. Ms. Drever cannot afford to purchase a power wheelchair. She will not only have difficulty at work but will also likely have difficulty with many activities of daily living (ADLs) and instrumental activities of daily living (IADLS).

    > **Box 15.10 Teaching Tip**
    > Have students consider the way that insurance coverage of adaptive equipment is similar to and different from insurance coverage for medication. Have students especially reflect on why insurance companies may be more likely to cover a treatment than adaptive equipment.

(b) **How does insurance coverage influence the ability of a person with disability? Is this influence the same as a person without disability?**

Insurance companies can deny people with disability access to needed equipment and therefore accommodation, based on the insurance plan in which the patient or family is enrolled. Without accommodation, people with disability can experience many negative outcomes, especially limited abilities in daily life and access to the environment and community.

Insurance coverage for a person without a disability can dramatically affect their health and wellness, and especially their access to healthcare services. This is also true for people with disability. The main difference is that for people with disability, it may *also* impact their access to the rest of their life and activities, not only their access to health care.

Insurance and insurance coverage are an incredibly complex issue, and policy in the United States related to health insurance has been shifting in the past several years. It is important to recognize that insurance coverage has a significant impact on the lives of all people.

Dr. Anker is divorced. She lives alone, but has a 22-year-old son who attends college and lives with her in the summer. Dr. Anker also has a 28-year-old daughter who is married, has one young child, and is pregnant with her second. Dr. Anker's daughter has talked about moving closer to home, but Dr. Anker does not want her to feel obligated to do so. Dr. Anker's parents moved to Sweden, their country of origin, after they retired; they are now in their 80s and find it difficult to travel.

Ms. Drever lives with her husband, who is 54 years old and has chronic back pain from his job. He helps her with many daily living tasks, but she does not want to feel like a burden. Ms. Drever realizes that her husband will not be able to continue to be a primary caregiver for her as her MS progresses, due to his own health problems. Her parents are in their late 80s and live in Florida, while her husband's parents are deceased. They have no other close family members.

15. **What options do Dr. Anker and Ms. Drever have for daily care?**

Dr. Anker may be able to pay for a caregiver or personal care attendant to come to her home to assist with tasks of daily living. She may also be able to move into an assisted living community. She also has the option to rely more heavily on her children, especially as her ability changes. She may be able to employ a combination of these strategies, such as living closer to her daughter for some aspects of her care but also having a home care assistant help her with other tasks of daily living, such as shopping, cooking, or transportation. Many of these options are possible for Dr. Anker because of her income level as a physician, which will allow her to pay for home care assistance and more easily afford other services.

Ms. Drever may be able to work with a community organization or access social services to find home health care, but her access to these services may depend on her income, employment, and insurance (as some insurance plans offer case management or home care services, while others do not). Ms. Drever may need to rely on her husband for assistance in daily living tasks despite his medical issues. It is also possible that if Ms. Drever begins experiencing more functional impairment, she may need to move to a long-term care facility earlier than might otherwise be necessary if she had other care resources available to her.

Likely the best place to start is for Ms. Drever to work with local community organizations to explore affordable home care to supplement the assistance she receives from her husband, so that the work for him is lessened. She may also consider legal counsel to help determine her and/or her husband's eligibility for Social Security Administration programs.

The best outcome for Ms. Drever is for her to be able to continue to work at her job, with accommodation, until she is 65. Her employment allows her to have stable income and to be covered under her company's insurance. When she is 65, she can qualify for Medicare and either retire or continue to work. However, if this is not possible due to functional limitations from her MS, Ms. Drever will need to pursue one of the following options.

If Ms. Drever is unable to continue working due to her MS, before she turns 65 and qualifies for Medicare, there are three main options available for her to receive supplemental income and maintain insurance coverage (as of 2018):

1. Apply for SSDI. Twenty-four (24) months after receiving SSDI benefits, she will qualify for Medicare.
2. "Spend down" or reduce her wealth (including real estate, savings, and investments) in order to qualify for Medicaid.
3. If her household income (the combined income of Ms. Drever and her husband) is below 138% of the Federal Poverty Level, and her state expanded Medicaid under the Affordable Care Act (ACA), then she and her husband will qualify for Medicaid without the need to reduce their wealth.

The most financially reasonable is Option 1, which allows her to have supplemental income through SSDI and health insurance coverage through Medicare. Option 1 would also allow her to continue working, at least part time, if she chooses (her income must be below the level that is considered "substantial" by the Social Security Administration). Medicare coverage is important for Ms. Drever, because it covers not only her medical and treatment expenses but also some home health care, which may decrease the amount of assistance for which Ms. Drever relies on her husband.

- For a clear explanation of SSDI and SSI for health professionals, review information at ▶ https://www.ssa.gov, and search for "Answers for Doctors and Other Health Professionals"
  ▶ https://www.ssa.gov/disability/professionals/answers-pub042.htm.
- For information about Social Security for people with disability, visit
  ▶ https://www.ssa.gov, search for "Benefits Planner," and select "Disability" from the top bar menu
  ▶ https://www.ssa.gov/planners/disability/index.html.
- For information about SSDI for individuals who are working, visit:
  ▶ https://www.ssa.gov, and search for "Disability Planner: Your Continuing Eligibility"
  ▶ https://www.ssa.gov/planners/disability/work.html.

Home care or home nursing care is often expensive and may not be available at a reasonable price in Ms. Drever's community. Long-term residential care can vary dramatically by state and by whether it is a private room and the level of care provided (skilled nursing facility or residential only). Prices on average in 2016 were between $3628 and $7698 per month depending on these factors [24]. Visit ▶ https://longtermcare.acl.gov/costs-how-to-pay/costs-of-care.html for more information of the cost of long-term care.

Medicare is insurance paid for by the federal government for individuals over 65 and those with qualifying disabilities such as kidney dialysis or who have qualified for SSDI. In most cases, people with disability are eligible for Medicare coverage 24 months after they receive SSDI benefits. Among other things, Medicare covers some home health services and skilled nursing facility services, but not long-term care facilities:

- For more information about what is covered by Medicare, visit ▶ https://www.medicare.gov/, and search "What Medicare Covers"
  ▶ https://www.medicare.gov/what-medicare-covers/.

- For more information on Medicare eligibility, visit ▶ https://www.hhs.gov/, and search "Who is eligible for Medicare?"
  ▶ https://www.hhs.gov/answers/medicare-and-medicaid/who-is-elibible-for-medicare/index.html.

Medicaid is insurance paid for by the state government and subsidized by the federal government that is provided for individuals below a certain level of income or poverty level (usually determined in relationship to the Federal Poverty Level). Medicaid coverage varies by state, and not all health practitioners will bill to Medicaid. For patients, this means that even with insurance coverage through Medicaid, they may not be able to find a healthcare practitioner who will take their insurance. Among other things, Medicaid covers home health services and nursing facility services. ▶ Chapter 8 explores the healthcare system in more detail.

- For more information on eligibility for Medicaid, visit: ▶ https://www.medicaid.gov/, and search "Eligibility"
  ▶ https://www.medicaid.gov/medicaid/eligibility/index.html.
- For more information about what is covered by Medicaid, visit: ▶ https://www.medicaid.gov/, search "List of Benefits," and select "List of Medicaid Benefits"
  ▶ https://www.medicaid.gov/medicaid/benefits/list-of-benefits/index.html.
- For more information on the Federal Poverty Level (FPL), visit: ▶ https://www.healthcare.gov/, search "FPL," and select "Federal Poverty Level (FPL)" at the ▶ HealthCare.gov Glossary
  ▶ https://www.healthcare.gov/glossary/federal-poverty-level-fpl/.

Insurance and insurance coverage is an incredibly complex issue, and policy in the United States related to health insurance has been shifting in the past several years. Insurance policy is subject to change based on many factors. This is especially true due to the complex funding streams for Medicare and Medicaid. It is important to recognize that insurance coverage has a significant impact on the lives of all people and varies on a case-by-case basis.

(a) **How do each of these women's situations influence the options available to them?**
Income plays a significant role in this decision for both women. Dr. Anker may be able to afford services that Ms. Drever cannot. Dr. Anker also has more options for family support from her children, despite her concerns about being a burden. Ms. Drever may need to consider government programs, which might affect her ability to continue to work, or the wealth of herself and her husband as they age.

✅ 16. **What responsibility do Dr. Anker's and Ms. Drever's families have to care for them?**
There are several difficulties that each of the families may encounter as caregivers:
- Ms. Drever's husband will likely have worsening back pain if he continues to be her primary caregiver. Furthermore, although the Family and Medical Leave Act (FMLA) guarantees people working for employers with 50 or more employees 12 weeks of unpaid leave per year to care for an immediate family member with a serious health condition, he may not be able to afford unpaid leave [25]. It is important to note that unpaid leave under FMLA would allow Mr. Drever to maintain his insurance if he has it through his employer. For more information on the Family and Medical Leave Act (FMLA), visit ▶ https://www.dol.gov/general/topic/benefits-leave/fmla.
- Dr. Anker's daughter is pregnant and has a young child. She likely does not have the time and energy to care for her mother in terms of ADLs but may be able to support her with IADLs, such as taking her grocery shopping. However, she would need to live closer to her mother in order

to become more involved with her care. Dr. Anker's son is focused on his education and then beginning his career. Since both he and his sister are relatively young, taking time away for their careers and family lives may impact their lives not only now but into the future.

Currently, in Western society, there is no legal responsibility for families to care for older adults as they age and often acquire disability. There is also relatively little cultural value placed on caring for older adults, and many older adults often do not want to feel dependent on family members. There is great variation in this value between cultural groups in the United States, and it is possible that individuals from different cultural backgrounds may feel strongly about the responsibility to care for elders. This is something that should be explored and never assumed.

To some extent, there is also little societal value placed on caring for people with disability, although there is a much stronger sense of cultural responsibility for parents to care for children with disability.

(a) **How might Ms. Drever's husband's medical conditions influence her decision regarding daily care?**
Ms. Drever might try to find care support from a community organization or, as her needs increase, consider moving to a long-term care facility earlier than might otherwise be necessary, out of concern for her husband's health. She may also experience more stress and conflicted feelings as she and her husband work to maintain their mutual health and well-being.

(b) **How might Dr. Anker's concern about being a burden influence her decisions?**
Dr. Anker may not choose to have her family as her primary care assistants. If her children do play a role in her care, Dr. Anker might feel conflicted or even embarrassed about that role. Changes in role are one aspect of changes in ability. Changing roles within the family may be challenging both for family members experiencing changes in ability and other family members, who may take on more of a caregiver role. A good primary care physician will get to know Dr. Anker and her family, if they are amenable, and help support the whole family through the process.

With respect to her medical practice, Dr. Anker is no longer participating in procedures but is still seeing patients. She is feeling some pressure from her colleagues to decrease her role further and is frustrated with feeling like they are constantly looking over her shoulder.

✅ 17. **Are there minimum physical requirements for being a physician?**

Box 15.11 Teaching Tip

Prompt by asking what assumptions learners are making in answering this question. For example, you might ask if potential modifications (such as amplified stethoscopes for healthcare practitioners with hearing impairment, or height-adjustable exam tables that would allow a healthcare professional using a wheelchair to conduct a physical exam) change the learner's opinion on what the minimum physical requirements are. Be sure that learners consider not only physical disability but also mental illness and other disability.

Learners may be interested in a discussion of whether requirements should be different for individuals just beginning their training and those who acquire a disability later in their career. They may raise issues of discrimination against people with a disability. Another issue that might be raised is the difference in how accommodation is considered for individuals whose ability changes later in their career compared to those who would like to enter training.

One important consideration is technical standards and essential requirements. These standards are used in the admission process for healthcare professionals, as explored in Background Question 8. These should be considered along with adaptive equipment that can allow some disabled physicians and healthcare practitioners to practice, such as amplified stethoscopes or height-adjustable exam tables. Some disabilities may be more easily accommodated than others, depending on adaptive equipment and other factors.

18. **Who has the responsibility to monitor physician ability?**
    This is a controversial topic, and research in this area is ongoing. Although many ideas have been proposed, a consensus has not been reached on the assessment of physician ability.
    (a) **Self-reflection: Do you trust yourself to be objective in assessing your own ability?**

> **Box 15.12 Teaching Tip**
> Have learners consider the ways that they are biased about their own abilities. One very important aspect of this question is that when your livelihood depends on your own reporting of your ability, it could affect your judgment. Be sure learners reflect on the ramifications for patients if they do not objectively assess their own ability but also the ramifications they may experience if they do. What is their obligation to objectively assess and respond to their assessment of their own ability?

(b) **Self-reflection: Do you trust your peers to be objective in assessing their own ability? Would you feel differently if your peers were responsible for assessing your ability?**
(c) **What are some potential misuses of this responsibility? What biases might come into play?**

> **Box 15.13 Teaching Tip**
> Ask learners to suggest some potential misuses of this responsibility, and how they might impact the physician being assessed. It may also be an opportunity for learners to consider their own biases regarding physician assessment.

One potential problem with physician assessment is the concept that there is a single standard against which all physicians can be measured [26]. There is also the possibility that peer assessment might include inaccurate reporting of competency to address personal disagreements or differences. Health professionals who dislike one of their colleagues might be able to utilize assessment of ability to threaten the employment and reputation of individuals they dislike. There is also the potential for assessment to be a mechanism by which healthcare practitioners are asked to leave organizations, which could allow for institutional changes to be made.

Physician assessment can take many forms including psychometric testing, realist and ethnographic methods of assessment [27], cognitive testing [26], multisource feedback [28], and continuing medical education [29]. However, some biases, especially ageism, have been identified [26]. It is important that all healthcare practitioners are able to provide high-quality care for their patients. Creating systems that allow for assessing and accommodating changes in ability at any age is necessary [30] in order to appropriately value health practitioner experience and to avoid stigmatizing certain disabilities or groups [26].

Assessment bias is an area of ongoing research. OSCEs (objective structured clinical exams) provide an opportunity to investigate bias in evaluation of medical students [31]. One study found in a sample of 33 students that females rated their own performance on self-assessments lower than peer assessment, while males rated their performance the same as peer assessment [32]. Although this evidence is very specific, it demonstrates the importance of exploring bias in self and peer assessment.

Some studies have shown that physicians do not accurately assess their own abilities [33]. More current research is necessary to explore the potential for bias in physician self-assessment and to explore possible solutions to maintain patient safety.

As in many other situations, bias is a factor that affects physician assessment. It is important for health practitioners to recognize their own bias and the possible biases of their peers. Recognizing and addressing bias is necessary to ensure patient safety in the case of changes in physician ability.

✓ 19. **Creative problem-solving:** How should ability be assessed in cases related to aging or medical events? [34].

> **Box 15.14 Teaching Tip**
> Encourage learners to consider some potential misuses of the approach they propose. What biases might come into play? This is a controversial topic, and research in this area is ongoing. Although many ideas have been proposed, a consensus has not been reached on the use of assessment of healthcare practitioner ability [34]. Learners should consider the perspective of physicians, who would have additional checks in place, as well as the perspective of patients, who could be put at risk if a physician was unable to perform the essential functions of their position and did not recognize their limitations.

✓ 20. **Self-reflection:** Did you approach aspects of this case differently because the first patient was a physician? Do you think physicians treat physician-patients differently from non-physician patients? How so? [35].

> **Box 15.15 Teaching Tip**
> Encourage learners to share experiences that they have had, since entering medical or health professions school, as patients, or family members of patients. There has been little research on physician-patients. However, one study has described some challenges and strategies identified by clinicians:
>
> » "Three of the challenges most commonly discussed by physician participants were: (1) maintaining boundaries between relationships with colleagues or between roles as physician/colleague/friend, (2) avoiding assumptions about patient knowledge and health behaviors, and (3) managing physician-patients' access to informal consultations, personal test results, and opinions from other colleagues. We were able to identify three main strategies clinicians use in addressing these perceived challenges: (1) Ignore the physician-patient's background, (2) Acknowledge the physician-patient's background and negotiate care, and (3) Allow care to be driven primarily by the physician-patient" [35].
>
> It is worth noting that challenge 3 is increasingly true for non-physician-patients as well, as there is now widespread access to medical information online and to opinions from patients and families with similar concerns via social media.

Acknowledgments The authors and editors wish to thank Burk Jubelt, who provided expert knowledge on multiple sclerosis; Martha Wojtowycz, who provided expert knowledge on Medicare, Medicaid, and the health system; and Michael Ioerger, who provided expert knowledge related to self-other overlap and disability; as well as Rebecca Garden, for her contributions to earlier cases which inspired this chapter.

References

1. Centers for Disease Control and Prevention. Key findings: prevalence of disability and disability type among adults, United States – 2013. In: Disability and health [Internet]. Atlanta (GA): CDC; 2015 [cited 2018 Jun 20]. Available from: https://www.cdc.gov/ncbddd/disabilityandhealth/features/key-findings-community-prevalence.html.
2. Okoro CA, Hollis ND, Cyrus AC, Griffin-Blake S. Prevalence of disabilities and health care access by disability status and type among adults — United States, 2016. MMWR Morb Mortal Wkly Rep. 2018 [cited 2018 Dec 31];67:882–7. https://doi.org/10.15585/mmwr.mm6732a3.
3. Information and technical assistance on the Americans with Disabilities Act. Introduction to the ADA [Internet]. Washington, D.C.: United States Department of Justice Civil Rights Division. Available from: https://www.ada.gov/ada_intro.htm.
4. Merriam-Webster. Definition of disability [Internet]. 2018 [cited 2018 Jun 20]. Available from: https://www.merriam-webster.com/dictionary/disability.
5. Johnson WG. There is a difference between differences and disparities. Disabil Health J. 2008;1:181–3.
6. Krahn GL, Walker DK, Correa-De-Araujo R. Persons with disabilities as an unrecognized health disparity population. Am J Public Health. 2015:e1–9. https://doi.org/10.2105/AJPH.2014.302182.
7. U.S. Department of Justice: Civil Rights Division. Americans with Disabilities Act: access to medical care for individuals with mobility disabilities [Internet]. Washington, D.C.: U.S. Department of Health and Human Services: Office for Civil Rights; 2010 [cited 2018 Jan 10]. Available from: https://www.ada.gov/medcare_mobility_ta/medcare_ta.pdf.
8. Iezzoni LI, Pendo E. Accessibility of medical diagnostic equipment — implications for people with disability. N Engl J Med. 2018;378:1371–3.
9. Practical recommendations for enhancing the care of patients with disability. In: Web-based continuing education [Internet]. Syracuse (NY): Upstate Medical University. [cited 2018 Jun 20]. Available from: http://www.upstate.edu/pmr/education/disability/index.php.
10. Disabilities, opportunities, internetworking, and technology (DO-IT). What is universal design? [Internet]. Seattle (WA): University of Washington; 2017 Jun 29 [cited 2018 Jun 12]. Available from: https://www.washington.edu/doit/what-universal-design-0.
11. Centers for Disease Control and Prevention. Social ecological model [Internet]. Atlanta (GA): CDC; 2015. Available from: https://www.cdc.gov/cancer/crccp/sem.htm.
12. Turk MA, Mudrick NR. Rehabilitation interventions. Los Angeles: SAGE Publishing; 2013.
13. Olek MJ. Clinical course and classification of multiple sclerosis [Internet]. In: González-Scarano F, editor. UpToDate. Waltham (MA): UpToDate Inc. [cited 2018 Jun 5]. Available from: http://www.uptodate.com.
14. McKee M, Case B, Fausone M, Zazove P, Ouellette A, Fetters MD. Medical schools willingness to accommodate medical students with sensory and physical disabilities: ethical foundations of a functional challenge to "organic" technical standards. AMA J Ethics. 2016;18:993–1002.
15. Kezar LB, Kirschner KL, Clinchot DM, Laird-Metke E, Zazove P, Curry RH. Leading practices and future directions for technical standards in medical education. Acad Med. 2018; https://doi.org/10.1097/ACM.0000000000002517.
16. Institute of Medicine. In: Pope AM, Brandt Jr EN, editors. Enabling America: assessing the role of rehabilitation science and engineering. Washington, DC: National Academies Press; 1997.
17. LaPierre TA, Zimmerman MK, Hall JP. "Paying the price to get there": motherhood and the dynamics of pregnancy deliberations among women with disabilities. Disabil Health J. 2017;10(3):419–25.
18. UpToDate. About us [Internet]. Waltham (MA): UpToDate, Inc. [cited 2018 Jun 20]. Available from: https://www.uptodate.com/home/about-us?&redirect=true.
19. Cifu DX. Braddom's physical medicine and rehabilitation. 5th ed. Philadelphia, PA: Elsevier; 2016.
20. Frumovitz M. Invasive cervical cancer: epidemiology, risk factors, clinical manifestations, and diagnosis. In: Goff B, Dizon DS, editors. UpToDate. Waltham (MA): UpToDate Inc.; updated 2018 Dec 7 [cited 2018 Jun 19]. Available from: http://www.uptodate.com.
21. Centers for Disease Control and Prevention. Understanding health literacy [Internet]. Washington, D.C.: U.S. Department of Health and Human Services; 2016 [cited 2018 Jan 9]. Available from: www.cdc.gov/healthliteracy/learn/Understanding.html.
22. Agency for Healthcare Research and Quality. Health literacy measurement tools (revised) [Internet]. Rockville (MD): AHRQ; 2016 Feb [cited 2018 Jun 20]. Available from: http://www.ahrq.gov/professionals/quality-patient-safety/quality-resources/tools/literacy/index.html.
23. Ward AL, Sanjak M, Duffy K, Bravver E, Williams N, Nichols M, Brooks BR. Power wheelchair prescription, utilization, satisfaction, and cost for patients with amyotrophic lateral sclerosis: preliminary data for evidence-based guidelines. Arch Phys Med Rehabil. 2010;91:268–72.
24. U.S. Department of Health & Human Services. Costs of care [Internet]. Washington, D.C.: Administration on Aging; 2017 [cited 2018 Jun 20]. Available from: https://longtermcare.acl.gov/costs-how-to-pay/costs-of-care.html.
25. United States Department of Labor. FMLA (Family & Medical Leave) [Internet] Washington, D.C.: DOL. [cited 2018 Jun 20]. Available from: https://www.dol.gov/general/topic/benefits-leave/fmla.
26. Devi G. Alzheimer's disease in physicians — assessing professional competence and tempering stigma. N Engl J Med. 2018;378:1073–5.
27. Whitehead CR, Kuper A, Hodges B, Ellaway R. Conceptual and practical challenges in the assessment of physician competencies. Med Teach. 2014;37:245–51.
28. Donnon T, Ansari AA, Alawi SA, Violato C. The reliability, validity, and feasibility of multisource feedback physician assessment. Acad Med. 2014;89:511–6.

29. Cervero RM, Gaines JK. The impact of CME on physician performance and patient health outcomes: an updated synthesis of systematic reviews. J Contin Educ Health Prof. 2015;35:131–8.
30. Kupfer JM. The graying of US physicians: implications for quality and the future supply of physicians. JAMA. 2016;315:341–2.
31. Jefferies A, Simmons B, Regehr G. The effect of candidate familiarity on examiner OSCE scores. Med Educ. 2007;41:888–91.
32. Madrazo L, Lee CB, McConnell M, Khamisa K. Self-assessment differences between genders in a low-stakes objective structured clinical examination (OSCE). BMC Res Notes. 2018;11(1):393.
33. Davis DA, Mazmanian PE, Fordis M, Harrison RV, Thorpe KE, Perrier L. Accuracy of physician self-assessment compared with observed measures of competence. JAMA. 2006;296:1094.
34. Levinson W, Ginsburg S. Is it time to retire? JAMA. 2017;317(15):1570–1.
35. Domeyer-Klenske A, Rosenbaum M. When doctor becomes patient: challenges and strategies in caring for physician-patients. Fam Med. 2012;44(7):471–7.

Further Reading on This Topic

Iezzoni LI, Pendo E. Accessibility of medical diagnostic equipment — implications for people with disability. N Engl J Med. 2018;378:1371–3.

Practical recommendations for enhancing the care of patients with disability. In: Web-based continuing education. Syracuse (NY): Upstate Medical University. [cited 2018 Jun 20]. Available from: http://www.upstate.edu/pmr/education/disability/index.php.

U.S. Department of Justice: Civil Rights Division. Americans with Disabilities Act: access to medical care for individuals with mobility disabilities [Internet]. Washington, D.C.: U.S. Department of Health and Human Services, Office for Civil Rights; 2010 July. Available from: https://www.ada.gov/medcare_mobility_ta/medcare_ta.pdf.

Young, Stella. Stella Young: I'm not your inspiration, thank you very much. TEDxSydney [web streaming video]. 2014 Apr. Available from: https://www.ted.com/talks/stella_young_i_m_not_your_inspiration_thank_you_very_much.

"It Runs in the Family"

Robert Roger Lebel

16.1 Background Questions – 316

16.2 Additional Case Information and Discussion Questions – 317

16.3 Answers to Background Questions – 318

16.4 Responses to Discussion Questions – 321

References – 330

- **Glossary**

Chromosomal diseases - Occur when the entire chromosome, or large segments of a chromosome, is missing, duplicated, or otherwise altered. Down syndrome is a prominent example of a chromosomal abnormality [1].

Multifactorial disorders - Occur as the result of mutations in multiple genes, frequently coupled with environmental causes. An example of a multifactorial disorder is diabetes [1].

Mitochondrial disorders - Rare disorders caused by mutations in nonchromosomal DNA located within the mitochondria. These disorders can be found to affect any part of the body including the brain and the muscles [1].

Single-gene disorders - Occur when an alteration occurs in a gene causing one gene to stop working. An example of a single-gene disorder is Huntington disease [1].

Learning Objectives

1. Apply principles of confidentiality to complex situations and identify exceptions.
2. Describe the risks and benefits of genetic testing for heritable diseases.
3. Analyze ethical, legal, and medical implications of diagnosing a family member with a disease that is known to be heritable.

Case Background

You are a clinical geneticist seeing a couple, Cristina and Mario Fontana, who were referred to you by their primary care physician to discuss testing for Huntington disease. Cristina is currently 8 weeks into their first pregnancy.

You learn that Mario had been the primary caregiver for his maternal aunt, who died last week from complications associated with Huntington disease (HD). Mario's maternal grandmother also had Huntington disease. Both women were in their mid-50s before they manifested signs and symptoms of the disease. Mario's mother, Paula, is in her late 40s and does not have any signs of Huntington disease at this time. When you ask if Paula has been tested for the disease, Mario tells you that she decided many years ago not to undergo genetic testing and that she recently reiterated that she is not open to reconsidering the matter.

The Fontanas' relationship has been under stress for the past several months, in part because of the uncertainty about the risk of HD. Mario has considered getting tested to learn about his status but changed his mind several times over the years. Cristina is very concerned about the risk of transmission of HD to their future child and recently asked her obstetrician, Dr. Albano, for testing to determine if the fetus has an HD mutation. Dr. Albano assured her that genetic testing for HD could be done by either chorionic villus sampling (CVS) or amniocentesis.

16.1 Background Questions

1. What is Huntington disease (HD)?

2. Under what circumstances is genetic testing for inheritable diseases generally recommended?

3. What is *definitive* prenatal genetic testing and what is *exclusion* prenatal testing? How are they different?

4. What is preimplantation genetic diagnosis (PGD)? Does your state require that insurance companies cover PGD?

5. What types of testing are routinely offered as part of standard prenatal care? How were these tests chosen?

6. Tay-Sachs disease (TSD) is an autosomal recessive condition that leads to death in early childhood. What ethnic groups were historically most likely to be affected by TSD? Is there evidence that prenatal testing for TSD has impacted the incidence in live births?

7. Several religious traditions have specific positions on the use of certain assisted reproductive technologies, such as in vitro fertilization (IVF). Choose one of these and research the basis and specifics of the position.

16.2 Additional Case Information and Discussion Questions

You explain to the Fontanas that because Mario does not have a living relative with diagnosed HD, they are not candidates for exclusion prenatal testing. Definitive prenatal testing is technically an option but is extremely rarely performed if neither parent has been diagnosed with HD nor is known to have the genetic mutation that causes it.

1. Given this information, should Mario be tested first? Why or why not?

2. Was it appropriate for Dr. Albano to offer genetic testing? Why or why not?
   (a) Should he have involved any other expert consultants prior to making the offer? If so, whom?
   (b) If you disagree with Dr. Albano's offer, what would you say to Cristina and Mario?
   (c) Would you contact Dr. Albano to tell him of your differing opinion? Why or why not? What are the implications for Dr. Albano's future patients?

After further discussion, you encourage the Fontanas to try to talk with Paula again since the most efficient approach to testing would be to determine her status—which, if negative, would put an end to the entire discussion since it would mean that Mario is not at risk himself. You also inform them that if Paula does not change her mind about testing and they decide that they would like to pursue a prenatal diagnosis, your recommendation will be for Mario to be tested first, since both CVS and amniocentesis are invasive and carry a risk of miscarriage. With the assistance of your genetic counselor, you encourage them to consider all of the new information that has been shared, the implications of genetic testing, and the resources and support systems that are available to them.

3. Is it appropriate to perform testing if a positive finding will implicate Paula as the source of the mutation? Why or why not?
   (a) Is Mario obligated to consider her wishes?
   (b) Does Paula have a right *not* to know?
   (c) Does the principle of respect for autonomy help to resolve these questions? How so?
   (d) What other principles or concepts factor into your analysis?

At the next visit, both Cristina and Mario seem tense and emotional. Mario states that he did talk with Paula. Initially she became more sympathetic toward their reasons for wanting to know his status; however, while they were discussing the reasons for testing, Cristina shared that she would want to terminate the pregnancy if further testing identified an HD mutation in the fetus. Paula became extremely upset about the possibility that Cristina might terminate the pregnancy and adamantly refuses to get tested. Cristina feels that Mario is letting his mother dictate their lives and angrily says that she has decided simply to have CVS done, since it can be done as early as the 11th week of pregnancy, allowing her to terminate her pregnancy earlier, when the procedure has fewer risks.

4. Given this new information, how should you proceed?
   (a) What are your obligations to Cristina? To Mario? To the fetus?
   (b) Do you have any obligations to Paula or to other family members?
   (c) To whom do you owe the duty of confidentiality?

5. Self-reflection: What do you think that you would choose for yourself, if you were in Cristina's or Mario's position?

6. If Cristina and Mario had seen an obstetrician prior to conception, should they have been offered in vitro fertilization (IVF) and preimplantation genetic diagnosis (PGD)? Why or why not?
   (a) Would your answer be different if there was a family history of Tay-Sachs disease?

(b) What about sickle cell disease?
  (c) Does the severity of the disease change your opinion about the morality of assisted reproduction or abortion? Should it?

At one point during your visits, Mario mentions that he overheard one of the staff members in the obstetrician's office make a derogatory comment, suggesting that people who have genetic diseases should adopt.

❓ 7. How would you respond to this remark if it came from one of your colleagues?

❓ 8. Should insurance cover IVF? PGD? Why or why not?
  (a) If not, should these services be available only to those who can afford them?
  (b) What role does the principle of justice play?
  (c) Should insurance premiums be higher for people who are identified as at risk for a genetic disease? Why or why not?

After several genetic counseling sessions with Mario and Paula, Mario decides that he does not want to get tested either. Although this is very upsetting to Cristina, she eventually agrees to support Mario and Paula in their decisions, and she decides against pursuing prenatal testing for HD.

Seven years later, Cristina returns to see you. She tells you that 3 years ago, Paula was diagnosed with HD. She also shares that she and Mario were recently divorced. She is here today because she wants to discuss testing her now-6-year-old daughter, Amelia, for the disease.

❓ 9. Should you test Amelia for HD? Why or why not?
  (a) Would your decision be different if Cristina wanted to test Amelia for a *BRCA1* gene mutation? Why or why not?
  (b) Are parents entitled to know medical information about their children that will not impact their children's health until adulthood?

Cristina also tells you that Mario learned that his mother had placed a child for adoption before he was born. Mario's older biological brother recently contacted the family and they have exchanged a few e-mails. She is unsure if the family has disclosed its history to the brother. She asks you to contact the brother because she feels he has a right to know.

❓ 10. Do you have any ethical or legal obligations to this sibling?

❓ 11. What if, instead of Huntington disease, the family had a history of something like familial adenomatous polyposis (FAP), in which early intervention can prevent cancer?

16.3 Answers to Background Questions

✅ 1. **What is Huntington disease?**
Huntington disease is a neurodegenerative disorder inherited in an autosomal dominant fashion; children of affected individuals have a 50% chance of inheriting the disease. The disease is caused by an expansion in the number of CAG repeats in the *HD* gene. Age of onset of symptoms is variable, with an average age of 40 years. Symptoms include involuntary choreiform movements, significant changes in mood and behavior, and cognitive impairment, all of which are progressive. There is no known treatment to prevent or ameliorate symptoms [2].

✅ 2. **Under what circumstances is genetic testing for inheritable diseases generally recommended?**
There are several circumstances in which genetic testing is recommended. The most common genetic testing is done as part of routine newborn screening to identify certain types of disease such as phenylke-

tonuria (PKU) and inheritable congenital hypothyroidism. In addition, genetic testing may be recommended for [3]:
- Diagnostic purposes for individuals who have symptoms consistent with an inheritable disease such as cystic fibrosis or HD
- Predictive purposes for individuals who have a family history of an inheritable disease [4]:
   - "Presymptomatic" testing for diseases in which the inherited condition has a delayed outcome; the outcome is predetermined by genes but the timing is not, such as HD
   - "Predisposition" testing to help assess the *risk* of a disease developing in an individual, such as *BRCA1* and *BRCA2* testing if there is a family history of breast and ovarian cancer
- Pharmacologic management in some circumstances such as testing for *CYP2C9/VKORC1* genes to reduce adverse drug events associated with warfarin
- Prenatal risk assessment:
   - Diagnosis when there is family history of an inheritable disease (e.g., sickle cell, HD)
   - Screening for genetic abnormalities (e.g., Down syndrome, trisomy 18)
- Preimplantation genetic diagnosis (discussed further in Discussion Question 4, later in this section)

Genetic testing for diseases that present in adulthood is generally not recommended in minors. The standard approach is to allow young people to reach adulthood and decide for themselves whether and when they wish to know their status. The exceptions to this rule are (1) when the minor is manifesting suspicious signs and symptoms so that the testing becomes diagnostic rather than predictive or (2) when there are preventative measures that can be taken in childhood or adolescence to reduce the risk or severity of the disease or disorder in question [5]. With respect to HD, no treatment is yet available [5, 6]. Fewer than 15% of people with a family history of Huntington disease choose predictive testing in the asymptomatic period [7].

3. **What is *definitive* prenatal genetic testing and what is *exclusion* prenatal testing? How are they different?**
Definitive testing applies to situations in which the exact genetic marker is known, and the question of its transmission is asked and answered directly through testing. For example, in this case, if Mario decided to get tested, he would know whether or not he had a genetic mutation that causes HD.

Exclusion testing can be done when the at-risk parent prefers not to be tested or is unavailable for testing, but another affected family member is known. In exclusion testing, the fetus is tested for the presence of a linked marker that identifies one grandparent (rather than the other) as the source of genetic material in the chromosome region of interest. Results of exclusion testing do not specify whether the gene in question has typical or atypical conformation, but rather yield an assessment of risk, based on which grandparent was affected. For example, assuming the disease is found in the paternal family, as in this case: If the paternal grandfather was unaffected and the relevant chromosome is shown to have been inherited from the paternal grandfather, then the conclusion is "low risk." If the relevant chromosome has been inherited from the affected paternal grandmother, then the conclusion is that the fetus has "half the risk of the father" [8–11]. Exclusion testing can also be performed as part of preimplantation genetic diagnosis, as discussed in Discussion Question 4 [12].

4. **What is preimplantation genetic diagnosis (PGD)? Does your state require that insurance companies cover PGD?**
Preimplantation genetic diagnosis is a process, requiring in vitro fertilization, to identify certain heritable diseases

before an embryo is even transferred to the uterus. An ovum is fertilized in a dish and allowed to reach the 8-cell stage. One cell is then removed and studied for the genetic marker in question. If deemed favorable, the pre-embryo can then be transferred to a uterus. Thus, the diagnosis has been accomplished prior to implantation.

Insurance coverage for PGD varies from state to state and from insurance company to insurance company. Infertility treatment is not one of the essential benefits covered under the Affordable Care Act. In the case of PGD, even if infertility is not the diagnosis, because it requires in vitro fertilization, it is unlikely to be covered [13].

5. **What types of testing are routinely offered as part of standard prenatal care? How were these tests chosen?**
First-trimester screening (between 10 and 13 weeks): An ultrasound exam evaluates nuchal "translucency" (thickness), which may indicate increased risk for Down syndrome or another aneuploidy, as well as defects of the heart, abdominal wall, and skeleton. All pregnant patients are offered the opportunity to learn their carrier status for cystic fibrosis (CF), hemoglobinopathies (such as sickle cell disease and thalassemia), and spinal muscular atrophy (SMA). Any who are found to be carriers are encouraged to have their partners tested, and if both are carriers, they are offered prenatal testing to determine if the fetus is affected.

Second-trimester screening (between 15 and 22 weeks): The quadrivalent or "quad" screen measures alpha-fetoprotein (AFP), uE3 (unconjugated estriol), human chorionic gonadotropin (hCG), and inhibin A, which screens for Down syndrome, trisomy 18, and neural tube defects.

The screens were decided based upon safety (noninvasive), risk of chromosomal abnormalities, sensitivity, specificity, and low false-positive rates. Each test has limitations in terms of sensitivity, specificity, and predictive value, which should be discussed with the patient pretest and posttest. If any screening test is positive, the patient may pursue diagnostic tests and/or cell-free testing starting at 10 weeks of pregnancy to test for aneuploidy. A positive screen can be followed by chorionic villus sampling (CVS) during the first trimester or amniocentesis during the second trimester to determine aneuploidy as well as neural tube defects in the case of amniocentesis [14–17].

6. **Tay-Sachs disease (TSD) is an autosomal recessive condition that leads to death in early childhood. What ethnic groups were historically most likely to be affected by TSD? Is there evidence that prenatal screening has impacted the incidence of TSD in live births?**
Historically, TSD is most common in Ashkenazi Jews (Central and Eastern European descent), with carrier frequency of 1 in 25 to 1 in 30 and disease incidence 1 in 3,600. Due to the founder effect, French, Canadians, and Louisiana Cajuns are also considered high risk, with the same carrier rates at Ashkenazi Jews. Screening protocols since the 1970s, specifically for Ashkenazi Jews, increased the number of cases of TSD detected prenatally three to fourfold. In Orthodox Jewish communities, the *Dor Yeshorim* program administers an anonymous carrier-testing program so that couples at high risk to conceive a child with Tay-Sachs disease or another genetic disorder might avoid marrying one another. The program is possible only because in many Orthodox Jewish communities, rabbis make the final decisions regarding marriages and can maintain confidential records that allow "appropriate" matches to prevail [18]. The American College of Medical Genetics and Genomics (ACMG) recommends routinely offering carrier testing to Ashkenazi Jewish patients for nine disorders including Tay-Sachs disease, while the American College of Obstetricians and Gynecologists (ACOG) Committee on Genetics recommends routinely offering Ashkenazi Jews carrier testing for four disorders, again including Tay-Sachs

disease [19, 20]. There is evidence that the incidence of TSD in live births has decreased, which is presumed due to such recommendations [21, 22].

✓ 7. **Several religious traditions have specific positions on the use of certain assisted reproductive technologies, such as in vitro fertilization (IVF). Choose one of these and research the basis of the position.**
Many different faiths, including but not limited to Christian, Jewish, and Muslim denominations, have objections to some of the methods employed in assisted reproduction. These objections frequently relate to such matters as (1) separating reproduction from person-to-person interactions in committed, exclusive relationships and (2) the disposal of zygotes and pre-embryos as though they represent something less than new persons [23]. It is important to note that many people do not accept the specific position of the faith to which they subscribe. For example, the Roman Catholic Church unequivocally prohibits the use of in vitro fertilization, yet patients identifying as Roman Catholic are well represented in published studies of IVF [24, 25].

> **Box 16.1 Teaching Tip**
> Although it is potentially more challenging for instructors, asking learners to select which faith they wish to research in response to this question avoids the risks of singling out specific religious beliefs for scrutiny and permits them to decide for themselves whether to discuss their own beliefs. The article cited above, by Joseph Schenker, gives a detailed, though not comprehensive, overview of potential religious objections [23].

16.4 Responses to Discussion Questions

You explain to the Fontanas that because Mario does not have a living relative with diagnosed HD, they are not candidates for exclusion prenatal testing. Definitive prenatal testing is technically an option but is extremely rarely performed if neither parent has been diagnosed with HD or is known to have the genetic mutation that causes it.

✓ 1. **Given this information, should Mario be tested first? Why or why not?**

> **Box 16.2 Teaching Tip**
> Ask learners to describe the risks to Mario, to Cristina, and to the fetus and to compare them. They should be able to articulate and defend their reasoning. How does one compare very different types of risk, such as physical versus psychological harm?

In most situations, Mario should be tested first, if he is willing. The risks of testing to Mario include (1) the potential for psychological harm from learning that he has Huntington disease and (2) the potential for future financial harm and loss of privacy.

Regarding the former, patients who learn that they have a life-limiting diagnosis, particularly one as devastating as HD, may experience depression, anxiety, and other adverse psychological symptoms. Thus, it is important that patients receive appropriate counseling prior to testing. It has long been assumed that individuals who undergo testing voluntarily (compared with those who learn their status through a relative's test result) are less likely to suffer such adverse effects, although this has not been proven [7, 26]. Approximately half of individuals who choose to undergo predictive testing for HD are adults aged 35 years or younger, and some researchers have found that those who test positive express more urgency in reaching key milestones of adulthood and that knowing one's status has positive effects on life choices, including choices related to education, career, romantic relationships, and family planning [27].

Regarding the latter, the Genetic Information Nondiscrimination Act of 2008 (GINA) bars employers and health insurers from discriminating against those with genetic risk (e.g., hiring/firing decisions, charging higher insurance premiums, etc.). A major omission from the law is that life, disability, and long-term care insurers are exempted [28, 29]. Thus, there is fear among patients that their long-term needs might not be covered if a genetic test is done. It should be noted that, as Mario's partner, Cristina might experience some of these consequences herself.

The risks of testing to Cristina include potential physical and psychological harm from undergoing an invasive procedure. Risks of amniocentesis and CVS include pain and discomfort, infection, amniotic fluid leak, Rh sensitization, and spontaneous abortion [30, 31]. Spontaneous abortion might be misconstrued as a risk to the fetus only; however, in addition to the obvious psychological cost of pregnancy loss, which may also affect Mario, there are physical risks to Cristina. The risk of spontaneous abortion after amniocentesis is approximately 0.2%. It is higher when amniocentesis is performed prior to 15 weeks of pregnancy; however, termination of pregnancy is safer when performed earlier in the pregnancy, so there is often tension between minimizing the risk of miscarriage, if the fetus is healthy, and minimizing the risks of the procedure to the mother, if the fetus is affected [32]. The risk after chorionic villus sampling is approximately 0.7% [31].

Some ethicists have argued that prenatal testing should not be pursued unless Cristina has already decided that she would want to terminate an affected pregnancy. The reasons for this relate to the right of the future child, once born, to make his or her own decisions about genetic information. (Rights of the future child are discussed further later in this section.) Many others, however, believe that no such constraining promise is appropriate, since real people might be inclined to one or another course of action at an early stage of the process but reach another opinion when confronted by the actuality of test results. No one is a perfect predictor of her own future behavior, and healthcare professionals should not place themselves in the position of coercing a course of action like pregnancy termination through some pseudo-contractual agreement in advance of actual testing.

2. **Was it appropriate for Dr. Albano to offer genetic testing? Why or why not?**
Dr. Albano is not a geneticist, and his previous experience with counseling patients in this situation is unknown. This may well be outside his scope of practice. The scope of practice is defined "as the activities that an individual health care practitioner is permitted to perform within a specific profession. Those activities should be based on appropriate education, training, and experience. Scope of practice is established by the practice act of the specific practitioner's board, and the rules adopted pursuant to that act" [33]. Three questions are pertinent to determining the scope of practice in this situation [34, 35].

First, does Dr. Albano have sufficient education, including on-the-job training, to adequately counsel patients such as Cristina regarding prenatal testing for Huntington disease? This question is the most pertinent to the case, as the information Dr. Albano provided to Cristina appears to be somewhat limited or even inaccurate, suggesting that he perhaps does *not* have the requisite knowledge.

Second, does the licensing or governing body (e.g., medical board) that oversees Dr. Albano as a practicing physician permit him to perform the diagnostic procedures in question? Most states license physicians and other healthcare practitioners (e.g., nurse practitioners, physician assistants) generally, without attention to the scope of practice defined by each specialty. However, several national medical boards certify physicians within specialties (e.g., the American Board of Internal Medicine, American Board of Surgery, American Board of Pediatrics), and professional organizations may set guidelines for scope of practice within specialties (e.g., American College of Obstetrics and Gynecology, American College of Medical Genetics and Genomics) and sub-specialties.

Third, does his institution (hospital or clinic) permit him to offer the service in question, in this case both genetic counseling and CVS or amniocentesis? If Dr. Albano is in independent private practice, he still must disclose to his malpractice insurance carrier what procedures he performs.

Ethically, even if Dr. Albano meets conditions 2 and 3, if he does not meet condition 1, he should not offer the service. Physicians have a professional and ethical obligation to know the limitations of their own knowledge and skills and to refer patients when these are exceeded.

(a) **Should he have involved any other expert consultants prior to making the offer? If so, whom?**

Dr. Albano's advice to Cristina suggests a limited understanding of genetic testing for Huntington disease. It would likely have been helpful for Dr. Albano to offer Cristina a referral to a geneticist and/or genetic counselor up front. As the field of medicine has advanced, it has simultaneously become more difficult for physicians to maintain broad expertise and much easier to access sophisticated, specialized testing.

(b) **If you disagree with Dr. Albano's offer, what would you say to Cristina and Mario?**

There is an important tension here to consider: the physician in this case is pulled between being honest with the patients and avoiding potentially inappropriate or unprofessional criticism of another colleague. The patient's welfare is primary: if the physician felt that the patient was in jeopardy because of ongoing decisions by the primary physician, she would be obligated to speak up. However, consultants may not have a full understanding of what was communicated between a referring physician and a patient, so it is important not to rush to making a judgment. In this situation, no physical harm has been done or is at imminent risk of being done, and criticizing Dr. Albano cannot repair psychological harm.

(c) **Would you contact Dr. Albano to tell him of your differing opinion? Why or why not? What are the implications for Dr. Albano's future patients?**

> **Box 16.3   Teaching Tip**
> If learners state that they would *not* contact Dr. Albano, even after considering the needs of future patients, ask them to reflect on what they would want for themselves or their family members, if they were Dr. Albano's patients.

Educating the referring healthcare practitioner is part of the professional role and responsibilities of consultants and subspecialists. The situation in this case presents a potential teaching opportunity for the geneticist and a learning opportunity for the obstetrician. By approaching Dr. Albano now, miscommunication with future patients can be avoided.

> **Box 16.4   Teaching Tip**
> If learners readily agree, encourage them to think about how they might speak to a fellow physician when they disagree with the management of a mutual patient. How can such information be communicated most effectively?

3. **Is it appropriate to perform testing when a positive finding will implicate Paula as the source of the mutation?**
"Appropriate" has many potential meanings. The physician in this case does not have any legal or ethical obligation to Paula and *does* have such obligations to Cristina and Mario. It is both ethically and legally permissible to perform testing

on a consenting adult, regardless of others who may be implicated as carrying the same disease. However, it is still worthwhile to consider the views of the key stakeholders in this situation, and Cristina and Mario should be counseled to consider how the results would affect their family. Such counseling is not directive: it is not intended to deter them from testing but rather to help prepare them so that they can effectively manage the consequences of testing.

(a) **Is Mario obligated to consider Paula's wishes?**

Legally, he is not. No laws restrict individuals' access to their own genetic information based on family members' interests in *not* having such information. However, as a member of a (presumably) loving, functional family, Mario may feel that he has other obligations to his mother, which should certainly be explored. For example, his desire to protect and not harm her might inspire restraint from actions that will impose on his mother information that she prefers not to have. Unfortunately, it will be challenging for him to know his HD status without his mother somehow finding out. If the test is negative, it may not be difficult to protect her from unwanted information (even if she has the HD mutation, he has a 50% chance of having a negative result), but positive results would be much more difficult to hide, since the implied source is inescapable.

(b) **Does Paula have a right *not* to know?**

One often hears of people having a right to know something about themselves and their health, and in fact this is the foundation of the notion of informed consent as applied to any medical or surgical procedure (e.g., the right to know the risks of adverse effects of a proposed treatment). In the genetics community, the right *not* to know is considered an essential aspect of the right to know. Logically, if one does not have the right to *refuse* specific knowledge or information, then that information can be forced upon the individual by someone else. Patients may have numerous personal reasons for not wanting to have this burdensome knowledge. They may want to avoid significant psychological consequences that include depression, anxiety, and suicide. To force a patient to know or not to know information regarding health would violate patient autonomy, beneficence, non-maleficence, and justice, and it would be paternalistic [36, 37].

(c) **Does the principle of respect for autonomy help to resolve these questions?**

From a physician's perspective, all legal and ethical obligations are toward the individual patient, although there may be important practical reasons to consider family members' interests. Cristina, as an autonomous actor, can choose which prenatal testing options she wishes to pursue. While Mario's interests are certainly relevant, they are not a constraint on her autonomy to consent to prenatal testing. However, she cannot compel Mario to be tested. Mario is autonomous with regard to his own body. This is a case in which the autonomy of one individual stakeholder constrains the options available to other stakeholders: Cristina may prefer to limit to physical risks to herself, but that is only a choice if Mario concurs. Paula clearly wishes not to know her status, but she cannot prevent Cristina from consenting to prenatal testing. The principle of respect for autonomy helps to clarify the available options for each stakeholder but does *not* resolve the inherent conflict.

"It Runs in the Family"

(d) **What other principles or concepts factor into your analysis?**
The other core ethical principles—beneficence, non-maleficence, and justice—are relevant to the case, although not overriding. Beneficence and non-maleficence might guide the physician to encourage Mario and Cristina to think through the effects of testing on other family members, in order to maximize benefits and minimize harms to the family unit. Justice is discussed in more detail in Discussion Question 8.

> **Box 16.5 Teaching Tip**
> Depending on the interests and backgrounds of the learners, it may be appropriate to delve more deeply into justice at this point in the case, or educators may choose to ask learners to "hold their thoughts" temporarily until additional events and information are revealed.

The biology of HD mutations is also deeply pertinent to the analysis by the parties involved. Specifically, the CAG triplet repeat usually remains stable when transmitted by a carrier female but is prone to expansion (sometimes dramatically) when transmitted by a male. An example is a family in which a man had 42 repeats with onset of symptoms in his 30s, his daughter inherited it as exactly 42 repeats and similar age of onset, and she transmitted it to her son who had 42 with onset in 30s; however, he passed along an expansion to 119 repeats to his son whose symptoms appeared at age 5 and he died at 11. If Mario does transmit HD in this pregnancy, it is quite possible that the next affected individual will experience early onset of symptoms and death, rather than the more common adult-onset presentation [38].

4. **Given this new information (regarding family dynamics), how should you proceed?**
This new information changes neither the available options nor the physician's legal and ethical obligations. However, practically, it may be helpful to ask to speak to each partner alone. Although they are a couple, Cristina and Mario each have their own values, beliefs, interests, and goals to consider, and careful evaluation of these might be best undertaken separately. In many situations, patients and family members may be more open and comfortable and better able to ask the questions of greatest concern to them in a private space. It might also be appropriate to refer them to individual and couples counseling.

(a) **What are your obligations to Cristina? To Mario? To the fetus?**
In order to determine the physician's ethical and legal obligations, it is crucial to understand who can be considered the patient. In this situation, Cristina and Mario have both presented to the geneticist for consultation. To them, the physician is obligated to provide the best possible medical advice, based on current standards of care. The physician is responsible for describing the available options (which, as noted previously, are different for each of them), assessing their understanding of these options, supporting them in making their choice or choices, ordering any testing desired (in the case of prenatal testing, this will probably require referral to a maternal-fetal medicine specialist), interpreting the results of the testing, and counseling them regarding options based upon test results (including referral for termination if necessary).

It is also worth considering this question from Dr. Albano's perspective as an obstetrician. Regardless of her personal opinions about the moral status of the fetus and about Mario's right to participate in decision-making for his potential child, Cristina is the patient, and

Dr. Albano's obligations are to her alone. From both Dr. Albano's perspective and the geneticist's perspective, the fetus is only a patient at this point insofar as Cristina presents the fetus as a patient. She cannot be compelled to be tested or not be tested, nor to terminate or not terminate. The potential for conflicting goals and interests in obstetric medicine is explored further in ▶ Chapter 7.

(b) **Do you have any obligations to Paula or to other family members?**
Except in the sense of fully counseling Cristina and Mario regarding the implications of their choices for other family members, the physician does not have ethical, professional, or legal obligations to non-patients. The circumstances might be different if there were family members who did not know of their risk for the disease and if there were potential treatments. This is discussed further in the Response to Discussion Question 10a. Here (1) Paula already knows she may carry the disease, and (2) the disease cannot be treated.

(c) **To whom do you owe the duty of confidentiality?**
The physician owes confidentiality to both Cristina and Mario, since both are presenting as patients in this scenario. However, that confidentiality applies to each member of the couple individually: If Cristina asks to withhold information about her decision-making from Mario, the physician would be obligated to do so. This requires great sensitivity to navigate in practice. The physician is prohibited from informing Paula, or other relatives, of the results of any genetic testing that Cristina and/or Mario receive.

✓ 5. **Self-reflection: What do you think that you would choose for yourself, if you were in the position of Cristina or Mario?**

> **Box 16.6 Teaching Tip**
> This question allows learners to reflect on the challenges of predicting one's own future actions in hypothetical circumstances, since such prediction is at best dubious. "I would never…" and "I would always…" statements are intrinsically dangerous. Further, temptations to guide patients with one's own opinions on such matters may lead into places that are not psychologically or ethically appropriate. Encourage learners to consider these issues and to realize that recognizing one's personal reactions to issues is the first step in becoming a more compassionate healthcare practitioner.

✓ 6. **If Cristina and Mario had seen an obstetrician prior to conception, should they have been offered in vitro fertilization (IVF) and preimplantation genetic diagnosis (PGD)? Why or why not?**
IVF with PGD is extremely expensive and may or may not be covered by insurance. It is also not guaranteed to be successful; however, it may be their only option for having shared biological children without a risk of passing on HD, if Mario has the disease, while also avoiding abortion. Other options might include adoption or use of a sperm donor (or an egg donor, if Cristina were the carrier).

(a) **Would your answer be different if there was a family history of Tay-Sachs disease?**
Tay-Sachs disease (TSD) is a metabolic disorder that causes progressive neurologic damage and death in early childhood. The disease has several characteristics that make it particularly devastating for children and families: Symptoms develop early, in the first year of life; infants who have appeared to develop normally for the first 6–9 months gradually regress; there is no treatment to alter the course of the disease; and death typically occurs by age 6, even with supportive care such as feeding tubes and tracheostomy placement. In at least one sample, more than 95% of patients with a prenatal diagnosis of Tay-Sachs disease terminated the affected pregnancy. It is considered by many in medicine

and public health to be a model for screening and prevention of devastating genetic diseases [20].

> **Box 16.7 Patient Perspective**
> Author Emily Rapp has written about her son's life with and death from Tay-Sachs disease:
> "If I had known Ronan had Tay-Sachs, I would have found out what the disease meant for my then unborn child; I would have talked to parents who are raising (and burying) children with this disease, and then I would have had an abortion. Without question and without regret, although this would have been a different kind of loss to mourn and would by no means have been a cavalier or uncomplicated, heartless decision. I'm so grateful that Ronan is my child. I also wish he'd never been born; no person should suffer in this way—daily seizures, blindness, lack of movement, inability to swallow, a devastated brain—with no hope for a cure. Both of these statements are categorically true; neither one is mutually exclusive" [39].

(b) **What about sickle cell disease?**
Sickle cell disease (SCD) is a disorder of hemoglobin production in which the abnormal hemoglobin results in sickle-shaped red blood cells that cluster together, causing severe pain crises as well as life-threatening respiratory problems (acute chest syndrome), stroke, avascular necrosis of bone, and increased susceptibility to sepsis and other infections. Like many inherited diseases, SCD is tremendously variable in its phenotype, and it is not possible to predict the severity prenatally (except insofar as certain combinations of variant hemoglobins, such as SC, are typically less severe than classical SS disease). For some children, SCD is a manageable chronic illness with little impact on quality of life; for others, it means frequent debilitating pain crises, hospitalizations, and other complications, including strokes and lung disease. Life expectancy is shorter for patients with SCD, but most live well into adulthood. Medication (hydroxyurea) and blood transfusion (simple or exchange) can reduce the severity of symptoms. Bone marrow transplant offers a possibility of cure but carries considerable morbidity and mortality itself [40].

Thus, SCD differs from HD and TSD in several key ways: it can be life-limiting but is not definitively fatal; it does not cause cognitive impairment, except in some patients with stroke; and there are variably effective treatments, including a potential cure for some patients.

> **Box 16.8 Teaching Tip**
> The purpose of this question is not to educate learners about the pathology but to encourage them to consider how we evaluate "quality of life." Who, and on what basis, decides if a life is worth living? What kind of information is needed to make that assessment? Is it even possible?

(c) **Does the severity of the disease change your opinion about the morality of assisted reproduction or abortion? Should it?**
This is a complicated question, requiring, first, ethical and moral evaluation of assisted reproduction and abortion and, second, consideration of whether that position changes based on a prenatal diagnosis. However, it is important to remember that while healthcare practitioners are free to have a personal objection to certain legal medical procedures, they are legally and ethically obligated to provide nonjudgmental and unbiased counseling to patients and to refer them to a qualified provider who is willing to provide the service in question [41]. The first step in being able to meet this ethical standard is being aware of one's own beliefs and assumptions. In this case, what matters are Cristina's (primarily) and Mario's beliefs and values. Conscientious objection is also discussed in ▶ Chapter 23.

> **Box 16.9 Teaching Tip**
> This question may be an opportunity to build bridges between learners who disagree about the morality of abortion. For some who are supportive of access to abortion, the idea of termination of pregnancy due to specific disease or a disability violates their personal beliefs about who should decide what kind of life is worth living. For some others who are generally opposed to abortion for other reasons, termination for a lethal or severely disabling condition might yet be morally acceptable.

At one point during your visits, Mario mentions that he overheard one of the staff members in the obstetrician's office make a derogatory comment, suggesting that people who have genetic diseases should adopt.

7. **How would you respond to this remark if it came from one of your colleagues?**
   Healthcare practitioners, perhaps especially physicians, are role models in the healthcare setting. They have an obligation to speak up when colleagues disparage patients. Ignoring and thus effectively tolerating such remarks can erode trust between patients and providers and among members of the healthcare team, negatively affecting the work environment and institutional culture. Furthermore, allowing some patients and families to be labeled as less deserving of care (or difficult or demanding) may, consciously or unconsciously, lead to bias in the care they receive, potentially resulting in errors or other undesirable outcomes.

8. **Should insurance cover IVF and PGD?**
   (a) **If not, should these services be available only to those who can afford them?**
   There are several parts to this question.
   - First, should insurance cover IVF in general? In some states (e.g., New York), state law requires insurers to cover treatment for infertility but specifically exempts IVF from coverage. Private insurance may or may not cover it but is more likely to do so when infertility is involved. Medicaid does not cover it [42–45].
   - Second, should insurance cover IVF with PGD for the purpose of selecting embryos free of genetic diseases? If so, all genetic diseases or specific ones?
   - Third, if IVF in general is covered, should it be covered or permitted for other purposes aside from the prevention of disease, such as sex selection?

   One approach is to consider individual rights. Is access to health care a right? Does that include access to *preventive* health care? IVF with PGD might be considered "preventive" care, but such a conclusion is in part predicated on beliefs that people are entitled to have biological children "no matter what."
   Another approach is to consider costs and societal resources. If provision of IVF and PGD to all who want these services leads to a healthier population, then it would be cost-effective to do so. However, other interests may operate at the societal level that also affect costs, such as the need for foster and adoptive parents or concerns regarding the impact of overpopulation upon available resources and climate change. If increased coverage for reproductive technology encourages childbearing among people who would otherwise have chosen to build families through other means, such an outcome may not be desirable at the societal level [46, 47].

   (b) **What role does the principle of justice play?**
   The high cost of these technologies can translate into these procedures only being accessible to those who can afford them. There is some concern that this will lead to further stratification of society, with wealthier families also being able to produce genetically privileged offspring. Furthermore, some people fear that making such services more widely available will lead to patients using them to select for certain "desirable" characteristics in their offspring (such as sex, eye color, or athletic ability,

based on current knowledge but perhaps extending to intelligence or personality in the future).

> **Box 16.10 Teaching Tip**
> It can be difficult to facilitate discussions of justice vis-à-vis healthcare costs within the context of the U.S. healthcare system. Encourage learners to consider this question in the broader context of the distribution of healthcare services in the USA. The Affordable Care Act was designed to ensure coverage of basic healthcare needs through essential services (e.g., pregnancy, maternal and child care, hospitalization, mental health), but not all healthcare services are covered, such as dental or eye care for adults. For example, consider whether it is "just" to cover IFV/PGD for a few when there are many things that are not covered for a much larger percentage of the population [42].

(c) **Should insurance premiums be higher for people who are identified as at risk for a genetic disease? Why or why not?**
"Actuarial fairness" would dictate that the patient is likely to cost more and thus should have higher premiums. However, this would violate the principle of justice, since the patient has no control over her or his genetic predisposition to a disease. Further, such policies might deter patients from seeking predictive testing for manageable diseases (due to fears of premiums increasing), even when such testing might in fact improve their health outcomes and reduce costs.

After several genetic counseling sessions with Mario and Paula, Mario decides that he does not want to get tested either. Although this is very upsetting to Cristina, she eventually agrees to support Mario and Paula in their decisions, and she decides against pursuing prenatal testing for HD.

Seven years later, Cristina returns to see you. She tells you that 3 years ago, Paula was diagnosed with HD. She also shares that she and Mario were recently divorced. She is here today because she wants to discuss testing her now-6-year-old daughter, Amelia, for the disease.

9. **Should you test Amelia for HD? Why or why not?**

The physician in this case should *not* test the daughter unless she is symptomatic. (As noted earlier, HD typically presents in adulthood, but it is possible this child has acquired a more severe form with earlier onset.) As noted previously, in the case of a disease for which there is no childhood intervention, the decision to undergo predictive (presymptomatic) genetic testing rightfully belongs to the child in question, when she reaches the age of majority [48]. Although ethical principles support this approach, it is more often described by the framework of a "right to an open future"— that is, parents and other caregivers should avoid making decisions that constrain the possible futures available to the child, except when such decisions provide important immediate benefits or prevent serious harms [49–51]. This framework has been critiqued as unrealistic—parents regularly make choices that have both positive and negative consequences for the children's futures—but are well suited for consideration of genetic testing in cases like this one.

(a) **Would your decision be different if Cristina wanted to test Amelia for a BRCA1 gene mutation? Why or why not?**
Adequate clinical knowledge of diseases is often critical to ethical decision-making. There are no interventions to prevent or ameliorate symptoms of Huntington disease in children, nor does the disease manifest in childhood or adolescence. BRCA mutations are primarily associated with adult-onset cancers, and there are not currently any recommendations for screening for cancer in children with a BRCA mutation [52]. Therefore, *not* testing for these genes would avoid harm to Cristina's daughter now as she would have neither any discomfort associated with the test itself nor the potential psychological consequences of living with the threat of oncoming disease. Furthermore, there is no benefit lost if testing is postponed until she is an adult and can consent for herself.

(b) **Are parents entitled to know medical information about their children that will not impact their children's health until adulthood?** Many ethicists think not, arguing that future adults are entitled to confidentiality in this regard, unless (as noted previously) their parents need to be able to consent to screening tests or other interventions while they are still children [48, 53]. However, several studies have reported that parents themselves often feel quite differently and believe that they should retain decisional authority. Other research suggests that this may be influenced by parents placing greater weight on the likelihood of (1) identifying diseases that can be treated or prevented and (2) receiving results that reassure them that their children are at low risk for diseases [54].

Cristina also tells you that Mario learned that his mother had placed a child for adoption before he was born. Mario's older biological brother recently contacted the family, and they have exchanged a few e-mails. She is unsure if the family has disclosed its history to the brother. She asks you to contact the brother because she feels he has a right to know.

10. **Do you have any ethical or legal obligations to this sibling?**
    The physician in this case does not have a legal obligation to this sibling but may, in some cases, have an ethical obligation to encourage patients to notify affected relatives. In this case, the physician is seeing Cristina and her daughter, and has not had any contact with Mario for several years, so it is difficult to perceive an ethical obligation.

11. **What if, instead of Huntington disease, the family had a history of something like familial adenomatous polyposis, in which early intervention can prevent cancer?**
    In this case, the physician's ethical obligation is stronger because early knowledge of the genetic predisposition can meaningfully change medical outcomes. (Arguably, early knowledge of Huntington disease can change psychological outcomes, allowing some with the disease to accomplish life goals earlier, knowing that early death is likely.) Confidentiality, however, still poses a problem. Does the physician have a "duty to warn" potentially affected relatives and therefore a *right* to violate confidentiality? Initially, the U.S. courts held that this was not the case, because it was not the patient's actions that put the relative "at risk." However, later decisions indicated that physicians might, in some cases, have such a duty [55, 56].

> **Box 16.11   Teaching Tip**
> Consider assigning students to argue for or against a law *requiring* physicians to disclose genetic information to potentially affected relatives. What are the implications of such a law?

References

1. WHO Genomics Resource Center. Genes and human disease [Internet]. Geneva: World Health Organization. [cited 2017 Dec 4]. Available from: http://www.who.int/genomics/public/geneticdiseases/en/.
2. Myers RH. Huntington's disease genetics. NeuroRx. 2004;1(2):255–62.
3. Mayo Foundation for Medical Education and Research (MFMER). Genetic testing-why it's done [Internet]. Rochester (MN): Mayo Clinic. [cited 2017 Dec 4]. Available from: https://www.mayoclinic.org/tests-procedures/genetic-testing/details/why-its-done/icc-20325416.
4. McPherson E. Genetic diagnosis and testing in clinical practice. Clin Med Res. 2006;4(2):123–9.
5. Fenwick A, Plantinga M, Dheensa S, Lucassen A. Predictive genetic testing of children for adult-onset conditions: negotiating requests with parents. J Clin Genet. 2017;26:244–50.
6. Keenan KF, McKee L, Miedzybrodzka Z. Help or hindrance: young people's experiences of predictive testing for Huntington's disease. Clin Genet. 2014; https://doi.org/10.1111/cge.12439.
7. Quaid KA, Eberly SW, Kayson Rubin E, Oakes D, Shoulson I. Factors related to genetic testing in adults at risk for Huntington disease: the prospective Huntington at risk observational study (PHAROS). Clin Genet. 2017;91(6):824–31.
8. Fulda KG, Lykens K. Ethical issues in predictive genetic testing: a public health perspective. J Med Ethics. 2006;32(3):143–7.
9. Adam S, Wiggins S, Whyte P, Bloch M, Shokeir MH, Soltan H, et al. Five year study of prenatal testing for

Huntington's disease: demand, attitudes, and psychological assessment. J Med Genet. 1993;30:549–56.
10. Losekoot M, van Belzen MJ, Seneca S, Bauer P, Stenhouse SA, Barton DE, European Molecular Genetic Quality Network (EMQN). EMQN/CMGS best practice guidelines for the molecular genetic testing of Huntington disease. Eur J Hum Genet. 2013;21:480–6.
11. Fahy M, Robbins C, Bloch M, Turnell RW, Hayden MR. Different options for prenatal testing for Huntington's disease using DNA probes. J Med Genet. 1989;26:353–7.
12. Sermon K, De Rijcke M, Lissens W, De Vos A, Platteau P, Bonduelle M, et al. Preimplantation genetic diagnosis for Huntington's disease with exclusion testing. Eur J Human Genet. 2002;10(10):591–8.
13. Schattman GL. Preimplantation genetic diagnosis. Waltham: UpToDate; 2016.
14. ACOG Committee on Practice Bulletins. ACOG Practice Bulletin No. 77: screening for fetal chromosomal abnormalities. Obstet Gynecol. 2007;109(1):217–27.
15. Northwestern University, General Obstetrics and Gynecology. Genetic screening and testing during pregnancy [Internet]. Chicago (IL): Northwestern Medicine® and Northwestern Memorial HealthCare. Available from: http://nfwh.nm.org/genetic-screening-and-testing-during-pregnancy.html.
16. American College of Obstetrics and Gynecology. Prenatal genetic screening tests. Available from: https://www.acog.org/Patients/FAQs/Prenatal-Genetic-Screening-Tests.
17. American College of Obstetrics and Gynecology. Standard prenatal care. Last accessed: 3 Nov 2017. Available from: https://www.acog.org/~/media/For%20Patients/faq133.pdf.
18. Ekstein J, Katzenstein H. The Dor Yeshorim story: community-based carrier screening for Tay–Sachs disease. Adv Genet. 2001;44:297–310.
19. Kaback M, Lim-Steele J, Dabholkar D, Brown D, Levy N, Zeiger K. Tay-Sachs disease--carrier screening, prenatal diagnosis, and the molecular era. an international perspective, 1970 to 1993. The International TSD Data Collection Network. JAMA. 1993;270:2307–15.
20. Kaback MM. Screening and prevention in Tay–Sachs disease: origins, update, and impact. Adv Genet. 2001;44:253–65.
21. Lew RM, Proos AL, Burnett L, Delatycki M, Bankier A, Fietz MJ. Tay Sachs disease in Australia: reduced disease incidence despite stable carrier frequency in Australian Jews. Med J Aust. 2012;197(11):652–4.
22. Nazareth SB, Lazarin GA, Goldberg JD. Changing trends in carrier screening for genetic disease in the United States. Prenat Diagn. 2015;35(10):931–5.
23. Schenker JG. Assisted reproduction practice: religious perspectives. Reprod Biomed Online. 2005;10(3):310–9.
24. Schover LR, Thomas AJ, Falcone T, Attaran M, Goldberg J. Attitudes about genetic risk of couples undergoing in-vitro fertilization. Hum Reprod. 1998;13(4):862–6.
25. Oskarsson T, Dimitry ES, Mills MS, Hunt J, Winston RM. Attitudes towards gamete donation among couples undergoing in vitro fertilization. BJOG. 1991;98(4):351–6.
26. Lucassen A, Kaye J. Genetic testing without consent: the implications of the new human tissue act 2004. J Med Ethics. 2006;32(12):690–2.
27. Gong P, Fanos JH, Korty L, Siskind CE, Hanson-Kahn AK. Impact of Huntington disease gene-positive status on pre-symptomatic young adults and recommendations for genetic counselors. J Genet Couns. 2016;25(6):1188–97.
28. Hudson KL, Holohan MK, Collins FS. Keeping pace with the times—the genetic information nondiscrimination act of 2008. New Engl J Med. 2008;358(25):2661–3.
29. Slaughter L. Genetic information non-discrimination act. Harv J Legislation. 2013;50:41.
30. Mayo Clinic. Amniocentesis: risks. https://www.mayoclinic.org/tests-procedures/amniocentesis/basics/risks/prc-20014529.
31. Mayo Clinic. Chorionic villus sampling: risks. https://www.mayoclinic.org/tests-procedures/chorionic-villus-sampling/basics/risks/prc-20013566.
32. Bartlett LA, Berg CJ, Shulman HB, Zane SB, Green CA, Whitehead S, et al. Risk factors for legal induced abortion-related mortality in the United States. Obstet Gynecol. 2004;103(4):729–37.
33. Federation of State Medical Boards. Assessing scope of practice in health care delivery: critical questions in assuring public access and safety. 2005. http://www.fsmb.org/Media/Default/PDF/FSMB/Advocacy/2005_grpol_scope_of_practice.pdf. Accessed 12/15/17.
34. Sherwood GD, Brown M, Fay V, Wardell D. Defining nurse practitioner scope of practice: expanding primary care services. Internet J Adv Nurs Pract. 1996;1(2)
35. Gunnar WP. The scope of a physician's medical practice: is the public adequately protected by state medical licensure, peer review, and the National Practitioner Data Bank? Ann Health Law. 2005;14:329.
36. Shaw MW. Testing for the Huntington gene: a right to know, a right not to know, or a duty to know. Am J Med Genet. 1987;26:243–6.
37. Andorno R. The right not to know: an autonomy based approach. J Med Ethics. 2004;30:435–9.
38. Andrew SE, Goldberg YP, Kremer B, Telenius H, Theilmann J, Adam S, Starr E, Squitieri F, Lin B, Kalchman MA, Graham RK. The relationship between trinucleotide (CAG) repeat length and clinical features of Huntington's disease. Nat Genet. 1993;4(4):398–403.
39. Rapp E. Rick Santorum, meet my son. Slate. Feb 27 2012. http://www.slate.com/articles/double_x/doublex/2012/02/rick_santorum_and_prenatal_testing_i_would_have_saved_my_son_from_his_suffering_.html.
40. Yawn BP, Buchanan GR, Afenyi-Annan AN, Ballas SK, Hassell KL, James AH, Jordan L, Lanzkron SM, Lottenberg R, Savage WJ, Tanabe PJ. Management of sickle cell disease: summary of the 2014 evidence-based report by expert panel members. JAMA. 2014;312(10):1033–48.
41. Card RF. Conscientious objection and emergency contraception. Am J Bioeth. 2007;7(6):8–14.
42. Lebel RR, Guzauskas GF. The new eugenics: testing the limits of prenatal diagnosis. Proc Greenwood Genet Center. 2005;24:50–7.
43. Devine K, Stillman RJ, DeCherney A. The affordable care act: early implications for fertility medicine. Fertil Steril. 2014;101(5):1224.

44. Adashi EY, Dean LA. Access to and use of infertility services in the United States: framing the challenges. Fertil Steril. 2016;105(5):1113–8.
45. Ho JR, Aghajanova L, Mok-Lin E, Hoffman JR, Smith JF, Herndon CN. Public attitudes in the US toward insurance coverage for IVF and the provision of infertility services to lower income patients. Fertil Steril. 2017;108(3):e9.
46. Rieder TN. Procreation, adoption and the contours of obligation. J Appl Philos. 2015;32(3):293–309.
47. Rieder TN. Toward a small family ethic: how overpopulation and climate change are affecting the morality of procreation. Switzerland: Springer International Publishing; 2016.
48. Ross LF, Saal HM, Anderson RR, David KL, American Academy of Pediatrics Committee on Bioethics. Ethical and policy issues in genetic testing and screening of children. Pediatrics. 2013;131(3):620–2.
49. Feinberg J. The child's right to an open future. Princeton: Princeton University Press; 1992.
50. Davis DS. Genetic dilemmas and the child's right to an open future. Hast Cent Rep. 1997;27(2):7–15.
51. Millum J. The foundation of the child's right to an open future. J Soc Philos. 2014;45(4):522–38.
52. Bradbury AR, Dignam JJ, Ibe CN, Auh SL, Hlubocky FJ, Cummings SA, et al. How often do BRCA mutation carriers tell their young children of the family's risk for cancer? A study of parental disclosure of BRCA mutations to minors and young adults. J Clin Oncol. 2007;25(24):3705–11.
53. Bredenoord AL, de Vries MC, van Delden H. The right to an open future concerning genetic information. Am J Bioeth. 2014;14(3):21–3.
54. Tercyak KP, Hensley Alford S, Emmons KM, Lipkus IM, Wilfond BS, McBride CM. Parents' attitudes toward pediatric genetic testing for common disease risk. Pediatrics. 2011;127(5):e1288–95.
55. Knoppers BM, Strom C, Clayton EW, Murray T, Fibison W, Luther L. Professional disclosure of familial genetic information. Am J Hum Genet. 1998;62:474–83.
56. Weaver M. The double helix: applying an ethic of care to the duty to warn genetic relatives of genetic information. Bioethics. 2016;30(3):181–7.

Further Reading on This Topic

Nance MA. Genetic counseling and testing for Huntington's disease: a historical review. Am J Med Genet. 2016;174B:75–92.

Rapp E. The still point of the turning world. New York: Penguin; 2014.

UNESCO. Universal declaration on the human genome and human rights, Paris 1997. http://portal.unesco.org/en/ev.php-URL_ID=13177&URL_DO=DO_TOPIC&URL_SECTION=201.html.

Wexler A. Mapping fate: a memoir of family, risk and genetic research. Berkeley: University of California Press; 1996.

Willis M. Not this pig: dignity, imagination, and informed consent. In: Van Cleve JV, editor. Genetics, disability, and deafness. Washington, DC: Gallaudet University Press; 2004. p. 174–85.

Medical Mistakes and Patient Safety

Contents

Chapter 17　"I Know Something Is Wrong" – 335

Chapter 18　"I'm in Pain!" – 349

"I Know Something Is Wrong"

Kendra Harris

17.1 Background Questions – 336

17.2 Additional Case Information and Questions for Discussion – 337

17.3 Answers to Background Questions – 338

17.4 Responses to Discussion Questions – 340

References – 346

© Springer Nature Switzerland AG 2019
A. E. Caruso Brown et al. (eds.), *Bioethics, Public Health, and the Social Sciences for the Medical Professions*, https://doi.org/10.1007/978-3-030-03544-0_17

◼ **Glossary**

Systems approach - Encourages a focus on the organization (i.e., inputs, processes) rather than the individual when seeking an understanding of categories of outcomes (i.e., adverse events). In a healthcare context, this approach suggests that because most errors reflect predictable human failings in the context of a poorly designed system, focusing on changing the underlying systems is the best way to reduce harm.

Swiss cheese model - A model of error causation, attributed to James Reason, that suggests that medical errors cause adverse patient outcomes in complex healthcare systems when flaws in the safeguards at multiple care levels align in a way that allows an error to reach a patient.

Root-cause analysis - A multidisciplinary process of study or analysis that uses a detailed, structured process to examine factors contributing to a specific outcome, such as an adverse event.

Just culture - A framework for balanced accountability for frontline operators and the organization in which they work that emphasizes learning over blaming and emphasizes system-level changes to reduce bad outcomes.

Full disclosure - In a medical context, this is often meant to include an admission of a mistake, discussion of the error, and a link from the error to harm.

Second victim phenomenon - Refers to the psychological harm, trauma, and distress that healthcare providers may experience as a result of their involvement in an unanticipated adverse patient event, medical error, and/or patient-related injury.

Learning Objectives

1. Identify factors contributing to poor outcomes and be able to distinguish between known complications and negligence and between human error and systems error.
2. Discuss the ethical, legal, and personal impacts medical mistakes can have on patients and providers.
3. Recognize the role and value of disclosing mistakes, and begin to be prepared to disclose your own mistakes in the future.
4. Describe common perceptions and misperceptions about medical malpractice.

Case Background

You are an intern, and you have just received sign-out from your co-intern on the general medicine service. Soon after your co-intern departs, you are urgently called to the room of William Begay, a 49-year-old man with newly diagnosed leukemia. His heart rate and respirations are elevated, while his blood pressure is extremely low. His skin is cool and clammy and pulses are weak. He is minimally responsive. You attempt to resuscitate him with fluids and call a code. He is soon intubated, placed on a ventilator, and started on norepinephrine for blood pressure support, but he decompensates in the intensive care unit (ICU) a few hours later and dies despite the team's best efforts. When you offer condolences to his wife, she is devastated but livid and furiously explains that she had been telling the nurse and intern all day that something was not right with her husband.

17.1 Background Questions

1. What is a medical error? What is the difference between an adverse event and a near-miss? What is a sentinel event?

2. How common are medical errors in the United States? How many people die each year due to medical errors?

3. What is the difference between medical error and medical malpractice? What is negligence? How are known complications different from adverse events attributable to error or negligence?

4. What is the process for identifying when medical errors occur in a hospital? Is there a way to compare one hospital to another in terms of the number and rate of medical errors?

5. Are physicians legally required to disclose errors in your state?

6. Are patients and/or families compensated for medical errors? If so, how?

7. Does Medicare reimburse hospitals for services provided to treat complications of medical errors or preventable compli-

"I Know Something Is Wrong"

cations? Find an example of how this is handled in another country's healthcare system.

8. Describe the evidence that patient- and family-centered care makes hospitals safer.

17.2 Additional Case Information and Questions for Discussion

When you review the chart, you notice that Mr. Begay was tachycardic and hypotensive throughout the day and repeatedly reported abdominal pain. The nursing notes state that your co-intern was called to assess but said he was not concerned because the patient did not have a fever. The intern recalled that constipation had been an issue for the patient earlier in the week and attributed the abdominal pain to that. He prescribed a laxative and did not re-examine the patient.

1. What do you think happened in this case?
   (a) Was Mr. Begay's death preventable? How so?
   (b) What could or should have been done differently? By your co-intern? You? Your attending physician? The nurse? The hospital?
   (c) Why didn't the intern and nurse listen to the patient's wife?

In the morning, when you tell your chief resident about the overnight events, he looks stricken and asks if you can wait for a few minutes while he sees the program director "about something." Two hours later, your chief resident comes to relieve you and says that your co-intern has taken a leave of absence.

2. Self-reflection: How do you think you would feel if you made a mistake of this magnitude? How would you feel about a colleague who did?

3. What should be said to Mr. Begay's wife and other family? Who should say it?
   (a) Are physicians or other healthcare practitioners ethically required to tell families when mistakes have been made? Why or why not?
   (b) Does it matter how serious the consequences for the patient were? That is, should errors that do not cause harm and near-miss events be disclosed?
   (c) Creative problem-solving: Should healthcare practitioners be required to do so? By whom? Consider the ethical and practical implications of your response.

Two years later, you are a senior resident at the same hospital, and you hear that your former co-intern has returned to complete his training. You offer but he declines to discuss the events surrounding Mr. Begay's death. He says only that it was "the worst day of [his] life."

4. Self-reflection: What worries you about saying, "I'm sorry" as a doctor?
   (a) What would you need to feel comfortable saying it?

You eventually begin an oncology fellowship and are working on a patient safety effort at the same hospital. One of your colleagues reintroduces you to Elena Begay, Mr. Begay's wife, who now sits on a family advisory committee for the hospital and advocates for patient safety. You learn that after repeated requests for her husband's medical records, the hospital eventually admitted responsibility for the patient's death and apologized to Ms. Begay.

5. Do you think that the legal system negatively impacts physicians' and other healthcare practitioners' willingness to discuss errors openly?
   (a) What would you recommend to address this? From a healthcare practitioner perspective? From a patient perspective?

6. Are there other possible reasons that the hospital took so long to tell Ms. Begay the truth about her husband's death?

The hospital has taken many steps to prevent this type of error in the future, including using "early warning scores" to identify patients whose vital signs are worsening, sometimes very subtly, and allowing families to call for a "rapid response" (assessment from a dedicated team, usually including a nurse or physician from the critical care unit) themselves. The residency program now includes training in cognitive errors and biases and how they impact clinical decision-making. Trainees learn about the importance of reconsidering assumptions when assessing and reassessing patients.

7. **What do you think about these changes? Are there others you would suggest?**
   (a) How does the hierarchical structure in medicine impact the occurrence of errors?

Ms. Begay generally feels positively about the direction the hospital has taken, although she tells you that the former intern was never willing to meet with her. He has completed his residency and is now working in a private practice.

8. **Should the intern have been required to meet with her before resuming training or otherwise directly address what happened? Why or why not?**

17.3 Answers to Background Questions

1. **What is a medical error? What is the difference between an adverse event and a near-miss? What is a sentinel event?**
   A **medical error** is a preventable adverse effect of care, which can include acts of omission or commission, contributing to an unintended result, whether or not it is evident or harmful to the patient [1]. Broadly defined, medical errors encompass situations in which healthcare practitioners choose inappropriate or incomplete diagnostic tests or therapeutic interventions, fail to perform appropriate tests and interventions, or improperly execute such interventions (including administering care to the wrong patient). Medical errors came into sharp focus as a serious public health problem after the landmark *To Err is Human* report from the Institute of Medicine (IOM), which estimated medical error deaths on a scale equivalent to a fiery jumbo jet crash occurring every day [2]. Since its publication in 2000, numerous systems and taxonomies have been developed to better classify medical errors [3, 4]. Many of these taxonomies distinguish between those medical errors that result in patient harm and those that do not result in patient harm.

   An **adverse event** is a medical error that results in patient harm. Adverse events include all unintended injuries or outcomes resulting in death, disability, or prolonged hospital stays and arising from medical care, including those that arise due to the absence of indicated treatment.

   A **near-miss**, on the other hand, is an error that has the potential to cause an adverse event (such as patient harm) but fails to do so because of chance or because it is intercepted before affecting the patient. According to the IOM, a near-miss is "an act of commission or omission that could have harmed the patient but did not cause harm as a result of chance, prevention, or mitigation" [5].

   A **sentinel event** is a medical error resulting in a very serious adverse event and is defined by The Joint Commission as any unanticipated event in a healthcare setting resulting in death or serious physical or psychological injury to a patient or patients and not related to the natural course of the patient's illness [6].

2. **How common are medical errors in the United States? How many people die each year due to medical errors?**
   The numbers are difficult to obtain since medical records can be inaccurate and providers may not want to disclose medical errors in charts. A study in 2013 suggested the numbers range from 210,000 to 440,000 deaths per year. The latter number would make it the third leading cause of death after heart disease and cancer [7].

3. **What is the difference between medical error and medical malpractice? What is negligence? How are known complications different from adverse events attributable to error or negligence?**

As defined above, medical error is a preventable, unintended outcome of care, whether or not it is evident or harmful to the patient. This might include an inaccurate or incomplete diagnosis or treatment of a disease, injury, syndrome, behavior, infection, or other ailment. Medical malpractice is a legal course of action that occurs when a medical or healthcare professional deviates from standards in his or her profession, thereby causing injury to a patient. Medical malpractice deals with professional negligence.

Negligence is the failure to provide a standard level of care or, in other words, the delivery of substandard care [8]. By way of contrast, a known complication is a recognized, potential harm that sometimes results from specific medical care and/or procedures, despite reasonable provider conduct and precautions. A known complication, for example, might involve an allergic reaction after receiving an antibiotic for the first time. In such cases, patients should be made aware that they may develop an allergy to a new antibiotic, but since they have not received the antibiotic before, it is not known whether they will, in fact, have an allergy to that specific antibiotic. Another example of a known complication is an adverse reaction to a vaccine. Such reactions occur at very low rates and are generally not predictable in advance of administration [9].

It is important to differentiate between known complications resulting in poor outcomes and poor outcomes resulting from negligence, because punishing poor outcomes per se could have the unintended effect of discouraging the treatment of complex conditions or the performance of difficult procedures, such as liver transplants or neurosurgery [5–7]. It might also dis-incentivize providers to care for high-risk patients with multiple comorbidities.

4. **What is the process for identifying when medical errors occur in a hospital? Is there a way to compare one hospital to another in terms of the number and rate of medical errors?**

There are a number of ways to compare hospitals to one another in terms of the number and rates of medical errors. For example, the U.S. Centers for Medicare and Medicaid Services (CMS) runs Hospital Compare, a consumer-oriented service that reports a number of performance measures, including complication rates, rates of readmissions and death, and measures of timely and effective care, stemming from a number of mandatory reporting measures [13]. The Leapfrog Group is a national nonprofit organization that assigns letter grades to hospitals based on their record of patient safety and incorporates consideration of the number and rate of medical errors into its assessments [14].

5. **Are physicians legally required to disclose errors in your state?**

Although no federal laws exist mandating disclosure of medical error, since 2002, 10 states have passed laws or implemented policies that mandate certain types of errors be disclosed: Tennessee, Pennsylvania, Maryland, Connecticut, Nevada, New Jersey, Oregon, Florida, South Carolina, and Washington. More than two-thirds of states have adopted laws that preclude some or all information contained in a physician's apology from being used in a malpractice lawsuit. These protections, dubbed "safe harbor" legislation, appear to alter malpractice liability outcomes in only about 1% of claims [15].

6. **Are patients and/or families compensated for medical errors? If so, how?**

There is no specific system in place for compensating patients and/or families for medical errors in the United States. However, some hospital systems are pioneering programs for taking responsibility for errors and offering appropriate compensation when necessary. For example, in 2001, the University of Michigan Health System risk management team implemented a model for handling medical errors and improving patient safety that was built on a premise of openness with

patients and families [16]. This model includes:
- Early reporting and analysis of adverse events
- Full explanations for patients and families
- Emotional support for healthcare professionals involved in the mishap
- Apologies and compensation to patients when the hospital is at fault

☑ 7. **Does Medicare reimburse hospitals for services provided to treat complications of medical errors or preventable complications? How is this handled in other countries' healthcare systems?**
In 2008, Medicare stopped reimbursing U.S. hospitals for the treatment of eight preventable medical errors [17]. For example, hospitals were no longer able to collect payment related to additional surgeries required for the retrieval of retained objects, such as sponges or tools left in patients after surgery; to collect for services required to treat bedsores developed while in the hospital; or to bill for interventions related to injuries from hospital falls and hospital-acquired infections. To give a sense of financial impact of this change, the Society of Actuaries estimated that in the United States, medical errors cost $19.5 billion, with a rate of roughly $93,000 per error [18].

Other countries' healthcare systems handle the treatment of complications related to medical errors differently, largely because most other high-income countries have some form of single-payer healthcare system (including Canada, the United Kingdom, France, Italy, Germany, Australia, Denmark, and Sweden, among others). Although these systems do not generally deny payment related to fixing medical mistakes, it is notable that many of these systems do have comprehensive patient compensation programs. Sweden, for example, handles malpractice claims administratively. Patients who are determined, by a panel of experts, to have either experienced a negative outcome due to care that could have been better or to have experienced a rare severe medical event are entitled to and provided with appropriate compensation [19].

☑ 8. **Describe the evidence that patient- and family-centered care makes hospitals safer.**
As discussed in ► Chapter 1: Introduction, patient- and family-centered care is "an approach to the planning, delivery, and evaluation of health care that is grounded in mutually beneficial partnerships among health care providers, patients, and families." In patient- and family-centered care, patients and families are involved in every aspect of their health care. Many hospitals have addressed this by encouraging or requiring bedside rounds. That is, the patient discussion happens at the bedside, and patients and/or families participate in rounds and have a voice in the plan of care. Evidence increasingly shows that this is good for patient safety and quality of care [20].

Examples of evidence that patient- and family-centered care makes hospitals safer come from a variety of sources. Assessment of the impact of such partnering activities, including patient education and discharge planning, has been shown to reduce the likelihood of recurrent chest pain among patients admitted for myocardial infarction [21]. Patient-centered communication in the outpatient setting has been shown to improve patients' perceptions of the quality of their care, while also reducing referrals and diagnostic testing [20].

17.4 Responses to Discussion Questions

When you review the chart, you notice that Mr. Begay was tachycardic and hypotensive throughout the day and repeatedly reported abdominal pain. The nursing notes state that your co-intern was called to assess but said he was not concerned because the patient did not have a fever. The intern recalled that constipation had been an issue for the patient earlier in the week and attributed the abdominal pain to that. He prescribed a laxative and did not re-examine the patient.

"I Know Something Is Wrong"

1. **What do you think happened in this case?**
   (a) **Was Mr. Begay's death preventable? How so?**
   Yes, this was likely to have been a preventable death. The patient had signs of evolving sepsis (tachycardia, hypotension) during the day that were apparently unrecognized and unaddressed. The intern failed to examine the patient when called by the nurse with concerns and made an assumption about the diagnosis (constipation), based on earlier information.

   Premature closure is a failure to consider alternative diagnoses once an initial impression is formed [22, 23]. Holding on too tightly to this initial impression when making subsequent decisions is called anchoring, a type of cognitive bias that can lead to adverse outcomes in patient care and is highly correlated with diagnostic errors [24]. This physician made both types of common errors, leading him to fail to recognize the signs of sepsis in Mr. Begay. It should be noted that there is exceedingly strong evidence that early recognition and intervention for patients with sepsis dramatically improve outcomes; this data suggests that, had the physician intervened, Mr. Begay's life may have been saved [25].

   (b) **What could or should have been done differently? By your co-intern? You? Your attending physician? The nurse? The hospital?**
   - **The co-intern**: Could have been more attentive to vital sign changes ("vital signs are vital"); assessed the patient in person more frequently during the day and when asked; been cognizant of cognitive biases that lead to assumptions (such as the bias from a preexisting diagnosis such as anxiety); recognized abdominal pain as a potential sign of a life-threatening condition; listened to the wife's concerns; and reached out to the attending physician if he was uncertain about what to do next.
   - **"You"**: Little is known about role of "you," the night intern, prior to the start of the call shift during which the patient died. If the night intern was caring for the patient more frequently during the proceeding days than the (day) co-intern, better handoffs might have helped the co-intern be better prepared to anticipate and manage possible complications of the patient's disease and treatment. During the handoff, the night intern should also have asked about the sickest patients and immediately examined those patients, though this is unlikely to have made a difference in the outcome.
   - **Attending physician**: It is possible that the attending physician should have followed up more closely during the day, as well as provided more education to the co-intern regarding this patient's diagnosis, treatment, and possible complications. If the risk of sepsis and possible early signs had been reviewed on rounds, the co-intern might have been more likely to recognize them.
   - **The nurse**: The bedside nurse might have noted the changes in vital signs and informed the co-intern; informed the attending when the co-intern failed to assess the patient in person; insisted on face-to-face assessment when the patient developed a new complaint; and/or listened to the wife's concerns with an open mind and helped relay them more effectively to the team.
   - **The hospital**: The hospital could have implemented early warning systems to call attention to patients with changes in vital signs that might make them high risk for acute events; implemented systems that allow families to call

for a rapid response (or similar); and/or created a culture in which (1) nurses feel comfortable calling physicians and expressing disagreement, (2) trainees feel comfortable calling their senior residents and/or attending physicians for input, and (3) families' voices are respected.

(c) **Why didn't the intern and nurse listen to the patient's wife?**
Family members may be subjected to the same biases as patients. Medical staff may dismiss their concerns as the products of (understandable) anxiety about their loved ones' health and well-being. Staff members may feel they do not have time to listen or that families do not have the medical knowledge to really understand what is happening.

There also may be other factors at play. For example, reducing interns' work hours has been shown to reduce serious medical errors in intensive care units; clinician fatigue affects the safety of patients [26, 27]. (This is discussed further in ▶ Chapter 20). There are also programmatic and peer impacts related to a clinician's willingness to listen to concerned family members. Ultimately, self-regulation plays an important role.

In the morning, when you tell your chief resident about the overnight events, he looks stricken and asks if you can wait for a few minutes while he sees the program director "about something." Two hours later, your chief resident comes to relieve you and says that your co-intern has taken a leave of absence.

2. **Self-reflection: How do you think you would feel if you made a mistake of this magnitude? How would you feel about a colleague who did?**

Making a serious mistake that results in patient harm impacts all team members. Physicians report feelings of devastation and embarrassment [28, 29]. There is evidence that they sometimes have difficulty acknowledging healthcare mistakes and worry about malpractice claims and incriminating themselves or others [30–33]. There is also evidence that reluctance to seek help, lack of time to do so, concerns about negative consequences, and stigma all impact the willingness of physicians to get help for the trauma that can result from these experiences [34]. The potentially catastrophic effect that making a mistake can have on healthcare practitioners has led to the development of the "second victim" concept [35–38]. That is, in addition to the victim of the medical error, there are often serious consequences for the involved healthcare practitioner.

> **Box 17.1   Teaching Tip**
> Encourage learners to recognize that good people and good healthcare practitioners do sometimes make mistakes—even serious ones. What is important is understanding (1) what led to the mistake and (2) what happens after.
> Was there a systems error, such as writing an order on the wrong patient in the electronic medical record? Was the physician who made the mistake impaired (under the influence of drugs, alcohol, or serious mental illness)? Was the physician truly negligent or even malicious? That some good physicians make serious mistakes or struggle with issues such as addiction does not negate the possibility that a minority of physicians are incompetent or even criminal in their behavior.

3. **What should be said to Mr. Begay's wife and other family? Who should say it?**
Patients and their families want to be informed of errors, regardless of severity, and they prefer to be informed by their physicians [39–42]. Generally, most

patients report, when asked, that their desire in the aftermath of a medical error is for the "doctor to tell [them] that he or she was sincerely sorry" [43]. That is to say, patients report that the most important feature of an apology from them in the context of a medical error is a physician's *acceptance of responsibility* for having caused harm [44–46]. Expressions of sympathy are also helpful, but generally not to the same degree [47].

(a) **Are physicians or other healthcare practitioners ethically required to tell families when mistakes have been made? Why or why not?**
Yes, they are ethically required to do so. Veracity is part of the physician's obligation to the patient, and it is something that patients expect from their physicians, as implied above. Patients cannot have true autonomy, or make voluntary, informed decisions, if they have inaccurate information about their health and treatment. Furthermore, mistakes often "come out" eventually, and patients can experience significant harm when they become aware that they were deceived by a trusted healthcare professional.

Numerous professional organizations, including the Institute of Medicine, The Joint Commission, and the American Society for Healthcare Risk Management, recommend full disclosure when a medical error occurs [48]. Hospitals that have policies in this regard can build cultures of trust, openness, and safety by creating an expectation of disclosure and developing appropriate support systems for physicians to do so. Educators can also play a role, ensuring that physicians-in-training understand that telling the truth to patients and their families when errors have occurred is part of their ethical duty.

(b) **Does it matter how serious the consequences for the patient were? That is, should errors that do not cause harm and near-miss events be disclosed?**
Yes. Although the principle error in this case was a failure to act, this omission likely contributed to the death of the patient, making this medical error particularly serious and therefore worthy of disclosure.

There is more controversy about prioritizing the disclosure of errors that do not cause harm. These errors include things such as a patient receiving an incorrect dose or type of medication that does not have any adverse effects. Most patients want to be told about these types of errors.

In the case of near-miss events—mistakes that do not reach the patient, sometimes due to safety systems in place, but may be tracked in order to further improve patient safety—there are concerns that informing patients of all these events would overwhelm hospital staff and frighten patients and families, rather than enhance trust.

(c) **Creative problem-solving: Should they be required to do so? By whom? Consider the ethical and practical implications of your response.**

Two years later, you are a senior resident at the same hospital, and you hear that your former co-intern has returned to complete his training. You offer but he declines to discuss the events surrounding Mr. Begay's death. He says only that it was "the worst day of [his] life."

✓ 4. **Self-reflection: What worries you about saying, "I'm sorry" as a doctor?**
   (a) **What would you need to feel comfortable saying it?**

> **Box 17.2   Teaching Tip**
> Consider asking learners to answer this question with various stakeholders in mind. For example, what do learners think they need *from patients or families* in order to feel comfortable apologizing?
> Responses might include forgiveness, patience, compassion, or a willingness to listen with an open mind. Other examples are given below:
> - *From their peers?* Empathy and imagination (this could have happened to anyone);
> - *From faculty and supervisors?* Trust (supervisors will continue to trust the trainee in clinical situations); openness and dialogue (supervisors share their own mistakes and responses)
> - *From their institutions?* A nonpunitive approach; emphasis on identifying causes and errors and implementing systems to prevent them; a culture of openness and disclosure, rather than blame and secrecy
> - *From society?* Acknowledgment that healthcare practitioners are human and that good healthcare practitioners can make mistakes; broader understanding of the role of systems in safety

You eventually begin an oncology fellowship and are working on a patient safety effort at the same hospital. One of your colleagues reintroduces you to Elena Begay, Mr. Begay's wife, who now sits on a family advisory committee for the hospital and advocates for patient safety. You learn that after repeated requests for her husband's medical records, the hospital eventually admitted responsibility for the patient's death and apologized to Ms. Begay.

5. **Do you think that the legal system negatively impacts physicians' and other healthcare practitioners' willingness to discuss errors openly?**
   This seems to generally be the case, even today. Many physicians believe that apologizing will result in being held legally responsible for the error. A 2008 survey found that "admissions of fault" were not admissible in court in only eight states. Thus, in 42 states, physicians' words could be used against them in court. That survey also found that most states did not allow "expressions of sympathy" to be admissible. However, data also indicate that patients are, in fact, "less likely to consider filing suit" when physicians apologize and fully disclose errors [49].

   When malpractice claimants are surveyed about what prompted them to actually file suit, more than 90% indicated that they wanted to prevent the same thing from happening to someone else; to get an explanation for what had happened; and/or for the doctors to realize what they had done (causing the error and its consequences) [49]. A substantial minority of respondents in the above study (40%) also reported that, if they *had* received an honest explanation and a sincere apology, they would not have filed suit. There are numerous studies reproducing these findings and supporting the assertion that apologies often make it *less* likely that a physician will be sued [46]. A famous example of this relationship in practice is the experience of the University of Michigan Health System, described in part above, which reported a 50% decrease in the frequency and cost of litigation in the 5 years following the implementation of an open-disclosure program that included "full disclosure of adverse events, appropriate investigations, implementation of systems to avoid recurrences, and rapid apology and financial compensation when care is deemed unreasonable" [50].

   (a) **What would you recommend to address this? From a healthcare practitioner perspective? From a patient perspective?**
   Education and culture change are important. Culture change should start from the top, contributing to a movement toward a just culture related to medical errors [51]. That is, physicians ought to feel comfortable that responses to medical errors will be reasonable and appropriate (e.g., that they will not be punished for an

"honest mistake" that "could have happened to anyone"). They, of course, also want to avoid being sued—sometimes as much for the negative experience of going to court as for the potential financial penalties. In a just culture, medical errors are recognized as opportunities for organizational learning, such that the risk for future errors is systematically reduced. This approach, shifting from a punitive legal approach to a systems-based and prevention-focused approach, may help healthcare practitioners to feel more comfortable acknowledging and apologizing for errors while also addressing patients' unambiguous preferences for knowing when errors have occurred [42, 52].

> Box 17.3 Teaching Tip
> In their book *Nudge*, behavioral economists Richard Thaler and Cass Sunstein propose a controversial idea: allow patients to waive the right to sue their doctor in order to receive cheaper health care (e.g., by reducing their health insurance premiums) [52]. They say this would allow fully informed patients to have more choice (and thus be more autonomous about their healthcare decisions).
>
> Others have argued that the typical patient does not have enough information to make a fully informed choice in this regard, and short-term monetary gains might lead them to underestimate the long-term risk of medical malpractice. There is the potential for quality of care to suffer if a secondary tier of medical care eventually developed, raising issues of justice.
>
> Learners may find this proposal provocative—consider offering it to stimulate discussion.

6. **Are there other possible reasons that the hospital took so long to tell Ms. Begay the truth about her husband's death?**

   There might be reputational reasons that a hospital would not want to disclose the information. Hospital care is a competitive business, and negative publicity is a worrisome prospect to hospital administrators. That said, they also might receive negative publicity if they look like they are withholding information from grieving families.

The hospital has taken many steps to prevent this type of error in the future, including using "early warning scores" to identify patients whose vital signs are worsening, sometimes very subtly, and allowing families to call for a "rapid response" (assessment from a dedicated team, usually including a nurse or physician from the critical care unit) themselves. The residency program now includes training in cognitive errors and biases and how they impact clinical decision-making. Trainees learn about the importance of reconsidering assumptions when assessing and reassessing patients.

7. **What do you think about these changes? Are there others you would suggest?**

   These changes are a good place to start. Other possibilities might include intensive education in error prevention for hospital staff; initiatives to address aspects of hospital culture (such as the medical hierarchy) that might have affected the willingness of some healthcare practitioners to speak up; or initiatives to promote family-centered care. The simple act of including the patient and his wife in daily rounds might have given them a better baseline relationship, allowing for better communication later.

   (a) **How does the hierarchical structure in medicine impact the occurrence of errors?**

   In a culture of safety, all members of the team are equally valued for their contributions and all have a voice in patient care. When a team member speaks, that person can expect to be heard and acknowledged. When hierarchies are deeply ingrained, lower-ranking members of the team may be reluctant to speak up, because they do not expect to be heard or because they fear potential reprisals from higher-ranking members. The importance of empowering all team members in order to care for patients safely has been widely reported and discussed in relation to Peter Pronovost's studies of a single checklist intervention to reduce catheter-related bloodstream infections in the ICU [53]. An important part of this project's rollout across the state of Michigan involved pairing nurse leaders with physician leaders; a second key aspect was the use of data to increase the comfort of all team members in speaking up whenever they saw a physician begin to place a central line without the appropriate infection-control measures in place.

Ms. Begay generally feels positively about the direction the hospital has taken, although she tells you that the former intern was never willing to meet with her. He has completed his residency and is now working in a private practice.

✓ 8. **Should the intern have been required to meet with her before resuming training or otherwise directly address what happened? Why or why not?**

> **Box 17.4 Teaching Tip**
> One way to approach this with learners is ask how they would feel about the former intern treating one of their family members. Would they feel differently if they knew he had met with the patient's wife, apologized, and worked with her on patient safety? Why? Does it say something about the character or moral strength of the physician? Does that kind of character matter in medicine?

There are many possible answers here. Given the magnitude of the error, it is difficult not to be concerned that the physician is still unwilling to apologize to Ms. Begay. This is particularly true since he is apparently aware that Ms. Begay has worked closely with the hospital over the years (i.e., he knows that her intentions are generally good and that she is not planning to berate or condemn him). The error clearly had a significant emotional impact on him—he quit medicine for some time—and his reluctance to face the patient's wife seems to suggest that he has not been able to fully reconcile it himself [54].

> **Box 17.5 Patient Perspective**
> Josie King was an 18-month-old girl who died after a medical error. Her mother, Sorrel King, became a leading advocate for patient safety and created a foundation in Josie's name. The following is an excerpt from Josie's story [54]:
>
> » I told him it was all wrong. None of what he was saying was making it any easier to take the money. 'We don't want their money,' I reiterated.
>
> 'What *do* you want?' he asked.
>   I told him that I wanted them to remember Josie, to learn something from her and to never let this happen again. 'I want every hospital in the country to know her name and why she died. I want them *all* to learn something,' I said angrily.

Acknowledgments This case was inspired by the stories of Alyssa Hemmelgarn, Josie King, and other patients whose families have worked to create safer hospitals and health care after their deaths.

References

1. Grober ED, Bohnen JM. Defining medical error. Can J Surg. 2005;48(1):39.
2. Institute of Medicine (US) Committee on Quality of Health Care in America. In: Kohn LT, Corrigan JM, Donaldson MS, editors. To err is human: building a safer health system. Washington, DC: National Academies Press; 2000.
3. Chang A, Schyve PM, Croteau RJ, O'Leary DS, Loeb JM. The JCAHO patient safety event taxonomy: a standardized terminology and classification schema for near misses and adverse events. Int J Qual Health Care. 2005;17(2):95–105.
4. Pace WD, Fernald DH, Harris DM, M Dickinson L, Araya-Guerra R, Staton Rebecca VanVorst EW, Bennett L. Developing a taxonomy for coding ambulatory medical errors: a report from the ASIPS collaborative. Rockville, MD: Agency for Healthcare Research and Quality; 2005.
5. Sheikhtaheri A. Near misses and their importance for improving patient safety. Iran J Public Health. 2014;43(6):853.
6. The Joint Commission [Internet]. Sentinel events. Oakland Terrace (IL): The Joint Commission; 2013 Jan [cited 2018 Jul 31]. Available from: https://www.jointcommission.org/assets/1/6/CAMH_2012_Update2_24_SE.pdf.
7. Carver N, Hipskind JE. Medical error. StatPearls [Internet]. StatPearls Publishing LLC; 2018 [cited 2018 Jul 31]. Available from: https://www.ncbi.nlm.nih.gov/books/NBK430763/.

8. Sohn DH. Negligence, genuine error, and litigation. Int J Gen Med. 2013;6:49.
9. Agha R. Complications of immunizations. Pediatr Rev. 1997;18(2):66.
10. Leape LL. Reporting of adverse events. N Engl J Med. 2002;347(20):1633.
11. New York patient occurrence reporting and tracking system (NYPORTS). Albany (NY): New York State Department of Health. [cited 2018 Jul 31]. Available from: www.health.ny.gov/facilities/hospital/nyports/.
12. The Joint Commission [Internet]. Sentinel events: policy and procedures [Internet]. Oakland Terrace (IL): The Joint Commission; 2017 Jun 29 [cited 2018 Jul 31]. Available from: https://www.jointcommission.org/sentinel_event_policy_and_procedures/.
13. Centers for Medicare and Medicaid Services. Hospital compare [Internet]. Baltimore (MD): CMS. [cited 2018 Jul 31]. Available from: www.medicare.gov/hospitalcompare.
14. The Leapfrog Group [Internet]. Washington, D.C. [cited 2018 July 31]. Available from: www.leapfroggroup.org.
15. Kachalia A, Little A, Isavoran M, Crider LM, Smith J. Greatest impact of safe harbor rule may be to improve patient safety, not reduce liability claims paid by physicians. Health Aff. 2014;33(1):59–66.
16. Brin DW. The best response to medical errors?: Transparency. AAMC News [Internet]. Washington, D.C.: American Academy of Medical Colleges; 2018 Jan 16 [cited 2018 Jul 31]. Available from: https://news.aamc.org/patient-care/article/best-response-medical-errors-transparency/.
17. Brooks J. US Medicare will stop paying for preventable errors. CMAJ. 2007;177(8):841–2.
18. Chmieleski S, Dekker M, Scott B, Shapiro SM, Siegel S, Elstein A, et al. The economic measurement of medical errors [seminar presentation]. Arlington County (VA): Casualty Acturial Society; 2010 Sep 20 [2018 Jul 31]. Available from: https://www.casact.org/education/clrs/2010/handouts/HC1-VanDenBos.pdf.
19. Pukk-Härenstam K, Ask J, Brommels M, Thor J, Penaloza RV, Gaffney FA. Analysis of 23 364 patient-generated, physician-reviewed malpractice claims from a non-tort, blame-free, national patient insurance system: lessons learned from Sweden. BMJ Quality & Safety. 2008;17(4):259–63.
20. Oates J, Weston WW, Jordan J. The impact of patient-centered care on outcomes. Fam Pract. 2000;49(9):796–804.
21. Fremont AM, Cleary PD, Hargraves JL, Rowe RM, Jacobson NB, Ayanian JZ. Patient-centered processes of care and long-term outcomes of myocardial infarction. J Gen Intern Med. 2001;16(12):800–8.
22. Scott IA. Errors in clinical reasoning: causes and remedial strategies. BMJ. 2009;338:b1860.
23. Norman GR, Eva KW. Diagnostic error and clinical reasoning. Med Educ. 2010;44(1):94–100.
24. Schiff GD, Hasan O, Kim S, Abrams R, Cosby K, Lambert BL, Elstein AS, Hasler S, Kabongo ML, Krosnjar N, Odwazny R. Diagnostic error in medicine: analysis of 583 physician-reported errors. Arch Intern Med. 2009;169(20):1881–7.
25. Jones DA, DeVita MA, Bellomo R. Rapid-response teams. N Engl J Med. 2011;365(2):139–46.
26. Landrigan CP, Rothschild JM, Cronin JW, Kaushal R, Burdick E, Katz JT, Lilly CM, Stone PH, Lockley SW, Bates DW, Czeisler CA. Effect of reducing interns' work hours on serious medical errors in intensive care units. N Engl J Med. 2004;351(18):1838–48.
27. Gaba DM, Howard SK. Fatigue among clinicians and the safety of patients. N Engl J Med. 2002;347(16):1249–55.
28. Christensen JF, Levinson W, Dunn PM. The heart of darkness: the impact of perceived mistakes on physicians. J Gen Intern Med. 1992;7:424–31.
29. Newman MC. The emotional impact of mistakes on family physicians. Arch Fam Med. 1996;5(2):71.
30. Robinson AR, Hohmann KB, Rifkin JI, Topp D, Gilroy CM, Pickard JA, Anderson RJ. Physician and public opinions on quality of health care and the problem of medical errors. Arch Intern Med. 2002;162(19):2186–90.
31. Kaldjian LC, Jones EW, Wu BJ, Forman-Hoffman VL, Levi BH, Rosenthal GE. Disclosing medical errors to patients: attitudes and practices of physicians and trainees. J Gen Intern Med. 2007;22(7):988–96.
32. Gallagher TH, Waterman AD, Ebers AG, Fraser VJ, Levinson W. Patients' and physicians' attitudes regarding the disclosure of medical errors. JAMA. 2003;289(8):1001–7.
33. Wu AW, Cavanaugh TA, McPhee SJ, Lo B, Micco GP. To tell the truth— ethical and practical issues in disclosing medical mistakes to patients. J Gen Intern Med. 1997;12:770–5.
34. Hu YY, Fix ML, Hevelone ND, Lipsitz SR, Greenberg CC, Weissman JS, Shapiro J. Physicians' needs in coping with emotional stressors: the case for peer support. Arch Surg. 2012;147(3):212–7.
35. Seys D, Wu AW, Gerven EV, Vleugels A, Euwema M, Panella M, Scott SD, Conway J, Sermeus W, Vanhaecht K. Health care professionals as second victims after adverse events: a systematic review. Eval Health Prof. 2013;36(2):135–62.
36. Wu AW. Medical error: the second victim: the doctor who makes the mistake needs help too. BMJ. 2000;320(7237):726.
37. Wu AW, Sexton J, Pham JC. Health care providers: the second victims of medical error. In: Croskerry P, Cosby KS, Schenkel SM, Wears RL, editors. Patient safety in emergency medicine, vol. 338-406. Philadelphia, PA: Lippincott Williams & Wilkins; 2008. p. 7.
38. Folkman S, McPhee SJ, Lo B. Do house officers learn from their mistakes? JAMA. 1991;265:2089–94.
39. Hobgood C, Tamayo-Sarver JH, Elms A, Weiner B. Parental preferences for error disclosure, reporting, and legal action after medical error in the care of their children. Pediatrics. 2005;116(6):1276–86.
40. Blendon RJ, DesRoches CM, Brodie M, Benson JM, Rosen AB, Schneider E, Altman DE, Zapert K, Herrmann MJ, Steffenson AE. Views of practicing physicians and the public on medical errors. N Engl J Med. 2002;347(24):1933–40.
41. Mazor KM, Simon SR, Yood RA, Martinson BC, Gunter MJ, Reed GW, Gurwitz JH. Health plan members' views

about disclosure of medical errors. Ann Intern Med. 2004;140(6):409–18.
42. Witman AB, Park DM, Hardin SB. How do patients want physicians to handle mistakes?: a survey of internal medicine patients in an academic setting. Arch Intern Med. 1996;156(22):2565–9.
43. Hobgood C, Peck CR, Gilbert B, Chappell K, Zou B. Medical errors—what and when: what do patients want to know? Acad Emerg Med. 2002;9(11):1156–61.
44. Beckman HB, Markakis KM, Suchman AL, Frankel RM. The doctor-patient relationship and malpractice. Lessons from plaintiff depositions. Arch Intern Med. 1994;154:1365–70.
45. Hickson GB, Federspiel CF, Pichert JW, Miller CS, Gauld-Jaeger J, Bost P. Patient complaints and malpractice risk. JAMA. 2002;287(22):2951–7.
46. Levinson W, Roter DL, Mullooly JP, Dull VT, Frankel RM. Physician-patient communication. The relationship with malpractice claims among primary care physicians and surgeons. JAMA. 1997;277(7):553–9.
47. Robbennolt JK. Apologies and legal settlement. Mich Law Rev. 2003;102:460–516.
48. Fein SP, Hilborne LH, Spiritus EM, Seymann GB, Keenan CR, Shojania KG, Kagawa-Singer M, Wenger NS. The many faces of error disclosure: a common set of elements and a definition. J Gen Intern Med. 2007;22(6):755–61.
49. Vincent C, Young M, Phillips A. Why do people sue doctors? A study of patients and relatives taking legal action. Lancet. 1994;343(8913):1609–13.
50. Clinton HR, Obama B. Making patient safety the centerpiece of medical liability reform. N Engl J Med. 2006;354:2205–8.
51. Edwards MT. An assessment of the impact of just culture on quality and safety in US hospitals. Am J Med Qual. 2018; 1062860618768057.
52. Thaler R, Sunstein C. Nudge: the gentle power of choice architecture. New Haven, Conn: Yale University Press; 2008.
53. Pronovost. An intervention to decrease catheter-related bloodstream infections in the ICU. New Eng J Med. 2006;355:2725.
54. King S. Josie's Story: A Mother's inspiring crusade to make medical care safe. New York, NY: Grove/Atlantic, Inc; 2009.

Further Reading on this Topic

Berlinger N. After harm: medical error and the ethics of forgiveness. Baltimore, MD: Johns Hopkins University Press; 2005.
Berwick DM. Escape fire: designs for the future of health care. San Francisco, CA: John Wiley & Sons; 2010.
Crisp DH. Anatomy of medical errors: the patient in room 2. Indianapolis, IN: Sigma Theta Tau International; 2017.
Dukhanin V, Edrees HH, Connors CA, Kang E, Norvell M, Wu AW. Case: A Second Victim Support Program in Pediatrics: Successes and Challenges to Implementation. J Pediatr Nurs. 2018;pii: S0882–5963(17)30626–7.
Gawande A. Complications: a surgeon's notes on an imperfect science. New York, NY: Profile Books; 2010.
Gawande A. The checklist manifesto: how to get things right. New York, NY: Henry Holt and Company; 2009. p. 13.
Institute for Healthcare Improvement [Internet]. Boston (MA): IHI. Available from: http://www.ihi.org.

"I'm inPain!"

Theresa Baxter

18.1 Background Questions – 350

18.2 Additional Case Information and Questions for Discussion – 351

18.3 Answers to Background Questions – 352

18.4 Responses to Discussion Questions – 356

References – 365

- **Glossary**

Addiction - Defined by the American Society of Addiction Medicine (ASAM) as a "primary, chronic disease of brain reward, motivation, memory and related circuitry." Like other chronic diseases, it involves cycles of relapse and remission and without treatment is progressive, resulting in disability or premature death. Addiction is characterized by an "inability to consistently abstain, impairment in behavioral control, craving, diminished recognition of significant problems with one's behaviors and interpersonal relationships, and a dysfunctional emotional response" [1].

Medication-assisted treatment (MAT) - Defined by the Substance Abuse and Mental Health Services Administration (SAMHSA) as "the use of FDA-approved medications, in combination with counseling and behavioral therapies, to provide a 'whole-patient' approach to the treatment of substance use disorders" [2].

Opioid use disorder - Defined in the *Diagnostic and Statistical Manual of Mental Disorders*, 5th edition, as the repeated occurrence within a 12-month period of at least 2 of 11 problems, including withdrawal, giving up important life events in order to use opioids, and excessive time spent using opioids; a cluster of 6 or more indicates a severe condition [3].

Pain - "Unpleasant sensory and emotional experience associated with actual or potential tissue damage, or described in terms of such damage" [4].

Sickle cell disease (SCD) - Inherited disorder resulting in abnormal hemoglobin molecules that distort the red blood cells into a sickle shape; characterized by hemolytic anemia and intermittent vaso-occlusive events that cause tissue ischemia, leading to acute and chronic pain as well as organ damage.

Stigma - According to Oxford it is a "mark of disgrace associated with a particular circumstance, quality, or person".

Learning Objectives

1. Understand the social, ethical, and public health context of opioid use disorder.
2. Recognize the importance of prevention and treatment of this disorder.
3. Begin to understand the ethical and social challenges in caring for patients with chronic pain, including people who have sickle cell disease and those on chronic opioid therapy.
4. Identify ways in which physicians and other healthcare practitioners have contributed to the opioid epidemic and how they can help to address it in their communities.

Case Background

You are an opioid care specialist working in a tertiary academic medical center. Today, you have been asked to consult on the care of three different patients recently admitted to your hospital.

Your first patient is Derek Brandon, a 26-year-old man who has an opioid use disorder and who recently overdosed on fentanyl at home. His father was able to revive him with naloxone, and he is currently waiting for a bed for inpatient rehabilitation. Next, you have been asked to see Brianna Meredith, a 21-year-old woman with sickle cell disease (SCD) and a history of chronic opioid use. She was recently admitted with a sickle cell pain crisis. Finally, your third patient will be Jeremy Rosenberg, a 37-year-old man with a history of heroin use who presented with fever and chills and was found to have bacterial endocarditis. Notes in the electronic medical record indicate that this is his third admission for endocarditis and that he underwent an aortic valve replacement during his last admission 4 months ago.

18.1 Background Questions

1. Describe trends in opioid-related deaths over the past 20 years. Is opioid use an "epidemic?"

2. How is pain assessed? When and why was "pain" introduced as the "fifth vital sign?"

3. When should patients be given opioids for acute pain? When should patients be maintained on chronic opioid therapy for chronic pain? What are the risks and benefits of opioids in acute and chronic settings?

4. How has the pharmaceutical industry contributed to the opioid use epidemic?

5. Describe the trends in physician prescribing of opioids over the past 20 years. How have physicians and healthcare practitioners contributed to the opioid use epidemic?

6. What are the current approaches to treating opioid use disorders and how effective are they? How available and accessible is addiction treatment? What is being done to address this shortage?

7. What is naloxone and how available is it nationally?

8. Find and describe two examples of laws that address the opioid epidemic.

18.2 Additional Case Information and Questions for Discussion

You decide that you will go meet all three patients in the morning, review their records, and develop their treatment plans in the afternoon. You go to see Mr. Brandon first. His parents, who are both college professors, are in the room with him. His parents have been aware of his substance use for many years and have tried to help him. Over the years, despite outpatient counseling programs, Mr. Brandon's drug use has escalated.

1. How would you prepare for the consultation?

2. What do you want to know? What would you ask in your first meeting?

3. Would you include his parents in the meeting, or friends if they're visiting? If so, how would you do so? If not, what would you say?

4. Self-reflection: How do you envision yourself with respect to treating patients with substance use disorders?

You learn two important pieces of history. Mr. Brandon has a strong family history of alcohol and substance use. His mother and grandfather have both been treated for alcohol use disorders, although his mother has not used alcohol since before he was born. You also learn that when he was 14 years old, he was in a serious car accident and had multiple fractures, some of which required surgery. He was prescribed narcotics during his hospital stay and his recovery.

5. Does this change your perception of responsibility? How so?
   (a) How do you balance the role that past events played in his current condition with the need for accountability today and in the future?
   (b) How might his parents' status be advantageous to him?
   (c) What biases or disadvantages might less-educated parents encounter in seeking care for their son?
   (d) Does it matter that his mother and grandfather have a history of alcohol use disorder? Why or why not?

6. How can you advocate for your patient at this point in his treatment?

7. What struggles do you think he is presently facing?

After your initial consultation with Mr. Brandon, you go see Ms. Meredith. Ms. Meredith was diagnosed with sickle cell disease on newborn screening. Her first admission was for dactylitis (painful swelling of the hands and feet) at the age of 10 months. Since then, she has been in and out of the hospital frequently with pain crises and other complications of sickle cell disease. She has prescriptions for both short- and long-acting opioid medications at home, yet she presents to the emergency department (ED) frequently for intravenous pain management when she feels her pain is intolerable.

When Ms. Meredith arrived in the ED for this admission, she specifically requested hydromorphone, telling the staff that morphine never works for her. She also asked for intravenous diphenhydramine to help with nausea and itching. Not for the first time, she became irate when the medications were not administered quickly.

8. Should you be concerned about Ms. Meredith's behavior? Why or why not?

9. Why might she be angry?

Ms. Meredith reports "10/10" pain, despite objectively appearing comfortable and conversational, often eating snacks and using her cell phone to play games and text her friends when the inpatient team comes in to examine her.

10. Self-reflection: Do you believe Ms. Meredith's assessment of her own pain?

11. How would you approach assessing her pain?

12. Ms. Meredith is a young adult with a chronic illness. What are some of the challenges she might be facing that are particular to her age?

13. Creative problem-solving: Both adults with SCD and caregivers of children with SCD may feel they are labeled as "drug-seeking," and indeed, clinicians have been found to erroneously think that opioid addiction is more common among patients with SCD than among those with other chronic pain syndromes [5–7]. How might you address the possibility of bias with Ms. Meredith's other healthcare practitioners?

14. A common criticism of the response to the opioid use epidemic has been that it illuminates the depth of racism in the United States—that active efforts to address the stigma of substance use, to treat it as a disease rather than a moral failing, and to change mandatory sentencing laws are happening because opioid use has impacted U.S.-born whites more than previous illicit substances. Do you agree or disagree? Why?

After completing your initial assessment on Ms. Meredith, you go to Mr. Rosenberg's room. As soon as you enter the room, he immediately asks, "Am I going to get my surgery?" He tells you that the cardiothoracic surgery team rounded this morning and demanded an ethics consultation before the surgeon would agree to replace his valve a second time. It appears, from the electronic medical record, that Mr. Rosenberg is medically a candidate for valve replacement.

15. What are the ethical issues at stake here?

16. Can you think of a situation in which treatment could ethically be withheld from a patient with another disease (not a substance use disorder) on the basis of nonadherence to other therapy?
    (a) Are substance use disorders truly regarded as diseases by the health professions?

17. What would you say to Mr. Rosenberg?
    (a) What more would you like to know?

18. Considering the socio-ecological model, your background research, and these three patients, identify key factors in each sphere that have contributed to the opioid epidemic.

18.3 Answers to Background Questions

1. Describe trends in opioid-related deaths over the past 20 years. Is opioid use an "epidemic?"
   From 1999 to 2016, there was a dramatic increase in the number of opioid-related deaths. The number of deaths from opioid overdoses (including both prescription and illegal opioids) in the United States was five times higher in 2016, compared to 1999. An estimated 40% of these were related to prescription opioids [8, 9]. The U.S. Centers for Disease Control and Prevention (CDC) describes three waves of increased opioid overdose deaths in the past 20 years. The first wave began around or before 1999 and involves deaths associated with natural and semisynthetic opioids, the second began in 2010 and involves deaths associated with heroin, and the third began in 2013 and involves deaths associated with synthetic opioids (fentanyl) [8].
   According to the CDC, the term "epidemic" refers to an increase in the number of cases of a disease above what is expected in a defined population, often occurring suddenly [10]. If overdose deaths serve as a proxy for opioid use (the proverbial "tip of the iceberg"), the aforementioned statistics demonstrate that this is an epidemic.

2. **How is pain assessed? When and why was "pain" introduced as the "fifth vital sign?"**

There are many different types of pain, but it is typically classified as coming from the nerves (neuropathic) or from the tissues (nociceptive) and as being acute or chronic. Pain is assessed using a wide variety of assessment scales, many of which are not specific to the aforementioned qualities. For adults, the most relevant and commonly used is the numerical pain scale, which measures pain from 0 (no pain) to 10 ("worst pain of your life"). For infants and preverbal children, the FLACC (face, legs, activity, cry, consolability) scale is preferred, and for young, verbal children, the Wong-Baker FACES scale is used (depicting faces with varying degrees of frowns or smiles). There are about 20 different scales, designed to meet the needs of a multitude of diverse patient populations [11, 12].

In 1996, the American Pain Society introduced the concept of pain as the "fifth vital sign" to address the widespread problem of underassessment and undertreatment of pain [13]. A public campaign was launched to introduce this to patients and providers. It became part of standard practice in healthcare settings to assess pain concomitantly with vital signs such as blood pressure and temperature.

In 2000, as part of a national effort, The Joint Commission: Accreditation, Healthcare, Certification (formerly the Joint Commission on Accreditation of Health care Organizations [JCAHO]) introduced standards to help organizations to improve care for patients with pain and began to assess organizations based on these new standards [14]. Patients were given surveys that were used to gauge their satisfaction with pain management, and some aspects of reimbursement were tied to this satisfaction. Consequently, healthcare organizations initiated policies to improve patient satisfaction related to pain relief measures. In recent years, after growing recognition of the contribution these standards and policies had to the opioid epidemic, The Joint Commission and the Centers for Medicare and Medicaid Services (CMS) reversed these standards and policies.

3. **When should patients be given opioids for acute pain? When should patients be maintained on chronic opioid therapy for chronic pain? What are the risks and benefits of opioids in acute and chronic settings?**

Opioids are one of many options that healthcare providers can consider for management of acute pain, but they should be avoided whenever possible. In some cases of severe acute pain, opioids may be indicated, usually in conjunction with adjuvant pain medications such as acetaminophen and nonsteroidal anti-inflammatory drugs (NSAIDs). In such situations, the dosage should be as low as possible to achieve adequate pain relief, and they should be prescribed for the shortest duration possible. The CDC, for example, recommends 3 days or less for most situations. Physiological dependence on and tolerance to opioids can develop in as little as 1 week [15, 16].

Box 18.1 Patient Perspective

Firoozeh Dumas' essay "After Surgery in Germany, I Wanted Vicodin, Not Herbal Tea" contrasts with the author's experience undergoing a laparoscopic hysterectomy in Munich with her expectations as an American patient [16]. She writes:

» I brought up the subject of painkillers with my gynecologist weeks before my surgery. She said that I would be given ibuprofen. 'Is that it?' I asked. 'That's what I take if I have a headache. The removal of an organ certainly deserves more.'

'That's all you will need,' she said, with the body confidence that comes from a lifetime of skiing in crisp, Alpine air.

> Ms. Dumas then asks the same questions of her surgeon and anesthesiologist. The anesthesiologist advises her:
>
> » Pain is a part of life. We cannot eliminate it nor do we want to. The pain will guide you. You will know when to rest more; you will know when you are healing. If I give you Vicodin, you will no longer feel the pain, yes, but you will no longer know what your body is telling you. You might overexert yourself because you are no longer feeling the pain signals. All you need is rest. And please be careful with ibuprofen. It's not good for your kidneys. Only take it if you must. Your body will heal itself with rest.
>
> Reflecting on this advice, she observes:
>
> » His gentle suggestion to trust my body almost brought me to tears. It reminded me of the poster in my doctor's waiting room, the one informing us that herbal tea is the first remedy to try when we have a cold. The first remedy I try is the decongestants I bring with me from the United States. I can't find those in Germany, nor can I find the children's cough medicine that makes my child drowsy. I also import that.
>
> After the surgery, she describes her recovery thus:
>
> » After a week, I took the tram to the doctor's office to have my stitches removed. My doctor, with her usual cup of chamomile tea in hand, remarked on my progress. 'I rested,' I told her. Normally, I would have said, 'I did nothing,' but I didn't say that. I had been healing, and that's something.

Use of long-term opioid therapy to address chronic pain is controversial (excluding its role in the management of cancer-related pain and terminal illness). As with all medical decision-making, healthcare practitioners and patients need to carefully weigh the risks and benefits of any intervention. At this time, there is not a sufficient evidence of opioids' long-term efficacy, yet there *is* evidence of their risks, which are death, addiction, dependence, tolerance, and opioid-induced hyperalgesia, among others. Still, there are some circumstances under which healthcare practitioners might appropriately consider opioids for long-term use, as part of a multimodal approach to care. The CDC recently released guidelines to assist healthcare practitioners in determining if opioids are indicated for chronic pain. If indicated, the guidelines also discuss dosage, duration, follow-up plans, and discontinuation, as well as how to assess the risks and address harms of opioid use [17].

It should be noted that many chronic pain patients who have been maintained on opioids for years are—not surprisingly—distressed about these findings and recommendations. Some patients have begun to organize their own grassroots advocacy groups to challenge this approach.

4. **How has the pharmaceutical industry contributed to the opioid use epidemic?**
There are many facets to the pharmaceutical industry's role in the opioid use epidemic. In 1980, the *New England Journal of Medicine* (NEJM) published a single-paragraph letter, written by Jane Porter and Dr. H. Jick. The authors briefly summarized a study of 11,000 postsurgical patients who received opioid pain medications at least once and concluded that, despite widespread use of narcotic drugs in hospitals, "development of addiction [was] rare in medical patients with no history of addiction" [18]. This letter was then used in marketing strategies by pharmaceutical companies that produced opioid medications. Some 37 years ago, the same prestigious journal published a follow-up letter, describing an analysis of the consequences of this letter and arguing that the 1980 letter helped to "shape a narrative that allayed prescribers' concerns about the risk of addiction associated with long-term opioid therapy," ultimately contributing to the North American opioid crisis [19].

As of this writing, there are numerous lawsuits pending against pharmaceutical companies for their role in the epidemic. Several consumer-initiated lawsuits allege that some pharmaceutical companies "failed to adequately warn about addiction risks on drug packaging and in promotional activities"; "deliberately withheld information about their products' dangers, misrepresenting them as safer than alternatives"; and "failed to include safety mechanisms" (e.g., tamper-resistant packaging) [20]. Similarly, lawsuits brought by both state and federal governments allege that companies "unreasonably interfered with the public's health by oversaturating the market with drugs and failing to implement controls against misuse and diversion, thereby creating a public nuisance"; "made false representations about their products' addictiveness and effectiveness, all calculated to mislead the state, prescribers, and the public"; and had "lax monitoring of suspicious opioid orders" [20].

5. **Describe the trends in physician prescribing of opioids over the past 20 years. How have physicians and healthcare practitioners contributed to the opioid use epidemic?**
Between 1999 and 2010, the sales of "opioid pain relievers" quadrupled [21]. In a more detailed follow-up study, analyzing data from 2006 to 2015, the CDC concluded: "In the United States, annual opioid prescribing rates increased from 72.4 to 81.2 prescriptions per 100 persons from 2006 to 2010, were constant from 2010 to 2012, and then decreased by 13.1% to 70.6 per 100 persons from 2012 to 2015" [22]. Although prescribing levels are still approximately three times higher today than they were in 1999, these data are encouraging. The report highlighted the fact that there are significant differences in prescribing rates by county; the CDC has since started providing more granular data on its website that demonstrates this geographic variation [22]. By 2016, the overall prescribing rate had fallen to the lowest it had been in more than 10 years (66.5 prescriptions per 100 persons); however, some counties had rates more than seven times that rate [23].

In addition to geographic differences, there is also significant variation in prescribing patterns between individual healthcare practitioners. Organizations such as the American Pain Society and the American Academy of Pain Medicine encouraged opioid use as part of aggressive campaigns to reduce pain beginning in the mid-1990s. Variation between healthcare practitioners might be in part due to differences in exposure to these campaigns, training era, and financial consequences of compliance with standards to address pain. (This is also explored in Background Question 2.) Inter-practitioner variability is almost certainly also related to the extent to which clinicians have been influenced by incentives from pharmaceutical companies.

6. **What are the current approaches to treating opioid use disorders and how effective are they? How available and accessible is addiction treatment? What is being done to address this shortage?**
According to the American Society for Addiction Medicine (ASAM), medication-assisted treatment (MAT)—a combination of counseling and behavioral therapies and medication—is the most effective intervention to treat opioid use disorder and is more effective than either behavioral interventions or medication alone [24]. However, more research is necessary to determine which medications are most efficacious and whether specific populations benefit from certain approaches. ASAM has also published a clinical practice guideline for opioid use disorder [24]. Presently, two medications are considered to be opioid replacement therapy: (1) buprenorphine, a partial opioid agonist and partial antagonist, and (2) methadone, an extremely long-acting opioid agonist.

Only about 1 in 10 people with a substance use disorder receives any type of specialty treatment, according to a report released by the U.S. Surgeon General [25]. One major limitation is the shortage of inpatient beds, outpatient programs, and credentialed medical providers able to prescribe MAT. Many initiatives seek to address the shortage, and laws have been proposed or passed along to increase state and federal funding for treatment. Recently, the U.S. Department of Health and Human Services (HHS) announced that $485 million (U.S. dollars) in Grants would be provided as part of the twenty-first century cures act to achieve these goals. This funding will be provided through the state-targeted response to the opioid crisis Grants program, administered by the Substance Abuse and Mental Health Services Administration [26].

✓ 7. **What is naloxone and how available is it nationally?**
Naloxone is a nonaddictive *mu*-opioid antagonist. It can reverse opioid overdose and prevent fatalities, with safety and efficacy well established [27]. Although naloxone has been shown to save thousands of lives, it is important to remember that it is still a rescue drug, not a treatment for opioid use disorder.

In 2015, the CDC recommended that naloxone access be expanded to reduce deaths associated with opioid overdoses [28]. Availability has increased widely across the country, with many states passing laws to increase access. Mechanisms for this increased access have included "allowing prescribers to write naloxone prescriptions for third parties (e.g., friends and family members of those who are at risk of an opioid overdose) and permitting pharmacists to dispense or distribute naloxone without a patient-specific prescription" as well as requiring naloxone to be available to certain emergency personnel (e.g., police officers) and in certain settings (e.g., public libraries) [29, 30].

To further reduce barriers to utilizing naloxone, many health insurance plans have eliminated co-payments for this medication, and some states have programs to assist patients and families with the costs. For example, in New York, the Naloxone Co-Payment Assistance Program (N-CAP) covers insurance co-payments up to $40. Many organizations across the country, including local public agencies and community-based organizations, have hosted naloxone training events. Some events have not only provided education but also distributed the medication itself to the public.

✓ 8. **Find and describe two examples of laws that address the opioid epidemic.**
On July 22, 2016, President Barack Obama signed the Comprehensive Addiction Recovery Act (CARA). According to the ASAM, this law addresses "the full continuum of care from primary prevention to recovery support, including significant changes to expand access to addiction treatment services and overdose reversal medications. It also includes criminal justice and law enforcement-related provisions" [31]. One provision allows nurse practitioners (NPs) and physician assistants (PAs) to obtain buprenorphine waiver training, in order to increase access to MAT.

Another example is New York's 911 Good Samaritan Law, which went into effect in 2011. The law seeks to encourage people to call 911 when witnessing or experiencing an alcohol or other drug overdose by providing some protections against drug-related charges and prosecution associated with the call [32]. It should be noted that this is not the case in every state, and in some states, friends or relatives who were sharing drugs at the time of a fatal overdose have been charged with homicide [33].

18.4 Responses to Discussion Questions

You decide that you will go meet all three patients in the morning, review their records, and develop their treatment plans in the afternoon. You go to see Mr. Brandon first. His parents, who are both college professors, are in the room with him. His parents have been aware of his substance use for many years and have tried to help him. Over the years, despite outpatient counseling programs, Mr. Brandon's drug use has escalated.

✓ 1. **How would you prepare for the consultation?**
In order to be prepared for the meeting, the consultant should enter the room having reviewed the details of Mr. Brandon's case, including his history of present illness, past medical history, events during this

hospitalization, pertinent laboratory results, and current treatment plan. His medications, both past and current, should also be reviewed. All U.S. states, with the exception of Missouri, have a functioning prescription drug monitoring program, although program participation is voluntary in 10 states, rather than mandated [34, 35]. These programs support databases in which clinicians can verify any controlled substances prescribed in any outpatient settings, including those from different prescribers.

2. **What do you want to know? What would you ask in your first meeting?**
It is important to take a thorough history; although the electronic medical record may contain some of this information, the key points should be verified. The consultant may want to ask more details about the circumstances of his recent overdose and other aspects of his past medical (including psychiatric), surgical, and psychosocial history. Clinicians may want to ask specifically whether he has been tested for hepatitis C virus and human immunodeficiency virus (HIV).

When inquiring about substance use, questions should include:
- When did he first use?
- What has been his drug of choice?
- What route does he use (e.g., intravenous, oral, and inhaled)?
- Has he used any other substances, including alcohol and tobacco?
- Has he had any other overdoses?
- What does he understand about his treatment plan?
- To what extent is he a ready and willing participant in treatment?

It is also important for healthcare practitioners to disclose their role upon entering the room. There is a great deal of mistrust due to perceived discrimination against people with substance use disorders. Many patients have felt judged by healthcare practitioners, and stigma remains high. History taking is an opportunity to build trust, before moving on to the physical examination, and certainly before making recommendations to the patient.

3. **Would you include his parents in the meeting, or friends if they're visiting? If so, how would you do so? If not, what would you say?**
As with any patient, the consultant should ask to speak with Mr. Brandon alone and then establish with him whether he would like to include his parents or friends in the meeting. It is essential to protect patient privacy. If Mr. Brandon so desires, any family and friends can be involved in the discussion at any point, and they can be a useful source of collateral information. Patients can also give permission for healthcare practitioners to speak to their family members alone.

4. **Self-reflection: How do you envision yourself with respect to treating patients with substance use disorders?**

> **Box 18.2 Teaching Tip**
> It is well-documented that many healthcare practitioners have negative attitudes about treating patients with substance use disorders and have fears that they may contribute to dependence or addiction. Explore this with the learners.

You learn two important pieces of history. Mr. Brandon has a strong family history of alcohol and substance use. His mother and grandfather have both been treated for alcohol use disorders, although his mother has not used alcohol since before he was born. You also learn that when he was 14 years old, he had been in a serious car accident and had multiple fractures, some of which required surgery. He was prescribed narcotics during his hospital stay and his recovery.

5. **Does this change your perception of responsibility? How so?**
It may and, to some extent, it probably should. Some healthcare practitioners will feel more sympathetic if they feel that the healthcare system contributed to the substance use disorder; others will find it easier to empathize when blame for the patient's illness is shifted away from the

patient. However, it is important to cultivate empathy and compassion regardless of the extent to which the patient can be considered responsible for the illness.

(a) **How do you balance the role that past events played in his current condition with the need for accountability today and in the future?**

Healthcare practitioners have an obligation to try to understand the root causes of any medical condition in order to optimize management of that condition. In cases of opioid use disorder, recognizing what role, if any, forces beyond the patient's control (including, but not limited to, societal forces, the pharmaceutical industry, other healthcare practitioners, etc.) had in the evolution of the disorder may be helpful to the patient and the patient's family. It is also important for healthcare providers to encourage and empower patients to take responsibility for the ongoing management of their disease, regardless of the factors that may have led to it.

(b) **How might his parents' status be advantageous to him?**

The Patient Protection and Affordable Care Act (ACA) required that health insurance companies cover treatment for substance use disorders as part of essential benefit packages [36]. Even so, many barriers to treatment remain. These include awareness of treatment options, stigma, non-covered costs, and others (discussed further below).

Mr. Brandon's parents are highly educated and are presumably quite affluent, likely with good health insurance. With his parents' support and resources, he will have more choices of treatment facilities, potentially including access to holistic facilities with added amenities, such as yoga, art therapy, nature activities, acupuncture, and exercise facilities. They may also be more likely to help him with his personal and financial responsibilities while he receives treatment, thereby reducing the burden of those concerns on him.

On the other hand, some parents from similar socioeconomic backgrounds might feel that their status increases the shame or stigma of having a child with a substance use disorder; such feelings might hinder their ability to provide optimal support.

(c) **What biases or disadvantages might less-educated parents encounter in seeking care for their son?**

As indicated previously, there are many barriers to accessing treatment for a substance use disorder. Some of the barriers exist at the level of the individual or the individual's family. Examples of such barriers include:

- Perceptions that the opioid use disorder is not a problem.
- Unfamiliarity with the concept of addiction as a treatable disease or with treatment options.
- Lack of awareness of the ACA's mandated essential benefits (leading them to believe they cannot afford treatment).
- Concern about what others may think of them if they seek treatment (as noted earlier, this may apply to any parent, regardless of socioeconomic status).
- Inability to take time away from other responsibilities (e.g., childcare, employment) to help the patient secure treatment.
- Non-medical costs, such as lost wages or transportation.

In addition, barriers within the healthcare system include eligibility for, and availability of, treatment options within the patient's community, as well as financial barriers even with the ACA coverage requirements. Less-educated individuals and their families are disproportionately impacted by these barriers and have fewer resources to circumvent them.

(d) **Does it matter that his mother and grandfather have a history of alcohol use disorder? Why or why not?**

It does. While not everyone with a family history of addiction will develop an addiction and while some people with no family history of addiction will still develop one, there is evidence that addictions are "moderately to highly heritable" [37]. In this case, it is noted that Mr. Brandon's mother has not used alcohol since before he was born—this is important because growing up in a household with an adult who has a substance use disorder is also a risk factor. These aspects of family history represent individual and interpersonal risk factors in the socio-ecological model.

Individual factors (e.g., genes), interpersonal factors (e.g., friends' and family's attitudes about substance use), community factors (e.g., density of alcohol outlets, availability of drugs, availability of effective community prevention programming), and policy factors (e.g., policies regarding substance use education, screening programs within healthcare systems, funding and resources for early identification and referral for treatment) all influence the likelihood of developing and recovering from a substance use disorder. (The socio-ecological model is also discussed in Discussion Question 18.)

6. **How can you advocate for your patient at this point in his treatment?**

The consultant can ensure that:
- A representative from the addiction treatment team rounds on him every few days over the course of his admission.
- His medications are appropriate for his treatment plan.
- He has access to peer support while he awaits transfer to an inpatient facility.
- In addition, healthcare practitioners can reduce stigma by ensuring that all team members have some training in substance use disorder.

7. **What struggles do you think he is presently facing?**

Mr. Brandon likely is still facing the stigma of having opioid use disorder. He may be feeling like a "failure" for continuing to use despite prior treatment, and he may still be struggling with other barriers to treatment, as described above.

After your initial consultation with Mr. Brandon, you go see Ms. Meredith. Ms. Meredith was diagnosed with sickle cell disease on newborn screening. Her first admission was for dactylitis (painful swelling of the hands and feet) at the age of 10 months. Since then, she has been in and out of the hospital frequently with pain crises and other complications of sickle cell disease. She has prescriptions for both short- and long-acting opioid medications at home, yet she presents to the emergency department (ED) frequently for intravenous pain management when she feels her pain is intolerable.

When Ms. Meredith arrived in the ED for this admission, she specifically requested hydromorphone, telling the staff that morphine never works for her. She also asked for intravenous diphenhydramine to help with nausea and itching. Not for the first time, she became irate when the medications were not administered quickly.

8. **Should you be concerned about Ms. Meredith's behavior? Why or why not?**

The consultant should be concerned, but Ms. Meredith's risk of opioid use disorder may not be the primary concern right now. The consultant might consider that Ms. Meredith's home pain management regimen is not working effectively, given her frequent ED visits for pain relief. On the other hand, it is possible that her sickle cell disease is poorly controlled or is simply very difficult to control, resulting in frequent acute pain crises.

Regardless, this is a frustrating situation for any patient. Furthermore, it is not uncommon for young adult patients with sickle cell disease to require frequent hospitalization for complications of SCD. Such complications include fever, infection, acute chest syndrome, and stroke, in addition to vaso-occlusive pain episodes, and several of these can be life-limiting and life-threatening [38, 39]. Current best-practice guidelines recommend that patients with SCD receive immediate and potent intravenous opioid medication and intravenous fluids for sickle cell crisis pain. These medications should be administered within 30 minutes of triage or 60 minutes of registration, and response to pain should be reassessed frequently, which may pose a challenge in a busy ED [40].

Unfortunately for clinicians, no biomarkers or imaging studies can validate pain or assess its severity, and the above guidelines recommend that management be guided by patient report of pain severity.

9. **Why might she be angry?**
   Ms. Meredith might be angry because she is in pain and her expectation of immediate relief has not been met. In addition, she truly *is* the expert in understanding the history of her disease and what works and does not work for her. She may not feel that her voice is being heard or respected. As a high utilizer of healthcare services, she likely recognizes that her clinicians are concerned about the possibility of addiction, and she may be reacting negatively to perceptions of a label she does not feel is accurate.

Ms. Meredith reports "10/10" pain, despite objectively appearing comfortable and conversational, often eating snacks, and using her cell phone to play games and text her friends when the inpatient team comes in to examine her.

10. **Self-reflection: Do you believe Ms. Meredith's assessment of her own pain?**

11. **How would you approach assessing her pain?**
    Ms. Meredith has had this painful disease her entire life, and it is likely that she has adjusted her coping mechanisms in order to allow her to continue to function despite intolerable pain. Healthcare practitioners should assess Ms. Meredith's pain using their institution's policy, such as employing the standard pain rating scale; however, it is important for practitioners to keep in mind that "pain is what the patient says it is" and it is her perception of pain that will guide her experience. Other questions can be very helpful, especially to clarify what her pain complaints are: What are the characteristics of her pain? Is this pain different from her usual sickle cell pain? Is what is being done helping? What has helped in the past? These answers might also help the consultant to consider the real cause of the pain. Does it seem musculoskeletal or neuropathic? Could it be related to stiffness from lying in bed?

    It is also helpful to frame questions for the patient. For example, asking if the patient is having "the worst pain of her life" serves a very different purpose (situational) than asking them to rate their pain based on ability to conduct activities of daily living (functional). Reeducation on rating pain in terms of functionality can assist the patient in returning to her baseline state of health and well-being.

    In the case of sickle cell disease, the pain is real, but a patient such as Ms. Meredith may also be worried about other, life-threatening complications of her disease. For instance, chest pain can be from vaso-occlusive pain or it can indicate acute chest syndrome, a very serious condition. If laboratory and imaging results are negative and if there is evidence that this episode is not an acute life-threatening condition, reassurance can be important for the patient.

12. **Ms. Meredith is a young adult with a chronic illness. What are some of the challenges she might be facing that are particular to her age?**
    Young adults with chronic illnesses face some particular challenges. They may

struggle with feeling different than their peers. At a time when their friends are developing more autonomy, perhaps going away to college or starting careers, they may feel that their diagnoses and treatments are interfering with their pursuit of similar goals. They may be learning to navigate their sexuality in the context of illness. They may also be searching for the right balance of involvement from their families, particularly if their parents were closely involved in managing their health and health care when they were younger.

Ms. Meredith is at the age where discussions about transitioning from pediatric SCD healthcare providers to adult healthcare practitioners should occur if they have not already. Young adults with SCD have increased ED utilization and rates of rehospitalization compared with other age groups [41]. Pediatric health care is more comprehensive than adult care, and many patients with SCD express dissatisfaction about the transition process. If Ms. Meredith has recently transitioned and now sees an adult hematologist for the first time, this stress may be contributing to her pain, frequent visits, and frustrations with care.

Every healthcare encounter presents an opportunity for healthcare practitioners to intervene and assist Ms. Meredith with developing and maintaining effective coping mechanisms. In optimal situations, a multidisciplinary team, including psychologists and social workers, will help support Ms. Meredith with developing and maintaining effective coping mechanisms. Every healthcare encounter presents an opportunity for healthcare practitioners to intervene in this way.

13. **Creative problem-solving: Both adults with SCD and caregivers of children with SCD may feel they are labeled as "drug-seeking," and indeed, clinicians have been found to erroneously think that opioid addiction is more common among patients with SCD than among those with other chronic pain syndromes [5–7]. How might you address the possibility of bias with Ms. Meredith's other healthcare practitioners?**

14. **A common criticism of the response to the opioid use epidemic has been that it illuminates the depth of racism in the United States—that active efforts to address the stigma of substance use, to treat it as a disease rather than a moral failing, and to change mandatory sentencing laws are happening because opioid use has impacted US-born whites more than previous illicit substances. Do you agree or disagree? Why?**

There are several aspects to this issue. A *Kaiser News Briefing* report summarized the concern that the societal response to the crack cocaine epidemic in 2000 that disproportionately affected blacks (84% of "offenders") compared to the more recent opioid epidemic, in which overdose rates are higher in whites than blacks. That report concluded: "Many see race as a crucial factor in how Congress and health officials have focused on prevention and education rather than punishment" [42].

Ironically, with respect to the first wave of the opioid epidemic (see Background Question 1), Andrew Kolodny, founder of Physicians for Responsible Opioid Prescribing, has argued that racial stereotyping may have protected minority populations from the addiction epidemic, in part because healthcare providers were not as willing to prescribe to these patients, wrongly fearing greater risks of addiction or diversion relative to white patients. White patients complaining of the same pain were prescribed opioid pain medication in abundance, increasing the risk of addiction [43].

After completing your initial assessment on Ms. Meredith, you go to Mr. Rosenberg's room. As soon as you enter the room, he immediately asks, "Am I going to get my surgery?" He tells you that the cardiothoracic surgery team rounded this morning and demanded an ethics consultation before the surgeon would agree to replace his valve a second time. It appears, from the electronic medical record, that Mr. Rosenberg is medically a candidate for valve replacement.

✅ 15. **What are the ethical issues at stake here?**

Healthcare practitioners have a duty and obligation to provide medical care to patients regardless of circumstances, irrespective of any perceived moral failings or lawbreaking behavior. They are not charged with the duty of punishing patients for self-destructive or socially undesirable behavior by withholding treatment [44].

In Mr. Rosenberg's case, the patient is clearly invested in having a second valve replacement. This appears to be his autonomous decision, assuming he has been appropriately advised of the risks, benefits, and alternatives and that he understands them. (The ethics of informed consent is explored in-depth in ► Chapter 3). In this context, performing the surgery is consistent with the principles of autonomy, as well as with beneficence. Assessment of the latter is based on the statement that he is indeed a candidate for this procedure.

In the U.S. and increasingly throughout the western world, autonomy is prioritized over other ethical principles when they conflict. However, one potential dilemma in this situation is the "duty to be good stewards of medical care and not to squander resources where they are unlikely to provide much benefit" [44]. In the case of substance abuse disorder, the question of futility (inability to provide benefit) is in part related to the patient's likelihood of recidivism. If Mr. Rosenberg continues to use drugs, he may again be at risk of endocarditis, thus putting his replacement valve at risk, and a third surgery may be impossible to perform safely and effectively. Performing futile procedures not only leads to worse patient outcomes, which violate the ethical principle of non-maleficence, but may also contribute to rising healthcare costs, possibly limiting access to care, which violates the principle of distributive justice [44].

Because of the fragmented nature of the U.S. healthcare system, questions regarding just distribution of finite healthcare resources are extremely difficult to answer. If many patients with opioid use disorder need and receive repeat valve replacements, and many of those patients subsequently relapse and have poor outcomes, then there might be a measurable impact to other patients, either by increasing healthcare costs or by limiting access to surgeons in medically underserved areas.

✅ 16. **Can you think of a situation in which treatment could ethically be withheld from a patient with another disease (not a substance use disorder) on the basis of nonadherence to other therapy?**

It is important to understand the reasons for nonadherence to therapy in patients, as it is multifactorial. (It should be noted that the term "nonadherence" is preferred over "non-compliance.") According to the CDC, the reasons for "medication nonadherence" are variable and include, but are not limited to, affordability, lack of comprehension regarding the importance of the medications, and adverse side effects [45]. The consequences of nonadherence include increased morbidity and mortality, costing billions of dollars per year. Hospital admission rates increase for nonadherent patients with chronic illness by as much as 69% [45].

In general, treatment is not withheld from patients based on nonadherence to therapy. For example, a patient with type 2 diabetes mellitus who continues to have uncontrolled blood glucose as a result of nonadherence to diet, exercise, and/or oral medications would not be denied dialysis for chronic kidney disease due to diabetes. Clinicians continue to provide chemotherapy for tobacco-related lung cancer or vascular diseases, even when patients are unable to follow smoking cessation recommendations or choose not to do so.

Rare exceptions include (1) situations in which continued treatment in the face of the nonadherence could compromise the patient's health and (2) situations that involve finite resources. An example of the former is a patient with diabetes who is given an insulin pump but who routinely overestimates how much he will eat at meals, resulting in life-threatening hypoglycemic events. If this issue cannot be addressed, the patient might switch to another regimen that is perhaps less optimal for blood sugar control but less likely to result in hypoglycemia. The principal example of the latter is organ transplantation. Demand for organs greatly exceeds supply; thus, candidates are evaluated for their ability to comply with complex posttransplant regimens to prevent immune rejection of the transplanted organ. Patients who are unable to comply for a variety of reasons are not considered candidates—nor are patients with untreated substance use disorders that are likely to compromise outcomes of transplant.

(a) **Are substance use disorders regarded as diseases by the health professions?**

The fact that the surgical team is considering denying this patient a needed surgery suggests the possibility that, in this case, these disorders are not truly regarded as diseases or at least are perceived as distinct from other medical conditions. This is not uncommon. The same can be said for laws that criminalize substance use during pregnancy [46]. A recent *New York Times* interview noted the unique challenges that healthcare practitioners who treat opioid use disorders face, including regulatory hurdles that are higher than those required to prescribe opioids themselves:

"Dr. Gastala has to follow strict federal requirements and live with the possibility that the Drug Enforcement Administration might inspect her office with no warning. Insurers require her to jump through constant hoops to get the medication approved for her patients." [47]

Furthermore, despite numerous international and national guidelines recommending that addiction be treated as a chronic disease, there are still situations in which patients may feel like healthcare practitioners consider their addiction to be a "choice" or a moral failing. A recent survey of emergency physicians at Johns Hopkins University revealed that a significant proportion of physicians had "low regard for patients with substance use" [48]. The Harm Reduction Coalition is a national advocacy group for persons with substance use disorder. They have noted that healthcare professionals can play a vital role in the care of people with addiction and that when persons with substance use disorder perceive disdain or rejection by healthcare workers, they may reject the care offered [49].

When patients with substance use disorders present to hospitals with other medical conditions, they may be wrongly perceived as "drug seekers" or labeled as "frequent flyers." At other times, people with substance use disorders may also be treated for presenting medical conditions without recognition of the substance use disorder as an underlying and causative condition. For these patients, failure to recognize their addiction also compromises their health outcomes.

17. **What would you say to Mr. Rosenberg? What more would you like to know?**

At this point, it is important to have more information about Mr. Rosenberg's understanding of his current situation. What does he want? Would he like to proceed with another major surgery? Does he understand the gravity of the situation? What are his plans postoperatively regarding treatment? It is important to ensure he understands the risks, benefits, and alternatives of the procedure and that he recognizes his risk for relapse and the importance of engagement in a relapse prevention program. In addition, if the decision is made to proceed with surgery, it is critical to address adequate pain management in the perioperative setting and proactively develop a plan for transitioning off opioid pain medications after surgery.

At the end of the discussion, the consultant may want to reassure Mr. Rosenberg of some of the aforementioned themes: that the consultant is not judging him and is not a judge of patients; that the care team is there to support Mr. Rosenberg in making the best choices for himself; and that the consultant will advocate for the patient with the surgical team, if necessary. In some such cases, it may be appropriate to involve an ethics consultation service, team, or committee, particularly if the ethical question arises regularly and if there is disagreement between different specialty services on how best to proceed.

✓ 18. **Considering the socio-ecologic model, your background research and these three patients, identify key factors in each sphere that have contributed to the opioid epidemic.**

Key factors at each level include:

Individual: Genetics, including epigenetics (briefly discussed in ▶ Chapter 22)

Interpersonal: Family values and preferences, including stigma, assumptions about substance use disorders, social support networks, bias from clinicians, clinicians' knowledge and attitudes about substance use and treatment

Organizational: Lack of availability and accessibility of prevention, support, and treatment programs, access to primary health care

Community: Media representation, density of alcohol outlets, access to drugs

Policy: Inadequate oversight of the pharmaceutical industry, inadequate coverage for prevention and treatment of substance use disorders, lack of recognition of addiction as a disease [50]

Box 18.3   Patient Perspective

The photographer Nan Goldin has written about her experiences with opioid addiction and her subsequent efforts to hold pharmaceutical manufacturers accountable. In one essay, she describes how it began:

» My relationship to OxyContin began several years ago in Berlin. It was originally prescribed for surgery. Though I took it as directed I got addicted overnight. It was the cleanest drug I'd ever met. In the beginning, forty milligrams was too strong but as my habit grew there was never enough. At first, I could maintain. Then it got messier and messier. I worked the medical field in Berlin for scripts. When they shut me out I turned to FedEx. That worked until it didn't. The drug, like all drugs, lost its effect, so I picked up the straw [50].

Later, she writes:

» My life revolved entirely around getting and using Oxy. Counting and recounting, crushing and snorting was my full-time job. I rarely left the house. It was as if I was Locked-In. All work, all friendships, all news took place on my bed. When I ran out of money for Oxy I copped dope. I ended up snorting fentanyl and I overdosed.

» I wanted to get clean, but I waited a year to go into treatment because of my fear of withdrawal. Then in January I went into rehab for two and a half months. I was one of the fortunate ones who could afford an excellent hospital, which isn't an option for most people. I've stayed clean for almost a year. Getting off drugs and staying off drugs are two different things, each painful in their own way. But going back is not an option. My endowment was cut off and I regret the money I wasted. I regret the time I lost, which is irretrievable. Now I find the world hard to navigate, but I have a sharpened clarity and a sense of purpose [50].

References

1. American Society of Addiction Medicine. Public policy statement: definition of addiction [Internet]. Rockville (MD): ASAM; 2011 Apr 12 [cited 2018 Jul 30]. Available from: https://www.asam.org/resources/definition-of-addiction.
2. Substance Abuse and Mental Health Services Administration. Medication-Assisted Treatment (MAT). Rockville (MD): SAMH. [cited 2018 Jul 30]. Available from: https://www.samhsa.gov/medication-assisted-treatment.
3. American Psychiatric Association. Diagnostic and statistical manual of mental disorders. 5th ed. Washington, DC: American Psychiatric Association; 2013.
4. International Association for the Study of Pain. IASP terminology [Internet]. Washington, D.C.: IASP. [cited 2018 Jul 30]. Available from: http://www.iasp-pain.org/Education/Content.aspx?ItemNumber=1698.
5. Zempsky WT. Treatment of sickle cell pain: fostering trust and justice. J Am Med Assoc. 2009;302:2479–80.
6. Royal CD, Jonassaint CR, Jonassaint JC, De Castro LM. Living with sickle cell disease: traversing "race" and identity. Ethn Health. 2011;16:389–404.
7. Wesley KM, Zhao M, Carroll Y, Porter JS. Caregiver perspectives of stigma associated with sickle cell disease in adolescents. J Pediatr Nurs. 2016;31(1):55–63.
8. Centers for Disease Control and Prevention. Understanding the epidemic. https://www.cdc.gov/drugoverdose/epidemic/index.html. Accessed 30 July 2018.
9. Department of Health and Human Services. What is the opioid epidemic. https://www.hhs.gov/opioids/about-the-epidemic/index.html. Accessed 30 July 2018.
10. Centers for Disease Control and Prevention. Lesson 1: Introduction to epidemiology. https://www.cdc.gov/ophss/csels/dsepd/ss1978/lesson1/section11.html. Accessed 31 July 2018.
11. Krebs EE, Carey TS, Weinberger M. Accuracy of the pain numeric rating scale as a screening test in primary care. J Gen Intern Med. 2007;22(10):1453–8.
12. Williamson A, Hoggart B. Pain: a review of three commonly used pain rating scales. J Clin Nurs. 2005;14:798–804.
13. Morone NE, Weiner DK. Pain as the 5th vital sign: exposing the vital need for pain education. Clin Ther. 2013;35(11):1728–32.
14. Baker DW. History of the joint Commission's pain standards. Lessons for Today's prescription opioid epidemic. JAMA. 2017;317(11):1117–8.
15. Centers for Disease Control and Prevention. Opioids for acute pain: what you need to know [Internet]. Atlanta (GA): CDC. [cited 2018 Jul 31]. Available from: https://www.cdc.gov/drugoverdose/pdf/patients/Opioids-for-Acute-Pain-a.pdf.
16. Dumas F. After surgery in Germany, I wanted Vicodin, not herbal tea. The New York Times [Internet]. 2018 Jan 27 [cited 2018 Aug 2]. Available from: https://www.nytimes.com/2018/01/27/opinion/sunday/surgery-germany-vicodin.html.
17. Centers for Disease Control and Prevention. CDC guideline for prescribing opioids for chronic pain [Internet]. Atlanta (GA): CDC. [cited 2018 Jul 31]. Available from: https://www.cdc.gov/drugoverdose/prescribing/guideline.html.
18. Porter J, Jick H. Addiction rare in patients treated with narcotics. N Engl J Med. 1980;302:123.
19. Leung P, Macdonald EM, Dhalla IA, Juurlink DN. A 1980 letter on the risk of opioid addiction. N Engl J Med. 2017;376:2194–5.
20. Haffajee RL, Mello MM. Drug companies' liability for the opioid epidemic. N Engl J Med. 2017;377:2301–5.
21. Centers for Disease Control and Prevention. Vital signs: overdoses of prescription opioid pain relievers --- United States, 1999—2008. MMWR. 2011;60(43):1487–92.
22. Guy GP Jr, Zhang K, Bohm MK, et al. Vital signs: changes in opioid prescribing in the United States, 2006–2015. MMWR Morb Mortal Wkly Rep. 2017;66:697–704.
23. Centers for Disease Control and Prevention. US prescribing rate maps [Internet]. Atlanta (GA): CDC. [cited Jul 2018]. Available from: https://www.cdc.gov/drugoverdose/maps/rxrate-maps.html.
24. American Society of Addiction Medicine. National practice guideline for the use of medications in the treatment of addiction involving opioid use. Rockville (MD): ASAM; 2015 [cited 2018 Jul 31]. Available from: https://www.asam.org/docs/default-source/practice-support/guidelines-and-consensus-docs/asam-national-practice-guideline-supplement.pdf.
25. Office of the Surgeon General. Facing addiction in America: the surgeon general's report on alcohol, drugs, and health. Washington, D.C.: U.S. Department of Health & Human Services. [cited 2018 Jul 31]. Available from: https://addiction.surgeongeneral.gov/.
26. U.S. Department of Health & Human Services. Trump administration awards grants to states to combat opioid crisis [Internet]. Washington, D.C.: HHS. https://www.hhs.gov/about/news/2017/04/19/trump-administration-awards-grants-states-combat-opioid-crisis.html. Accessed 31 July 2018.
27. American Society of Addiction Medicine. Public policy statement on the use of naloxone for the prevention of opioid overdose deaths [Internet]. Rockville (MD): ASAM. [cited 2018 Jul 31]. Available from: https://www.asam.org/docs/default-source/public-policy-statements/use-of-naloxone-for-the-prevention-of-opioid-overdose-deaths-final.pdf.
28. Centers for Disease Control and Prevention. Expanding Naloxone use could reduce drug overdose deaths and save lives [Internet]. Atlanta (GA): CDC; 2015 Apr [cited 2018 Jul 31]. https://www.cdc.gov/media/releases/2015/p0424-naloxone.html.
29. Association of State and Territorial Health Officials. 2018 State legislation addressing naloxone. Arlington (VA): Astho; 2018 Apr [cited 2018 Jul 31]. Available from: http://www.astho.org/StatePublicHealth/2018-State-Legislation-Addressing-Naloxone/04-05-18/.
30. New York State Library. Combatting the opioid epidemic: New York libraries save lives [Internet]. Albany: New York State Department of Education; updated 2018 Sep 4 [cited 2018 Aug 2]. Available from: http://www.nysl.nysed.gov/libdev/opioid/index.html.

31. American Society of Addiction Medicine. Summary of the comprehensive addiction and recovery act [Internet]. Rockville (MD): ASAM. [cited 2018 Jul 31]. Available from: https://www.asam.org/advocacy/issues/opioids/summary-of-the-comprehensive-addiction-and-recovery-act.
32. Drug Policy Alliance. 911 good Samaritan: explaining NY's fatal overdose prevention law [Internet]. New York: Drug Policy Alliance. [cited 2018 Jul 31]. Available from: http://www.drugpolicy.org/sites/default/files/911_Good_Samaritan_Informational_Brief.pdf.
33. Goldensohn R. They shared drugs, someone died: does that make them killers? The New York Times [Internet]. 2018 May 25 [cited 2018 Aug 2]. Available from: https://www.nytimes.com/2018/05/25/us/drug-overdose-prosecution-crime.html.
34. National Alliance for Model State Drug Laws. Prescription drug monitoring program (PMPM/PMP) basics [Internet]. Harrisburg (PA): NAMSDL; 2018 [cited 2018 Aug 1]. Available from: http://namsdl.dynamicwebware.com/News%20Tab/Highlighted%20Issues/PDMP%201-Pager%20(FINAL).pdf.
35. The Pew Charitable Trusts. When are prescribers required to use prescription drug monitoring programs? [Internet]. Philadelphia (PA): Pew Trusts; 2018 Jan 24 [cited 2018 Aug 1]. Available from: http://www.pewtrusts.org/en/research-and-analysis/data-visualizations/2018/when-are-prescribers-required-to-use-prescription-drug-monitoring-programs.
36. U.S. Department of Health & Human Services. Mental health and substance abuse coverage [Internet]. Baltimore (MD): U.S. Centers for Medicare & Medicaid Services. [cited 2018 Aug 1]. Available from: https://www.healthcare.gov/coverage/mental-health-substance-abuse-coverage/.
37. Bevilacqua L, Goldman D. Genes and addictions. Clin Pharmacol Ther. 2009;85(4):359–61.
38. Bender MA. Sickle cell disease. In: Adam MP, Ardinger HH, Pagon RA, et al., eds. GeneReviews®. Seattle: University of Washington, Seattle; 1993–2018. [2003 Sep 15; updated 2017 Aug 2017]. Available from: https://www.ncbi.nlm.nih.gov/sites/books/NBK1377/.
39. Brousseau DC, Owens PL, Mosso AL, Panepinto JA, Steiner CA. Acute care utilization and Rehospitalizations for sickle cell disease. JAMA. 2010;303(13):1288–94.
40. Department of Health and Human Services, National Heart, Lung and Blood Institute. Expert Panel: evidence based management of sickle cell disease. Rockville (MD): National Institute of Health; 2014 [cited 2018 Aug 1]. Available from: https://www.nhlbi.nih.gov/sites/default/files/media/docs/sickle-cell-disease-report%20020816_0.pdf.
41. Porter JS, Wesley KM, Zhao MS, Rupff R, Hankins JS. Pediatric to adult care transition: perspectives of young adults with sickle cell disease. J Pediatr Psychol. 2017;42(9):1016–27.
42. Kaiser Health News. Two drug epidemics, decades apart: why government's response to opioid epidemic different than crack crisis [Internet]. San Francisco (CA): Kaiser Family Foundation; 2018 May 7 [cited 2018 Aug 1]. Available from: https://khn.org/morning-breakout/two-drug-epidemics-decades-apart-why-governments-response-to-opioid-epidemic-different-than-crack-crisis/.
43. Kolodny A, Frieden TR. Ten steps the Federal Government should take now to reverse the opioid addiction epidemic. JAMA. 2017;318(16):1537–8.
44. Hull SC, Jadbabaie F. When is enough enough? The dilemma of valve replacement in a recidivist intravenous drug user. Ann Thorac Surg. 2014 May;97(5):1486–7.
45. Centers for Disease Control and Prevention. Overcoming barriers to medication adherence for chronic diseases [web streming video]. Atlanta (GA): CDC; presented 2017 Feb 21 [cited 2018 Jul 30]. Available from: https://www.cdc.gov/grand-rounds/pp/2017/20170221-medication-adherence.html.
46. Hui K, Angelotta C, Fisher CE. Criminalizing substance use in pregnancy: misplaced priorities. Addiction. 2017 Jul;112(7):1123–5.
47. Goodnough A. When an Iowa family doctor takes on the opioid epidemic. The New York Times [Internet]. 2018 Jun 23 [cited 2018 Aug 2]. Available from: https://www.nytimes.com/2018/06/23/health/opioid-addiction-suboxone-treatment.html.
48. Mendiola CK, Galetto G, Fingerhood M. An exploration of emergency physicians' attitudes toward patients with substance use disorder. J Addict Med. 2018;12:132–5.
49. Bartlett R, Brown L, Shattell M, Wright T, Lewallen L. Harm reduction: compassionate care of persons with addictions. Medsurg Nurs. 2013;22(6):349–58.
50. Goldin N. I survived the opioid crisis. Artforum International Magazine. 2018 Jan; 56(5). Available: https://www.artforum.com/print/201801/nan-goldin-73181.

Further Reading on this Topic

Bergman EJ, Diamond NJ. Sickle cell disease and the "difficult patient" conundrum. Am J Bioeth. 2013;13(4):3–10.

Quinones S. Dreamland: The true tale of America's opiate epidemic. New York, NY: Bloomsbury Press; 2016.

Reading N. Methland: the death and life of an American small town. New York, NY: Bloomsbury; 2009.

Verghese A. The tennis partner: a doctor's story of friendship and loss. New York, NY: HarperCollins; 1998.

Conducting Research

Contents

Chapter 19 "I Don't Want to Be a Guinea Pig" – 369

Chapter 20 "Wait, I'm a Research Subject?" – 389

"I Don't Want to Be a Guinea Pig"

Gregory L. Eastwood

19.1 Background Questions – 370

19.2 Additional Case Information and Questions for Discussion – 370

19.3 Answers to Background Questions – 372

19.4 Responses to Discussion Questions – 376

References – 386

© Springer Nature Switzerland AG 2019
A. E. Caruso Brown et al. (eds.), *Bioethics, Public Health, and the Social Sciences for the Medical Professions*, https://doi.org/10.1007/978-3-030-03544-0_19

Learning Objectives

1. Analyze the ethical conduct of contemporary human subjects research in light of historical events in research involving humans.
2. Describe the process of drug development and approval in the United States.
3. Identify potential conflicts of interest and reasons why conflicts of interest occur.
4. Appreciate the views that patients and families may have about being participants in a research trial.

Case Background

You are a neurologist caring for a family with Duchenne muscular dystrophy (DMD), an X-linked recessive disorder that causes muscle degeneration, eventually leading to paralysis and death from respiratory or cardiac failure. There is no effective treatment for DMD. Your patients include Arun, age 11; his brother Ajay, age 14; and their maternal uncle, Satish, age 22. Another maternal uncle, Praveen, died from complications of DMD last year at age 24.

Two years ago, Arun Rajan was enrolled in a Phase II study of an experimental drug for DMD. At the time Arun was enrolled, Ajay did not qualify for the study because he already required a wheelchair for mobility.

19.1 Background Questions

1. What was the 1947 Nuremberg "Doctors' Trial," and what were the main stipulations of the Code that resulted from it?

2. What are the important elements of consenting to participate in a research study? What must the person who is the human subject understand for consent to be valid? How does consent apply to people under age 18?

3. What were the main recommendations of the 1964 Declaration of Helsinki and the 1979 Belmont Report of the U.S. Congress? Who are "vulnerable subjects," and what protections must be in place in a research study to safeguard these people?

4. How are new drugs developed and approved in the United States? What are the differences between Phase I, Phase II, Phase III, and Phase IV trials? How long does the process usually take to bring a drug to the market?

5. What are compassionate use (expanded access) programs? Some researchers, pharmaceutical companies, and the U.S. Food and Drug Administration (FDA) leaders argue that compassionate use programs can interfere with the conduct of clinical trials that would conclusively prove whether a drug works. What are their arguments?

6. The study on which this case is based has been criticized for its small sample size and use of historic controls. Why do these two features of study design matter?

7. What are "orphan diseases," and what federal entities are in place to encourage the development of treatments for these diseases?

8. What is a conflict of interest? What are some potential conflicts of interest in research studies?

19.2 Additional Case Information and Questions for Discussion

Over the past 2 years, Arun has tolerated the experimental drug without any significant problems. Furthermore, the rate of progression of his disease has markedly slowed down. Arun's parents credit the experimental drug with his continuing ability to walk with minimal assistance.

1. What are some of the key ethical questions in this case?

2. Who are the key stakeholders? What are the concerns, values, and preferences of each stakeholder?

3. Whose interests conflict in this case? What are some of the causes of these conflicts?

4. What can be done to address conflicts of interest in research? In medical practice?

Two years ago, before enrolling Arun in the study, his parents asked you what you would do if he were your child.

5. How would you answer this question? Is it acceptable to answer it directly? Why or why not?

6. How do you weigh the risks and benefits of study participation?

In order to consent to participate in a study, people need to have the mental capacity to understand and consent, have sufficient relevant information about the study, and understand the likely outcomes, benefits, and risks. Even when potential study participants meet these criteria, it is important for the person who is obtaining the consent to be aware of, and if indicated, discuss some potential pressures that study participants or their families may feel.

7. What interests may interfere with someone's ability to make a fully informed and voluntary decision to participate in research?

When you initially discussed the study with the family, you asked to speak with Arun without his parents in the room to get a better sense of how he felt about the situation. When you talked with him, he said, "I don't want to be a guinea pig!"

8. How do you respond to his concerns?

9. Are children entitled to know that they are participating in research? What about adults with cognitive impairment (e.g., a 25-year-old with fragile X syndrome and autism or a 65-year-old with dementia from Parkinson's disease)?

10. To what extent should children be able to decide for themselves whether to participate in a research study? What about adults who do not have decision-making capacity?

The Rajans are in your office today for a follow-up appointment and share that they have tried to obtain the experimental drug through compassionate use for both Ajay and Satish. The drug company has denied the request, stating that it has only enough drug for patients currently enrolled in its clinical trials. The family asks you to participate in a social media campaign to persuade the company to change its position.

11. Self-reflection: Would you participate, as the family requests, in a social media campaign to convince the drug company to change its position? Why or why not?
    (a) What conclusions do you think would be drawn from your participation by the family? By your other patients? By the public?
    (b) Do you think that this might lead other patients to make similar requests? What would be your obligation to those patients?

Reflecting on a similar situation in which a young child's family advocated on social media for access to a drug, bioethicist Arthur Caplan wrote: "If Josh were 67 instead of 7, he would already be out of luck. Those who are not very cute get less attention in their pursuit of unproven drugs. If Josh had parents who did not understand how to use social media, he would already be out of luck. If Josh did not have sharp, well-connected doctors, he would already be out of luck. But he is not in any of these categories, so he may yet get the drug" [1].

12. Does social media "level the playing field" between families and pharmaceutical companies, or does it perpetuate injustice?

13. How is justice important in this case?
    (a) With respect to medical care, to what, if anything, are Arun, Ajay, and Satish entitled?

14. What are the interests of the public at large, if public funds, such as grants from the National Institutes of Health (NIH), are being used to support the research?

While some people feel that participating in research is a privilege because it may provide access to drugs or procedures that are otherwise not available, as in this case, others may share Arun's feeling of "being a guinea pig" and thus choose not to participate. However, ensuring that participants have that choice has not always been the standard. Henrietta Lacks was an African-American woman who lived in Baltimore. In February 1951, at the age of 30, she was told she had epidermoid carcinoma of the cervix. She was treated with radium implants but died of metastatic cancer about 6 months later. Cervical biopsies of her cancerous cells, taken during her treatment, were given to laboratory researchers without her consent or knowledge. These cells grew into an immortal cell line, the first of its kind to grow successfully in cell culture, called HeLa cells, which have been used since that time by researchers around the world [2, 3].

15. **Henrietta Lacks' story is different, but it is important to our understanding of the history of medicine and medical research. What lessons apply to this case?**
    (a) **Clinical trials have historically under-enrolled patients from minority groups. Why?**
    (b) **Why is this problematic?**
    (c) **What should be done to address this?**

19.3 Answers to Background Questions

1. **What was the 1947 Nuremberg "Doctors' trial," and what were the main stipulations of the Code that resulted from it?**
   After World War II, a series of military trials was held in Nuremberg, Germany, to prosecute the leaders of Nazi Germany who were held responsible for war crimes. In 1947, 20 doctors and three administrators were tried for murder under the guise of euthanasia and atrocities that had occurred during Nazi medical research in the 1930s and 1940s, in which prisoners of war, civilians, and those deemed insane, aged, deformed, or incurably ill were subjected to experiments that involved freezing, shooting, infecting, exposing to high altitude, sterilizing, and torturing [4, 5].
   Current standards of rights of research subjects have their origins in the stipulations of the Nuremberg Code, which was developed as a consequence of the trial [6]. Current standards of these rights include:
   - To understand the purposes, risks, and benefits of the research before they consent to participate and to do so without coercion
   - To be protected against possibilities of injury, disability, or death
   - To have such research be conducted by qualified individuals
   - To ask questions at any time
   - To withdraw from participation at any time without repercussions
   - To be represented, if they are unable to understand the conditions or consequences of the research, by surrogates who can consent and otherwise act on their behalf and make decisions for them according to their wishes or best interests

2. **What are the important elements of consenting to participate in a research study? What must the person who is the human subject understand for consent to be valid? How does consent apply to people under age 18?**
   The human subject must:
   1. Have the mental capacity to understand and consent
   2. Have sufficient relevant information about the study
   3. Understand the likely outcomes, benefits, and risks

   This applies to adults 18 years and above. With regard to children and teenagers under age 18, the concept of graduated participation and assent by the subject applies. The law recognizes that parents or guardians of children most often make decisions in the best interests of the child and, without evidence to the contrary, the law accords parents this right. However, the ability of a 5-year-old to comprehend and contribute to the decision of whether to participate in a research study would be quite different from the ability of a 15-year-old to do the same. Thus, as a child matures and gains the ability to help make rational decisions, that child's participation in and assent to the decision should be considered. Children who have experienced prolonged or chronic illness may have a stronger grasp of their health and of medicine than older children and may therefore be more capable of meaningful participation [7].

✓ 3. **What were the main recommendations of the 1964 Declaration of Helsinki and the 1979 Belmont Report of the U.S. Congress? Who are "vulnerable subjects," and what protections must be in place in a research study to safeguard these people?**

The Declaration of Helsinki acknowledged the need to study people who cannot give consent and may be vulnerable but still need to be involved in research studies, such as infants, children, those who are mentally or developmentally disabled, and those who are critically ill. It introduced the notion that such individuals should have a surrogate or guardian decision-maker to make decisions on their behalf according to their wishes or best interests [8, 9].

The Belmont Report defined the distinctions between research and clinical practice [10]. It said that in the practice of medicine, interventions are intended to benefit the patient, whereas research is intended to test a question and the subject may or may not benefit. The report also affirmed several ethical principles.

- **Respect for persons**: People are autonomous, and if autonomy is diminished, the person is entitled to protection. (This echoes a recommendation of the Declaration of Helsinki.)
- **Beneficence**: Researchers should make every effort to ensure the subject's well-being.
- **Justice**: The report raised questions about who should benefit or bear the burdens of research. For example, should certain socially, politically, and/or economically disadvantaged groups of people bear the burden of being research subjects while others benefit?

✓ 4. **How are new drugs developed and approved in the United States? What are the differences between Phase I, Phase II, Phase III, and Phase IV trials? How long does the process usually take to bring a drug to market?**

The preclinical phase of drug development includes the search for new drugs, understanding the physicochemical properties of potential drugs, and testing them in vitro and in animals to define therapeutic actions, side effects, and toxicity. Results may be submitted to the Food and Drug Administration (FDA) as an Investigational New Drug (IND) application. Approval of the IND leads to the clinical phase [11, 12].

The clinical phase includes three or four stages [13, 14].
- **Phase I**: The drug is tested in small numbers of healthy volunteers to determine safety and appropriate dosing.
- **Phase II**: The drug is tested in small groups of people having the disease for which the drug is intended to determine whether it is effective and to provide further information regarding safety.
- **Phase III**: The drug is tested in large numbers, usually hundreds, of people with the disease to further determine efficacy and safety. If the drug is deemed safe and effective, it may be approved for the status of New Drug Application (NDA). In the NDA stage, the FDA considers all the information from the manufacturer, including the results of the Phase 1–3 trials, the reliability of the formulation of the drug, and prescribing information that will be included on the product label and elsewhere.
- **Phase IV**: Sometimes the FDA requires a post-approval trial after the drug is on the market as a condition of approval.

The search for new drugs and testing in vitro and in vivo may require several years before the IND application. It may then take 6–8 years for a drug to go from IND application to market. Approximately 10% of candidate drugs eventually make it to market [15]. The pricing of new drugs is a controversial issue. It is related in part to costs of development, but also—more significantly—to what the market will bear, advertising, and whether the drugs are sold in the United States or somewhere else. If the drug is sold in the United States, pricing will further depend upon whether it is sold on the free market or sold by negotiation to large organizations, such as the Veterans Administration, insurers, or groups of healthcare institutions [16, 17].

5. **What are compassionate use (expanded access) programs? Some researchers, pharmaceutical companies, and the U.S. Food and Drug Administration (FDA) leaders argue that compassionate use programs can interfere with the conduct of clinical trials that would conclusively prove whether a drug works. What are their arguments?**
Compassionate use programs allow some patients to use a drug that currently is in a clinical trial, before the drug has been FDA-approved, without entering as a subject in the trial [18]. Such patients usually are not eligible for entry into a clinical trial of the drug because they do not meet the entry criteria, but nevertheless their physician thinks they might benefit from the drug.

Regarding requirements for expanded access, the FDA states that "[u]nder the Federal Food, Drug, and Cosmetic Act (FD&C Act), a patient may seek individual patient expanded access to investigational products for the diagnosis, monitoring, or treatment of a serious disease or condition if the following conditions are met [18]:
- The patient and a licensed physician are both willing to participate;
- The patient's physician determines that there is no comparable or satisfactory therapy available to diagnose, monitor, or treat the patient's disease or condition;
- The probable risk to the person from the investigational product is not greater than the probable risk from the disease or condition;
- The FDA determines that there is sufficient evidence of the safety and effectiveness of the investigational product to support its use in the particular circumstance;
- The FDA determines that providing the investigational product will not interfere with the initiation, conduct, or completion of clinical investigations to support marketing approval;
- The sponsor (generally the company developing the investigational product for commercial use) or the clinical investigator (or the patient's physician in the case of a single patient expanded access request) submits a clinical protocol (a document that describes the treatment plan for the patient) that is consistent with FDA's statute and applicable regulations for INDs or investigational device exemption applications (IDEs), describing the use of the investigational product; and
- The patient is unable to obtain the investigational drug under another IND or to participate in a clinical trial" [18].

Compassionate use programs have the potential to interfere with the conduct of clinical trials that would prove whether a drug works by diminishing the number of study subjects, thus reducing the data needed to make statistical or observational conclusions and possibly lengthening the duration of the study to accrue enough subjects to achieve statistical significance. This would be particularly relevant to diseases and disorders that are rare and for which there are few potential study participants. Further, if patients receive study drugs under compassionate use programs, such stories may encourage other patients who otherwise would participate in the drug trial to seek compassionate use instead. It is also important to note that because the efficacy of investigational drugs is unproven, patients seeking compassionate use may not benefit from the drug.

✓ 6. **The study on which this case is based has been criticized for its small sample size and use of historic controls. Why do these two features of study design matter?**

In general, a small sample size makes it more difficult to detect statistically significant differences between study groups. During the process of designing and planning the study, researchers should do a power calculation that depends on both sample size and the expected difference in response rates between the control and treatment groups. In general, a small sample size increases the chance of a Type II error (failing to find a difference when one really exists). However, if the anticipated difference in outcomes is great, a smaller sample size may be acceptable [19].

Historic controls are considered weaker than concurrent control groups but are often used for practical purposes. There are several limitations of using historical controls. First, at least some conditions of treatment are likely to be different between the historic controls and the current study patients. Even if the disease or condition being studied lacks effective therapy, other standards for related health care (such as management of comorbid conditions) change and may confound the results. Second, the study patients should be matched to the historic controls as closely as possible with consideration to such characteristics as age, sex, duration of disease, treatments, comorbid conditions, and ethnicity. Third, when a study uses historic controls, all study participants receive the intervention and therefore cannot be blinded. Thus, there may be subtle differences between patients treated in the past, who did not choose to participate in a clinical trial, and patients in the present, who are choosing to participate [20, 21].

✓ 7. **What are "orphan diseases" and what federal entities are in place to encourage the development of treatments for these diseases?**

An orphan disease is a rare disease, often genetic in origin, for which there are few, if any, effective treatments. The Office of Rare Diseases Research, within the National Institutes of Health, recommends a national research agenda, coordinates the research, and maintains a centralized database on rare disease research. The Office provides research grants for the study of rare diseases, encourages investigators to collaborate and share data, and facilitates support groups for patients with rare diseases [22]. Also, the U.S. Food and Drug Administration (FDA) has an Office of Orphan Products Development (OOPD) that advances the development of products that demonstrate promise for the diagnosis or treatment of rare diseases [23].

✓ 8. **What is a conflict of interest? What are some potential conflicts of interest in research studies?**

In clinical medicine, a conflict of interest is anything that threatens to come between the physician (or nurse, or other healthcare practitioners) and his/her primary duty to the welfare of the patient. In research a conflict of interest may take two forms: (1) anything that threatens to come between the investigator and his/her primary duty to the welfare of the human research subject and (2) anything that threatens to come between the investigator and the integrity of the research.

Stated another way, "A conflict of interest in research exists when the individual has interests in the outcome of the research that may lead to a personal advantage and that might therefore, in actuality or appearance, compromise the integrity of the research" [24].

Examples of conflicts of interest include:

- Possibility of financial gain by receiving payment for the use of products related to the research or for making presentations on behalf of a company, owning an equity interest in a relevant company, or having any other financial connection to an entity that

might affect one's judgment regarding the human subject or the proper conduct of the research [25]
- Influence by gifts or other considerations by other interested parties, such as industry representatives [26]
- Lure of prestige, fame, or enhanced reputation
- Desire for promotion or advancement in the university or company in which the investigator works
- Genuine interest in science, i.e., finding answers to questions, by the investigator, which may override the primary duty to the patient or research subject
- Desire to get good grades (students) or evaluations (trainees or other subordinates)
- Pressures by the university or company to complete the research quickly
- Personal, family, and social interests and obligations that conflict with responsible conduct of research

Some conflicts can be acknowledged and managed so that they do not interfere with the primary duty to the welfare of the human subject or the integrity of the research. Other conflicts cannot be managed and must be relinquished, such as the requirement in some institutions that investigators either do not enter into or give up certain relations with industry.

19.4 Responses to Discussion Questions

Over the past 2 years, Arun has tolerated the experimental drug without any significant problems. Furthermore, the rate of progression of his disease has markedly slowed down. Arun's parents credit the experimental drug with his continuing ability to walk with minimal assistance.

✓ 1. What are some of the key ethical questions in this case?
   Issues and questions in this case include:

> **Box 19.1 Teaching Tip**
> Specific issues will be explored in detail in the questions that follow; however, there are advantages to beginning with a broad, open-ended question such as this. It allows from advanced learners a chance to analyze the case "from scratch," without being led in a particular direction. It also allows instructors to gain a better sense of what knowledge and experience learners are bringing to the course. If the response is silence, use the subheadings below to guide follow-up questions or move on and return to this question later in the discussion.

- *The primacy of the welfare of the patient or research subject.* All ethical principles in current health care derive from and relate in some relevant, applied way to the fundamental commitment of healthcare professionals—physicians, nurses, physical therapists, respiratory therapists, social workers, and others—to the welfare of the person who is a patient. In contemporary medical oaths, this is phrased something like, "The welfare of my patient will be my primary concern." This principle can be traced to about 2500 years ago at the time of the Greek physician Hippocrates. The primacy of the welfare of the patient not only drives physicians and other healthcare professionals, but it also is the reason we perform biomedical research, educate students to be practitioners and researchers, promote social actions, and even engage in healthcare administrative activities—all to foster the welfare of patients.
- *The balance between beneficence and non-maleficence.* Beneficence means to do good, and non-maleficence means to avoid or minimize harm. Non-maleficence is not synonymous with the maxim "first do no harm." If the first principle were to avoid harm, physicians would not prescribe anti-cancer drugs or perform open-heart surgery. In fact, we may be paralyzed into inaction, because almost everything we do has the potential

to cause some harm. Healthcare practitioners accept that some harm may occur but always do so with the expectation that in the balance between beneficence and non-maleficence, the benefit will exceed the harm or the possibility of harm.

- *Justice.* Justice is a complex bioethical principle. In its several connotations, justice has broad applications not only to individuals but also to groups of people, even to whole populations. With respect to individuals, justice has implications of fairness and equality. For example, justice provokes questions such as:
  - Do I have access to a healthcare system that is comparable to what is available to other people?
  - When I engage the healthcare system, will I be treated fairly?
  - Will I suffer from racial, ethnic, social, economic, gender, age, or some other bias?
  - Will I be given the same access to diagnostics and treatments as others who are in a similar situation?

With respect to groups of people, concerns of individuals about age, ethnicity, and so on also may apply. This relates to the notion of distributive justice, which considers the fair distribution of resources, some of which may be in short supply. Will all children receive the necessary immunizations or just some who can afford them? Will all people with diabetes have access to proper care? Are cardiac angiography, cancer therapy, preventative diagnostics and treatments, and so on available to all who need them?

2. **Who are the key stakeholders? What are the concerns, values, and preferences of each stakeholder?**
   - **The parents:** They credit Arun's ability to walk with minimal assistance to the drug. They have the dilemma of feeling that they are not doing enough to provide effective care to Ajay. The extent to which the drug has helped Arun is not known objectively. Should their impression about the drug's effectiveness be generalized and applied to Ajay? They may also feel guilt for having children who are affected with such a devastating disorder, particularly the mother, because DMD is X-linked. Ultimately, they want what is best for Arun and Ajay, in the absence of any information to suggest otherwise, and they are hopeful that eteplirsen will be effective.
   - **Arun:** He has an interest in maintaining his mobility to the greatest extent possible and in living as long as possible with a good quality of life; however he might define this for himself. He may have some guilt that his brother has not been able to maintain the same degree of mobility. He may have fears about mortality. He almost certainly is aware that his disease likely will kill him at a young age, given his uncle Praveen's death, and he probably assumes that improved motor function translates to increased longevity. In some diseases, children with life-limiting diseases are more likely than their parents to express a desire to participate in research that will benefit other children in the future. Arun may be motivated by this type of altruism. On the other hand, he might have been fearful initially about participating in the study or resentful of the burdens of participation. We will explore Arun's initial reaction to the study below.
   - **Ajay:** His interests are similar to Arun's. He may have some resentment that his brother has more mobility than him, possibly through the use of a drug that he has not been allowed to take, and he might envy Arun's good luck in this regard.
   - **The neurologist:** Should the neurologist be involved, as the family requests, in a social media campaign?

To do so would be to declare the neurologist's acceptance that the drug is working. Furthermore, doing so would send a message that the drug trial, and possible interests of other patients with DMD who may benefit from a properly conducted trial, should be disregarded. This may also put the neurologist in a difficult position when other patients with DMD or other conditions make similar requests to appeal for compassionate use of a drug. If the neurologist does it for one patient, it sets a precedent. Why not then advocate for other patients who do not fit the criteria for admission as a study subject? Like the parents, the neurologist wants to do what is best for Arun and Ajay.

- **The drug maker and study investigators**: They also want to find an effective treatment but may have conflicts of interest that need to be reconciled with what is best for the patient(s) and research subjects. These are discussed further in the next question.
- **The FDA**: The U.S. Food and Drug Administration is a federal public health agency. As such, its interests should be in achieving the best health of a population by ensuring that (1) approved drugs are both safe and effective prior to use in the general population but that (2) they are approved in a timely and efficient manner. This is a difficult balance to achieve. Ultimately, the FDA is staffed by individuals who are at risk for conflicts of interest. During the discussion of a recent nominee for FDA Commissioner, it was noted that nominees are typically criticized for being "too close to [the pharmaceutical] industry," implying that they are likely to have conflicts of interest in favor of the industry, or for "not being close enough," implying that they do not understand enough about the pharmaceutical industry to collaborate and regulate effectively [27].

Compassionate use exceptions formally require "sufficient evidence of the safety and effectiveness" for the patient and that such use "not interfere with the initiation, conduct, or completion of clinical investigations to support marketing approval" [18]. The death of a patient who receives an investigational new (experimental) drug will usually halt open Phase I and II studies, until the association between the drug and the patient's death can be fully investigated, even if the patient was receiving the drug through an expanded access, or compassionate use, program.

- **The insurance company**: Insurance companies are conflicted because they have a duty, if they are a for-profit company, to make a profit, and if a not-for-profit company, not to lose money. They need to be prudent in determining which treatments and procedures they cover and in respecting the protocols established by the FDA.
- **Other patients with DMD, their families, and their healthcare professionals**: Their interests may align or at least overlap with the interests of the Rajans and the neurologist. However, as noted previously, there may be negative consequences for patients currently enrolled in clinical trials, or for future patients, if a patient receives eteplirsen through an expanded access program and has an adverse response, even if the drug is, in fact, safe and effective, and the adverse reaction was coincidental. This is one reason that most trials have strict entry criteria. Whereas it is true that enrolling all patients, regardless of comorbidities or disease progression, might best reflect the real population with the disease, patients who are sicker are more likely to experience serious adverse events (SAEs) that are unrelated to the drug under study, and thus they may be restricted from

the study. Furthermore, eteplirsen targets a mutation that is found only in about 14% of patients with DMD, meaning most patients are not likely to benefit from it and may be at a disadvantage if resources are directed away from other potential treatments and toward one that is receiving media attention [28].

3. **Whose interests conflict in this case? What are some of the causes of these conflicts?**
As discussed earlier, the drug maker, study investigators, and insurance company may stand to benefit financially in different ways from the outcome of this case, the trial, and the drug itself in the long run. This interest stands in potential conflict with the individual best interests of the patients. In some situations, the treating physician (such as the neurologist in this case) may also be a study investigator, and even if the physician does not stand to benefit financially, he or she may stand to benefit professionally from the successful completion of the trial, publication of the results in an academic journal, and eventual approval of the drug.

4. **What can be done to address conflicts of interest in research? In medical practice?**
Most institutions in which research is done have policies that address the ethical and proper conduct of research. Included usually are requirements to declare financial conflicts, such as stock ownership in relevant companies and receipt of payment from industry for speeches or consultative services. The institution then has processes to determine how to manage the conflict, including sometimes withdrawing from participation in the research or divesting the relationship with industry.

Such policies also may prohibit gifts from research sponsors, regulate reimbursement for travel from sponsors, describe appropriate use of intellectual property, and prohibit participation on institutional committees that award contracts to industry, make purchasing decisions, and the like.

Most research institutions also have a conflict of interest officer, research integrity officer, ethics officer, or similarly named individual or office. These officers typically promulgate programs that educate researchers about institutional policies and should be consulted for questions and guidance. Medical practice in institutions and groups may have similar policies to which clinicians are expected to adhere.

> **Box 19.2   Teaching Tip**
> Consider reviewing with learners the relevant policies of your institution.

Medical and graduate education curricula also commonly include instruction in the proper conduct of research and of medical practice. Although the medical profession and the research community are subject to governmental and institutional oversight and regulation, they also are expected to be self-regulating, meaning that individuals are expected to behave in a professional and ethical manner and to take responsibility for their peers in this regard [29].

Two years ago, before enrolling Arun in the study, his parents asked you what you would do if he were your child.

5. **How would you answer this question? Is it acceptable to answer it directly? Why or why not?**
The physician has three options:
1. Answer directly, based on what she/he believes she/he would do.
2. Decline to answer because she/he has never been in the position of making a similar decision and therefore cannot truly say.
3. Deflect the question in some way, such as asking more questions (e.g., "How do you think that would help you to make this decision?"), offering further information (including, if

possible, contact with other families who have made similar decisions), or using reflective language to summarize the family's thought processes so far.

Should the physician exercise option 1 and answer directly? Physicians often are asked this question in one context or another, so it is something that should be expected. Sometimes the stakes are low (e.g., a question related to a common procedure or treatment), or the physician in fact does have direct experience with a condition comparable to the one that the patient has, so a direct response could be justified. In this situation, the stakes seem higher, and one could argue that unless the physician has a child with DMD, the physician should not use option 1. Even in such a case, there may be important differences in personal beliefs and values, family circumstances, or medical history that differ; however, the physician would be speaking honestly from experience, rather than hypothesizing.

In many cases, option 2 is the most appropriate response for a healthcare professional. Option 3 allows the physician to gather more information and perhaps to understand what is behind the patient's concerns. In practice, with both options 2 and 3, it is possible that the parents will refuse to accept those answers and will demand a "real answer."

Patients come to a physician because of the physician's expertise in medicine and should expect to receive expert opinion and advice about how to manage their health. The physician needs to balance the autonomy of the patient to make his/her own decisions with the physician's appropriate role of expert advisor to the patient. Patients often ask physicians "What would you do?" simply because physicians are the experts. We consult lawyers, auto mechanics, and plumbers for their expert advice and direction, so do we consult physicians for the same reasons. However, even when physicians think they "know" what they would do, that does not make their hypothetical choice the right choice in general, or for this family. Regarding research, a healthcare practitioner's "professional intuition" regarding the potential benefits of study participation does *not* mean that this study is more likely to produce positive results than another study.

✓ 6. **How do you weigh the risks and benefits of study participation?**

> **Box 19.3   Teaching Tip**
> Questions such as this one can be daunting for more junior learners, particularly those with limited background in study design. If learners have been assigned background questions to research prior to class, it may be helpful to note which learners should have a stronger foundation in certain topics. Referring back to those questions can help them build connections between content areas, as well as creating space for quieter learners to speak up and share their deeper knowledge with the group.

Understanding a few aspects of this clinical trial is helpful. This is a Phase II study of eteplirsen. This means that the drug has gone through Phase I trials in a small number of people and thus has been deemed safe, a safe dosage range has been determined, and some side effects may have been observed. During the Phase II study, the drug is given to a larger number of patients to see whether it is effective and monitor for side effects. The genetic defect in DMD patients leads to the production of inadequate amounts of functional dystrophin because of an X-linked genetic mutation. This causes muscle degeneration, which progresses to paralysis and death from respiratory or cardiac failure. The presumed mechanism of action of eteplirsen is to remove the exon near the DMD mutation so that the downstream reading frame for dystrophin can be corrected and functional dystrophin can be made. Whether this truly improves symptoms in DMD patients is the intent of the Phase II study [12, 20].

Risks: Common side effects of eteplirsen include loss of balance, headaches, vomiting, and proteinuria, none of which were dose-related, dose-limiting, or interrupted treatment [28]. Side effects may impair quality of life, and the requirements of study participation may be burdensome.

Benefits: The major benefit is improved physical health, *if* indeed eteplirsen is effective, or some improvement might also be achieved due to enhanced supportive care as a research participant. Similarly, improved quality of life might be appreciated if eteplirsen works or if there is a secondary psychological benefit of research participation. Some patients will experience an enhanced sense of agency, associated with the decision to participate, or an enhanced sense of self-worth and altruism (if the participant is acting from a desire to help future patients) [30].

In order to weigh risks and benefits, both healthcare practitioners and prospective participants (and their families) need to understand the probability of each, the magnitude or severity, and any factors that might ameliorate or exacerbate the risks and benefits. A typical consent form might report "common but not serious" risks (affecting 5–20% or more of patients receiving the intervention but typically minor, such as temporary nausea and vomiting), "uncommon but serious" or potentially serious (affecting 1–10% of patients), and "rare but life-threatening" (affecting fewer than 1% of patients).

At the time of the events of this case, eteplirsen had not been shown to have any serious or life-threatening side effects, but this was less certain earlier in clinical trials. Data for efficacy remained limited. One potential problem, common in early-phase trials, is the use of a surrogate biologic endpoint—in this case, levels of the dystrophin protein—rather than a clinically relevant one, such as the ability to walk [20, 31].

In order to consent to participate in a study, people need to have the mental capacity to understand and consent, have sufficient relevant information about the study, and understand the likely outcomes, benefits, and risks. Even when potential study participants meet these criteria, it is important for the person who is obtaining the consent to be aware of, and if indicated, discuss some potential pressures that study participants or their families may feel.

✓ 7. **What interests may interfere with someone's ability to make a fully informed and voluntary decision to participate in research?**
   The patient and family may also have conflicting interests that interfere with the ability to make a fully informed and voluntary decision to participate in research. Such reasons may include:
   - Desire to be viewed as a compliant or as a "good" patient.
   - Beliefs by the parents that "doing everything" is part of being a good parent and that enrolling in a clinical trial is part of "doing everything" [32].
   - Therapeutic misconception—a phenomenon in which potential research participants conflate the discussion of research participation by the physician as a therapeutic recommendation to participate [33].
   - Unrealistic optimism—a phenomenon in which potential research participants or their surrogates overestimate the likelihood that the individual patient will benefit from participation [34, 35].
   - Fears—side effects and randomization.
   - Concerns about the costs of being in a clinical trial [36].
   - Erroneous or misleading information from family and friends or media (including social media and Internet sources), which may contribute to all of the above.

When you initially discussed the study with the family, you asked to speak with Arun without his parents in the room to get a better sense of how he felt about the situation. When you talked with him, he said, "I don't want to be a guinea pig!"

✓ 8. **How do you respond to his concerns?**
A good start would be to explore what he means when he says "guinea pig." Many people understand, admire, and respect science and have realistic notions of what science can do and cannot do. Many people do not feel that way at all but rather are suspicious of science, the scientific method, and scientists. It is important to understand why Arun said this and explore his understanding of the situation. What does he imagine being a research subject will involve? Is he envisioning that he will be poked and prodded in a laboratory or, less fancifully but perhaps realistically, that he will have more frequent medical appointments and additional blood tests? How much does he understand about his disease and prognosis, the lack of effective treatment, and the study? Does he believe that effective treatments exist outside the study? Can he articulate more sophisticated reasons for not wanting to participate?

The patient's objection might be approached differently, depending on the responses to the questions above. For example, irrational fears that are not grounded in truth can be allayed with age-appropriate information, perhaps including meetings with other children who are participating in research. On the other hand, practical concerns about potential pain and suffering, even impaired quality of life if the child and his parents must make frequent trips to a distant hospital, should be addressed seriously, especially for a child with a condition that will significantly limit his life span. The child, along with his parents and the medical team, will have to weigh carefully the potential benefits against the burdens of participation. If the child does not have "veto power," he should be told this honestly, but he should still have a voice in the decision-making process.

✓ 9. **Are children entitled to know that they are participating in research? What about adults with cognitive impairment** (e.g., a 25-year-old with fragile X syndrome and autism or a 65-year-old with dementia from Parkinson's disease)?
In most cases, children old enough to understand the basic principles of scientific exploration should be given an age-appropriate explanation of the study in which they are enrolled. In some cases, parents have expressed concerns that disclosure about research participation would result in their children losing trust in their physicians or confidence in the efficacy of the treatment and have asked that this information be withheld. Assuming the specific details of such a case placed it outside the bounds of standards set by institutional policy, the final decision would likely involve a compromise between parent and physician, and perhaps an ethics consultation would be helpful. The physician might work with the parents and other team members, including members of the psychosocial support team, such as a child life specialist, social worker, and psychologist, to agree upon an acceptable way to disclose the information to the child. Depending on the likelihood of benefit from the research, the physician might refuse to enroll the patient in the study. The younger the child, the more likely the physician might acquiesce to the parents' request. Similar considerations apply to adults with cognitive impairment, but are related to the degree that the physician perceives the patient can understand the research protocol and its implications for the patient.

✓ 10. **To what extent should children be able to decide for themselves whether to participate in a research study? What about adults who do not have decision-making capacity?**
In most cases, as indicated in the previous question, children old enough to understand the basic principles of scientific exploration should be given an age-appropriate explanation of the study in which they are enrolled. Most institutional review boards require that

investigators produce an assent form and describe an assent process, which the board then reviews for approval; some have institution-wide policies regarding assent. A typical policy might require a child's verbal assent at age 6 or 7 years and written assent around age 10–14 years with an option for a parent and physician both to sign if they agree that the child does not have the capacity to provide assent. Many institutions and investigators will not enroll a child in a study that does not provide direct benefit without that child's assent, even if the study is minimal risk. Studies of children must provide "direct" benefit if they are "more than minimal risk." Risk assessments usually are based upon the type of risk a child will encounter in his or her daily life [37, 38].

Regarding adults who do not have decision-making capacity, a category that includes many vulnerable subjects, some of the concerns are comparable to those present when children are involved. However, adults may be less likely to have a surrogate decision-maker who is able to fully prioritize their best interests. On the other hand, they may have previously expressed wishes, goals, and desires that shed light upon whether they would choose to participate in a study. Historically, adults with conditions that impair decision-making, such as dementia or intellectual impairment, have been subjected to unethical and risky research with limited benefits. In recent years, the pendulum has swung, and they probably have been excluded from research that would have benefited them, individually or as a group.

The Rajans are in your office today for a follow-up appointment and share that they have tried to obtain the experimental drug through compassionate use for both Ajay and Satish. The drug company has denied the request, stating that it has only enough drug for patients currently enrolled in their clinical trials. The family asks you to participate in a social media campaign to persuade the company to change its position.

11. **Self-reflection: Would you participate, as the family requests, in a social media campaign to convince the drug company to change its position? Why or why not?**
    (a) **What conclusions do you think would be drawn from your participation by the family? By your other patients? By the public?**
    If the physician participates in a social media campaign on behalf of the parents, it is likely that the family would be grateful. But by doing so, intentionally or not, the physician would be declaring to the family, other patients, and the public, acceptance that the drug is working. Furthermore, the physician is sending the message that the drug research study, and possible interests of other patients with DMD who may benefit from a properly conducted trial, should be disregarded. This is an example of having to balance the interests of an individual patient with the interests of other patients with DMD. By acquiescing to the request of the parents to participate in a social media campaign, are you working against the interests of other patients with DMD and their families?
    (b) **Do you think that this might lead other patients to make similar requests? What would be your obligation to those patients?**
    Depending on the physician's practice, this may be unlikely because of the rarity of the disorder. A primary care pediatrician may have only one patient with DMD. Conversely, a specialist who has many such patients, would be more likely to have more patients with DMD and this could be a serious concern. Justice requires that physicians do not favor or privilege some patients over others. The decision to advocate for one particular patient, in this way, requires that the physician would be willing to do so for others.

Reflecting on a similar situation in which a young child's family advocated on social media for access to a drug, bioethicist Arthur Caplan wrote: "If Josh were 67 instead of 7, he would already be out of luck. Those who are not very cute get less attention in their pursuit of unproven drugs. If Josh had parents who did not understand how to use social media, he would already be out of luck. If Josh did not have sharp, well-connected doctors, he would already be out of luck. But he is not in any of these categories, so he may yet get the drug" [1].

✓ 12. **Does social media "level the playing field" between families and pharmaceutical companies or does it perpetuate injustice?**
It seems to have the potential to do both, as Caplan's quote notes. On one hand, it can provide a platform for patients and families to reach a much larger audience and overcome some of the hierarchy inherent in doctor-patient interactions. Sometimes attention, and attendant public outrage, is effective and necessary to correct injustices. On the other hand, certain aspects of privilege may persist even in social media spheres, bestowing more attention upon the highly educated, motivated, eloquent, or attractive.

✓ 13. **How is justice important in this case?**
Justice is relevant here because, as noted earlier, other patients with DMD, and their families, may lose out on benefiting from properly conducted trials of treatment for DMD. Also, it is possible that the patient with DMD and his family may be treated unfairly or inequitably. That means that the patient, having skirted the usual rules for receiving the drug in a trial, may either benefit from receiving the drug after the drug company caves to pressure from you and others, while other patients do not have that benefit, or be harmed if it turns out that the drug being studied is worse than standard treatment (▶ Chapter 20 also explores this issue.)

Box 19.4   Teaching Tip
If learners did not discuss the interests of other patients and families with DMD, ask them to do so here.

(a) **With respect to medical care, to what, if anything, are Arun, Ajay, and Satish entitled?**
Arun, Ajay, and Satish are entitled to medication that is both safe and effective, if such a treatment is available. However, it is not clear that the medication in question meets those criteria. So in the absence of a safe and effective treatment for DMD, Arun, Ajay, and Satish are entitled to safeguards, either as patients or research subjects, that respect their wishes and protect their interests.

Box 19.5   Teaching Tip
Ask learners to consider the risks of drawing rapid conclusions from limited data. Can they think of an example of something that was rapidly and widely adopted by biomedicine and subsequently disavowed?

✓ 14. **What are the interests of the public at large, if public funds, such as grants from the National Institutes of Health (NIH), are being used to support the research?**
The public contributes through tax dollars to the federal funds and thus has an interest in how those funds are allocated. Also, the public by voting selects governmental representatives and officials who establish policy and make decisions about the allocation of public funds. So the public has an interest in having funds used to support studies that are designed and conducted responsibly and in such a way that their results are likely to advance the practice of medicine and improve human health. At the same time, the public also has an interest in assuring that pharmaceutical companies' profit margins are reasonable, since many drugs begin in laboratories that receive at least some federal funding. Finally, the public has an interest in having access to trials and to treatments, once approved, in a reasonable timeframe, while still being able to trust that those treatments are safe and effective. Further, the NIH makes available to the public all manuscripts that have been accepted for publication of research supported by the NIH [39].

"I Don't Want to Be a Guinea Pig"

While some people feel that participating in research is a privilege because it may provide access to drugs or procedures that are otherwise not available, as in this case, others may share Arun's feeling of "being a guinea pig" and thus choose not to participate. However, ensuring that participants have that choice has not always been the standard. Henrietta Lacks was an African-American woman who lived in Baltimore. In February 1951, at the age of 30, she was told she had epidermoid carcinoma of the cervix. She was treated with radium implants but died of metastatic cancer about 6 months later. Cervical biopsies of her cancerous cells, taken during her treatment, were given to laboratory researchers without her consent or knowledge. These cells grew into an immortal cell line, the first of its kind to grow successfully in cell culture, called HeLa cells, which have been used since that time by researchers around the world [2, 3].

✅ 15. **Henrietta Lacks' story is different, but it is important to our understanding of the history of medicine and medical research. What lessons apply to this case?**

> **Box 19.6 Teaching Tip**
> Encourage learners to reflect on the opposing but coexisting views of research participation as (1) something desirable and sought-after by some and (2) something undesirable, inflicted on the vulnerable without their consent.

(a) **Clinical trials historically have under-enrolled patients from minority groups. Why?**
Patients who are more affluent and educated seem more likely to make their own inquiries on the Web or elsewhere regarding treatment options, to request second opinions, and to have the resources to travel to other cities and medical centers. Patients who belong to ethnic or racial minority groups may be less likely to receive care at research institutions that could recruit them or may not seek out such institutions.
    Researchers may feel that low-income or disadvantaged patients are less reliable as research subjects because in the clinics such patients may have more trouble keeping appointments and following treatment plans for reasons that may be out of their control. Patients from historically marginalized groups also sometimes have distrust of research institutions due to legitimate concerns about being studied without appropriate informed consent (e.g., Henrietta Lacks' case, the Tuskegee syphilis study discussed in Background Question 6, the Kennedy Krieger lead abatement study discussed in ▶ Chapter 6) [3, 40, 41]. Further, members of disadvantaged groups are themselves underrepresented among researchers, and researchers generally are more likely to recruit subjects who look like them and speak like them [42].

(b) **Why is this problematic?**
Without good data in these populations, we cannot reliably predict how the treatment will work for them. There may be variation in the distribution of disease variants or of polymorphisms, for instance, involved in drug metabolism [43]. Although 80–90% of genetic variation is found within racial groups, rather than between, variants that are rare in one group might not be included within a small study population [44]. There also may be critical socioeconomic or cultural reasons that treatments work in one population but not another. This may lead to treatment that is not effective or even harmful to some patients [42].
    The Jehovah's Witness community offers an interesting example for contrast: their refusal of blood transfusions has driven research and innovation in "bloodless medicine" and "bloodless surgery" with benefits far beyond the Jehovah's Witness community [45].

(c) **What should be done to address this?**
Researchers and entities associated with research, such as institutional

review boards and funding agencies, need to make more effort to recruit from underrepresented groups of people. Requirements that NIH-funded studies report race and ethnicity are part of this. Specific grant opportunities that target such research also might help, as well as increasing programs to develop and support minority researchers [46]. See ▶ Box 19.7 for case conclusion [47, 48].

> **Box 19.7 Case Conclusion**
> *Eteplirsen received accelerated approval for use in a subset of patients with DMD in 2016 [47]. However, some families have encountered difficulty getting their insurance companies to pay for the therapy for children who are wheelchair-bound. Excellus BCBS, for example, argued that the treatment was only approved for children who were still walking [48].*

Acknowledgments This case was adapted from an earlier case written by Amy Caruso Brown and inspired by the stories of Josh Hardy and the many children and parents who advocated for the approval of eteplirsen.

References

1. Cihelka P. #SaveJosh? Maybe, but what about the rest? NBC News [Internet]. 2014 Mar 11 [cited 2018 Feb 5]. Available from: https://www.nbcnews.com/health/kids-health/bioethicist-savejosh-maybe-what-about-rest-n50051.
2. Masters JR. HeLa cells 50 years on: the good, the bad and the ugly. Nat Rev Cancer. 2002;2(4):315–9.
3. Skloot R. The immortal life of Henrietta lacks: Crown Publishing Group; 2010.
4. Shuster E. Fifty years later: the significance of the Nuremberg Code. New Engl J Med. 1997;337(20):1436–40.
5. Reicher H. Medicine in the Third Reich: the 65th anniversary of the Doctors' Trial at Nuremberg. Penn Medicine. 2012;23(2):28–37.
6. The Nuremberg Code. Trials of war criminals before the Nuremberg military tribunals under Control Council Law No 10. Washington, D.C.: U.S. Government Printing Office; 1949. Vol. 2, pp. 181–2. [cited 2018 Feb 5]. Available from: https://history.nih.gov/research/downloads/nuremberg.pdf.
7. Unguru Y. Pediatric decision-making: informed consent, parental permission, and child assent. In: Diekema DS, Mercurio MR, Adam MB, editors. Clinical ethics in pediatrics: A Case-Based Textbook. Cambridge, MA: Cambridge University Press; 2011.
8. Office of History and Stetten. World Medical Association Declaration of Helsinki (1964, 1975, 1983, 1989) [Internet]. Bethesda (MD): NIH. [cited 2018 Feb 5]. Available from: https://history.nih.gov/about/timelines/helsinki.html.
9. Human Experimentation: Declaration of Helsinki. Ann Intern Med. 1966;65(2):367–369.
10. The National Commission for the Protection of Human Subjects of Biomedical and Behavioral Research. The Belmont Report: ethical principles and guidelines for the protection of human subjects of research [Internet]. Rockville (MD): Office for Human Research Protections; 1979 [cited 2018 Feb 5]. Available from: https://www.hhs.gov/ohrp/regulations-and-policy/belmont-report/index.html.
11. U.S. Food and Drug Administration. How drugs are developed and approved: investigational new drug application [Internet]. Silver Spring (MD): FDA; updated 2017 Oct 5 [cited 2018 Feb 5]. Available from: https://www.fda.gov/Drugs/DevelopmentApprovalProcess/HowDrugsareDevelopedandApproved/ApprovalApplications/InvestigationalNewDrugINDApplication/default.htm.
12. U.S. Food and Drug Administration. The drug development process [Internet]. Silver Spring (MD): FDA; updated 2018 Jan 4 [cited 2018 Feb 5]. Available from: https://www.fda.gov/ForPatients/Approvals/Drugs/default.htm.
13. Wright B. Clinical trial phases. In: A comprehensive and practical guide to clinical trials. London, UK: Academic Press; 2017. p. 11–5.
14. U.S. Food and Drug Administration. Step 3. clinical research [Internet]. Silver Spring (MD): FDA; updated 2018 Jan 4 [cited 2018 Feb 5]. Available from: https://www.fda.gov/ForPatients/Approvals/Drugs/ucm405622.htm.
15. Hay M, Thomas DW, Craighead JL, Economides C, Rosenthal J. Clinical development success rates for investigational drugs. Nat Biotechnol. 2014;32(1):40–51.
16. Picavet E, Morel T, Cassiman D, Simoens S. Shining a light in the black box of orphan drug pricing. Orphanet J Rare Dis. 2014;9(1):62.
17. Terry M. Duchenne muscular dystrophy, genetics, the FDA and drug pricing. J Assoc Genetic Technologists. 2017;43(2):53.
18. U.S. Food and Drug Administration. Expanded access (compassionate use) [Internet]. Silver Spring (MD): FDA. [cited 2018 Feb 5]. Available from: https://www.fda.gov/NewsEvents/PublicHealthFocus/ExpandedAccessCompassionateUse/default.htm.
19. Gordis L. Epidemiology. 4th ed. Philadelphia, PA: Saunders; 2009.
20. Mendell JR, Goemans N, Lowes LP, Alfano LN, Berry K, Shao J, et al. Eteplirsen study group and Telethon Foundation DMD Italian network. Longitudinal effect of eteplirsen versus historical control on ambulation in Duchenne muscular dystrophy. Ann Neurol. 2016;79(2):257–71.
21. Viele K, Berry S, Neuenschwander B, Amzal B, Chen F, Enas N, et al. Use of historical control data for assessing treatment effects in clinical trials. Pharm Stat. 2014 Jan-Feb;13(1):41–54.
22. NIH Office of Rare Diseases Research. ORDR Brochure. Bethesda (MD): National Center for Advancing

23. U.S. Food & Drug Administration. Developing products for rare diseases and conditions. Silver Spring (MD): FDA. [cited 2018 Feb 5]. Available from: https://www.fda.gov/ForIndustry/DevelopingProductsforRareDiseasesConditions/default.htm.
24. Integrity in Scientific research: creating an environment that promotes responsible conduct. Washington, DC: The National Academies Press; 2002 p. 38.
25. Angell M. Industry-sponsored clinical research. A broken system. JAMA. 2008;300:1069–71.
26. Katz D, Caplan AL, Merz JF. All gifts large and small. Toward an understanding of the ethics of pharmaceutical industry gift-giving. Am J Bioeth. 2003;3(3):39–46.
27. Tavernise S. FDA nominee Califf's ties to drug makers worry some. The New York Times [Internet]; 2015 Sep 20 [cited 2018 Feb 5]. Available from: https://www.nytimes.com/2015/09/20/health/fda-nominee-califfs-ties-to-drug-industry-raise-questions.html?_r=0.
28. Lim KR, Maruyama R, Yokota T. Eteplirsen in the treatment of Duchenne muscular dystrophy. Drug Design Dev Ther. 2017;11:533.
29. AMA Code of Ethics. Opinion 9.031: reporting impaired, incompetent, or unethical colleagues. Chicago (IL): AMA; 2009 [cited 2018 Feb 5]. Available from: http://www.lb7.uscourts.gov/documents/08-42541.pdf.
30. Godskesen T, Nygren P, Nordin K, Hansson M, Kihlbom U. Phase 1 clinical trials in end-stage cancer: patient understanding of trial premises and motives for participation. Support Care Cancer. 2013;21(11):3137–42.
31. Mendell JR, Rodino-Klapac LR, Sahenk Z, Roush K, Bird L, Lowes LP, et al. Eteplirsen study group. Eteplirsen for the treatment of Duchenne muscular dystrophy. Ann Neurol. 2013;74(5):637–47.
32. Hinds PS, Oakes LL, Hicks J, Powell B, Srivastava DK, Spunt SL, et al. "Trying to be a good parent" as defined by interviews with parents who made phase I, terminal care, and resuscitation decisions for their children. J Clin Oncol. 2009 Oct 5;27(35):5979–85.
33. Miller FG, Brody H. A critique of clinical equipoise: therapeutic misconception in the ethics of clinical trials. Hastings Cent Rep. 2003;33(3):19–28.
34. Jansen LA, Appelbaum PS, Klein WM, Weinstein ND, Cook W, Fogel JS, et al. Unrealistic optimism in early-phase oncology trials. IRB. 2011;33(1):1–8.
35. Crites J, Kodish E. Unrealistic optimism and the ethics of phase I cancer research. J Med Ethics. 2013;39(6):403–6.
36. Silverman E. Most Americans would avoid clinical trials due to worries over safety and costs. Pharmalot. STAT [Internet]. 2016 May 25 [cited 2018 Feb 5]. Available from: https://www.statnews.com/pharmalot/2016/05/25/clinical-trials-cancer-safety/.
37. Whittle A, Shah S, Wilfond B, Gensler G, Wendler D. Institutional review board practices regarding assent in pediatric research. Pediatr. 2004;113(6):1747–52.
38. Ross L. Phase I research and the meaning of direct benefit. J Pediatr. 2006;149(1):S20–4.
39. National Institutes of Health. NIH Public Access Policy [Internet]. Rockville (MD); updated 2016 Mar 25 [cited 2018 Feb 5]. Available from: https://publicaccess.nih.gov/policy.htm.
40. Freimuth VS, Quinn SC, Thomas SB, Cole G, Zook E, Duncan T. African Americans' views on research and the Tuskegee syphilis study. Soc Sci Med. 2001;52(5):797–808.
41. Buchanan DR, Miller FG. Justice and fairness in the Kennedy Krieger Institute lead paint study: the ethics of public health research on less expensive, less effective interventions. Am J Public Health. 2006;96(5):781–7.
42. George S, Duran N, Norris K. A systematic review of barriers and facilitators to minority research participation among African Americans, Latinos, Asian Americans, and Pacific islanders. Am J Public Health. 2014;104(2):e16–31.
43. Ortega VE, Meyers DA. Pharmacogenetics: implications of race and ethnicity on defining genetic profiles for personalized medicine. J Allergy Clin Immunol. 2014;133(1):16–26.
44. Kaplan JM, Winther RG. Prisoners of abstraction? The theory and measure of genetic variation, and the very concept of "race". Biological Theory. 2013;7(4):401–12.
45. Shander A, Goodnough LT. Objectives and limitations of bloodless medical care. Current Opinion Hematol. 2006;13(6):462–70.
46. NIH National Institute of Minority Health and Health Disparities [Internet]. Washington, D.C.: U.S. Department of Health & Health Services. [cited 2018 Feb 5]. Available from: https://www.nimhd.nih.gov.
47. Kesselheim AS, Avorn J. Approving a problematic muscular dystrophy drug: implications for FDA policy. JAMA. 2016;316(22):2357–8.
48. Thomas K. Insurers battle families over costly drug for fatal disease. The New York Times [Internet]. 2017 Jun 22 [cited 2018 Feb 5]. Available from: https://www.nytimes.com/2017/06/22/health/duchenne-muscular-dystrophy-drug-exondys-51.html.

Further Reading on This Topic

Caplan AL, Ray A. The ethical challenges of compassionate use. JAMA. 2016;315(10):979–80.

Collaborative Institutional Training Initiative. Basic course in human subjects research [Internet]. Fort Lauderdale (FL): CITI Program. Available from: https://about.citiprogram.org/en/homepage/.

Goldacre B. Trial sans error: how pharma-funded research cherry-picks positive results [Excerpt]. Scientific American [Internet]; 2013 Feb 13. Available from: http://www.scientificamerican.com/article/trial-sans-error-how-pharma-funded-research-cherry-picks-positive-results/.

Goldacre B. Bad pharma: how drug companies mislead doctors and harm patients. London (UK): Fourth Estate; 2012.

"Wait, I'm a Research Subject?"

Gregory L. Eastwood

20.1 Background Questions – 390

20.2 Additional Case Information and Questions for Discussion – 391

20.3 Answers to Background Questions – 392

20.4 Responses to Discussion Questions – 395

References – 404

© Springer Nature Switzerland AG 2019
A. E. Caruso Brown et al. (eds.), *Bioethics, Public Health, and the Social Sciences for the Medical Professions*, https://doi.org/10.1007/978-3-030-03544-0_20

Glossary

Duty hours - Defined by the Accreditation Council for Graduate Medical Education (ACGME) as all clinical and academic activities related to a graduate medical training program (e.g., internship, residency, fellowship), including time spent in direct patient care in any setting, administrative duties related to patient care, arranging for transfer of patient care, physically in the healthcare institution while on call (but not call taken from home), in scheduled academic activities, and in research activities; does not include time spent at home while not engaged in any of the above (including during "home call"); and does not include time spent reading or studying.

Learning Objectives

1. Apply study design and basic research principles to a complex question.
2. Articulate why failure to perform research that produces valid and reliable results is as much an ethical problem as a scientific one.
3. Analyze a recent study in which some researchers and ethicists think that participants' rights were violated, and discuss why this might have occurred.
4. Recognize some of the obligations and barriers to "speaking up" in healthcare institutions and as a member of a health profession.

Case Background

You are a preliminary year intern at an academic medical center. Next year you will begin residency training in radiology. Soon after arriving for orientation, you learned that the internal medicine and surgery residency programs at the hospital, in which you will spend much of your year, are part of a multicenter, randomized controlled trial (RCT) to evaluate the effect of resident work-hour restrictions on patient outcomes, measured primarily by mortality rates 30 days after admission.

When you applied to residency, you knew nothing about this study. You expected that your shifts as an intern would be capped at 16 consecutive hours, a policy that was implemented in 2011. Therefore, you are surprised to learn that you will be working up to 28 hours at a stretch when you are on surgery and medicine rotations and that you will need to record your duty hours. You also find out that your consent is not needed as a participant in the study because it was deemed "exempt" or "not human subject research" by the hospital's IRB.

20.1 Background Questions

1. What types of studies must be approved by an institutional review board (IRB) in the United States? What categories of research are exempt from IRB approval and generally do not require informed consent from participants?

2. What is the Common Rule? What are the relevant rules for research under the Common Rule?

3. What are patient handoffs? What role do patient handoffs play in the occurrence of medical errors?

4. What is a confounder? What factors might confound the relationship between resident duty hours and 30-day patient mortality?

5. Who was Libby Zion, and what role did her death play in the development of resident duty-hour restrictions?

6. Identify a historically important research study in which the participants' rights were violated. Discuss what happened and how the violations have been redressed or not.

7. What is comparative effectiveness research? What are pragmatic randomized controlled trials? How are these different from other types of research, with regard to ethical concerns?

"Wait, I'm a Research Subject?"

20.2 Additional Case Information and Questions for Discussion

1. What are the ethical issues in this case?

2. Who are the key moral stakeholders?
   (a) Who are the research subjects in this study—the hospitals, the residency programs, the residents, or the patients treated at the participating hospitals?

3. Should you have been informed of the study and the implications regarding duty hours earlier? Why or why not?
   (a) What are your options in this situation?
   (b) Self-reflection: Which one would you choose? Why?
   (c) Would you pursue legal action? Why or why not?

4. Was the IRB correct to deem this study exempt and not involving human subject research? Why or why not?
   (a) If the IRB had not classified the study as exempt from review, a waiver of informed consent or informed *written* consent would have been an option. In that case, should informed written consent have been required from the prospective interns?
   (b) Many medical practices and policies are rooted more in tradition than evidence, including some practices deemed "standard of care." (Comparative effectiveness research, as addressed above, seeks to redress this.) Should scientific experimentation be held to a higher level of scrutiny than widespread implementation of an untested practice? Why or why not?

The Libby Zion case is important for historical reasons. Her father's anger led to investigations, legal charges, and the Bell Commission, which resulted in the 405 (Bell) Regulations that limited resident duty hours in the state of New York. Similar restrictions have been adopted throughout the country [1].

5. Did the Libby Zion case prove that long duty hours by the residents were the cause of Libby's death?
   (a) What role did the lack of supervision of the residents play in her death?
   (b) What role did lack of awareness regarding the interaction of meperidine and phenelzine, causing serotonin syndrome, play?
   (c) What is the relationship between work-hour restrictions and increased supervision of residents?

6. Why aren't the duty hours of attending physicians restricted? Should they be?

Early in the morning during your first 24-hour shift, you find yourself talking with some of the nurses about their schedules. One of your nursing colleagues tells you that she read a study in which most Americans said they wanted to know if their physician has been working for more than 24 hours and, in that case, wanted to be able to request a different physician [2].

7. If longer duty hours are shown to be safe, what should physicians do to address patient concerns?
   (a) Are physicians obligated to inform patients about the number of hours they are working?
   (b) Are physicians obligated to honor patients' requests for another physician who might be less tired?

Six weeks after starting your internship, you realize that you are sometimes in violation of duty-hour restrictions. Although the study does not require you to log your duty hours, you also have rotations at other hospitals, where routine logging is required. When you ask your senior resident what happens when you log duty hours that exceed the limit, she laughs and says that she always records the maximum permitted number of hours, even when she exceeds it. She explains that the system will prompt you to choose an acceptable explanation for the violation (for instance, staying later to care for a patient who is actively dying) and you will be called to meet with the program director. If the program accrues too many violations, it can be penalized by the ACGME.

❓ 8. **Self-reflection:** Would you conceal the violation and log fewer hours than you actually worked? Why or why not?
   (a) What resources are available in your institution for you to ask questions or "speak up" when you see something that troubles you ethically or legally?

Two years later, the study has concluded. Based on the results of this study and others, duty hours are somewhat more flexible than in the recent past, but questions remain about the impact of fatigue on health care. You have been selected to be a chief resident and have an opportunity to propose a new study to address the question.

❓ 9. What are some of the problems with the original study design?
   (a) Why do you think it was designed this way?
   (b) **Creative Problem-Solving:** How would you design a new study to answer the questions related to differences in duty hours?
      (i) What endpoints would you choose?
      (ii) Are patient-related outcomes the only ones of interest?

❓ 10. Why is good study design imperative to the responsible conduct of research?

20.3 Answers to Background Questions

✅ 1. **What types of studies must be approved by an institutional review board (IRB)? What categories of research are exempt from IRB approval and generally do not require informed consent from participants?**
An Institutional Review Board for the Protection of Human Subjects, or IRB, is a federally mandated institutional committee whose purpose is to protect the safety, confidentiality, and other interests of human subjects in research studies [3].
   In general, all research that involves human subjects must be reviewed and approved by the institution's IRB, with some exceptions. The exceptions include [4]:

1. Research conducted in commonly accepted educational settings, involving normal educational practices, such as the effectiveness of regular and special education instructional strategies or comparisons among instructional techniques, curricula, or classroom management methods.
2. Research involving the use of educational tests (e.g., cognitive, diagnostic, aptitude, achievement), survey procedures, interview procedures, or observation of public behavior. Information must be obtained in such a manner that human subjects cannot be identified and any disclosure of the human subjects' responses outside the research should not place the subjects at risk of criminal or civil liability or be damaging to the subjects' financial standing, employability, or reputation.
3. Research involving the collection or study of existing data, documents, records, pathological specimens, or diagnostic specimens, if these sources are publicly available or if the information is recorded by the investigator in such a manner that subjects cannot be identified.
4. Research and demonstration projects, which study public benefit or service programs, their benefits, procedures for obtaining benefits, and possible changes or alternatives to the programs.
5. Studies of taste and food quality and consumer acceptance, if wholesome foods without additives are consumed or if a food is consumed that contains a food ingredient at or below the level and for a use found to be safe by governmental agencies.
   The decision that a particular research study is exempt is made by the IRB, or its representative, not by the study investigator. That means that investigators must inform the IRB and provide some information about the study before it is initiated.

2. **What is the Common Rule? What are the relevant rules for research under the Common Rule?**
The Common Rule is shorthand for federal policy for protection of human subjects in research. It has been in place since 1981 and governs IRB and the like. It relies on the ethical principles of the Declaration of Helsinki and the Belmont Report (see ▶ Chapter 19, Background Question 3). The Common Rule regards a human subject as a living individual about whom the researcher gathers data through "intervention or interaction" or gathers "identifiable private information."

3. **What are patient handoffs? What role do patient handoffs play in the occurrence of medical errors?**
Patient handoffs usually occur at the junction between shifts. The residents on the shift that is ending meet with the incoming residents who are on the next shift. During handoffs, an outgoing resident reviews the patients for whom an incoming resident will be responsible, providing relevant background information and highlighting concerns and things that need to be done. Any time information is transferred from one person to another, there is potential for miscommunication and error. The shorter the shifts, the more frequent the handoffs; thus theoretically more chance for error or patient mishap. However, in some studies of duty-hour limitations, programs and hospitals also implemented changes to the handoff process and other efforts (e.g., improvement in teamwork) that may attenuate the risks of more frequent handoffs [5, 6].

4. **What is a confounder? What factors might confound the relationship between resident duty hours and 30-day patient mortality?**
A confounder, or confounding variable, is anything other than the thing being studied that could be causing the results in the study. In other words, it is a third variable that may make it appear, falsely, that a particular exposure is associated with a given outcome. It is an unobserved exposure that is associated with both the exposure of interest and the outcome of interest.

The more controlled the criteria for participation in the research, the fewer the confounders. In research involving human subjects, and particularly research involving complex real-world phenomena, such as duty hours, it is difficult to identify, control, or eliminate all confounders, and studies that attempt to do so may not translate well to real-world practice. For example, some clinical trials have stringent eligibility criteria that eliminate the sickest patients, despite the drug or intervention being most applicable to precisely those patients who were excluded from the trials. Some confounders in studying the relationship between resident duty hours and 30-day patient mortality might be differences among hospitals within the study with regard to patient population, nosocomial infections, incidence of bed sores, etc.; differences among patients with regard to age, disease and disability, ethnicity, etc.; and differences among residents with regard to capability, mood, and expectations.

5. **Who was Libby Zion and what role did her death play in the development of resident duty-hour restrictions?**
Libby Zion was an 18-year-old college freshman admitted the night of March 4, 1984, to New York Hospital, a large teaching hospital in Manhattan, because of a flu-like illness. She had been taking a prescribed antidepressant, phenelzine. Two residents on duty, in consultation with a senior physician by phone, treated her, prescribing meperidine (also called pethidine) to control Libby's jerking motions. Both residents were busy with many other patients. After Libby became more agitated, medical restraints and haloperidol were prescribed. Libby became hyperthermic (temperature 107 ° F), suffered a cardiac arrest, and could not be resuscitated [1].

Later, experts opined that the combination of her antidepressant phenelzine and meperidine led to the development of serotonin syndrome, which caused agitation and eventually cardiac arrest. Libby's parents were convinced that their daughter's death was due to overworked residents. Her father, a lawyer and former publishing executive, brought suit against the hospital and resident physicians, claiming among other things that the long shifts for residents impaired their judgment. The lawsuit and publicity led to changes in resident duty hours, promulgated by the New York State Department of Health and later by the ACGME [1]. In 1989, New York ruled that residents could not work more than 80 hours a week or more than 24 consecutive hours, and attending physicians need to be present in the hospital at all times. In 2003, the ACGME adopted similar regulations for all accredited residency programs in the United States [7].

✅ 6. **Identify a historically important research study in which the participants' rights were violated. Discuss what happened and how the violations have been redressed or not.**
Following are two examples of research that have been regarded as ethically flawed and that have influenced the conduct of subsequent research.

The *Tuskegee Syphilis Study* in Tuskegee, Alabama, began in 1932 under the auspices of the U.S. Public Health Service and concluded in 1972. In the study, 399 poor black men with syphilis, who were told they had "bad blood," were followed for many years to observe the natural history of syphilis, in comparison with 201 men without syphilis. The men with syphilis were not treated for the disease even after an effective treatment, penicillin, became available in 1947. In 1974, a $10 million court settlement included lifetime medical benefits and burial services to all living participants. A year later, those benefits were applied to wives, widows, and children of the study participants. In 1997, President Clinton apologized on behalf of the nation [8].

The *Willowbrook Hepatitis Study*, begun in 1956 and concluded in 1970, took place at a state hospital for disabled children on Staten Island, New York. In that study, children who were newly admitted to the facility were intentionally infected with hepatitis A to investigate the effect of gamma globulin on the disease and develop a vaccine. The deliberate infection of the children was rationalized because the historical experience with previous children at Willowbrook indicated that nearly all children contracted hepatitis A within a short time after admission, presumably from cross infection from children who already were there. Although the results were presented at scientific meetings and published in journals, the study was criticized because the children were not able to give consent and parents received inadequate information before they were asked to give consent [9].

✅ 7. **What is comparative effectiveness research? What are pragmatic randomized controlled trials? How are these different from other types of research, with regard to ethical concerns?**
Comparative effectiveness research (CER) refers to research that directly compares existing healthcare interventions in order to establish definitively which intervention is optimal. It has been driven, in part, by studies demonstrating that there is significant unwarranted variation in medical practices among different healthcare providers, among different institutions, and even among geographic areas of the United States. The National Academy of Medicine (formerly the Institute of Medicine) defines comparative effectiveness research as "the generation and synthesis of evidence that compares the benefits and harms of alternative methods to prevent, diagnose, treat, and monitor a clinical condition or to improve the delivery of care. The purpose of CER is to assist consumers, clinicians, purchasers, and policy makers to make informed decisions that will improve health care at both the individual and population levels." [10]

A pragmatic randomized controlled trial is one that is conducted in real-world settings that match the expected practice environment in which the intervention would be adopted. Rather than eliminating variation, the goal of pragmatic trial is to produce results that accurately reflect "variations between patients that occur in real clinical practice," thus better informing choice between treatments [11]. Both types of research lend themselves well to randomization of programs, institutions, or even communities, raising questions about consent. Further, because the interventions are typically part of the repertoire of accepted medical practice, it can be tricky to define precisely the risks of research participation [12].

20.4 Responses to Discussion Questions

 1. **What are the ethical issues in this case?**
    – *Welfare of the patient, the fundamental principle in medicine.* Whether longer or shorter duty hours are more satisfactory for the residents or more conducive to their education is of interest, of course, but the main purpose of raising the question of duty hours is in the context of patient welfare. Resident duty hours were restricted and regulated primarily in the interest of patient welfare, not simply to make residents feel better. As with most studies that test a new or different approach or therapy, one does not know whether longer or shorter duty hours will be better or worse for patients or make no measureable difference. That presumably is why the study is being conducted.
    – *Welfare and rights of the human subject, in this case, the residents.* Have the residents been accorded the rights of human subjects, the most relevant of which are to understand the purposes, risks, and benefits of the research before they consent to participate and to do so without coercion, to ask questions at any time, and to withdraw from participation at any time without repercussions? Clearly, in order to do a study such as this, it may not be possible to accord the interns all the rights of an ordinary human subject trial.
    – *Autonomy.* Do the interns and/or the patients have the right to make their own decisions about participation in the study, based on appropriate information? Again, there are situations in which autonomy is necessarily limited: for instance, when the autonomous choices of one individual infringe upon others' choices, or when other ethical principles take precedence. For example, a new diagnosis of human immunodeficiency virus (HIV) must be reported to the public health authorities. The patient with HIV may feel that this infringes on his autonomy as well as his right of privacy. However, the health of others is at risk, particularly those who have been sexual partners with the HIV patient or have been put at risk in some other way. The rights of those individuals to know of their potential disease exposure and perhaps to act on that information by seeking diagnosis and treatment for themselves must be balanced against the autonomy and privacy of the patient. In this case, the principle of the welfare of other people, perhaps many, takes precedence over the principle of autonomy for the patient with HIV.

 2. **Who are the key moral stakeholders?**
    – *The people who designed or permitted the study*, such as ACGME, professional organizations, faculty researchers, department chairs, residency program directors, associate deans for Graduate Medical Education (GME), hospital administrators, and the IRB.
    – *This intern, in particular; other interns participating in the study; and other trainees* who may be affected by the results of such research.
    – *Patients and their families.* They have a stake in the research. They have limited ability to alter the outcome, although they can voice support

or concern if they know about the research and, unwittingly, can die or not, develop complications or not, or in general do better or worse under the new duty hours.
- *The public through the news media.* This study, if made public, could shine a spotlight on teaching hospitals and large academic medical centers, for better or worse.

(a) **Who are the research subjects in this study—the hospitals, the residency programs, the residents, or the patients treated at the participating hospitals?**
The hospitals and residency programs clearly must participate in the study in order for the study to be done, so in that sense they are participants in the research, or research subjects broadly construed. However, conventionally the term "research subject" is taken to mean a human research subject and thus a person. It would seem that the residents are the most obvious candidates to be research subjects in this study. They are obligated to record their duty hours; the study variables of extended duty hours are applied to them. However, the patients also are research subjects, since the main endpoint of the study is their 30-day mortality. The IRB that approved the study did not agree: They argued that the "subjects" were the institutions, since randomization was occurring at the institutional level [13, 14].

3. **Should you have been informed of the study and the implications regarding duty hours earlier? Why or why not?**
This question relates to matters of autonomy, informed consent, and the rights of human subjects. Did the intern have the right to know about the study and the implications regarding duty hours? In this study, the surgical interns seem to be research subjects. The interns should have been told about the study and at least given the opportunity to provide input and ask questions. The change in duty hours has the chance of affecting the welfare of the human subjects (interns) and the patients. The latter is evident in the measurement of 30-day mortality.

(a) **What are your options in this situation?**
In this context, particularly given that the intern matched to this program several months before arriving for orientation and that there are no practical options now for pursuing alternative training, the intern's autonomy is limited. There still should be opportunity for input and discussion with institutional and residency program leaders, who should have an interest in resident well-being and satisfaction.

> **Box 20.2   Teaching Tip**
> Although the narrative states that the intern learned about the new duty hours at orientation and was surprised to learn that he/she would be working up to 28 hours at a time and that consent for this is not needed, we do not know how the intern feels. This is an opportunity for learners to imagine how they think they would feel in this situation. Surprise at the new rules does not necessarily mean disapproval or concern. The concerns, values, preferences, and personal considerations of the intern, and perhaps other stakeholders, may be quite different.

> **Box 20.1   Teaching Tip**
> Ask learners whether feasibility and logistics should matter to the determination of who is a "human subject." Contrast this scenario with emergency medical research, where informed consent is waived upfront, due to practical issues, but should be obtained as soon as possible once the patient is stable [14].

Practical options include:
- Accepting the situation as it is without comment (or perhaps with some venting to peers).

"Wait, I'm a Research Subject?"

- Asking the residency director to provide additional support and perhaps holding an open forum for interns who are concerned about their duty hours, the study, or simply the lack of prior informed consent.
- Contacting the study's principal investigator with concerns, which might include asking for information regarding how safety will be assessed during the study and how concerns will be addressed.
  - It should be noted that human subject research protocols must include a specific plan for how safety will be monitored during the study, frequently involving outside observers or a formal Data Safety Monitoring Board (DSMB) for drug trials, what will be done if unexpected adverse events occur, and who subjects should contact if they have concerns or feel they were harmed by the study. However, because this was deemed not to be "human subject research" by the approving IRB (see the Discussion Question 4, below), the investigators would not be required to have such a plan in place.
- Contacting the IRB who approved the study with concerns.
- Seeking legal advice or recourse.

(b) **Self-reflection: Which one would you choose?**

> **Box 20.3 Teaching Tip**
> Ask learners to explain their reasons for choosing one or another option. If learners are quiet, taking an informal poll or vote on the options can stimulate conversation and move the case forward.

(c) **Would you pursue legal action? Why or why not?**
Pursuing legal action likely will carry financial costs and, indeed, even raising questions within the institution, may have a reputational cost. There is a realistic chance that such a lawsuit might impact the intern's future career and employment prospects. Ethically, these individual concerns should be weighed against risks to both patients and participating interns.

✓ 4. **Was the IRB correct to deem this study exempt and not involving human subject research? Why or why not?**
Whether this is a human subjects study or not is a matter of perspective. Individuals and groups of individuals, such as an IRB, can come to different conclusions after consideration of the same facts and rules. The IRB that approved this study evidently felt that it was not human subject research. Conversely, one could mount a strong argument to support the contention that it is human subject research. As was explored above (see Discussion Question 2a), both the residents and patients, without whom the study could not be done, are human participants and thus seem to be in the position of human subjects. This seems evident because the research is studying variations in both resident and patient behaviors—with respect to residents, their duty hours; with respect to patients, their 30-day mortality.

(a) **If the IRB had not classified the study as exempt from review, a waiver of informed consent or informed *written* consent would have been an option. In that case, should informed written consent have been required from the prospective interns?**
Consider the following statement: An IRB may waive the requirements to obtain informed consent provided the IRB finds and documents that (1) the research involves no more than minimal

risk to the subjects; (2) the waiver or alteration will not adversely affect the rights and welfare of the subjects; (3) the research could not practicably be carried out without the waiver or alteration; and (4) whenever appropriate, the subjects will be provided with additional pertinent information after participation [15].

Does this research study conform to the first condition of involving no more than minimal risk to the subjects? If 30-day patient mortality is adversely affected by the conditions of this study, one could argue that death exceeds minimal risk to the subjects. Of course, since the IRB that approved this apparently did not see the patients as subjects, that would not be relevant under the circumstances of approval of the study. Whether the second condition pertains—i.e., the waiver will not adversely affect the rights and welfare of the subjects—is arguable. That, too, depends on whether residents and patients are regarded as research subjects. If they are, their rights and welfare, taking the form of whether to participate or not and, as mentioned regarding patients, their 30-day mortality, seem relevant. The third condition is probably valid; it would be difficult to carry out the research without the waiver. This raises the question: Is the likelihood of meaningful results from the study high enough to override the rights and welfare of residents and patients? The fourth condition could be satisfied by providing additional pertinent information to the residents and patients after participation. However, this seems impractical. It may be inconvenient and time-consuming and may encourage some patients to withdraw participation, which would be difficult for patients who needed to remain in hospital and would confound the study.

Should informed written consent have been required from the prospective interns? The time to explain the study to the interns would have been during the recruitment and application process, months before they actually began their internship. A formal consent document probably would be preferable, but consent could be included in the response of the applicant to an offer of acceptance for the intern position.

(b) **Many medical practices and policies are rooted more in tradition than evidence, including some practices deemed "standard of care." (Comparative effectiveness research, as addressed above, seeks to redress this.) Should scientific experimentation be held to a higher level of scrutiny than widespread implementation of an untested practice? Why or why not?**

One essential characteristic of scientific inquiry should be unbiased objectivity in framing the questions to be addressed, designing the experiments, observing the data and other information, and interpreting the results. This is not often possible or, if possible, not practical, in clinical practice. Evidence-based practice seeks to remedy this, usually prospectively in determining the usefulness of new diagnostics and treatments, but also sometimes in examining established diagnostic and therapeutic practices. For some practices that have been used for hundreds of years, such as the use of aspirin-like medications to treat pain, it seems unreasonable to subject them to renewed scientific scrutiny to establish effectiveness. Other accepted practices have been challenged, such as long durations of bed rest to treat acute myocardial infarction in a former era. In some medical disorders (e.g., prostate cancer, back pain, coronary artery disease), practitioners cannot agree on the best treatment, so large clinical trials of different treatments seem appropriate.

Sometimes medical societies and federal agencies, such as the National Institutes of Health, convene acknowledged experts to discuss medical disorders for which treatments may be controversial or unclear and then develop consensus statements on accepted approaches in the management of those conditions, taking into account what scientific evidence is available as well as what the experts deem to be reasonable clinical practices.

> **Box 20.4 Teaching Tip**
> Encourage learners to think about current clinical practices that are simply accepted, those that have been challenged and changed, and the practicality of requiring new interventions and practices to conform strictly to scientific scrutiny before they are implemented broadly. Learners might also think beyond diagnostics and therapeutics to question whether the structure and methods of healthcare delivery, both in the clinic and in the healthcare system, might benefit from scientific inquiry.

The Libby Zion case is important for historical reasons. Her father's anger led to investigations, legal charges, and also the Bell Commission, which resulted in the 405 (Bell) Regulations which limited resident duty hours in the state of New York. Similar restrictions have been adopted throughout the country [1].

5. **Were long duty hours by the residents the cause of Libby's death?**
   (a) **What role did the lack of supervision of the residents play in her death?**
   The investigation and legal case never proved that long duty hours by the residents were the cause of Libby's death. If anything, her death may have been a consequence of lack of supervision of the residents by someone who might have known that the addition of meperidine by the residents to Libby's already taking the antidepressant phenelzine could lead to serotonin syndrome and death [1].
   (b) **What role did lack of awareness regarding the interaction of meperidine and phenelzine, causing serotonin syndrome, play?**
   It certainly did play a role; however, at a subsequent hearing, most expert witnesses in toxicology, emergency medicine, and internal medicine were unaware of the interaction of meperidine and phenelzine before this case, suggesting that greater supervision would not *necessarily* have prevented the concurrent use of the two drugs [1].
   (c) **What is the relationship between duty-hour restrictions and increased supervision of residents?**
   As duty hours for residents became shorter and more regulated, the supervision of residents' work, including chart notes, orders, and clinical decisions, by attending physicians also became more explicit. Although some of this was inherent in the ACGME rules about resident duty hours, the main driver was the requirement by Medicare and other insurance providers that attending physicians take more responsibility for the care of patients and document that in the patients' records [16].

> **Box 20.5   Personal Perspective**
>
> Perri Klass' essay "Getting Through the Night" is a reflection on the author's nostalgia toward the long hours of her residency training [16]. She writes:
>
> » Most of all, and most indefensibly, I sometimes feel nostalgic for the sheer crazy, slightly sick intensity of it all—for the punch-drunk fatigue you felt when you hadn't slept, because you were doing the most important job in the world, awake while everyone else slept, you and the babies, getting through the night. For the kind of sleep-deprived connection you sometimes felt with particular patients all through that postcall day, going by to check on them, smiling proudly when the family members exclaimed that you were still there.
>
> But she goes on to say:
>
> » And yet. If we ignore the satisfactions and pleasures of that crazy schedule, I think we risk losing some aspects of what makes our jobs as good as they are, when they're good. That doesn't mean we should return to the bad old days, but I think we have to find ways to acknowledge and teach that doing this job properly sometimes feels painful or difficult or isolating. And sometimes being there for the patient is all and goes against all the other imperatives of your own best interests.

✓ 6. **Why aren't the duty hours of attending physicians restricted? Should they be?**

The past drivers for shorter resident duty hours may have been several, and many of these drivers did not affect attending physicians in comparable ways. In the 1970s, long before Libby Zion's case occurred, residency programs moved from every other night to every third night. Perhaps that was out of concern for residents' well-being, but it may also have been necessary to make programs more competitive, as the number of training programs and available slots increased. In the current climate, restricting the duty hours of attending physicians could be viewed as interfering with the personal freedom and rights of physicians in practice, rather than preventing mistreatment of trainees, who are in subordinate positions and less empowered to speak up regarding fatigue. Finally, it might also be more difficult to monitor physicians' duty hours and to enforce infractions.

Whereas the correlation between sleep deprivation and cognitive impairment is well established, a causative relationship between sleep deprivation and poor patient outcomes has not been definitively established. Furthermore, studies that have found higher rates of errors specifically evaluated trainee performance [17–19]. One study, titled "Outcomes of Daytime Procedures Performed by Attending Surgeons after Night Work," showed that "the risks of adverse outcomes of elective daytime procedures were similar whether or not the physician had provided medical services the previous night." [20]

Some studies have also indicated a protective effect of expertise on decisions made while sleep-deprived, which guided the 2011 duty-hour limitations, allowing upper-level residents, who were presumed to have more expertise, to work 8 hours longer than interns [7]. Another study of residents performing laparoscopic procedures did find reduced efficiency and safety after sleep deprivation. Notably, after controlling for expected performance based on training level, the performance of more experienced residents suffered less than the performance of novice residents [21].

Early in the morning during your first 24-hour shift, you find yourself talking with some of the nurses about their schedules. One of your nursing colleagues tells you that she read a study in which most Americans said they wanted to know if their physician has been working for more than 24 hours and, in that case, wanted to be able to request a different physician [2].

✓ 7. **If longer duty hours are shown to be safe, what should physicians do to address patient concerns?**

One possibility might be for physician professional organizations to carry out educational programs to increase public awareness of what is known about duty hours and patient safety. Hospitals may do this at a local level, educating patients about their hospital's policies and, perhaps, providing appropriate and rapid recourse for patients who have real concerns about their physician's level of fatigue. Many patients would likely be receptive to the idea that continuity of care, achieved by having longer duty hours and decreasing the frequency of handoffs, is a desirable goal that increases their safety—if supported by data.

(a) **Are physicians obligated to inform patients about the number of hours they are working?**
If patients ask directly, physicians are obligated to answer truthfully; that is, they cannot lie to their patients about how long they have been working. Attempts to evade answering this question will almost certainly increase mistrust and impair the patient-physician relationship. However, there is no requirement that physicians disclose information that is not relevant to the care provided. If a physician who has been working for 24 hours provides care equivalent to a physician who has been working for 6 hours, then this information most likely is not relevant.

(b) **Are physicians obligated to honor patients' requests for another physician who might be less tired?**
This depends on the specific situation, but in most cases, physicians are not obligated to honor requests for another physician. For routine, nonurgent, or elective visits and procedures, patients would of course be free to reschedule or transfer care to another provider. However, fatigue is most likely to be an issue with resident physicians and during the night, situations in which it is less likely that an alternative provider would be readily available to take over. Unless the provider is visibly impaired, such that colleagues and staff have also noted a problem, residency programs and hospitals may find it difficult to accede to every request for another physician. In that extraordinary case, the hospital would likely face some liability if the patient's request for a different, more alert physician was denied, and then the patient experienced a medical error or even a known complication that could be partly attributable to fatigue and inattention.

> **Box 20.6 Teaching Tip**
> Learners may wish to discuss other conditions that may affect professional competence, such as experience (discussed in ▶ Chapter 3), alcohol and drug abuse, illness, or cognitive decline (discussed in ▶ Chapter 15).

Six weeks after starting your internship, you realize that you are sometimes in violation of duty-hour restrictions. Although the study does not require you to log your duty hours, you also have rotations at other hospitals, where routine logging is required. When you ask your senior resident what happens when you log duty hours that exceed the limit, she laughs and says that she always records the maximum permitted number of hours, even when she exceeds it. She explains that the system will prompt you to choose an acceptable explanation for the violation (for instance, staying later to care for a patient who is actively dying) and you will be called to meet with the program director. If the program accrues too many violations, it can be penalized by the ACGME.

✓ 8. **Self-reflection: Would you conceal the violation and log fewer hours than you actually worked? Why or why not?**
Some aspects of the medical culture contribute to silence and acquiescence regarding work-hour violations. Residents may fear retaliation, especially in smaller programs where anonymity might be difficult or impossible. They may be concerned that their program will be placed on probation or even lose accreditation,

requiring them to seek a position in another program or even jeopardizing their careers.

(a) **What resources are available in your institution for you to ask questions or "speak up" when you see something that troubles you ethically or legally?**
Both the Liaison Committee on Medical Education (LCME) and the ACGME require that medical schools have policies regarding mistreatment and systems in place to monitor for mistreatment [22, 23]. These are typically found in a student or resident handbook or equivalent document. Course or clerkship syllabi or orientation packets for specific rotations for residents may also include instructions for what to do if a trainee experiences mistreatment or observes something that he or she thinks is ethically or legally problematic. Sometimes, students and residents observe questionable ethical behaviors on the part of residents, physicians, nurses, and others, such as disrespectful comments about patients, in their presence or not; disrespectful comments about other physicians or specialties; or discrimination against patients because of ethnicity, gender, or social class. Commonly, options to address concerns include discussing the situation with other students, trainees, or faculty mentors or advisors and bringing the matter to the attention of the course director, department chair, or other institutional person who has the responsibility for responding to trainees' concerns, such as the dean of students or designee, or the dean for Graduate Medical Education or designee. If the situation occurs in a clinical setting, an ethics consultation or referral to the ethics committee or, in unusual cases, consultation with hospital legal counsel may be considered.

✓ 9. **What are some of the problems with the original study design?**
The flaws in the design of the study are critically relevant to the ethical evaluation of the situation. First, if 30-day patient mortality is the only measure in the study, it is a gross one and probably will not provide much information. The main criticism is that the single outcome of 30-day mortality is too narrow and nonspecific to be the only endpoint used to rely on a definitive answer to an important question about patient safety. Thus, one could raise the question: Why go through such a study that requires so much organization, monitoring, and staff time and may cause significant unhappiness and complaints for so little likelihood of useful information?

A multisite study is at risk for many confounders related to the patient population they serve, and, in this case, the composition of the residents in the study. For instance, a reviewer of this study would probably ask what attempts were made to control for differences in how sick the patients were at admission and in their access to care and adherence to treatment after discharge. Programmatic efforts to promote good follow-up after discharge, assure safety at home, and so on may have a measurable effect on 30-day mortality. It would be important to know about any pilot efforts ongoing at the same hospitals participating in the duty-hour study. Further, some residency programs have had high rates of duty-hour violations, prior to the study, and at other programs there may be pressure to misreport hours (see Discussion Question 8). What efforts did the researchers make to assure that participants assigned to limited duty hours were in fact working restricted hours? Further, duty-hour restrictions are not synonymous with adequate sleep. Another confounding variable might be that some residents assigned to restricted hours might have gone home to young children or other personal demands that affected their sleep.

Finally, frequent handoffs are a known problem in patient safety and present an additional confounder (see Background Question 3 above). Shorter shifts mean more frequent handoffs, which may confound any attempt to study the effect of sleep deprivation on patient safety [5]. One view is that these two issues either need to be studied separately or perhaps the answer to the problem of handoffs is not to decrease the number of handoffs but to improve the way handoffs are managed. Conversely, one could argue that the real question of the study is about the effect of work-hour restrictions, which broadly includes both the effect on sleep and the effect on handoffs, not simply the effect of sleep deprivation alone, and that might justify the design of the study.

(a) **Why do you think it was designed this way?**

There are two possible explanations here: the pragmatic and the cynical.

- **Pragmatic**: Practical concerns most likely drove the study design. This type of study question is a difficult one to answer well, since there are so many potential confounders. To design a study that is feasible, sufficiently powered (large enough sample size), *and* one that can be done in a reasonable time frame and at a reasonable cost, is very challenging.
- **Cynical**: Conflict of interest is another possibility. Some of the study investigators may have felt that duty-hour restrictions were compromising medical education, particularly surgical training. This may have intentionally or subconsciously influenced the design of the study, biasing it toward a design that was likely to find no difference between differences in duty hours, because of the weakness of the 30-day mortality endpoint.

(b) **Creative Problem-Solving: How would you design a study to answer the questions related to differences in duty hours?**

> **Box 20.7   Teaching Tip**
> Prompt learners to clearly define the type of study they would choose (e.g., interventional), the design (e.g., RCT), the study population and setting (e.g., residents working in intensive care units of similar size and patient population), the endpoints (see below), and possible confounders. How would they control for those confounders? Depending on the level of familiarity with research methods, learners may benefit from working together in pairs or groups of three to four and then reporting their decisions back to the larger group. This question can also be used as a written assignment after the case discussion.

(i) **What endpoints would you choose?**

Thirty-day mortality should be one measure, but other measures might include patient infection rates and other complications, length of hospital stay, return to hospital within 30 days of discharge, and patient satisfaction.

(ii) **Are patient-related outcomes the only ones of interest?**

Whereas patient-related outcomes are very important, they are not the only relevant ones. The study also should include measurement of the effects on the interns and perhaps attending physicians of longer duty hours. These might include tests of judgment and cognitive ability by the interns after the longer vs. shorter duty hours and estimates by the attending physicians of intern performance. One significant challenge of this type of study, to be done well, would be interpreting multiple endpoints and prioritizing them.

Box 20.8 Teaching Tip

Provide learners with a hypothetical *conflicting* set of clinically relevant outcomes and changes in resident and patient satisfaction. For example, tell them to suppose that 30-day mortality (a clinical outcome) is unchanged between the 2 duty-hour conditions, but patient satisfaction increases (suggesting patients prefer the continuity of providers working longer hours) and resident satisfaction decreases (perhaps due to fatigue or lack of time to maintain personal relationships and interests outside the hospital). Which outcomes would they prioritize? Which are most important? Who should decide this type of question?

 10. **Why is good study design imperative to the responsible conduct of research?**

Poorly designed studies carry more of the risks of research but few of the benefits. If one cannot, for practical or logistical reasons, perform a study that will (1) answer the study question and (2) have a meaningful impact on the problem, it is not ethical to proceed with the study [24]. See ▶ Box 20.9 for case conclusion [7, 25–28].

Box 20.9 Case Conclusion

The real study upon which this case was based was approved by the ACGME, as well as by the IRB at Northwestern University, which classified it as exempt (not meeting the definition of human subject research and, therefore, not requiring informed consent) by the IRB. The results were reported in the New England Journal of Medicine on February 25, 2016 [25].

The study involved residents at 117 general surgery residency programs in the United States during the 2014–2015 academic year. Programs were assigned randomly to conform to current ACGME duty-hour policies or, under an ACGME waiver, to more flexible policies that waived rules on maximum shift lengths and time off between shifts. Both groups adhered to the ACGME duty-hour requirements of a maximum of 80 hours on duty per week, 1 day off per 7 days, and on-call duty no more than every third night. Outcomes included 30-day rate of postoperative death or serious complications, other postoperative complications, and resident perceptions and satisfaction regarding well-being, education, and patient care. The flexible, less restrictive duty-hour policies were associated with no worse patient outcomes and no significant difference in residents' satisfaction with overall well-being and education quality.

Despite the concerns about the ethics of the study, as well as its applicability to nonsurgical trainees, ACGME duty-hour restrictions were changed recently to allow first-year residents (interns) to work 24 consecutive hours [26, 27]. (Beginning in 2011, it was limited to 16 hours; prior to 2011, it was 30 hours in most states and 27 hours in New York [7].) Surveys of surgical residents have largely supported the idea that most would prefer to work longer shifts with fewer handoffs from resident to resident and that the 16-hour shift was not popular [26, 28].

Acknowledgments This case was adapted from an earlier case written by Amy Caruso Brown and was inspired by the FIRST and iCOMPARE trials cited in this chapter.

References

1. Lerner B. A case that shook medicine. Washington Post [Internet]; 2006 Nov 24 [cited 2018 Feb 5]. Available from: http://www.washingtonpost.com/wp-dyn/content/article/2006/11/24/AR2006112400985.html.
2. Blum AB, Raiszadeh F, Shea S, Mermin D, Lurie P, Landrigan CP, et al. US public opinion regarding proposed limits on resident physician work hours. BMC Med. 2010;8:33.
3. U.S. Food & Drug Administration. Institutional Review Boards (IRBs) and protection of human subjects in clinical trials [Internet]. Silver Spring (MD): FDA; updated 2015 Feb 3 [cited 2018 Feb 5]. Available from: https://www.fda.gov/AboutFDA/CentersOffices/OfficeofMedicalProductsandTobacco/CDER/ucm164171.htm.
4. University of California, San Diego. Human research protections program: exemption from IRB Review. La Jolla (CA): UCSD; 2018 Mar 1 [cited 2018 Feb 1]. Available from: https://irb.ucsd.edu/Exemption_fact_sheet.pdf.
5. Pucher PH, Johnston MJ, Aggarwal R, Arora S, Darzi A. Effectiveness of interventions to improve patient handover in surgery: a systematic review. Surgery. 2015;158(1):85–95.
6. Lin H, Lin E, Auditore S, Fanning J. A narrative review of high-quality literature on the effects of resident duty hours reforms. Acad Med. 2016;91(1):140–50.
7. Rosenbaum L, Lamas D. Residents' duty hours—toward an empirical narrative. New Engl J Med. 2012;367(21):2044–9.

8. Centers for Disease Control and Prevention. U.S. public health service syphilis study at Tuskegee [Internet]. Atlanta (GA): CDC; 2015 [cited 2018 Feb 5]. Available from: https://www.cdc.gov/tuskegee/timeline.htm.
9. DuBois JM, Bioethics Research Center. Hepatitis studies at the Willowbrook State School for Children [Internet]. Washington University in St. Louis, School of Medicine. [cited 2018 Feb 5]. Available from: https://bioethicsresearch.org/resources/case-studies/hepatitis-studies-willowbrook/.
10. Cassel C, Dickersin K, Garber A, Gatsonis C, Gottlieb G, Guest J, et al. What is comparative effectiveness research? [Chapter 2]. In: Initial national priorities for comparative effectiveness research. Washington, D.C.: The National Academies Press; 2009 [cited 2018 Feb 5]. Available from: https://www.nap.edu/read/12648/chapter/4.
11. Roland M, Torgerson DJ. Understanding controlled trials: What are pragmatic trials? BMJ. 1998;316(7127):285.
12. Lantos JD, Wendler D, Septimus E, Wahba S, Madigan R, Bliss G. Considerations in the evaluation and determination of minimal risk in pragmatic clinical trials. Clin Trials. 2015;12(5):485–93.
13. Minami CA, Odell DD, Bilimoria KY. Ethical considerations in the development of the Flexibility in Duty Hour Requirements for Surgical Trainees trial. JAMA Surg. 2017;152(1):7–8.
14. Ellis GB, Lin MH, Office for Protection from Research Risks. Informed consent requirements in emergency research [OPRR Letter, no. 97-01]. Rockville (MD): OHRP; 1996 Oct 31 [cited 2018 Feb 5]. Available from: https://www.hhs.gov/ohrp/regulations-and-policy/guidance/emergency-research-informed-consent-requirements/index.html.
15. Protection of human subjects, 45 C.F.R. § 46. Rockville: Office for Human Research Protections (OHRP).; last reviewed 2016 Feb 16 [cited 2018 Feb 5]. Available from: https://www.hhs.gov/ohrp/regulations-and-policy/regulations/45-cfr-46/index.html.
16. Klass P. Getting through the night. New Engl J Med. 2013;369(24):2279–81.
17. Williamson AM, Feyer A-M. Moderate sleep deprivation produces impairments in cognitive and motor performance equivalent to legally prescribed levels of alcohol intoxication. Occup Environ Med. 2000;57:649–55.
18. Landrigan CP, Rothschild JM, Cronin JW, Kaushal R, Burdick E, Katz JT, et al. Effect of reducing interns' work hours on serious medical errors in intensive care units. N Engl J Med. 2004;351:1838–48.
19. Bolster L, Rourke L. The effect of restricting residents' duty hours on patient safety, resident well-being, and resident education: an updated systematic review. J Grad Med Educ. 2015;7(3):349–63.
20. Govindarajan A, Urbach DR, Kumar M, Li Q, Murray BJ, Juurlink D, Kennedy E, et al. Outcomes of daytime procedures performed by attending surgeons after night work. N Engl J Med. 2015;373:845–53.
21. Tsafrir Z, Korianski J, Almog B, Many A, Wiesel O, Levin I. Effects of fatigue on residents' performance in laparoscopy. J Am Coll Surg. 2015;221(2):564–70.
22. Accreditation Council for Graduate Medical Education. Improving physician well-being, restoring meaning in medicine [Internet]. Chicago (IL): ACGME. [cited 2018 Feb 5]. Available from: https://www.acgme.org/What-We-Do/Initiatives/Physician-Well-Being.
23. Liaison Committee on Medical Education. Functions and structure of a medical school: Standard 3.6. LCME; 2015 Apr [cited 2018 Feb 5]. Available from: http://jabsom.hawaii.edu/docs/LCME/standards/standard3.pdf.
24. Ioannidis JP. Why most clinical research is not useful. PLoS Med. 2016;13(6):e1002049.
25. Bilimoria KY, Chung JW, Hedges LV, Dahlke AR, Love R, Cohen ME, et al. National cluster-randomized trial of duty-hour flexibility in surgical training. New Engl J Med. 2016;374:713–27.
26. Asch DA, Bilimoria KY, Desai SV. Resident duty hours and medical education policy—raising the evidence bar. N Engl J Med. 2017;376(18):1704–6.
27. Khoong EC, Linker AS. An appeal for evidence-based resident duty Hours Reform. JAMA Int Med. 2017;177(11):1555–6.
28. Bernstein L. Some new doctors are working 30-hour shifts at hospitals around the U.S Washington Post [Internet]. 2015 Oct 28 [cited 2018 Feb 5]. Available from: https://www.washingtonpost.com/national/health-science/some-new-doctors-are-working-30-hour-shifts-at-hospitals-around-the-us/2015/10/28/ab7e8948-7b83-11e5-beba-927fd8634498_story.html?utm_term=.01e39d5b46d7.

Further Reading on This Topic

Collaborative Institutional Training Initiative. Basic course in human subjects research [Internet]. Fort Lauderdale (FL): CITI Program. Available from: https://about.citiprogram.org/en/homepage/.

Ofri D. Singular intimacies: becoming a doctor at Bellevue. Boston (MA): Beacon Press; 2009.

Rothman DJ. Strangers at the bedside: a history of how law and bioethics transformed medical decision making. New York: Routledge; 2017.

XI

Stigma and Marginalization

Contents

Chapter 21   "I Need Blockers So I Don't Turn Into a Girl" – 409

Chapter 22   "You Don't Understand: He Needs That Bottle" – 429

"I Need Blockers So I Don't Turn Into a Girl"

Karen L. Teelin

21.1 Background Questions – 410

21.2 Additional Case Information and Questions for Discussion – 411

21.3 Answers to Background Questions – 413

21.4 Responses to Discussion Questions – 416

References – 426

- **Glossary**

Gender diverse or gender nonconforming - A person whose behavior or appearance does not match cultural and societal expectations for what is stereotypically expected for their assigned gender. Examples of gender-diverse identities include gender nonbinary (neither male nor female) and gender fluid (more male or more female at different times).

Gender dysphoria (GD) - Distress or discomfort that may occur when a person's internal sense of gender does not match the sex assigned at birth, based on anatomy and/or chromosomes.

Gender identity - A person's internal sense of being male, female, neither, or both.

Gender incongruence - A diagnosis for a transgender person, in other words, a person whose gender identity (internal sense of gender or asserted gender) does not match the gender assigned at birth. This is an International Classification of Diseases-11 (ICD-11) diagnosis, and it replaces the term "transexualism."

Gonadotropin-releasing hormone (GnRH) analog - A class of drugs used for suppression of precocious puberty and to treat hormone-responsive cancers, endometriosis, and uterine fibroids. It is also used for suppression of puberty in adolescents with gender dysphoria.

Transgender - A person whose sex assigned at birth (based on external anatomy and/or chromosomes) does not match their gender identity. It is important to note that transgender individuals may identify as gay, straight, bisexual, or something else. Sexual orientation and gender orientation are separate concepts.

Transgender female - A person whose sex assigned at birth was male, but who identifies as a female.

Transgender male - A person whose sex assigned at birth was female, but who identifies as a male.

Learning Objectives

1. Understand the basic language used to discuss gender, as well as how gender dysphoria can be approached in young adolescents entering puberty.
2. Begin to understand the ways minority stress and stigmatization contribute to health disparities.
3. Apply knowledge of ethical and legal principles to situations in which parents do not agree on treatment for a child.
4. Develop an approach to addressing discrimination, bias, and microaggressions in your professional role.

Case Background

You are a pediatric resident working with an adolescent specialist on an elective rotation. You are both involved in the care of an 11-year-old boy, Nik Chen, who was assigned a female sex at birth. Nik has identified as a boy for most of his life. He first expressed a male gender identity when he started talking at age 2, and he adopted his current male first name in kindergarten. He wears his hair cropped short, dresses in conventionally male clothes, participates in sports on boys' teams, and has a group of close friends who are boys. In school, he uses the boys' locker rooms and bathrooms.

21.1 Background Questions

1. Summarize typical physical and emotional development during puberty.

2. What is gender dysphoria? What is the current estimated prevalence of gender dysphoria and/or transgender identity in the United States? What are the limitations of these data?

3. What percentage of gender diverse children continue to identify with their affirmed gender in adulthood (i.e., what percentage grow up to be transgender adults)? What are the limitations of these data?

4. What is the recommended medical treatment for young adolescents (early puberty) with persistent gender dysphoria? How do gonadotropin-releasing hormone (GnRH) analog work? What are the potential side effects?

5. What are some of the health disparities (including those related to mental health) that potentially affect patients who are gender diverse or transgender? What role might health care play in exacerbating or mitigating these disparities?

6. How is decision-making for children by divorced or never-married parents approached? In your state, when, if ever, is consent from both parents required?

7. Gender identity is distinct from sexuality and sexual orientation but is also related, both historically and in individual lives. How have professional organizations in medicine, psychiatry, and psychology historically regarded (1) gender identity and (2) sexual orientation?

21.2 Additional Case Information and Questions for Discussion

Nik has been followed by a family therapist who specializes in the care of children with gender dysphoria. She referred Nik and his family to the adolescent specialist to discuss treatment to suppress puberty. Nik is eager to start this medication as soon as possible, because he has noted the beginning of breast growth and he is worried about being ostracized by his classmates and teammates. However, at the first visit, Nik's mother, Diana Chen, makes some comments that suggest Nik's father, Ryan Gerson, who is not present at the appointment, is not in favor of the treatment. In fact, Mr. Gerson has made comments to the therapist suggesting that he might try to prevent the treatment from occurring.

1. What ethical principles can guide your discussion with the child and family and ultimately the treatment decisions you make?
    (a) How does one weigh these factors when making these decisions?
    (b) How much weight do you put on Nik's preferences?
    (c) Do values, beliefs, and social norms play a role in approaching this quandary? What are some examples of values, beliefs, or social norms that may influence decision-making in this situation?

During the conversation, you feel that you have established a good rapport with Nik and his mom. However, when you hand them the printout from the medical record system after the visit, on which is printed a diagnosis of "gender dysphoria," you see Nik make a face.

2. How can you respond when a patient uses a term you have not heard previously?
    (a) How do labels both help and hurt patients? Can you think of other examples of this?
    (b) Does using a term like this medicalize or pathologize difference?
    (c) Self-reflection: How do you address difference respectfully without bias?

Nik's parents are not married but have had a good co-parenting relationship in the past. Nik primarily lives with his mother and visits his father regularly. Mr. Gerson has struggled more than Ms. Chen to adjust to Nik's affirmed gender. He still calls Nik his "daughter" at times, but he no longer uses Nik's feminine birth name.

3. What are Nik's and his mother's options if his father refuses to consent to the treatment?
    (a) Under what circumstances can healthcare practitioners override a parent's wishes?

You suggest requesting an ethics consultation to discuss the case.

4. How do you think the ethics consultant will approach this case? What would you do if you were the ethics consultant?

5. What are the benefits and the risks of overriding Nik's father's objections (proceeding with the treatment despite his objections)?
    (a) What are the ethical principles that would justify overriding Nik's father?
    (b) How might the risks be minimized?

6. What are the benefits and risks of overriding Nik's mother's consent for puberty-blocking treatment, i.e., not providing the treatment due to his father's objections?
   (a) What are the ethical principles that would justify this?
   (b) How might the risks be minimized?

7. Would your analysis of this case be different if Nik was very short for his age and his mother wants to treat him with growth hormone for height while his father did not?

The ethics consultant meets with Nik's mother and the adolescent medicine specialist and concludes that the decision to start a GnRH analog is in Nik's best interests and that the potential harm from withholding treatment is sufficiently serious and irreversible to justify overriding Nik's father's parental authority. Nik's father declines to meet in person with the consultant but does agree to a phone call. He says that he is "really worried" about the long-term effects of this medication, and furthermore he thinks Nik will "be okay" as he matures, because women who like football and hunting are well-liked and "popular" with men.

8. Does this change your view of the situation? How so?

9. Would more studies, producing better data and more information specific to this case, help you resolve this situation? How so?
   (a) How might you conduct a study to determine the mental health outcomes for adolescents with gender dysphoria who undergo puberty-suppressing medical treatment compared with those who don't? What are the pros and cons of different study designs (e.g., randomized controlled trial) for this study?
   (b) What kind of study is most appropriate for studying a rare event, such as GnRH analog therapy for pediatric gender dysphoria?
   (c) What biases are inherent in this type of study?

Nik's mother asks if she should seek legal advice, since the ethics consultant's recommendations are not binding. The adolescent medicine physician explains she cannot move forward over the objections of one parent, unless Nik's mother obtains a letter from a judge. The physician refers her to a local attorney who provides free legal services to support LGBTQ (lesbian, gay, bisexual, transgender, or queer/questioning) individuals. Ms. Chen also wants to switch Nik to her insurance company, so that his father will not see any bills or be involved in the payment. However, the case manager suspects that Nik's father's insurance company is more likely to pay for the treatment, based on prior experience with the companies involved.

10. Should insurance companies and public programs such as Medicare and Medicaid be required to cover this treatment?

11. GnRH agonist therapy is used for many other indications. For example, it is sometimes used to suppress puberty in young adolescents with cancer because it may help reduce the risk of infertility from chemotherapy. Should insurance coverage be different for different indications for the same treatment?

You decide to present this case and one of the journal articles you discovered at your program's monthly journal club. The journal article is emailed to the faculty, residents, and medical students on the rotation in advance. You overhear a resident and medical student laughing about the topic and comparing whose experiences with transgender patients in the emergency department (ED) were more outrageous.

12. What, if anything, would you do or say?
    (a) Do you have any obligation to address derogatory comments and microaggressions when you witness them?
    (b) Self-reflection: Would you feel differently if the comment were directed at a group to which you belong? Would you be more likely to speak up if a member of the group (e.g., a transgender person) was present?

"I Need Blockers So I Don't Turn Into a Girl"

(c) Creative problem-solving: How can we create a space in which any staff member feels comfortable speaking up when they witness mistreatment?

21.3 Answers to Background Questions

1. **Summarize typical physical and emotional development during puberty.**
   Pubertal development includes physical, cognitive, and psychosocial maturation. Physical changes are triggered by an increase in pulsatile release of GnRH from the hypothalamus. In girls (female gender assigned at birth), the first outward sign of puberty is breast development (thelarche), followed by a growth spurt and genital hair, and then menarche (first menstrual period), typically 2–2.5 years after the onset of puberty. In boys (male gender assigned at birth), the first detectable sign of puberty is an increase in testicular volume to 4 mL or greater, followed by penile growth, pubic hair, growth spurt, and spermarche. Testosterone contributes to muscle development, acne, and increased red blood cell counts. Both genders experience accrual of bone mineral density during puberty. Healthy adolescent development is also associated with brain changes, contributing to the development of abstract reasoning, empathy, self-awareness, and mature relationships.

2. **What is gender dysphoria? What is the current estimated prevalence of gender dysphoria and/or transgender identity in the United States? What are the limitations of these data?**
   Children are assigned a gender at birth based on genital anatomy or chromosomes. In most children, this gender assignment matches the child's innate sense of maleness or femaleness, or gender identity. In some children, the assigned gender does not match their gender identity, or their internal sense of being male or female. Gender dysphoria is the term for distress or discomfort that may be associated with this incongruence. "Gender incongruence" or "gender/body divergence" are other diagnostic terms used to describe transgender individuals, preferred by some because they may be seen as less pathologizing.
   The second part of the question is difficult to answer because the U.S. Census Bureau does not currently ask about gender identity, and furthermore gender identity may be fluid or hard to define. Current estimates for the prevalence of gender dysphoria range widely, with reported estimates increasing over the last decade, in association with cultural and media trends creating a slightly more favorable environment for those who identify as transgender. A study in 2018 of more than 2000 high school students found 2.7% of the studied population self-defined as transgender or gender nonconforming [1]. Another analysis 2 years earlier found that 0.6% of U.S. adults identify as transgender, implying that 1.4 million adults in the United States are transgender [2]. These data are limited by the definition used for transgender (e.g., some organizations include gender nonconforming within the definition of transgender), current lack of national survey data, and a reluctance of some to identify as transgender on official surveys. Current estimates are thought to underestimate the true prevalence.

3. **What percentage of gender diverse children continue to identify with their affirmed gender in adulthood (i.e., what percentage grow up to be transgender adults)? What are the limitations of these data?**
   Of children diagnosed with gender dysphoria, available data suggest that a minority will grow up to be transgender adults. For example, in one study of children ages 5–12 years (*n* = 77) referred for gender dysphoria, the gender dysphoria was persistent in 27% and no longer present in 43%, with 30% lost to

follow-up [3]. Studies are limited by an unclear definition for gender dysphoria and by the stigma associated with this diagnosis.

Despite reported low overall likelihood of the persistence of gender dysphoria, a subset of children with intense, consistent, persistent, and insistent gender dysphoria, especially when persisting or strengthening through puberty, are likely to be transgender adults [4].

4. **What is the recommended medical treatment for young adolescents (early puberty) with persistent gender dysphoria? How do gonadotropin-releasing hormone (GnRH) analog work? What are the potential side effects?**

If the gender dysphoria is persistent, consistent, and insistent and persists with the start of puberty, treatment with a GnRH analog to suppress endogenous puberty is recommended by the Endocrine Society Clinical Practice Guidelines and the World Professional Association for Transgender Health (WPATH) Standards of Care [5, 6].

GnRH analog provide constant agonist activity, blocking the effects of pulsatile release of GnRH, therefore decreasing luteinizing hormone and follicle-stimulating hormone, leading to suppression of estrogen (females) and testosterone (males) production. GnRH analog are approved for treatment of precocious puberty in pediatric patients, as well as treatment of prostate cancer, endometriosis, and fibroids. They are used off-label for several indications including breast cancer, delaying puberty during treatment of pediatric leukemia, and gender dysphoria. Potential adverse effects include local injection site pain or reaction in fewer than 20% of children and headache, emotional lability, and weight gain in 7% or fewer. Effects are not permanent—that is, if treatment is stopped, endogenous pulsatile activity of GnRH will resume.

Treatment with GnRH analog reversibly suppresses endogenous puberty, preventing the development of secondary sexual characteristics, e.g., breast development in natal females and facial hair, body hair, deepening voice, and Adam's apple in natal males. Development of these unwanted secondary sex characteristics may be devastating to adolescents with gender dysphoria [7]. GnRH analog are a reversible treatment that provides time for the family and patient to further explore gender identity and the process of transition. During treatment, adolescents who have been living in their affirmed gender can continue to do so. Those who have been treated with GnRH analog and who go on to take gender-affirming hormones will have improved outcomes in that they will be able to avoid surgical procedures, such as mastectomy for transgender males, creating an easier, less costly transition. In addition, several studies have shown that the use of GnRH analog is associated with decreased psychosocial distress, including decreased likelihood of depression, anxiety, and behavioral problems [7–10].

There is currently a lack of longitudinal data on the use of GnRH analog for children with gender dysphoria, and, as noted previously, the use of these medications for this indication does not have U.S. Food and Drug Administration (FDA) approval. Concerns include the effect on the pubertal growth spurt, effect on bone health, and effects on the developing brain. These effects likely resolve once steroid hormones are begun and puberty ensues. Emerging studies support the safety and efficacy of these medications for this indication [11]. Finally, two additional concerns with the use of these medications are (1) their high cost, coupled with lack of insurance coverage in many cases, and (2) fertility concerns for those who go on to take gender-affirming hormones, an important area of active research that is beyond the scope of this chapter.

5. **What are some of the health disparities that potentially affect patients who are gender diverse or transgender? What role might health care play in exacerbating or ameliorating these disparities?**

It is important to note that health disparities in patients who are transgender are not inevitable, but rather a consequence of minority stress, stigmatization, and marginalization. Mental health disparities have been well-documented in LGBTQ populations and include depression, suicide attempts and completed suicides, non-suicidal self-injury, and anxiety [12, 13]. These can be mitigated by a gender-affirming environment, especially an affirming home environment, but also an affirming school and community environment [11, 12]. Transgender youth are also at increased risk for homelessness, teen dating violence, and bullying [14–17]. The role of stigmatization, social rejection, and violence in contributing to poor health outcomes for transgender individuals has been reported across cultures [18].

Interestingly, autism is more common among transgender youth, and the converse is also true (i.e., children with autism are more likely to be transgender) [19, 20]. The reasons behind the discrepant rates are not well-understood and require further research.

From a broader perspective, patients who identify as LGBTQ may experience disparities that are multifaceted, with individual, institutional, and societal components. Overt discrimination in health care was reported by the National Transgender Discrimination Survey in which 19% of respondents reported being denied care due to their transgender or gender-nonconforming status and 28% postponed *necessary* medical care due to experience with, or fear of, discrimination and harassment [21]. However, providers may fail to meet the health needs of LGBTQ patients through omission, rather than overt discrimination, by not asking about gender identity or sexual identity or assuming heterosexuality and cis-sexuality. There are many reasons why a provider may not discuss gender or sexual identity with patients, including feeling it is not relevant to care, lack of time, personal discomfort, or lack of familiarity [13, 22–24].

6. **How is decision-making for children by divorced or never-married parents approached? In your state, when, if ever, is consent from both parents required?**
Generally, when parent or guardian consent is required, the consent of one parent or guardian is considered sufficient. However, in cases of divorce or separation of the parents or guardians, the healthcare practitioners must determine which parent(s) or guardian(s) have legal authority to consent to treatment. Typically, in cases of divorce, both parents are granted legal custody, and often "major medical decisions" require permission from both parents. Other exceptions, applicable to all parents, include participation in clinical trials in which there is more than minimal risk, other research studies at the discretion of the institutional review board, and prenatal interventions solely for the benefit of the fetus. Exceptions exist when one parent is deceased, unknown, incompetent, or temporarily incapacitated or does not have legal responsibility for the minor child [25].

7. **Gender identity is distinct from sexuality and sexual orientation but is also related, both historically and in individual lives. How have professional organizations in medicine, psychiatry, and psychology historically regarded (1) gender identity and (2) sexual orientation?**
Sexuality and gender are separate but related spectrums that are often confused or conflated. When clinicians do so, it is problematic for the health and wellness of transgender individuals as well as gay, bisexual, and lesbian individuals. Transgender individuals may identity as straight, gay, bisexual, or something else. Gender and sexuality are uniquely related for all individuals (cisgender or transgender), but gender identity is distinct from sexual orientation. It is important for clinicians to seek to understand how their patients identify and to respect those identities.

While this conflation does not respect how individuals themselves identify, it shows how they have been viewed throughout history by medical professionals, as well as many other groups. Both sexual and gender minority patients have historically been unfairly and inaccurately pathologized. Affirming clinicians recognize that gender diverse identities do not constitute pathology but are rather natural biologic variations that should be affirmed, understood, supported, and destigmatized.

The following timeline considers selected points in the history of medicine with regard to both gender identity and sexual orientation:

- 1930: Lili Elbe underwent the first known surgical transition (male-to-female) in Europe [26–28].
- 1947: Alfred Kinsey founded the Institute for Sex Research at Indiana University. His work on sexual behavior, though not initially widely recognized, would eventually contribute to changing attitudes toward same-sex attraction [29, 30].
- 1952: Publication of the first edition of the Diagnostic and Statistical Manual of Mental Disorders (DSM) by the American Psychiatric Association, which classified homosexuality as a mental illness, codifying beliefs that had grown throughout the twentieth century [31].
  - The existence of such diagnostic labels arose from and promoted medicalization of sexuality and promulgated efforts to "cure" homosexuality [32].
- 1952: Christine Jorgensen became the first American widely known as transgender [26–28].
  - The diagnostic criteria utilized by physicians, for what was then called "transsexualism," emphasized sexual and gender conformity. For example, biological males might be diagnosed as a "transsexual" if they demonstrated certain characteristics, such as a desire to be penetrated, a desire for domesticity, or deference to men. In these early cases, gender and sexual orientation were inextricably linked [33].
- 1957: Evelyn Hooker published her seminal paper demonstrating no difference in psychological testing between homosexual and heterosexual men and arguing that previous data had falsely indicated a link between homosexuality and mental illness by studying *only* homosexual men *with* mental illness [29].
- 1973: The American Psychiatric Association's board voted to remove the diagnosis of "sexual orientation disturbance" from the DSM, arguing that "same-sex orientation [was] not inherently associated with psychopathology" [32, 33]. This vote was not without controversy.
- 1980: "Gender identity disorder" was added to the DSM [34].
- 1987: "Ego-dystonic homosexuality" (essentially meaning that a patient was experiencing anxiety related to and a desire to change sexual orientation), the last sexual orientation-related diagnosis, was removed from the DSM [34, 35].
- 2002: Several organizations, including the American Psychiatric Association and the American Academy of Pediatrics (AAP), issued statements supporting the civil rights of LGBTQ individuals in marriage, adoption, and parenting [36].
- 2009: The American Psychological Association issued a report "conclud[ing] that efforts to change sexual orientation are unlikely to be successful and involve some risk of harm" [31].
- 2013: Gender identity disorder was reclassified as gender dysphoria, due to the efforts of transgender activists and advocates [34, 35].

21.4 Responses to Discussion Questions

Nik has been followed by a family therapist who specializes in the care of children with gender dysphoria. She referred Nik and his family to the adolescent specialist to discuss treatment to suppress puberty. Nik is eager to start this medication as soon as possible, because he has noted the beginning of breast growth and he is worried about being ostracized by his classmates and teammates. However, at the first visit, Nik's mother, Diana Chen, makes some comments that suggest Nik's father, Ryan Gerson, who is not present at the appointment, is not in favor of the treatment. In fact, Mr. Gerson has made comments to the therapist suggesting that he might try to prevent the treatment from occurring.

1. **What ethical principles can guide your discussion with the child and family and ultimately the treatment decisions you make?**
   - **Beneficence.** How can we maximize the benefits to Nik? What decision is in his best interests? How do his best interests intersect with the best interests of his family, as a unit, and of his individual family members?
   - **Non-maleficence.** How can we avoid doing harm? Are there decisions that would actively harm Nik? Are there others that might prevent or mitigate that harm?
   - **Respect for autonomy or respect for persons.** Nik is not a fully autonomous adult, but his autonomy is developing. How can we support and respect that development? His parents have decision-making authority: Not only do they have the right to make healthcare decisions, but they also have power over the way he has been raised (faith, values, etc.). How can we proceed in a way that respects their values as well as the ones that Nik may be developing independently?
   - **Justice.** Are we providing Nik with the same standard of care that we would for a patient whose gender conformed to his sex assigned at birth? Does Nik have access to the same care and services as any other patient?
   (a) **How does one weigh these factors when making these decisions?**
       In the United States, we generally value autonomy above the other ethical principles; this is complicated by the fact that Nik is 11 years old. In one sense, we have to weigh his parents' current authority over him against his future autonomy. Various approaches to pediatric decision-making have been proposed to help resolve this conflict (defined and discussed in greater detail in ▶ Chapter 10: "Our Son's Cancer Is Gone. Why Can't We Stop Treatment?"), including:
       - The *best interests standard*. Which decision is in Nik's best interests? [37].
       - The *harm principle*. Will the parents' decision cause serious or irreversible harm? Only in such cases should it be overridden [38].
       - The *right to an open future*. Are there certain decisions that close off Nik's future options? If possible, they should be avoided [39].
   (b) **How much weight do you put on Nik's preferences?**

> **Box 21.1 Teaching Tip**
> Consider different perspectives and how the learner's own background (family, education, culture, religion) informs responses.

This case would not have occurred if Nik—and other children—did not already have strong preferences. Unlike many other medical dilemmas, Nik has already *powerfully* demonstrated his commitment to the decision. Despite being assigned female gender at birth based on his anatomy, he has lived—day in and day out—as a boy for at least the past 6 years. Most healthcare practitioners—even those with limited experience with children—will understand that this is beyond simple exploration or experimentation. Historically, some have advocated for psychotherapy to attempt to reverse the gender dysphoria. This has not been shown to be successful, and may be harmful, and it is not the current standard of care [5, 6].

Children of Nik's age (11 years) or who are younger have not typically been considered "mature minors"—patients under the age of majority but thought to have adult-like capacity to make their own decisions (see ▶ Chapter 10). While some state laws do not specify a lower age limit for making reproductive and mental health decisions (a category typically protected, allowing for adolescents to give consent and retain confidentiality), it is controversial as to whether an 11-year-old should be allowed to make such decisions without parental input.

(c) Do values, beliefs, and social norms play a role in approaching this quandary? What are some examples of values, beliefs, or social norms that may influence decision-making in this situation?

> **Box 21.2 Teaching Tip**
> Ask learners to work in groups of three to four to brainstorm and produce their own lists. Learners should be able to recognize that values and beliefs vary in each person. Social norms are broader. They vary with different societies but also can vary within a society depending on race, gender, ethnicity, and geography. And social norms change over time, of course.

Some relevant examples include:
- Values:
  - Traditional, binary gender roles are good for families and for societies.
  - Some separation between men and women is inherently good as it values the unique contributions of each sex.
  - Variability and difference are normal aspects of human biology and are good for society.
  - Children should be allowed to express who they are and should be supported and affirmed in their identity.
  - Individuals' lived experiences should be respected.
- Beliefs:
  - Binary separation of roles between two genders is natural and inevitable.
  - Birth-assigned sex always aligns with lived experience of gender identity, and anything else is a voluntary behavior.
  - Different does not mean abnormal.
  - Efforts to minimize gender stereotyping among young children are confusing and ultimately harmful.
  - Unconditional love is more important for children than forcing them to conform to societal expectations.
  - A transgender person cannot live a good and productive life in today's society.
  - The suffering of a transgender individual attempting to deny their gender identity is greater than the discomfort a cisgender person feels witnessing transgender identity.
- Social norms (behavioral expectations):
  - Girls are not aggressive; boys are not nurturing.
  - Girls enjoy domestic activities (such as cooking, childcare, etc.) more than boys do; boys enjoy sports more than girls do.
  - All children should conform to the gender assigned at birth. (Note this is beginning to shift in some communities and societies to the opposite norm that all children should be allowed to explore their gender identity if they choose).

Nik's own values and beliefs, and those of his parents, are important in achieving a successful resolution.

Understanding the values, beliefs, and social norms that underpin his father's opposition to GnRH analog therapy might help resolve the case.

For many people, conceiving of gender identity as separate from sex assigned at birth feels foreign. There is an important and necessary role for education, especially for those with no prior experience with transgender individuals. Some find it helpful to consider intersex children who are assigned a gender at birth that may not be concordant with their anatomical genitalia or their sex chromosomes. These infants often grow up to express a strong gender identity that does not match their assigned social gender. Alternatively, a parent might believe that gender exists along a spectrum and yet have difficulty accepting hormonal therapy or surgical transition.

It is also worth noting that seemingly unrelated values and beliefs may play a role here. For instance, perhaps Nik's father values living a "natural" life and avoiding reliance on Western medical interventions, he might then be opposed to a course of action that seems to suggest his son will need regular medical treatment for the rest of his life.

In this case, while his parents may not agree on using GnRH analog therapy, it appears that they have both been somewhat supportive of his decision to live as a boy. He has also benefitted from having a therapist who is understanding and apparently knowledgeable regarding gender dysphoria. Other children may not be as fortunate.

During the conversation, you feel that you have established a good rapport with Nik and his mom. However, when you hand them the printout from the medical record system after the visit, on which is printed a diagnosis of "gender dysphoria," you see Nik make a face.

2. **How do labels both help and hurt patients? Can you think of other examples?**

"Gender dysphoria" is meant to place the focus on the person's experience of discomfort, rather than the gender identity itself as a problem. However, some transgender, genderqueer, and gender-nonconforming individuals do not experience "dysphoria" nor any mental health problems. There continues to be a controversy around the need for a diagnosis, particularly a mental health diagnosis. Consequently, this nomenclature continues to be in flux. "Gender incongruence" and "gender/body divergence" are newer terms preferred by some advocates.

During each encounter between a patient and healthcare practitioner, a diagnosis must be assigned in order to document correctly and bill for services. Giving a diagnosis is sometimes psychologically difficult or stigmatizing to patients. However, sometimes, practical benefits outweigh the risks of a potentially stigmatizing label. Healthcare practitioners should be aware of these issues. It is possible to address the significant shortcomings with the diagnostic terms by being open and upfront with patients about how the diagnosis can be problematic yet also sometimes necessary.

Other examples include the diagnostic labels of obesity and autism. Being "diagnosed" as obese may worsen the stigma already felt by obese individuals at the doctor's office and may promote bias. However, the diagnosis might be necessary to obtain insurance coverage for a referral to a nutritionist. In the case of autism, the label may provoke significant anxiety in parents but be necessary in order to get appropriate services provided in school. While some adults with autism have embraced the term (preferring to be called "autistic" rather than "person with autism"), others have advocated for descriptors like "neurodiverse." ▶ Chapter 15, "Please Look Beyond My Disability," explores the idea of health differences, in contrast to disparities, in more detail.

(a) **Does using a label such as gender dysphoria medicalize or pathologize difference?**
De In a sense, yes. It could be pathologizing to a person who does not view themselves as having a problem but who might be viewed by some in the medical community as such. Moreover, many experts question the inclusion of the diagnosis of gender dysphoria within the DSM-5, arguing that this furthers stigma and misunderstanding and that gender dysphoria should not imply a mental health condition. Gender dysphoria might better fit as an endocrine diagnosis rather than a psychiatric term. As noted earlier, there is some need to give a diagnosis in the medical culture. There also needs to be a diagnostic code in order to systematically assess and manage patients. Certainly, billing is a part of this, but there is also a need for different physicians and other healthcare practitioners to use consistent terminology in order to communicate effectively and provide optimal care. If the patient is not reporting dysphoria this term should not be used. Patients who have already transitioned also typically do not have gender dysphoria.

(b) **Self-reflection: How do you address difference respectfully?**

Nik's parents are not married but have had a good co-parenting relationship in the past. Nik primarily lives with his mother and visits his father regularly. Mr. Gerson has struggled more than Ms. Chen to adjust to Nik's affirmed gender. He still calls Nik his "daughter" at times, but he no longer uses Nik's feminine birth name.

3. **What are Nik's and his mother's options if his father refuses to consent to the treatment?**
The first option is to simply accept his father's decision for now (until Nik is 18, until he is a few years older and may be considered a mature minor or simply until he is more obviously entering puberty). This option has potentially serious consequences for Nik's physical and mental health.

Other options depend on whether Nik's parents have a custody agreement that specifies joint decision-making for health care or gives decision-making to one parent alone. It is also possible that, as unmarried parents, they may have only an informal agreement or that Nik's father has no legal rights, if he was not listed on the birth certificate. If Nik's mother has sole decision-making, she can legally consent to the treatment without the involvement of Nik's father and may not even be obliged to inform him.

If they have joint decision-making, it is likely that their agreement specifies joint decision-making for major decisions. In that case the healthcare practitioner would have to decide whether this is a major decision or a minor one (given its reversibility). Assuming it is a major decision, Nik's mother—with the assistance of the healthcare practitioner—could attempt to reach a compromise, perhaps through an ethics consultation (see below) or through outside mediation. If they cannot reach a compromise, Nik and/or his mother could hire an attorney and seek a court order to permit Nik's treatment. Nik would likely have a guardian ad litem (GAL) appointed to advocate on his behalf. GALs are special representatives for minors and legally incompetent adults who are selected by judges; they are typically charged with determining the designee's best interests.

(a) **Under what circumstances can healthcare practitioners override a parent's wishes?**
Healthcare practitioners are not obligated to provide any treatment requested. For example, there is no obligation to provide a treatment that would not be effective or would be more harmful than beneficial. Within the scope of recommended therapy, however, courts have historically ordered treatment when a child's life is at stake or when the

child is at risk for serious morbidity without treatment. In this case, both the child and one parent feel strongly that he should have the treatment, and the treatment is in-line with national and international standards of care for children with gender dysphoria.

You suggest requesting an ethics consultation to discuss the case.

4. **How do you think the ethics consultant will approach this case? What would you do if you were the ethics consultant?**
Most ethics consultants will want to meet separately or jointly with the various stakeholders—including both parents (likely separately), Nik, the family therapist, and the adolescent specialist—to better understand their values, concerns, and preferences. There are many sources of misinformation regarding gender dysphoria, and it would be helpful to determine what Nik's father's specific concerns are and what is at the root of his hesitation. Does he think the gender dysphoria will resolve as Nik goes through puberty? Is he concerned about adverse effects from GnRH analog therapy? Is he conflating this therapy with testosterone therapy (which is only partially reversible)? Is he worried about what others will think and his own social standing? Are his objections based in religious beliefs? Is he worried about Nik's future fertility? It would be important to acknowledge Nik's father's concerns. It is also important that someone is available to provide information and data to address his questions. If the health care team can foster a relationship with him, so that rather than feeling cornered he feels affirmed, he may be willing to listen, to return to clinic, and to consider alternate viewpoints. It also may be helpful to include Nik's mental health therapist, who may have a relationship with Mr. Gerson. The consultant might also want to understand who else is important to Nik and to his parents—grandparents, close family friends, religious group members, etc.

Some of these concerns may be amenable to educational interventions or to conversations with Nik in the presence of a therapist. The ethicist may act as a mediator, helping the parents reach a compromise, if possible. If Nik's mother does have sole decision-making, the ethics consultant can help support the decision to treat with only the mother's consent. While the ethicist can make a recommendation regarding treatment, ultimately, the legal circumstances described previously will still apply. If Nik's father rejects the recommendation, legal action might still be necessary.

5. **What are the benefits and the risks of overriding Nik's father's objections (proceeding with the treatment despite his objections)?**
The benefits are that Nik will be able to access a treatment that will allow him to more easily continue living as a boy. Children with gender dysphoria face additional burdens during puberty—both internal and external. Their bodies are changing in ways that are not compatible with their gender identity. These children may be ostracized from the peer groups that were accepting of the prepubertal child. GnRH agonist therapy prevents these physical changes. If Nik chooses to live as a male for the rest of his life, GnRH therapy will allow him to progress through normal male puberty once he is older and can be prescribed testosterone. He will not develop wider hips or larger breasts, and thus he will avoid the need for future gender affirming top surgery.

The main risk is that this will damage Nik's relationship with his father, although there is a chance that this will happen regardless. The risks of GnRH analog therapy itself are small, as noted in the background question above. Concerns regarding bone density, effects on the growth spurt, and effects on the brain seem to be unfounded as stated above. GnRH analog therapy does not have permanent physical effects. Another risk to note is that, with GnRH analog therapy,

Nik will go through puberty later than most of his peers. Both early and late puberty can be sources of social distress.

While the possibility that Nik may change his mind might seem like a risk, GnRH analog therapy is reversible and could be stopped at any point that Nik wished, allowing endogenous puberty to proceed.

(a) **What are the ethical principles that would justify overriding Nik's father?**
This would be justified if we agree that the benefits of intervention and harms of nonintervention substantially outweigh the benefits of nonintervention and harms of intervention. In other words, the benefits to Nik's present and future outweigh the risk of upsetting his father and possibly damaging his relationship with his father (which may happen eventually anyway.)

(b) **How might the risks be minimized?**
Assuring that Nik and both parents are receiving appropriate counseling and psychosocial support is important regardless of the treatment decision.

6. **What are the benefits and risks of overriding Nik's mother's consent for puberty blocking treatment, i.e., not providing the treatment due to his father's objections?**
Benefits might include the father's satisfaction with outcome and, perhaps, preservation of relationship with Nik (though the opposite could occur if Nik is resentful of the decision or emotionally harmed by the experience of puberty) and co-parenting relationship with Nik's mother. There is also a hypothetical chance that the hormonal changes of puberty will facilitate resolution of gender dysphoria; however, no data support this hypothesis. The benefits of this approach may accrue more to Nik's parents than to Nik himself.

Risks include significant distress at going through puberty and developing secondary sex characteristics (including menstruation) of the gender with which Nik does not identify, as well as further psychological harm if he is subsequently excluded from his male peer group and activities in which he previously participated as a boy. Nik is consequently at increased risk for mental health issues including depression and suicide. He will also face a greater need for surgical procedures later in life if he chooses to live as a male.

(a) **What are the ethical principles that would justify this?**
This would be justified if it were agreed that the benefits of nonintervention and harms of intervention substantially outweigh the benefits of intervention and harms of nonintervention.

(b) **How might the risks be minimized?**
As noted previously, good psychosocial support for Nik and his family is important, regardless of the final decision. As in many cases, it is important to consider the family interests even if they do not change the recommendation. Overriding one parent's objections may cause that parent significant psychological and/or moral distress. It may have other unforeseen consequences, such as if Nik's father decides to withhold child support or stopped visiting his son because of this decision.

The healthcare practitioner in this case may want to consider other scenarios in which adolescents seek to make autonomous decisions and the consequences of failing to respect autonomy and confidentiality. While many or even most adolescents might find that their parents are more supportive than they expected, in other cases, an adolescent's disclosure of sexual activity or sexual identity, for example, might lead to physical or emotional abuse or being thrown out of the family home.

"I Need Blockers So I Don't Turn Into a Girl"

✓ 7. **Would your analysis of this case be different if Nik was very short for his age and his mother wants to treat him with growth hormone for height while his father did not?**
Like the use of GnRH analog, the decision to use growth hormone for idiopathic short stature must be made within a specific timeframe, usually before the child is sufficiently mature to make the decision for her-/himself. Like GnRH analog therapy, the purpose of the intervention is to impact physical appearance in a way that improves quality of life and facilitates social acceptance.

Unlike GnRH analog, use of growth hormone is not reversible and not nearly as effective (children on average acquire two additional centimeters in final height compared with predicted final height). While the use of GnRH analog may spare a child or teen some procedures later in life, those procedures are still an option for a child who does not receive puberty blocking therapy; there are no options for enhancing adult height for an older teen or young adult who did not receive growth hormone. The risks of growth hormone therapy are also more significant. On the whole, growth hormone therapy has fewer benefits and more risks, compared with GnRH analog.

> **Box 21.3   Teaching Tip**
> Encourage learners to consider whether there is a difference in the social norms affecting the different decisions and to discuss how that affects their willingness to overrule the father in each case.

The ethics consultant meets with Nik's mother and the adolescent medicine specialist and concludes that the decision to start a GnRH analog is in Nik's best interests and that the potential harm from withholding treatment is sufficiently serious and irreversible to justify, overriding Nik's father's parental authority. Nik's father declines to meet in person with the consultant but does agree to a phone call. He says that he is "really worried" about the long-term effects of this medication, and furthermore he thinks Nik will "be okay" as he matures, because women who like football and hunting are well-liked and "popular" with men.

✓ 8. **Does this change your view of the situation? How so?**
It is helpful to better understand Nik's father's thinking. As in most cases involving treatment disagreements, he wants Nik to be happy and healthy, but he disagrees about what choices are most likely to produce those outcomes. Nik's father has real concerns about the safety of the intervention, which suggests that more time needs to be spent discussing the risks and benefits, and the science behind them. It might also be worthwhile to better understand the beliefs and social norms he has endorsed, on which he seems to be relying to help alleviate the risks of denying Nik treatment.

Although Nik's father has not said so explicitly, healthcare practitioners should also give some consideration to the twin roles of grief and fear. Both Nik's parents may experience a period of grieving for the loss of the daughter they once had and the future they had imagined for that child. Such grief does not mean they do not love and accept him just as he is. All parents have hopes, dreams, and expectations, spoken and unspoken, for their children. Concurrently, parents may be fearful for the challenges their children will face as transgender teens and adults, and such fears might contribute to Nik's father's concerns.

It is also important to note that rejecting parents tend to become less rejecting over time, particularly with access to accurate information about gender [40].

✓ 9. **Would more studies, producing better data, and more information specific to this case help you resolve this situation? How so?**
More studies would be helpful, particularly regarding long-term effects of GnRH analog used for gender dysphoria and also regarding the mental health impacts on children who choose to delay or not to delay puberty. More information specific to this case would also be helpful. Anecdotally, some parents have changed their initial thinking on

this topic once they better understand gender dysphoria. Nik's father may, for instance, not understand the potential negative impact of endogenous puberty on his son.

(a) **How might you conduct a study to determine the mental health outcomes for adolescents with gender dysphoria who undergo puberty-suppressing medical treatment compared with those who do not? What are the pros and cons of different study designs (e.g., randomized controlled trial) for this study?**

> **Box 21.4 Teaching Tip**
> Assure that learners can demonstrate understanding of the strengths, limitations, and ethical issues associated with each type of study.

- **Randomized controlled trial**: For the study to be feasible and ethical, a population of pre-teens and parents who are truly undecided about puberty-suppressing treatment and willing to be randomized is required. This will not be the case for most, and currently puberty-suppressing treatment is the standard of care and thus any healthcare practitioner who thinks a patient is a good candidate should be offering it.
- **Cohort study**: To conduct a cohort study, researchers enroll a cohort of patients with gender dysphoria who have decided against treatment and a cohort of those who have decided to be treated (or, alternatively, one large cohort, although there may be a significant imbalance in sample sizes between the two groups) and then compare mental health outcomes in those who choose puberty-suppressing treatment with those who do not. Studies of this nature would take 10 years or more to complete prospectively, while retrospective studies might suffer from recall bias. Both types can be affected by selection bias, as discussed below. If puberty-suppressing treatment is the standard of care, patients and families who choose not to receive it may be distinctly different from those who do. For instance, they may have more doubts or questions about their gender identity and be more likely to identify with their sex assigned at birth later in life. Furthermore, in practice, it is difficult to identify children with insistent gender dysphoria who decline puberty blocking treatment.
- **Case-control study**: To conduct a case-control study, researchers identify cases of young adults with gender dysphoria who had a specific poor mental health outcome (such as a suicide attempt) and match them with controls, also with gender dysphoria, who did not. They then assess whether one group was more likely to have received puberty-suppressing treatment.

(b) **What kind of study is most appropriate for studying a rare event, such as GnRH analog therapy for pediatric gender dysphoria?**
Rare events are usually best studied in a case-control design, because the rarity of the event may make it difficult to capture in a cohort study, requiring unrealistically large sample sizes or extended timeframes to conduct the study. Further, cohort studies may suffer from significant biases that affect validity of results. However, more common events or outcomes could be studied using a cohort design.

(c) **What biases and limitations are inherent in this type of study?**
- **Selection bias**: Adolescents and their families who did not use GnRH analog therapy might be less likely to participate (or,

simply, harder to identify for recruitment purposes) in a retrospective case-control study than young people who took GnRH analog therapy. If there are important differences between those who can be found and agree to participate and those who cannot be identified or do not agree to participate, this would result in selection bias.

- **Recall bias**: Cases may be more likely to recall the exposure of interest than controls. For instance, a young adult who received treatment and is now doing well might "gloss over" any struggles during puberty. Case-control studies may be particularly affected by recall bias, although that may be less likely for studies in which the exposure is a memorable medical intervention (i.e., participants might be less likely to forget receiving a hormone-blocking medication, compared with a vitamin supplement).
- **Confounding factors**: A confounding factor must be associated with both the treatment and the outcome but not in the causal pathway. Patients who do not receive puberty-suppressing treatment may have factors that independently contribute both to the nontreatment decision and also to poorer mental health outcomes. For example, these could include less supportive or less educated parents or families or communities, or fewer financial resources.

Nik's mother asks if she should seek legal advice, since the ethics consultant's recommendations are not binding. The adolescent medicine physician explains she cannot move forward over the objections of one parent, unless Nik's mother obtains a letter from a judge. The physician refers her to a local attorney who provides free legal services to support LGBTQ individuals. Ms. Chen also wants to switch Nik to her insurance company, so that his father will not see any bills or be involved in the payment. However, the case manager suspects that Nik's father's insurance company is more likely to pay for the treatment, based on prior experience with the companies involved.

✅ 10. **Should insurance companies and public programs such as Medicare and Medicaid be required to cover this treatment?**
Given the substantial potential benefits and the potential cost savings later (by minimizing the need for surgical procedures), this treatment should be covered by insurance companies. Coverage of treatment for conditions such as gender dysphoria, as well as mental illness, should not be different because of our inability to fully understand their pathophysiology. In practice, coverage often differs between private insurance companies and public programs such as Medicare and Medicaid, but this is not ethically justified. (Differences occur in both directions, particularly in pediatric medicine, with Medicaid sometimes covering services that private insurance companies do not.) If a treatment is medically indicated and effective, it is not relevant whether the patient receives Medicaid, due to the family's low income, or is insured through a parent's employer, for instance.

✅ 11. **GnRH agonist therapy is used for many other indications. For example, it is sometimes used to suppress puberty in young adolescents with cancer because it may help reduce the risk of infertility from chemotherapy. Should insurance coverage be different for different indications for the same treatment?**
In general, an insurance company would be acting reasonably to cover only less expensive, equally effective treatments for a given condition, even if a more expensive treatment option is covered for other conditions. So, if a less expensive equally effective option existed for children with gender dysphoria, it might be reasonable to cover it and not to cover GnRH analog for this indication, even while this class of medications is covered

for other indications. However, in this case there is no less expensive equally effective treatment available.

You decide to present this case and one of the journal articles you discovered at your program's monthly journal club. The journal article is emailed to the faculty, residents, and medical students on the rotation in advance. You overhear a resident and medical student laughing about the topic and comparing whose experiences with transgender patients in the ED were more outrageous.

☑ 12. **What, if anything, would you do or say?**
   (a) **Do you have any obligation to address derogatory comments and microaggressions when you witness them?**
   Healthcare practitioners should recognize that this is an obligation of a professional: Failure to speak up is de facto support for these types of comments and perpetuates biases, which lead to substandard care for transgender patients (among others). Physicians and healthcare leaders can lead the way for others to speak up for marginalized and stigmatized populations.
   (b) **Self-reflection: Would you feel differently if the comment was directed at a group to which you belong? Would you be more likely to speak up if a member of the group (e.g., a transgender person) was present?**
   (c) **Creative problem-solving: How can we create a space in which any staff member feels comfortable speaking up when they witness mistreatment?**
   See ▶ Box 21.6 for case conclusion.

> **Box 21.6   Case Conclusion**
> Nik's mother sought a court order to allow Nik to receive treatment. However, prior to the judge's ruling, Nik's father conceded and agreed to treatment. Nik began receiving a GnRH analog.

Acknowledgments The author wishes to thank Irene Sills for her review of this chapter. The author and editors also wish to thank Lilly Rizzo for her assistance with background research and Samuel Castleberry and Jordana Gilman for their editorial input.

References

1. Rider G, McMorris B, Gower A, Coleman E, Eisnebert M. Health and care utilization of transgender/gender non-conforming youth: a population based study. Pediatrics. 2018;14(3):e20171683.
2. Flores AR, Herman JL, Gates GJ, Brown TNT. How many adults identify as transgender in the United States? [Internet] Los Angeles (CA): The Williams Institute, UCLA School of Law; 2016 [cited 2018 Feb 16]. Available from: https://williamsinstitute.law.ucla.edu/wp-content/uploads/How-Many-Adults-Identify-As-transgeder-in-the-United-States.pdf.
3. Wallien MS, Cohen Kettenis PT. Psychosexual outcome of gender dysphoric children. J Am Acad Child Adolesc Psychiatry. 2008;47(12):1413–23.
4. Steensma TD, McGuire JK, Kreukels BP, Beekman AJ, Cohen-Kettenis PY. Factors associated with desistence and persistence of childhood gender dysphoria: a quantitative follow-up study. J Am Acad Child Adolesc Psychiatry. 2013;52(6):582–90.
5. Hembree WC, Coehn-Kettenis PT, Gooren L, Hannema SE, Meyer WJ, Murad MH, Rosenthal SM, Safer JD, Tangpricha V, T'Sjoen GG. Endocrine treatment of gender-dysphoric/gender-incongruent persons: an Endocrine Society Clinical Practice Guideline. J Clin Endocrinol Metab. 2017;102(11):3869–903.
6. Coleman E, Brockting W, Botzer M, Cohen-Kettenis P, DeCuypere G, Feldman J, Fraser L. Standards of care for the health of transsexual, transgender, and gender nonconforming people. 7th vol. East Dundee (IL): The World Professional Association for Transgender Health; 2012 [cited 2018 Aug 28]. Available from: https://www.wpath.org/publications/soc.
7. de Vries AL, Steensma TD, Doreleijers TA, Cohen-Kettenis PT. Puberty suppression in adolescents with

> **Box 21.5   Patient Perspective**
> "From a young age I knew I was a boy. I cried at night and wondered why God made a mistake. I prayed that I would wake up a cis-boy. At times I felt like I should not have been born. The mistake was too big and people wouldn't understand. I would fight with my mother when she tried to make me wear an Easter dress. I tried to explain to her that I was a boy. Once I learned that transgender is a thing, I knew right away that's what I am. Knowing there are other people out there like me made it bearable." (Paraphrased from a letter received by the author.)

gender identity disorder: a prospective follow-up study. J Sex Med. 2011;8(8):2276–83.
8. Delemarre-van de Waal HA, Cohen Kettenis PT. Clinical management of gender identity disorder in adolescents: a protocol on psychological and paediatric endocrinology aspects. Eur J Endocrinol. 2006;155(Suppl 1):S131.
9. de Vries AL, McGuire JK, Steensma TD, Wagenaar EC, Doreleijers TA, Cohen-Kettenis PT. Young adult psychological outcome after puberty suppression and gender reassignment. Pediatrics. 2014;134(4):696–704.
10. Turban JL, Ehrensaft D. Research Review: Gender identity in youth: treatment paradigms and controversies. J Child Psychol Psychiatry. 2018;59(12):1228–43.
11. de Vries AL, McGuire JK, Steensma TD, Wagenaar EC, Doreleijers TA, Cohen-Kettensi PT. Young adult psychological outcome after puberty suppression and gender reassignment. Pediatrics. 2014;134(4):696–704.
12. Eisenberg ME, Gower AL, McMorris BJ, Rider GN, Shea G, Coleman E. Risk and protective factors in the lives of transgender/gender nonconforming adolescents. J Adolesc Health. 2017;61(4):521–6.
13. Mayer KH, Bradford JB, Makadon HJ, Stall R, Goldhammer H, Landers S. Sexual and gender minority health: what we know and what needs to be done. Am J Public Health. 2008;98(6):989–95.
14. Russell ST, Ryan C, Toomey RB, Diaz RM, Sanchez J. Lesbian, gay, bisexual, and transgender adolescent school victimization: implications for young adult health and adjustment. J Sch Health. 2011;81(5):223–30.
15. Dank M, Lachman P, Zweig JM, Yahner J. Dating violence experiences of lesbian, gay, bisexual, and transgender youth. J Youth Adolesc. 2014;43(5):846–57.
16. Martin-Storey A. Prevalence of dating violence among sexual minority youth: variation across gender, sexual minority identity and gender of sexual partners. J Youth Adolesc. 2015;44(1):211–24.
17. Cochran BN, Stewart AJ, Ginzler JA, Cauce AM. Challenges faced by homeless sexual minorities: comparison of gay, lesbian, bisexual, and transgender homeless adolescents with their heterosexual counterparts. Am J Public Health. 2002;92(5):773–7.
18. Robles R, Fresán A, Vega-Ramírez H, Cruz-Islas J, Rodríguez-Pérez V, Domínguez-Martínez T, Reed GM. Removing transgender identity from the classification of mental disorders: a Mexican field study for ICD-11. Lancet Psychiatry. 2016;3(9):850–9.
19. De Vries AL, Noens IL, Cohen-Kettenis PT, van Berckelaer-Onnes IA, Doreleijers TA. Autism spectrum disorders in gender dysphoric children and adolescents. J Autism Dev Disord. 2010;40(8):930–6.
20. Strang JF, Kenworthy L, Dominska A, Sokoloff J, Kenealy LE, Berl M, Walsh K, Menvielle E, Slesaransky-Poe G, Kim KE, Luong-Tran C. Increased gender variance in autism spectrum disorders and attention deficit hyperactivity disorder. Arch Sex Behav. 2014;43(8):1525–33.
21. Grant J, Mottet L, Tannis J. Injustice at every turn: a report of the National Transgender Discrimination Survey. Washington, DC: National Center for Transgender Equality; 2011.
22. Kitts RL. Barriers to optimal care between physicians and lesbian, gay, bisexual, transgender, and questioning adolescent patients. J Homosex. 2010;57(6):730–47.
23. Committee on Lesbian, Gay, Bisexual, and Transgender Health Issues and Research Gaps and Opportunities, Board on the Health of Select Populations. Health of lesbian, gay, bisexual, and transgender people: building a foundation for better understanding. Washington, D.C.: National Academies Press; 2014.
24. Safer JD, Coleman E, Feldman J, Garofalo R, Hembree W, Radix A, et al. Barriers to health care for transgender individuals. Curr Opin Endocrinol Diabetes Obes. 2016;23(2):168–71.
25. Katz AL, Webb SA; Committee on Bioethics. Informed consent in decision-making in pediatric practice. Pediatrics. 2016;138(2). pii: e20161485.
26. Meyerowitz J. Sex change and the popular press: Historical notes on transsexuality in the United States, 1930–1955. GLQ. 1998;4(2):159–87.
27. Whittle S. A brief history of transgender issues. The Guardian [Internet]. 2010 Jun 2 [cited 2018 Aug 28]. Available from: https://www.theguardian.com/life-andstyle/2010/jun/02/brief-history-transgender-issues.
28. Goddard JC, Vickery RM, Terry TR. Development of feminizing genitoplasty for gender dysphoria. J Sex Med. 2007;4(4):981–9.
29. Kinsey AC, Pomeroy WR, Martin CE. Sexual behavior in the human male. Am J Public Health. 2003;93(6):894–8.
30. Bayer R. Chapter 2: Dissenting views. In: Homosexuality and American psychiatry: the politics of diagnosis. Princeton: Princeton University Press; 1987.
31. Glassgold JM, Beckstead L, Drescher J, Greene B, Miller RL, Worthington RL. Report of the American Psychological Association task force on appropriate therapeutic responses to sexual orientation. Washington, D.C.: APA; 2009.
32. Bayer R. Chapter 1: From abomination to disease. In: Homosexuality and American psychiatry: the politics of diagnosis. Princeton: Princeton University Press; 1987.
33. Herek GM. Beyond "homophobia": thinking about sexual prejudice and stigma in the twenty-first century. Sex Res Soc Policy. 2004;1(2):6–24.
34. Glicksman E. Transgender today. American Psychological Association; April 2013;44(4):36. [cited 2018 Aug 22]. Available from: http://www.apa.org/monitor/2013/04/transgender.aspx.
35. Drescher J. Queer diagnoses: parallels and contrasts in the history of homosexuality, gender variance, and the Diagnostic and Statistical Manual. Arch Sex Behav. 2010;39(2):427–60.
36. Perrin EC, Siegel BS, Committee on Psychosocial Aspects of Child and Family Health. Promoting the well-being of children whose parents are gay or lesbian. Pediatrics. 2013;131(4):e1374–83.
37. Kopelman LM. The best-interests standard as threshold, ideal, and standard of reasonableness. J Med Philos. 1997;22(3):271–89.
38. Diekema D. Parental refusals of medical treatment: the harm principle as threshold for state intervention. Theor Med Bioeth. 2004;25(4):243–64.

39. Mills C. The child's right to an open future? J Soc Philos. 2003;34(4):499–509.
40. Ryan c. Generating a revolution in prevention, wellness & care for LBGT children. Temple Political and Civil Rights Review. 2014;23(2):331–44.

Further Reading on This Topic

American Psychological Association. Guidelines for psychological practice with transgender and gender nonconforming people. Am Psychol. 2015;70(9):832–64.

Bornstein K. Gender outlaw: on men, women and the rest of us. New York: Routledge; 2013.

Levine DA. The committee on adolescence. Office-based care for lesbian, gay, bisexual, transgender, and questioning youth. Pediatrics. 2013;132(1):e297–313.

Lopez X, Stewart S, Jacobson-Dickman E. Approach to children and adolescents with gender dysphoria. Pediatr Rev. 2016;37(3):89–98.

Rosin H. A boy's life. The Atlantic monthly [Internet]. Boston (MA); 2008 Nov. Available from: https://www.theatlantic.com/magazine/archive/2008/11/a-boys-life/307059/.

"You Don't Understand: He Needs That Bottle"

Lauren Hall Mutrie and Janet H. Goode

22.1 Background Questions – 430

22.2 Additional Case Information and Questions for Discussion – 431

22.3 Answers to Background Questions – 432

22.4 Responses to Discussion Questions – 440

References – 447

© Springer Nature Switzerland AG 2019
A. E. Caruso Brown et al. (eds.), *Bioethics, Public Health, and the Social Sciences for the Medical Professions*, https://doi.org/10.1007/978-3-030-03544-0_22

Glossary

Built environment - The human-designed spaces in which people live, learn, work, and play, including homes, buildings, streets, public spaces, transportation systems, and overall infrastructure—all of which influence health and wellness [1].

Low income - Categorization of working families in the United States who earn less than twice the federal poverty threshold.

Medical-legal partnership - A collaborative partnership between healthcare workers and lawyers aimed at addressing health-harming social and legal needs of patients.

Obese - Having a body mass index (BMI) at or greater than the 95th percentile in children or a BMI greater than 30 kg/m^2 in adults.

Overweight - Having a BMI between the 85th and 95th percentiles in children or a BMI between 25 and 30 kg/m^2 in adults.

Poverty threshold - A federal poverty measure, also referred to as the poverty line, used by the U.S. Census Bureau to define and quantify poverty in the United States every year for families of different sizes. Individuals or families are classified as poor if their annual pretax income falls below a certain dollar amount (i.e., poverty threshold) [2, 3]. The federal poverty guidelines are a simplified version of the threshold and are used by government programs to determine eligibility.

Learning Objectives

1. Apply the socio-ecological model to a complex problem such as obesity
2. Identify trends in obesity and discuss how a public health approach may be more crucial to reversing these trends in the long term than a medical approach
3. Describe the complex relationship between food insecurity and obesity
4. Recognize weight bias and its negative health outcomes, including how it affects clinical care given by health professionals

Case Background

You are a family physician working in a busy urban clinic that serves primarily low-income patients, most of whom are Medicaid recipients. This morning, you are seeing Mayson Bustamante Valdez and his mother, Alicia Valdez García. Both Mayson and Alicia are obese, and Alicia has a history of hypertension and type II diabetes. Mayson is in your clinic for his 18-month checkup, and this is the first time you have met the family.

22.1 Background Questions

1. Define obesity in adult and pediatric populations, and describe epidemiologic trends and disparities within childhood obesity. Why is it important to identify children with obesity?

2. Find and describe an effective and evidence-based (nonsurgical) approach to helping patients lose weight. What kinds of interdisciplinary interventions are helpful in achieving weight loss for families who struggle with obesity?

3. Recent research challenges the concept that the key to a healthy weight is "calories in = calories out." What does determine obesity? How do prenatal exposure to glucose and epigenetic modification contribute to obesity?

4. What are some examples of secondary and tertiary prevention of obesity and obesity-related diseases? Why is it difficult to use secondary prevention for obesity in children?

5. How does weight bias affect health outcomes?

6. What benefits do the Supplemental Nutrition Assistance Program (SNAP); Special Supplemental Nutrition Program for Women, Infants, and Children (WIC); and the National School Lunch Program (NSLP) provide? How are WIC and NSLP benefits different from SNAP? What are the income thresholds for each? What are the residency and immigration status requirements to access these benefits?

7. What is food insecurity and how can healthcare practitioners screen for it? How is it related to obesity and health outcomes?

8. What are food deserts? What are other features of the built environment that contribute to obesity?

"You Don't Understand: He Needs That Bottle"

22.2 Additional Case Information and Questions for Discussion

Upon reviewing past notes in the electronic medical record, you note that weight loss has been discussed at the last two visits with other providers. However, both Mayson and his mother have gained weight in the past year, with Mayson crossing multiple percentiles on his growth curve, such that he is now at the 99th percentile of weight-for-length. The social history notes that Alicia works as a convenience store clerk and has two other children, aged 6 and 8.

1. **How would you approach this visit?**
   (a) What questions would you ask Alicia about her family's diet and lifestyle?

Alicia screened positively for food insecurity when she was first seen at the clinic, about 4 months into her pregnancy.

2. **How can you respectfully address food insecurity with Alicia?**

When you enter the exam room, you are frustrated to see that both your patient and his mother have bags of food from a fast-food chain restaurant. Alicia apologetically explains that it was a "special treat" because of the doctor's visit and the anticipation of Mayson needing immunizations. Indeed, Mayson is fussy during the interview, but he consoles immediately with the snacks.

3. **Self-reflection: How do you feel hearing this piece of the encounter? Have you ever felt judged or stigmatized by a healthcare provider?**

On examination, Mayson is playful and interactive. He is "pudgy" but actively walks around the room and is curious about your medical instruments. He has normal vital signs. Aside from visible decay of his front teeth, his physical exam is normal. You ask Alicia what he usually drinks. She replies that he prefers whole milk or juice, but when he sees her drinking soda, he wants to have some, too. At bedtime, he only wants milk and drinks two or three bottles of either leftover powdered formula or cow's milk during the night. You start to explain to Alicia that this is a major contributor to both his weight and his dental caries. Alicia becomes extremely upset.

4. **Why do you think she is reacting this way, and what would you do next?**

You learn that Alicia's family was evicted from the apartment that she shared with her three children, her mother, her teenage sister, and her 10-year-old niece, who was recently adopted after the child's own mother died from complications of diabetes. All seven have had to temporarily move into a two-bedroom apartment rented by her uncle and his adult son. Her uncle and cousin are not accustomed to living with young children, and they get up early to work at a construction site. If she does not give Mayson a milk bottle, he cries—sometimes for hours—and she is terrified that her uncle might throw them out. While listening to Mayson's mother, you notice that she has dark, velvety skin over her neck and axillae, and she reports that the entire family struggles with obesity.

5. **How does this change your view of the situation?**

6. **What are some of the typical recommendations that you have heard healthcare practitioners give overweight and obese patients regarding weight loss?**
   (a) How realistic do you think these recommendations are for Alicia, given what you know about her resources and responsibilities?
   (b) What additional resources can you offer her?

You ask the clinic social worker to meet with Alicia and help connect her with resources in the community. The social worker tells you that many families like Alicia's spend more than 80% of their income on rent, so it is understandable that Alicia prioritizes keeping her uncle happy so that she can stay in his home. It can take years of being on a waiting list to finally receive a housing voucher, and the list is often closed to new names. Even if she was willing to join the list, she might only qualify for pro-rated assistance if she is living in a mixed-status household that includes both legally documented and undocumented residents of the United States. (A few states, notably Massachusetts, have programs to help undocumented immigrants obtain housing [4].) The social worker also tells you that many families who do receive subsidized housing immediately redistribute their income toward more and better food for their families.

7. Considering families like Alicia's, why do you think primary prevention of obesity at the population level is challenging?
   (a) What has worked? What has not worked?
   (b) Creative problem-solving: Suppose you were given a small grant ($10,000) to address obesity in your community. What would you propose doing with the grant?

Alicia also tells you that Mayson is an extremely picky eater, so she only offers him things she knows he will eat. In addition to milk and formula, this includes French fries, hot chips, and macaroni and cheese. With frustration, she says that the residents rotating in the clinic always tell her to "put out a variety of healthy foods and let him pick and choose what to feed himself" but that they cannot afford to waste food like that. You recall telling other parents that toddlers often need to try a new food 8–15 times before they will accept it.

8. Do you think this is a realistic suggestion for Alicia and Mayson? Why or why not?

Your medical student is curious about the process of enrolling in WIC and SNAP and offers to help Alicia navigate it. However, Alicia refuses the assistance. You recall that Alicia's chart mentions that she was born in Guatemala and came to the United States at the age of 3. Alicia's younger sister, her niece, and her three children were born in the United States.

9. Based on what you know about WIC and SNAP eligibility, is Alicia's family eligible?
   (a) Why might Alicia be reluctant to register her children for benefits?
   (b) Should you ask explicitly about Alicia's immigration status? Why or why not?

10. Self-reflection: Do you feel differently about this patient now than when you first read the case?
    (a) Have you learned anything that changes the way you will approach overweight and obese patients in the future?

22.3 Answers to Background Questions

1. **Define obesity in adult and pediatric populations, and describe epidemiologic trends and disparities within childhood obesity. Why is it important to identify children with obesity?**
   Obesity can be defined as a complex, chronic, largely preventable, multifactorial disorder, characterized by excess body fat and significant comorbidities. Obesity has multiple contributing causes, including genetics, metabolic and neurobehavioral activity, diet, physical activity level, and sociocultural factors [5]. Body mass index is the most commonly used screening measure of obesity in adults and children over 2 years of age. It is a calculation that uses weight in kilograms and height in meters to indirectly assess adiposity, and it correlates with adverse health outcomes associated with increased body fat: BMI (kg/m^2) is typically calculated by taking the weight in kilograms divided by square of height or length in meters. Overweight adults have a BMI greater than 25 and less than 30 kg/m^2; obese adults have a BMI of 30 kg/m^2 or higher. In contrast with adults, BMIs in children are interpreted relative to other children of the same gender and age via percentiles calculated from U.S. Centers for Disease Control and Prevention (CDC) growth charts. Children with BMIs between the 85th and 95th percentiles are considered overweight while children with BMIs over the 95th percentile are considered obese. For children under 2 years of age, weight-for-length is more commonly used to measure nutritional status [6]. BMI-for-age growth charts are now available for children under 2, but the predictive ability of this measure on long-term health outcomes compared to the standard weight-for-length requires further investigation [7, 8].
   While obesity rates remain high in the United States, the rise of obesity on the whole has begun to stabilize after decades of alarming increase since the 1980s. Significant disparities exist across gender, age, race, and geographic region.

Rates vary between states, counties, and even neighborhoods: in general, more than 1 in 3 adults, 1 in 6 children (ages 2–19), and 1 in 11 young children (ages 2–5) are obese [9]. Childhood rates are highest in Mississippi (22%) and lowest in Oregon (10%). According to the 2017 State of Obesity Report (2011–2014 National Health and Nutrition Examination Survey [NHANES] data), childhood obesity rates have tripled since 1980, though overall prevalence has remained stable for the past decade around 17%. Rates of obesity among adolescents (ages 12–19) represent the fastest growing age group, quadrupling from 5% to 20.5% since 1980. Extreme obesity, having a BMI at or greater than 120% of the 95th percentile on CDC BMI-for-age growth charts, has increased in all age groups and is becoming particularly notable in children under 5 years old. In fact, rates of extreme obesity are 2% for children aged 2 to 5 years, 4.3% for 6 to 11 years, and 9.1% for teens [9].

Social inequities contribute to other disparities seen in childhood obesity. Obesity disproportionately affects low-income and rural communities as well as certain racial and ethnic groups, namely, African Americans, Latinos, and Native Americans. Prevalence rates are highest among Native American and Latino children, followed by African American, white, and Asian children. Extreme obesity among Latino and African American preschool children is nearly double that of white children, and overall rates of childhood obesity in these groups are higher at earlier ages with faster rates of increase. Families in impoverished environments often lack options for obtaining both adequate physical activity and healthy food options. Contributing factors for these observations in many neighborhoods include lack of safe outdoor spaces for play, paucity of safe walking routes to school, and inadequate recess time at school for exercise. Furthermore, grocery stores with healthy, affordable food choices may not exist within accessible range for families in poor communities; thus people obtain unhealthy, highly processed food from corner stores or fast-food chains at cheaper prices [9, 10].

Prevention and treatment of childhood obesity are of utmost importance due to the risk of persistence through adulthood with significant long-term health consequences, namely, hypertension, type II diabetes, hypercholesterolemia, cardiovascular disease, obstructive sleep apnea, osteoarthritis and chronic pain, nonalcoholic fatty liver disease, gastroesophageal reflux, cancer, and depression. Obese children experience school dysfunction, poor academic performance, low self-esteem, and bullying in addition to the aforementioned comorbidities. Obese children encounter these adverse metabolic and psychosocial outcomes by adolescence. On a broader level, obesity and its associated health problems also have significant economic impact on individuals and healthcare systems: The direct medical expenses of prevention, diagnostics, and treatment coupled with the indirect costs of decreased productivity, disability, and absenteeism are all costly to society. It is estimated that the obesity epidemic costs more than $150 billion in annual healthcare expenditures with billions of dollars more in lost productivity. Investing in primary obesity prevention and a healthier overall population therefore provides significant economic return [9].

2. **Find and describe an effective and evidence-based (nonsurgical) approach to helping patients lose weight. What kinds of interdisciplinary interventions are helpful in achieving weight loss for families who struggle with obesity?**
Sustainable weight loss strategies aim for long-term weight control, prevention of weight gain in obese adults, and decrease in velocity of weight gain in obese children. It is much easier to help individuals maintain a healthy weight and prevent obesity than to reverse it later once obesity has developed. Comprehensive, integrative, family-centered

interventions are required for successful weight control, namely, dietary therapy and education, exercise, and behavioral modification, all of which may be coupled with pharmacotherapy and/or surgery in cases of extreme obesity [11]. When treating or preventing obesity in children, it is important to emphasize healthy lifestyle choices for the entire family in order to maximize success and also to identify barriers to healthy living for that family. Understanding where the family lives, the financial and food resources available to them, their level of education and literacy, and the safety of their neighborhood are all important in developing a treatment strategy.

The socio-ecological model for public health interventions provides a helpful framework and systems approach for the prevention and treatment of obesity. In order to support and sustain healthy living, interventions guided by this model address the drivers of obesity at the individual, family, healthcare, community, and societal levels through primary and secondary prevention efforts. This kind of intervention incorporates the individual's knowledge, beliefs, and behaviors into a treatment plan, which also accounts for the interpersonal influence of family and social networks and the policies and institutional structures at community, state, and federal levels. Within this approach, obesity is understood to be a complex disorder by which food, physical activity behaviors, and psychosocial structures are not simply determined by personal choice but are also enabled or restricted by environmental and policy factors [12].

Medical-legal partnership (MLP) is a complementary interdisciplinary tool that can also be useful at both individual and community levels. Lawyers and social workers in an MLP can help gather information about family dynamics, resources, and living situations in order to identify and advocate for potential solutions on an individual level. An MLP can also identify recurring trends and pursue policy change at local, state, and national levels.

In order for families to be successful in their weight loss efforts, they need nutritious food and time for active play and exercise. They need a medical home for regular assessment with BMI screening, goal development, and empowerment by their healthcare provider. Local healthcare systems must support community-based programs and counseling in order to guide families through the lifestyle changes necessary to sustain weight loss. Thoughtful community design and land use policies to support green spaces and active living are important, particularly in blighted areas. Food and beverage distributors must be encouraged to produce and market healthy, affordable options with reduced sugar and fat in their products. Key strategies in the management of obesity at an individual level thus involve family education about healthy food and beverage choices, calorie restriction (elimination of sugar-sweetened beverages and fast-food consumption, transition from whole to low-fat milk), increased physical activity (mandatory physical education in school, classroom activity breaks, walking/biking to school), adjunctive behavioral therapy, and limitations on screen time. Community-level strategies include the promotion of accessible, affordable healthy food and beverage options at home, school, child-care settings, and work; coverage of obesity services through health insurance, including strong preventive health care; and policies that support healthy living [13].

3. **Recent research challenges the concept that the key to a healthy weight is "calories in = calories out." What does determine obesity? How do prenatal exposure to glucose and epigenetic modification contribute to obesity?**
While obesity was once simply viewed as a lifestyle "choice" to overeat and underexercise, it actually stems from a complex web of interactions between genetic, biologic, environmental, social, and economic factors [6, 10, 13, 14]. Furthermore, obesity can either be

enabled or prevented by the built and shared environments, which control the availability and accessibility of calories and safe spaces for physical activity [13]. The greatest risk factor for children is parental history of obesity [14].

There is emerging evidence to suggest that fetal nutrition and environmental exposures have lasting effects on later growth, adiposity, and energy regulation. The idea is that epigenetic modifications, or heritable changes in gene expression that produce changes in phenotype without changes in genotype, may increase risk of obesity in offspring and perpetuate obesity across generations [15]. Epigenetic modifications do not alter DNA sequence, but rather change DNA accessibility through mechanisms like DNA methylation and histone modification, thereby influencing gene expression [15]. Maternal dietary changes, such as folate or fat increases, can alter DNA methylation patterns. Research is ongoing into the effects of maternal hyperglycemia on epigenetic modification and how glucose, insulin, and leptin function as epigenetic factors. Prenatal exposure to a diabetic intrauterine environment appears to increase the risk of developing diabetes later in life, and within animal models, this "metabolic imprinting" can be intergenerational. Children who are exposed to a diabetic intrauterine environment are more likely to be obese and have a higher incidence of diabetes later in life. Of note, adjustment for maternal weight does not explain excess risk in offspring, suggesting that "nutrient-mediated developmental abnormalities in utero contribute independently to the development of obesity and diabetes in offspring of diabetic mothers" [16]. A comparison of siblings born before and after the development of maternal diabetes provides evidence for epigenetic modification, with the excessive risk observed in those children born after the development of maternal diabetes attributed to the hyperglycemic intrauterine environment [16].

4. **What are some examples of secondary and tertiary prevention of obesity and obesity-related diseases? Why is it difficult to use secondary prevention for obesity in children?**

Because it develops over time and is compounded by social determinants, obesity can be difficult to treat. Long-term healthy weight management can be difficult to achieve without a comprehensive strategy, and the health consequences of chronic obesity may not be fully reversible even with weight loss. Thus, primary prevention, especially in childhood, is of utmost importance.

Secondary prevention includes identification of a person in the early stages of obesity and intervention to treat the disease, thus preventing significant adverse outcomes. Routine surveillance of growth parameters and BMI during preventive visits can be used to screen children and adults with, or at risk for, obesity. Once overweight and obese individuals are identified through these simple screening measures, a more thorough assessment of risk factors should occur—namely, a dietary and exercise history, family history of obesity and related complications, and social history—so that a tailored, comprehensive intervention can be developed. Children and families with risk factors for obesity should be seen at more regular intervals and even referred to external groups and exercise programs for further accountability and support. Barriers to care should be addressed, and experiences, emotions, struggles, and challenges with weight should be explored.

A classic example of secondary prevention is screening for disease: BMI for obesity, cholesterol level for hypercholesterolemia, blood sugar or Hgb A1C for diabetes, and blood pressure for hypertension. For those who screen positive, treatment that prevents a negative outcome, such as a heart attack or stroke, or improves atherosclerosis should be started. One example of this kind of prevention is the Diabetes Prevention Program [17]. The original study enrolled people who screened positively as being overweight (by BMI) or having prediabetes (by Hgb A1C). An intensive diet and exercise program were then implemented and found to be more effective at preventing full-blown diabetes than metformin or placebo. This diet and exercise program is now offered by many community organizations.

The complexity of obesity makes secondary prevention difficult in children. There are few, if any, evidence-based strategies that help children lose weight. There are a number of environmental and social factors that influence a child's weight, all of which are often very difficult to address. It is also true that long-term consequences of complex disease such as obesity are discounted. Most families are not thinking about their children having heart attacks in 30 years.

Tertiary prevention involves interventions to limit the disability associated with a disease that has already been acquired. In the case of diabetes, tertiary prevention includes annual podiatric ("foot checks") and ophthalmologic examinations. These measures attempt to limit neuropathy, poor healing, and blindness, all complications of diabetes. Bariatric surgery can be considered a form of tertiary prevention. Bariatric surgery does not "cure" obesity but often prevents associated complications or disabilities, such as yeast infections or chafing from excess fat and joint problems that limit mobility.

5. **How does weight bias affect health outcomes?**
Weight bias is defined as "negative weight-related attitudes, beliefs, assumptions, and judgments toward individuals who are overweight and obese," and this definition may also be extended to individuals with low weight [18]. Weight bias stems from the faulty assumption that obesity is a choice that results from poor self-control and that individual noncompliance explains weight loss failure. However, while individual choice is an important component of weight management, the complexity of obesity lies beyond the individual. Treating people disrespectfully or discriminatorily because of their weight is harmful to physical and mental health and does not result in positive behavior change around weight loss.

Understanding weight bias and its adverse health consequences is important to achieving well-being across the spectrum of weight-related issues, especially obesity. Weight bias negatively impacts health by causing anxiety, stress, depression, poor self-esteem, and body image issues [18]. Rather than empowering positive behavior change around weight, shaming overweight or obese individuals may actually lead to disordered eating, including fasting, extreme dieting, restrictive and/or binge eating, compulsive exercise, or the avoidance of exercise for fear of being shamed. Weight bias may also become internalized, meaning that obese individuals believe that they deserve negative treatment because of their weight. This internalization is associated with psychological maladjustment, eating pathology, and avoidance of preventive health care. Weight bias can also lead to social inequity: People with obesity are vulnerable to unfair treatment at school (harassment or bullying by peers as early as preschool, biased attitudes from teachers, exclusion from play or social activities), at work (negative impact on wages, promotions, or hiring preferences), in interpersonal relationships, and within the healthcare system (negative impact on quality of health care) because of their size [19]. The culture of medicine around obesity is not always consistent or equitable, and facilitating healthy lifestyle change requires significant social restructuring in the built environment of health systems. It is crucial to improve knowledge on the multifactorial nature of obesity among healthcare providers, increase awareness about the harm of weight bias, and provide sensitivity training on the prevention and management of obesity [18, 19]. The best way to improve health is to enable healthy lifestyles in compassionate, non-judgmental ways.

✅ 6. **What benefits do the Supplemental Nutrition Assistance Program (SNAP); Special Supplemental Nutrition Program for Women, Infants, and Children (WIC); and the National School Lunch Program (NSLP) provide? How are WIC and NSLP benefits different from SNAP? What are the income thresholds for each? What are the residency and immigration status requirements to access these benefits?**
   1. **Supplemental Nutrition Assistance Program**: SNAP benefits are available for adult applicants and are provided on an electronic benefit transfer (EBT) card, which works like a debit card. The EBT card can be used to purchase food with some limitations. Income thresholds vary from state to state, but the federal government has set the eligibility floor at a net income of 100% of the federal poverty guidelines (FPG). Although a number of deductions and exceptions are considered when determining SNAP eligibility, it is also limited by household resources. If a family has trouble applying for or receiving SNAP benefits, a referral to a social worker is a good first step. Medical-legal partnerships (MLPs) can also be particularly helpful when SNAP benefits are erroneously denied.
   2. **Special Supplemental Nutrition Program for Women, Infants, and Children**: Pregnant and postpartum women and children up to the age of 5 are eligible to receive WIC benefits, which include supplemental nutritious foods (though the choice of foods included is not without controversy and notably does include juice), nutrition education and counseling at WIC clinics, and screening and referrals to other health, welfare, and social services. The federal income threshold is a net income of 100% and no more than 185% of the FPG.
   3. **National School Lunch Program**: This program is administered through local schools that receive cash subsidies and food for reimbursable meals served to low-income children. Children whose family incomes are below 130% of the FPG qualify for free lunch, and children whose family incomes are between 130% and 185% of the FPG qualify for reduced price school meals. Children may also qualify if they already receive other benefits, such as SNAP, or based on their status as a homeless, migrant, runaway, or foster child. Certain school districts, with more than 40% eligible students, may enroll the entire student body, and therefore avoid certain administrative costs.

   All three programs require that a person be a resident of the state in which that person is receiving the benefit. While WIC and NSLP do not have restrictions based on immigration status, SNAP is only available for citizens with several limited exceptions. The most common exceptions include those who have lived in the United States for at least 5 years, those who receive disability-related assistance or benefits, or children under the age of 18 [20].

✅ 7. **What is food insecurity and how can healthcare practitioners screen for it? How is it related to obesity and health outcomes?**
   Food insecurity is inconsistent access to adequate, healthy, and affordable food due to constrained resources [21]. Affecting nearly one in six U.S. households, food insecurity is associated with adverse health outcomes, health disparities, and limited healthcare access. People who are food insecure have worse physical and mental health, with higher BMIs; higher prevalence of diabetes, smoking, and depression; and significant unmet needs for chronic disease prevention for conditions such as diabetes and hypertension [22].

   Data from NHANES (2005–2010) indicated that 58% of those who identified as food insecure received no food assistance, 20.3% received SNAP benefits, 9.7% received food bank assistance, and 12% received both SNAP and food bank assistance. Interestingly, receipt of both SNAP and food bank assistance was associated with the poorest health, while receiving no assistance was associated with the best health [22]. Compared to people who have enough food, people who were food insecure were more likely to be young, female, single, Hispanic, or African American with less than a high school diploma. Children living in poverty are also more likely to experience food insecurity [22].

The following two-question Hunger Vital Sign™ screening tool is recommended by the American Academy of Pediatrics (AAP) and has 97% sensitivity (correctly detects 97% of patients who are food insecure) [23]:
1. Within the past 12 months, we worried whether our food would run out before we got money to buy more. Yes or no?
2. Within the past 12 months, the food we bought just didn't last, and we didn't have money to get more. Yes or no?

Though it may seem counterintuitive, food insecurity is associated with obesity in the United States in some populations. For example, rates of obesity are higher among women who are food-insecure than among those who are food-secure. For men, a correlation has generally not been identified, and for children, the data is mixed and may be mediated by the mother's obesity status since obesity in pregnancy is a predictor of childhood obesity. While evidence is not sufficient to show causality at this point, there are data showing that low-income women might be independently at risk for both obesity and food insecurity.

Possible reasons for this observation stem from the effects of poverty on access to healthy, affordable foods. Poor neighborhoods often lack grocery stores or farmers markets where residents can purchase quality fruits and vegetables, whole grains, and low-fat dairy products. They may be limited to small corner markets or convenience stores where fresh produce and low-fat items are limited, if available at all. If these stores lack refrigeration, they may only carry highly processed, calorie-dense food and beverages with long shelf-lives. Low-income households are also less likely to have a vehicle for food shopping, and food may be limited to what can be carried by walking or using expensive public transportation. Sometimes, low-income households may be dependent on family or friends for transport to grocery stores. If these families are able to get a ride to the grocery store, once monthly, for example, they can purchase the majority of their food for that month. With limited access to convenient or regular grocery store visits, they may limit perishable foods, such as fruits and vegetables, in favor of canned goods or highly processed foods that will last longer. When available, healthy food is often more expensive and has higher potential for waste due to spoilage if not eaten in a timely manner or if a household lacks refrigeration. Low-income households stretch their food budgets by purchasing highly processed grains, added sugars, and fatty foods, which are inexpensive, palatable, and readily available in low-income communities. The cheaper, processed, energy-dense foods have lower nutritional quality and are associated with obesity due to overconsumption. Low-income communities also have greater availability of fast-food restaurants, which carry high-calorie, low-nutrient food and are linked to obesity [24].

Obesity can arise from food insecurity in other ways based on cycles of food deprivation and overeating ("feast or famine"). In order to stretch food budgets and supplies, low-income families may eat less or skip meals at times, then overeat when food is more readily available. This disordered eating behavior can lead to weight gain and metabolic changes that actually promote lower metabolic rates and fat storage, particularly among mothers who are food-restricting in order to feed their children [24].

8. **What are food deserts? What are other features of the built environment that contribute to obesity?**
A "food desert" is defined by the U.S. Department of Agriculture (USDA) as a community having both low-median income and limited access to a grocery store [25]. Food deserts are neighborhoods where healthy food choices are inaccessible to the people who live there due to limited availability or cost. Food deserts are typically found in low-income, rural, and minority neighborhoods and contribute to obesity in those populations [26]. There are specific criteria used to map food deserts, which take into account the following basic indicators:
   - Healthy food accessibility, as measured by distance to source of healthy food or number of stores in an area
   - Individual-level resources affecting food accessibility, including family income and vehicle availability (access to a vehicle is a key determinant of access to food in many communities)
   - Neighborhood-level resources affecting food availability, including average neighborhood income and public transportation availability [26]

If the only nearby stores are convenience stores that do not sell affordable, readily available, healthful foods, then these foods are less likely to be purchased and consumed. Hence, people are more likely to consume unhealthy options because they lack other choices. Access to a vehicle is a key determinant of access to food in many communities. For example, one report noted that 88% of food retailers in Washington, DC, sell mostly "junk" or processed food. Those who own or manage grocery stores in urban areas often find it hard to keep healthy foods in stock due to lower demand. This demand is intricately tied to cooking skills in the home. People do not buy things they do not know how to cook [27].

Food availability, production, and consumption are public health concerns packed with environmental and social issues. Because there has been a decline in small farming and locally grown food sources, there is increased reliance upon highly processed, cheap foods—especially in food deserts—which has contributed to obesity. Transforming food deserts and the built environment into healthy environments requires multilevel interdisciplinary collaboration. Policy-level solutions include preservation of affordable farmland, improved market access for small farmers, improved access to healthy markets for low-income communities, transparent food labeling, mindful advertising, and more progressive food policies that make healthy choices easy, affordable choices. Individual-level solutions include reduction of overall calorie consumption with emphasis on less sugar and more whole foods, increase in home cooking and family meals, and family education about healthy food choices.

Other features of the built environment that may contribute to obesity involve the financial and emotional pressures of food insecurity, low-wage employment, limited access to health care, substandard housing, neighborhood violence, and inadequate transportation. Low-income neighborhoods have fewer parks, green spaces, and recreational facilities, making a physically active lifestyle difficult. If these areas are present, they may have unattractive features, including fewer trees, visible trash and disrepair, and more noise, making regular use less likely. Furthermore, safety in low-income areas may be challenged by crime, violence, and traffic, making families more likely to remain indoors where sedentary activities such as watching TV, video games, and overall screen time increase the risk for obesity [24].

22.4 Responses to Discussion Questions

Upon reviewing past notes in the electronic medical record, you note that weight loss has been discussed at the last two visits with other providers. However, both Mayson and his mother have gained weight in the past year, with Mayson crossing multiple percentiles on his growth curve, such that he is now at the 99th percentile of weight-for-height. The social history notes that Alicia works as a convenience store clerk and has two other children, aged 6 and 8.

1. **How would you approach this visit?**
   (a) **What questions would you ask Alicia about her family's diet and lifestyle?**
   Measurement of growth (height, weight, head circumference, and BMI if over age 2) and anticipatory guidance regarding diet and nutrition are crucial components of all pediatric preventive care visits. The growth values are plotted on World Health Organization (WHO) growth curves that show the distribution of growth across the pediatric population. From birth to 5 years, the WHO growth chart represents a standard based upon the growth of healthy, breastfed children from diverse geographical regions. This new standard, over the former CDC growth charts, reflects normal pediatric growth and an ethnically diverse sample appropriate for generalized use in multiethnic communities. Because Mayson is under 2 years of age, healthcare practitioners would trend the velocity of his own growth while comparing percentile values for his weight and his height, rather than using a BMI at this age; that is, if his height is at the 20th percentile and his weight is at the 80th percentile, healthcare practitioners should be concerned. The trend of his weight over time would also be important to note. There are also weight-for-length charts that serve as surrogates for BMI in children under 2.

   Anticipatory guidance around diet would generally include an assessment of what the child eats and what foods and beverages are available at home. It should not be assumed that access to WIC and SNAP benefits secures healthy food availability at home, as many foods available through these benefit packages are not healthy (e.g., juice) and parents may tend toward purchasing cheap, calorie-rich foods in order to make benefits last through the entire month. An 18-month-old child should be able to eat foods by himself and is starting to use utensils. The healthcare practitioner should ask the mother what he likes and does not like, as well as what she and the other children like, and whether she caters to the child, since it is generally recommended to provide a balanced meal and not prepare something special for a picky eater. An even more basic question is whether they eat prepared food at home or purchase pre-made food from a restaurant or store. Finally, it is important to ask about what the child drinks (preferably water and milk with limited juice and no soda).

   These questions should be asked in a nonjudgmental way and with the goal of working together to find easy, incremental ways to trade unhealthy habits for more healthy ones (not necessarily the healthiest). Dramatic changes in diet or lifestyle tend to be difficult to achieve and nearly impossible to sustain. For example, the healthcare practitioner can provide education about the health consequences of juice on weight and dentition, along with counseling to either dilute juice with water or offer a piece of fruit and water in its place. Parents often think of juice as healthy, even though it contains primarily sugar. Whole fruit contains natural sugar as well but also other nutrients and fiber.

> **Box 22.1 Teaching Tip**
> More anticipatory guidance for toddlers and health diets is available from the American Academy of Pediatrics Website. Share this link with learners:
> ▶ https://www.aap.org/en-us/advocacy-and-policy/aap-health-initiatives/HALF-Implementation-Guide/Age-Specific-Content/Pages/Toddler-Food-and-Feeding.aspx

> **Box 22.2 Teaching Tip**
> This is an opportunity for learners to be honest and self-critical. Many, if not most, pediatricians might feel a little critical or judgmental when seeing a patient who is overweight or obese eating junk food—even those who offer their children similar "treats" on such occasions.

Alicia screened positively for food insecurity when she was first seen at the clinic, about 4 months into her pregnancy.

✅ 2. **How can you respectfully address food insecurity with Alicia?**
It is important for healthcare providers to address food insecurity in a sensitive manner, to connect families with nutrition resources in the community as needed, and to support local and national policies that increase access to healthy food for children and families. Without judgment, families should be reassured that the screening is universal. Food insecurity and poverty are often associated with shame and stigma; thus, normalization of the screening process may help families feel more comfortable disclosing food insecurity and other social concerns. Because of the pervasiveness of food insecurity, healthcare providers may consider screening at all visits and during hospitalizations [28].

When you enter the exam room, you are frustrated to see that both your patient and his mother have bags of food from a fast-food chain restaurant. Alicia apologetically explains that it was a "special treat" because of the doctor's visit and the anticipation of Mayson needing immunizations. Indeed, Mayson is fussy during the interview, but he consoles immediately with the snacks.

✅ 3. **Self-reflection: How do you feel hearing this piece of the encounter? Have you ever felt judged or stigmatized by a healthcare provider?**

On examination, Mayson is playful and interactive. He is "pudgy" but actively walks around the room and is curious about your medical instruments. He has normal vital signs. Aside from visible decay of his front teeth, his physical exam is normal. You ask Alicia what he usually drinks. She replies that he prefers whole milk or juice, but when he sees her drinking soda, he wants to have some too. At bedtime, he only wants milk and drinks two or three bottles of either leftover powdered formula or cow's milk during the night. You start to explain to Alicia that this is a major contributor to both his weight and his dental caries. Alicia becomes extremely upset.

✅ 4. **Why do you think she is reacting this way and what would you do next?**
The healthcare practitioner needs to think both creatively and compassionately in this situation. However, many might find themselves perplexed by the intensity of this mother's response. In practice, they need to translate the initial confusion into recognition of the importance of exploring and understanding her reaction. This is a powerful clue that the physician is missing important information about this family's home life. Failure to ask questions—due to a lack of curiosity, lack of time, or discomfort with the subject or the patient's distress—will hinder the healthcare practitioner's ability to make an accurate diagnosis and provide effective recommendations. When questions and education are met with upset or defensiveness, these emotions should be explored. It is critical to understand this family's home life and social determinants.

The healthcare practitioner should thus ask, with empathy and without judgment, why she is upset and give her space to respond.

You learn that Alicia's family was evicted from the apartment that she shared with her three children, her mother, her teenage sister, and her 10-year-old niece, who was recently adopted after the child's own mother died from complications of diabetes. All seven have had to temporarily move into a two-bedroom apartment rented by her uncle and his adult son. Her uncle and cousin are not accustomed to living with young children, and they get up early to work at a construction site. If she doesn't give Mayson a milk bottle, he cries—sometimes for hours—and she is terrified that her uncle might throw them out. While listening to Mayson's mother, you notice that she has dark, velvety skin over her neck and axillae, and she reports that the entire family struggles with obesity.

5. **How does this change your view of the situation?**
   Many healthcare practitioners might have made assumptions about why Alicia is still giving her son a bottle of milk or formula at night. For instance, they may have assumed that she did not know any better or did not care enough to take care of his teeth, that it was easier for her (since teaching children who are accustomed to needing a bottle to sleep without it can be time-consuming and exhausting for parents), or that she would get more sleep if he was pacified with a bottle all night. Clearly, this family's home life is more complicated than it first appeared, and the "textbook" recommendations for parents of toddlers must be adapted to their particular situation. For instance, "crying it out" (method of teaching an older infant or toddler to sleep through the night independently—including without a bottle) can be effective in a short period of time, but even a few sleepless nights may not be an option for this family. Some parents are resistant to this method, not wanting to distress their child in such a way.

6. **What are some of the typical recommendations that you have heard healthcare practitioners give overweight and obese patients regarding weight loss?**
   Frequent recommendations, which will be familiar to most patients as well as most healthcare practitioners, include:

- Eat smaller portions.
- Eat healthier foods – mostly fruits and vegetables, avoid processed foods, red meat, sugars, and/or all carbohydrates.
- Exercise regularly.

However, what is recommended based on evidence and what is typically recommended in the clinic are probably not the same, and there is little to no evidence suggesting that these stereotypical recommendations lead patients to make meaningful changes that produce weight loss. The U.S. Preventive Services Task Force (USPSTF) "recommends that clinicians screen children aged 6 years and older for obesity and offer or refer them to comprehensive, intensive behavioral intervention to promote improvement in weight status." This is a B-level recommendation, meaning that the USPSTF recommends the service because there is high certainty that the net benefit is moderate or moderate certainty that the net benefit is moderate to substantial [29]. Similar recommendations exist for adults with BMIs greater than 30 kg/m^2 [30].

Yet in many geographic regions, options for "comprehensive, intensive behavioral intervention" are nonexistent. If available, such programs would generally involve a nutritionist or dietitian, exercise physiologist, physician, and a psychologist (i.e., comprehensive) and would incorporate frequent, sometimes prolonged visits to the program (i.e., intensive). They might also engage a bariatric surgeon to consult in cases where surgery is a consideration: gastric banding for adolescents or gastric banding or bypass surgery for adults. (The ethics of such procedures in children are beyond the scope of this chapter.) Families with limited resources may also have difficulty adhering to additional appointments and interventions.

In general, primary care physicians often do minimal, if any, counseling regarding weight outside of an intensive intervention. A study of clinical encounter data regarding obese adults found that only 29% received a diagnosis of obesity and fewer received counseling on weight reduction (17.6%), diet (25.2%), and exercise (20.5%) [31].

Surveys tend to report higher rates of counseling, which is probably due to (1) overreporting or recall bias on the part of physicians and (2) underreporting in chart reviews because "counseling" is not always documented well in medical charts. Counseling for children should include (1) physical activity for 30–60 minutes per day, (2) limiting screen time, (3) calorie reduction (i.e., limited juice for kids and transition to low-fat milk, etc.), and (4) eating together as a family. Additionally, there is an increasing body of literature indicating that inadequate sleep is a driver of obesity in both children and adults: thus, inquiry and counseling about sleep might also be warranted.

(a) **How realistic do you think these recommendations are for Alicia, given what you know about her resources and responsibilities?**

Alicia has significant social challenges that will affect her ability to adhere to these recommendations. She has financial stressors, food insecurity, multiple children, a limited support system, and a risk of homelessness if she disturbs her extended family's routine. Further, recommendations such as increased exercise or physical activity may be practically difficult in neighborhoods with limited safe outdoor space. Limiting screen time may also be difficult for busy parents with limited childcare or preschool resources. Calorie reduction, too, may present challenges, which will be explored further later in the case.

(b) **What additional resources can you offer her?**

Programs that would address relevant social determinants of health would be especially helpful. Make sure that she is enrolled in WIC and/or SNAP, which would give her more resources for food purchases, as well as further nutrition counseling (particularly true for WIC) [32, 33]. Local food banks also give food to families in need, often on a recurring basis, and have lots of resources for families in their situation. A patient can call the food bank, and staff will conduct an intake interview and direct that patient to appropriate resources. Healthcare practitioners should be aware of other members of the healthcare team with expertise in this area. Physicians are not expected to do everything and are often unaware of community programs that might help. Nutritionists, social workers, and case managers are all essential members of the healthcare team who can be helpful in this situation.

> **Box 22.4 Teaching Tip**
> This may be a good opportunity to introduce learners to interdisciplinary practice models, particularly medical-legal partnerships. Ask learners to identify the kinds of issues an MLP might help with and refer to resources on the website for the National Center for Medical-Legal Partnership to learn more about how those legal issues connect to healthier outcomes for patients and families: ▶ https://medical-legalpartnership.org

Many practitioners hesitate to probe further into complicated socioeconomic and interpersonal issues because such inquiries can be lengthy and lead to the identification of problems that doctors are not fully equipped to address. Referrals to social workers and medical-legal partnerships (MLP) allow practitioners to address social determinants of health as part of an interdisciplinary team. Practitioners may also use legal needs screening tools in the clinic.

> **Box 22.3 Teaching Tip**
> Ask learners if any have ever had to take public transportation to the grocery store or if they had to borrow a friend's car or hitch a ride. How do they feel it affected their choices? This can also be expanded into an out-of-class poverty exercise. Learners can be asked to shop, cook, and eat for a week on a typical SNAP budget, using only public transportation or walking, and then write or speak about their experiences.

For example, in Alicia's scenario, a social worker may perform a needs and benefits assessment with a family. An MLP may be able to challenge an eviction or help the family recover the security deposit from a landlord, which is often what enables a family to secure another apartment. An MLP may also be able to provide legal counseling about the family's current living situation. One issue may be the number and identity of people living in the uncle's apartment because leases often contain restrictions. An MLP may also ask further questions about the children and the support each is receiving i.e., whether the niece is receiving benefits as a result of her adoption and whether Alicia is receiving child support. Comprehensive screenings and assistance from health practitioners, social work, and MLP may help the family achieve a greater degree of financial and overall stability.

You ask the clinic social worker to meet with Alicia and help connect her with resources in the community. The social worker tells you that many families like Alicia's spend more than 80% of their income on rent, so it is understandable that Alicia prioritizes keeping her uncle happy so that she can stay in his home. It can take years of being on a waiting list to finally receive a housing voucher, and the list is often closed to new names. Even if she was willing to join the list, she might only qualify for prorated assistance if she is living in a mixed-status household that includes both legally documented and undocumented residents of the United States. (A few states, notably Massachusetts, have programs to help undocumented immigrants obtain housing [4].) The social worker also tells you that many families who do receive subsidized housing immediately redistribute their income toward more and better food for their families.

✓ 7. **Considering families like Alicia's, why do you think primary prevention of obesity at the population level is challenging?**
Primary prevention is outright prevention of a disease. Vaccination is a classic example. In terms of obesity, primary prevention is very difficult in our American society. Preventing weight gain goes against strong biologic and environmental cues. Much of the processed food available in our society has been specifically engineered to be both incredibly cheap and incredibly addictive. It is also backed by very strong corporate interests. In addition, the weight loss industry is worth billions of dollars, which gives it some misplaced incentives, counter to its purported goals. In addition, many processed foods are marketed as "healthy," which raises a host of concerns including who decides what "healthy" means and how much of it is "healthy." Furthermore, these foods and even traditionally healthy unprocessed foods can be part of an overall diet that leads to weight gain. For example, smothering carrots in ranch dressing full of sugar and saturated fat is unlikely to be a healthy snack, despite the good intentions of promoting vegetable consumption.

(a) **What has worked? What has not worked?**

> **Box 22.5 Teaching Tip**
> Encourage learners to draw heavily on their research for the background questions in considering their responses.

The CDC has found that the changes in the WIC program nutrition guidelines contributed to a decrease in obesity prevalence in many states between 2008 and 2011 [9]. Although this has not been shown for SNAP, this program has other benefits, such as improved diet quality and decreased food insecurity [34].

The CDC Community Guide recommends interventions for communities on a variety of topics (analogous to the USPSTF recommending interventions for physicians). They recommend multicomponent interventions in schools to provide healthier foods and beverages. These would include meals provided in the cafeteria as well as other food service options in the school (vending, snack bars, etc.). They specifically recommend increasing fruits and vegetables in school meals and snacks. These strategies have evidence showing their effectiveness at reducing prevalence of overweight and obesity (albeit only slightly) and increasing fruit and vegetable intake among students [35].

An interdisciplinary, comprehensive approach is needed to create measurable impact in the obesity epidemic. On an individual level, interdisciplinary partners such as social workers and MLP lawyers can help families with economic and family stability, thus enabling them to reallocate resources and have a meaningful opportunity to make healthier choices. Local, state, and federal agencies must create and support policies that benefit neighborhoods and families across the United States. MLP lawyers also help identify, propose, and advocate for those policies. Universities and public health agencies are involved in the research and development of obesity prevention strategies and nutritional standards for children and adults: This evidence base must be translated into policies and guidelines that can be easily implemented at the community, family, and individual levels. The most successful approaches are comprehensive, interdisciplinary, localized, and community-oriented. Community members must partner with public health workers and healthcare providers; with hospitals, schools, and universities; with grassroots, faith-based, and community development organizations; and with transportation and housing planners to leverage community resources and develop strategies to best meet the needs of that community. Aims to reduce obesity are to increase access to fresh, nutritious food; to make healthy eating and physical activity a normal part of people's daily routines; to improve school nutrition and access to open safe spaces; to incentivize healthy food purchases; to make food labeling transparent; and to limit advertising of unhealthy foods to children.

Sadly, most programs have not been able to show measurable impact on obesity at the population level or on health outcomes related to obesity. For instance, the "Healthy Corners" program (promoting fresh fruits and vegetables in corner stores) and other programs to introduce grocery stores into food deserts did not measurably reduce weight [27]. As previously noted, these outcomes are difficult to accurately measure for a number of reasons; however, this does not mean the efforts are not worthwhile and beneficial in less measurable ways.

(b) **Creative problem-solving: Suppose you were given a small grant ($10,000) to address obesity in your community. What would you propose doing with the grant?** [36].

> **Box 22.6   Teaching Tip**
> Consider assigning the Schwartz and Brownell report, "Actions Necessary to Prevent Childhood Obesity: Creating the Climate for Change" to stimulate ideas [39].

Alicia also tells you that Mayson is an extremely picky eater, so she only offers him things she knows he will eat. In addition to milk and formula, this includes French fries, hot chips, and macaroni and cheese. With frustration, she says that the residents rotating in the clinic always tell her to "put out a variety of healthy foods and let him pick and choose what to feed himself" but that they can't afford to waste food like that. You recall telling other parents that toddlers often need to try a new food 8–15 times before they will accept it.

✓ 8. **Do you think this is a realistic suggestion for Alicia and Mayson? Why or why not?**
A recent study has suggested that this is probably not realistic. Low-income parents are often aware of recommendations

for healthy eating but choose to offer their children the foods they know will be fully consumed and thus not thrown away, which are often calorie-dense and nutrient-poor, in order to minimize waste on a limited budget [37].

Your medical student is curious about the process of enrolling in WIC and SNAP and offers to help Alicia navigate it. However, Alicia refuses the assistance. You recall that Alicia's chart mentions that she was born in Guatemala and came to the United States at the age of 3. Alicia's younger sister, her niece, and her three children were born in the United States.

9. **Based on what you know about WIC and SNAP eligibility, is Alicia's family eligible?**
   Because Mayson is 18 months old, he is eligible for WIC. While Alicia is not eligible for SNAP in her own right, she can apply for and receive benefits on Mayson's behalf, and legally, the state cannot inquire about Alicia's own immigration status when she applies for him [38]. However, there is a "chilling effect" where undocumented parents are afraid to enroll their citizen children. Even legal permanent residents (LPRs) frequently do not apply because they fear, incorrectly, they will be considered a "public charge," which has negative implications for long-term residency. The citizen children of undocumented immigrants are eligible, but often parental fears around deportation are so strong that the children are not enrolled.
   (a) **Why might Alicia be reluctant to register her children for benefits?**
   Given the fear in immigrant communities about deportation of undocumented family members, Alicia and her mother may be reluctant to provide any identifying information to government and social service agencies. The healthcare practitioner in this case should encourage Alicia to consult with a lawyer to discuss her immigration status. Patients and families may not understand what lawyers do or that MLP lawyers represent clients and not the hospital. By explaining that lawyers are bound by confidentiality, much like doctors, and that the role of the lawyer is to represent the client's interests and wishes, Alicia may be more willing to consult with the MLP or another lawyer.
   (b) **Should you ask explicitly about Alicia's immigration status? Why or why not?**
   Learning more about Alicia's immigration status will help you determine which benefits she is eligible for and some of the barriers she is facing. Trust is crucial to successful treatment of patients, and evidence is mounting that patients will not seek medical care if they believe that their immigration status will be reported to authorities. Assuring Alicia that all questions asked are for Mayson's benefit, that doctors have no obligation to report a family's immigration status, and that doctors are bound by confidentiality may allay Alicia's fears [39, 40].
   Some advocacy groups recommend that healthcare practitioners not ask directly about immigration status but rather create opportunities for patients to discuss it if they choose. For example, clinicians can say, "Many of my patients are currently experiencing anxiety concerning immigration problems. Are you, someone in your family, or your friends having these kinds of problems?" [41]. Alternatively, they might ask "if a parent/other key family member has left or is potentially going to leave the family for any reason" [42].

10. **Self-reflection: Do you feel differently about this patient now than when you first read the case?**
    (a) **Have you learned anything that changes the way you will approach overweight and obese patients in the future?**

> **Box 22.7 Personal Perspective**
>
> "My greatest fear is that I will die and leave my daughter alone."
>
> In an MLP at an urban children's hospital, many of our clients have told us that our interdisciplinary team was among the only professionals to listen to them in a nonjudgmental way. One family included a mother and her 2-year-old daughter, both of whom were morbidly obese. At this young age, the daughter had acanthosis nigricans in addition to a BMI well over the 95th percentile. By simply listening without judgment and asking thoughtful questions, we learned that the mother was the sole caretaker for two adult siblings with intellectual disability, in addition to her elderly mother and her daughter. The mother had congestive heart failure with frequent hospitalizations, and the family had extremely limited income. Although these facts may seem irrelevant to her obesity at first, they provided a more complete picture of the obstacles faced by the family in addressing adult and childhood obesity. These facts allowed us to more completely assess the medical, legal, and social resources available to help the family achieve greater stability, improving their ability to address obesity and its causes in a comprehensive, sustainable, and holistic way.
>
> In our interactions with families who struggle with obesity and all of the physical, emotional, and social pathology associated with this complex disease, we continue to learn about the depth of its effect on families and children. This mother confided to us that her greatest fear was that she would die as a result of her disease and leave her daughter motherless. Sadly, 3 months into working with her, she died after suffering a heart attack at age 39.
>
> – Janet H. Goode and Lauren H. Mutrie

Acknowledgments The authors and editors wish to thank Rachel Fabi for her review of this chapter.

References

1. Centers for Disease Control and Prevention. The built environment assessment tool manual [Internet]. Atlanta (GA): CDC; 2017 [cited 2018 Apr 5]. Available from: https://www.cdc.gov/nccdphp/dnpao/state-local-programs/built-environment-assessment/index.htm.
2. United States Census Bureau. Poverty thresholds, 2018. Washington, D.C.: U.S. Department of Commerce. [cited 2018 Apr 28]. Available from: https://www.census.gov/data/tables/time-series/demo/income-poverty/historical-poverty-thresholds.html.
3. Institute for Research on Poverty. What are poverty thresholds and poverty guidelines? [Internet]. Madison (WI): University of Wisconsin-Madison; 2018 [cited 2018 Aug 29]. Available from: https://www.irp.wisc.edu/resources/what-are-poverty-thresholds-and-poverty-guidelines/.
4. Massachusetts Legal Help. Chapter 9: Immigrants and housing. In: Legal tactics: finding public and subsidized housing. 3rd ed.; 2009 [updated 2015 Mar; cited 2018 Jul 27]. Available from: https://www.masslegalhelp.org/housing/finding-public-and-subsidized-housing/immigrants.pdf.
5. Nammi S, Koka S, Chinnala KM, Boini KM. Obesity: an overview on its current perspectives and treatment options. Nutr J. 2004;3:3.
6. Centers for Disease Control and Prevention. About child and teen BMI [Internet]. Atlanta (GA): CDC; 2018 [cited 2018 Apr 5]. Available from: https://www.cdc.gov/healthyweight/assessing/bmi/childrens_bmi/about_childrens_bmi.html.
7. Furlong KR, Anderson LN, Kang H, Lebovic G, Parkin PC, Maguire JL, et al. BMI-for-age and weight-for-length in children 0 to 2 years. Pediatrics. 2016;138(1):e20153809.
8. Roy SM, Spivack JG, Faith MS, Chesi A, Mitchell JA, Kelly A, et al. Infant BMI or weight-for-length and obesity risk in early childhood. Pediatrics. 2016;137(5):e20153492.
9. The State of Obesity. Obesity rates and trends overview [Internet]. New Jersey: Robert Wood Johnson Foundation and Trust for America's Health; 2017 [cited 2018 May 5]. Available from: https://stateofobesity.org/obesity-rates-trends-overview/.
10. Roundtable on Obesity Solutions; Food and Nutrition Board; Institute of Medicine. The current state of obesity solutions in the United States: workshop summary. Washington, DC: National Academies Press; 2014; ch. 2: Current epidemiology of obesity in the United States. Available from: https://www.ncbi.nlm.nih.gov/books/NBK223173/.
11. NHLBI Obesity Education Initiative Expert Panel on the Identification, Evaluation, and Treatment of Obesity in Adults (US). Clinical guidelines on the identification, evaluation, and treatment of overweight and obesity in adults: the evidence report. Summary of evidence-based recommendations. Bethesda (MD): National Heart, Lung, and Blood Institute; 1998 Sep [cited 2018 May 5]. Available from: https://www.ncbi.nlm.nih.gov/books/NBK2009/.
12. Hoelscher DM, Butte NF, Barlow S, Vandewater EA, Sharma SV, Huang T, et al. Incorporating primary and secondary prevention approaches to address childhood obesity prevention and treatment in a low-income, ethnically diverse population: study design and demographic data from the Texas Childhood Obesity Research Demonstration (TX CORD) study. Child Obes. 2015;11(1):71–91.
13. Centers for Disease Control and Prevention. Health equity resource toolkit for state practitioners address-

14. Canadian Task Force on Preventive Health Care. Primary and secondary prevention of overweight/obesity in children and youth [Internet]. Ottawa: Public Health Agency of Canada; 2012 Jul 10 [revised 2013 Feb; cited 2018 Jul 10]. Available from: https://canadiantaskforce.ca/wp-content/uploads/2015/04/2015-obesity-children-protocol-en.pdf.
15. Dhasarathy A, Roemmich JN, Claycombe KJ. Influence of maternal obesity, diet and exercise on epigenetic regulation of adipocytes. Mol Asp Med. 2017;54:37–49.
16. Vrachnis N, Antonakopoulos N, Iliodromiti Z, Dafopoulos K, Siristatidis C, Pappa KI, et al. Impact of maternal diabetes on epigenetic modifications leading to diseases in the offspring. Exp Diabetes Res. 2012; 2012:538474.
17. National Institute of Diabetes and Digestive and Kidney Diseases. Diabetes Prevention Program (DPP) [Internet]. Bethesda (MD): National Institutes of Health. [cited 2018 Jul 27]. Available from: https://www.niddk.nih.gov/about-niddk/research-areas/diabetes/diabetes-prevention-program-dpp/Pages/default.aspx.
18. Alberga AS, Russell-Mayhew S, von Ranson KM, McLaren L. Weight bias: a call to action. J Eat Disord. 2016;4:34.
19. Obesity Action Coalition. Understanding obesity stigma brochure. Tampa (FL): OAC; 2018 [cited 2018 Jul 27]. Available from: https://www.obesityaction.org/get-educated/public-resources/brochures-guides/understanding-obesity-stigma-brochure/.
20. USDA, Food and Nutrition Service. Programs and services [Internet]. Washington, D.C.: U.S. Department of Agriculture; updated 2018 Mar 26 [cited 2018 Jul 27]. Available from: https://www.fns.usda.gov/programs-and-services.
21. Economic Research Service. Definitions of food security [Internet]. Washington, D.C.: U.S. Department of Agriculture; 2017 Sep 6 [updated 2017 Oct 4; cited 2018 Jul 27]. Available from: https://www.ers.usda.gov/topics/food-nutrition-assistance/food-security-in-the-us/definitions-of-food-security.aspx.
22. Pruitt SL, Leonard T, Xuan L, Amory R, Higashi RT, Nguyen OK, et al. Who is food insecure? Implications for targeted recruitment and outreach, National Health and Nutrition Examination Survey, 2005–2010. Prev Chronic Dis. 2016;13:160103. [cited 2018 May 5]. Available from: https://www.cdc.gov/pcd/issues/2016/16_0103.htm.
23. Council on Community Pediatrics; Committee on Nutrition. Promoting food security for all children. Pediatrics. 2015;136(5):e1431.
24. Food Research and Action Center. Why low-income and food-insecure people are vulnerable to poor nutrition and obesity [Internet]. Washington, D.C.: FRAC; 2018 [cited 2018 Jul 27]. Available from: http://frac.org/obesity-health/low-income-food-insecure-people-vulnerable-poor-nutrition-obesity.
25. Economic Research Service. Documentation [Internet]. Washington, D.C.: USDA; updated 2017 Dec 5 [cited 2018 Jul 27]. Available from: https://www.ers.usda.gov/data-products/food-access-research-atlas/documentation/.
26. Economic Research Service. Food access research atlas [Internet]. Washington, D.C.: USDA; updated 2017 May 18 [cited 2018 Apr 5]. Available from: https://www.ers.usda.gov/data-products/food-access-research-atlas/.
27. Khazan O. Why don't convenience stores sell better food? The Atlantic monthly [Internet]. Boston (MA); 2015 Mar 2 [cited 2018 Jul 27]. Available from: https://www.theatlantic.com/health/archive/2015/03/cornering-the-market/386327/.
28. FRAC. Addressing food insecurity: a toolkit for pediatricians. February 2017. http://www.frac.org/wp-content/uploads/frac-aap-toolkit.pdf. Accessed 27 July 2018.
29. U.S. Preventive Services Task Force. Final recommendation statement: obesity in children and adolescents, screening [Internet]. Rockville (MD): USPSTF; published 2013 Dec 30 [updated 2016 Oct; cited 2018 Jul 27]. Available from: https://www.uspreventiveservicestaskforce.org/Page/Document/RecommendationStatementFinal/obesity-in-children-and-adolescents-screening.
30. U.S. Preventive Services Task Force. Final recommendation statement: obesity in adults, screening and management [Internet]. Rockville (MD): USPSTF; published 2013 Dec 30 [updated 2016 Dec; cited 2018 Jul 27]. Available from: https://www.uspreventiveservicestaskforce.org/Page/Document/RecommendationStatementFinal/obesity-in-adults-screening-and-management.
31. Bleich SN, Pickett-Blakely O, Cooper LA. Physician practice patterns of obesity diagnosis and weight-related counseling. Patient Educ Couns. 2011;82(1):123–9.
32. USDA: Food and Nutrition Service. Women, Infants, and Children (WIC) [Internet]. Washington, D.C.: USDA; updated 2018 Oct 17 [cited 2018 Aug 14]. Available from: https://www.fns.usda.gov/wic/women-infants-and-children-wic.
33. United States Department of Agriculture. Supplemental Nutrition Assistance Program (SNAP). Washington, D.C.: USDA; 2018 Apr 25 [cited 2018 Aug 29]. Available from: https://www.fns.usda.gov/snap/supplemental-nutrition-assistance-program-snap.
34. Center for the Study of the Presidency and Congress (CSPC). SNAP and obesity: the facts and fictions of SNAP nutrition. SNAP to health [Internet]. Washington, D.C.: New America Foundation. [cited 2018 Jul 27]. Available from: https://www.snaptohealth.org/snap/snap-and-obesity-the-facts-and-fictions-of-snap-nutrition/.
35. The Community Preventive Services Task Force (CPSTF). Multicomponent interventions to increase availability of healthier foods and beverages. The Community Guide [Internet]. [cited 2018 Jul 27]. Available from: https://www.thecommunityguide.org/findings/obesity-multicomponent-interventions-increase-availability-healthier-foods-and-beverages.
36. Schwartz MB, Brownell KD. Actions necessary to prevent childhood obesity: creating the climate for change. J Law Med Ethics. 2007 Spring;35(1):78–89.
37. Daniel C. Economic constraints on taste formation and the true cost of healthy eating. Soc Sci Med. 2016; 148:34–41.

38. United States Department of Agriculture. SNAP: supplemental nutrition assistance program guidance on non-citizen eligibility [Internet]. Washington, D.C.: USDA; 2011 Jun [cited 2018 Jul 27]. Available from: https://fns-prod.azureedge.net/sites/default/files/snap/Non-Citizen_Guidance_063011.pdf.
39. Sconyers J, Tate T. How should clinicians treat patients who might be undocumented? AMA J Ethics. 2016;18(3):229–36.
40. Hoffman J. Sick and afraid, some immigrants forgo medical care. The New York Times [Internet]. 2017 Jun 26 [cited 2018 Aug 29]. Available from: https://www.nytimes.com/2017/06/26/health/undocumented-immigrants-health-care.html.
41. Mejias-Beck J, Kuczewski M, Blair A, Neiswanger Institute for Bioethics & Healthcare Leadership. Treating fear: sanctuary doctoring [Internet]. Maywood (IL): Loyola University Chicago. [cited 2018 Jul 27]. Available from: https://hsd.luc.edu/bioethics/content/sanctuary-doctor.
42. American Academy of Pediatrics. Immigrant child health toolkit: immigration status FAQs [Internet]. Rochester (NY): AAP. [cited 2018 Jul 27]. Available from: https://www.aap.org/en-us/advocacy-and-policy/aap-health-initiatives/Immigrant-Child-Health-Toolkit/Pages/Immigration-Status-FAQs.aspx.

Further Reading on This Topic

Miller DP Jr, Spangler JG, Vitolins MZ, Davis MS, Ip EH, Marion GS, Crandall SJ. Are medical students aware of their anti-obesity bias? Acad Med. 2013;88(7):978.

Muennig P. The body politic: the relationship between stigma and obesity-associated disease. BMC Public Health. 2008;8(1):128.

Puhl RM, Heuer CA. Obesity stigma: important considerations for public health. Am J Public Health. 2010;100(6):1019–28.

Schwartz MB, Brownell KD. Actions necessary to prevent childhood obesity: creating the climate for change. J Law Med Ethics. 2007;35(1):78–89.

Video

Kahan S. Why we eat the way we eat [web streaming video]. TEDxManhattan; published 2011 Jun 3. Available from: https://youtu.be/wD8Ndjn_350.

Nevins S, Hoffman J, Teale S, Chaykin D. Stigma: the human cost of obesity-bonus short from HBO's The Weight of the Nation series [web streaming video]. HBO; originally aired 2012 May. Available from: http://theweightofthenation.hbo.com/films/bonus-shorts/stigma-the-human-cost-of-obesity.

Books

Desmond M. Evicted: poverty and profit in the American city. New York: Broadway Books; 2016.

Guthman J. Weighing in: obesity, food justice, and the limits of capitalism. Berkeley/Los Angeles: University of California Press; 2011.

Holmes S. Fresh fruit, broken bodies: migrant farmworkers in the United States. Berkeley/Los Angeles: University of California Press; 2013.

Tirado L. Hand to mouth: living in bootstrap America. New York: Penguin; 2015.

Global Health

Contents

Chapter 23   "They Say My Baby's Head Is Too Small" – 453

Chapter 24   "I Just Want to Help People and See the World" – 477

"They Say My Baby's Head Is Too Small"

Amy E. Caruso Brown and Cynthia B. Morrow

23.1 Background Questions – 454

23.2 Additional Case Information and Questions for Discussion – 455

23.3 Answers to Background Questions – 457

23.4 Responses to Discussion Questions – 462

References – 473

- **Glossary**

Bedside rationing - Withholding of a medically beneficial service from an individual patient because of that service's cost to someone other than the patient. It is usually at the discretion of the treating physician. Rationing not involving an individual patient and physician is often described as resource allocation.

Conscientious objection - Refusal to perform a role or responsibility because of moral or other personal beliefs.

Emerging infection - An infection that has recently appeared within a population or one with a rapidly increasing incidence or geographic range.

Re-emerging infection - An infection that at one time was within a population, disappeared, and has now returned with a rapidly increasing incidence.

Vector-borne disease - Defined by the World Health Organization (WHO) as "human illnesses caused by parasites, viruses and bacteria that are transmitted by mosquitoes, sandflies, triatomine bugs, blackflies, ticks, tsetse flies, mites, snails and lice"; and distinguished from diseases that are transmitted directly from human to human (via respiratory or oral-fecal routes, for instance) and zoonoses, which are transmitted directly from animal to human.

Vertical transmission - Infection caused by pathogens that are transmitted from the mother to an embryo, fetus, or neonate during pregnancy or childbirth; human immunodeficiency virus (HIV) is a well-established example of this.

Learning Objectives

1. Identify sources of information and recommendations to address current global infectious disease outbreaks.
2. Apply ethical principles to global public health concerns, such as Zika.
3. Analyze the ethical and legal issues surrounding abortion in the setting of a potentially devastating congenital infection, with special attention to the concept of conscientious objection.
4. Explore the concepts of bedside rationing and resource allocation in health care and research in a global context.

> **Case Background**
>
> *You are an obstetrician in a busy clinic in Recife, Brazil. Your first appointment of the day is with a 36-year-old woman, Juliana Silva. She is 22 weeks pregnant and had symptoms of Zika virus infection early in her pregnancy. She is accompanied by her partner, Miguel Oliveira, and their three young children.*

23.1 Background Questions

1. What is the history of Zika virus? What happened with Zika virus activity in Brazil in 2015? What is the current status of Zika activity in the United States? In the world?

2. How is Zika transmitted? What are the symptoms of Zika infection? What percentage of infected individuals are symptomatic?

3. What are the known risks of pregnancy complications in women who are infected with Zika? How common are these complications?

4. Describe how the incidence of microcephaly has changed in Brazil over the past 5 years. What are some hypotheses regarding why the incidence of Congenital Zika Syndrome (CZS) after Zika infection is higher in Brazil than in other countries?

5. What are the current U.S. Centers for Disease Control (CDC) recommendations for women who are identified as being exposed to Zika while pregnant?

6. What are the current CDC recommendations for people who would like to travel to areas in which Zika is endemic?

7. How have religious groups, such as the Catholic Church, responded to the Zika epidemic? Contrast this response with the response to the HIV epidemic in the 1980s and 1990s. How have these epidemics affected U.S. policy with regard to provision of reproductive health services internationally?

8. What are some of the strategies proposed to reduce the risk of Zika infection in countries impacted by the disease? Which ones have been implemented? Were they effective?

23.2 Additional Case Information and Questions for Discussion

An ultrasound is concerning for microcephaly consistent with Ms. Silva's history of presumed Zika infection.

1. How would you break this news to Ms. Silva and Mr. Oliveira?

2. Self-reflection: How would you discuss the potential risks and options in an appropriate way? How do you avoid conveying your personal biases?

Ms. Silva and Mr. Oliveira are understandably devastated. The baby is a girl—a much hoped for daughter after three sons. Ms. Silva asks what you would do if it were you, or your significant other, who was in this situation.

Abortion law is complicated in Brazil. It is illegal except in cases of rape and where a woman's life is in danger, although the Supreme Court ruled in 2012 that it was permissible in cases of anencephaly and in late 2016 that it should not be illegal in the first trimester [1, 2]. Despite the laws, abortion is common, with a 2010 study indicating that 20% of Brazilian women had obtained an abortion illegally [3].

3. What are Ms. Silva's options?

4. Self-reflection: How does the legality of the procedure affect your response?
    (a) Is it ever appropriate to share your personal experiences (for instance, if you had been in a similar position or had a close friend or family member who had)?

One of the nursing assistants suggests that it is cruel to tell Ms. Silva about the microcephaly when there is nothing she can do to prevent it now.

5. How do you respond?

A 2003 survey included more than 4000 obstetrician-gynecologists (30% of those invited to participate) in Brazil and concluded that the closer they were to a patient, the more likely they were to support her decision to have an illegal abortion. Of all respondents, 41% had helped a patient to get an abortion and 49% had helped a relative. Among those who had personally experienced unplanned pregnancy, 78% of female physicians had obtained abortions for themselves, and 80% of male physicians had helped a partner to get an abortion. The figures were only slightly lower for physicians who self-described as "very religious," with 70% choosing abortion for themselves or their partners in the face of unwanted pregnancy [4].

6. Self-reflection: How do you react to these statistics? Do you think studies in the United States would yield similar findings?

7. Is this information useful in public policy debates? Why or why not?

Some countries, most notably Sweden, have imposed legal limits on conscientious objection by physicians and other healthcare practitioners, arguing that many have a monopoly on services, particularly in smaller communities, and that they should therefore offer all legal treatments and procedures within their scope of practice [5, 6].

8. Do you agree or disagree with this argument?
    (a) Is there an ethical difference between a pharmacist who refuses to dispense birth control and an OB/GYN who refuses to perform abortions? Why or why not?
    (b) Does it matter if the professional with the objection is practicing in a small community where patients do not have alternatives?

Ms. Silva's pregnancy is far too advanced for her to obtain an abortion in-country, even if she were able to afford it, and she does not have the means to go out of country. Three months later, she gives birth to a baby girl, Ana Lucia, who appears healthy but has a head circumference well below the 3rd percentile. At her follow-up visit with you, Ms. Silva asks, "But she could still be normal, right?"

9. What would you say now?

10. What are some ways in which Zika disproportionately affects poor women and their families?

(a) What are some of the challenges the family faces now?

As Ana Lucia grows, she has trouble feeding and eventually begins having seizures. The family lives several hours from the nearest clinic. Once a month, Ms. Silva makes the 8-hour journey (each way) by bus so that Ana Lucia can receive physical therapy. She performs the therapies at home between visits and has quit her job to do so. During those travel days, while Mr. Oliveira is working, their oldest son, Theo, aged 8, looks after his siblings, who are 3 and 4 years old.

Ms. Silva also participates in a support group called Maes de Anjos Unidos (United Mothers of Angels). They advocate for the Brazilian government to provide services in local hospitals, to compensate families for the care they must provide at home, and to offer psychological support for families under stress.

11. Are your obligations to advocate for this family different if you are the only healthcare provider for 20,000 people, rather than a private physician with a patient load of 2000?

12. What obligations does the government have to families of children like Ana Lucia?

In May of 2017, Brazil announced the end of the official public health emergency due to the Zika epidemic, although new infections continue to occur (at lower rates) in Brazil, as well as in the United States and many other countries. Many questions remain unanswered, particularly why CZS occurred in Brazil at much higher rates than in other countries with outbreaks.

The end of the emergency is attributed to a reduction in the number of susceptible hosts and to the success of mosquito eradication efforts, including fumigation of around 20 million homes, as well as the release of mosquitoes genetically modified to produce sterile offspring and the deliberate infection of mosquitoes.

13. What are some of the potential ethical concerns with the interventions used?
    (a) In what other ways are environmental concerns also ethical concerns?
    (b) The government was criticized for giving fumigators permission to forcibly enter any building, including private homes. Were they justified in doing so?

"Funding by crisis" is a critical way of describing how high-income countries address epidemics. That is, that many countries, including the United States, do not invest in adequate public health infrastructure (both domestically and globally) that could prevent public health emergencies through early detection but rather cobble together resources when an emergency arises [7].

The Ebola epidemic in West Africa in 2014 presented some very different ethical issues. Among these were the use of a still-experimental therapy called ZMapp. During the outbreak, ZMapp was in extremely limited supply and still in the experimental phase with little data available about its safety or its effectiveness [8, 9].

14. Given that very few of the thousands of patients with Ebola would have been able to get ZMapp, which criteria for selecting who gets ZMapp are ethically justifiable and which are not? Why?
    (a) What ethical or scientific concerns might exist if a similar experimental therapy existed for Zika?

Several months later, you see Mr. Oliveira outside the pediatric intensive care unit. Ana Lucia has been admitted with influenza and staphylococcal pneumonia. He is distraught that she may need to be intubated and mechanically ventilated, but all the ventilators are currently in use.

15. Creative problem-solving: Like the situation with ZMapp discussed above, this is a realistic example of a situation in which bedside rationing of care may occur. If you were working in this intensive care unit, how would you decide which children get placed on ventilators?
    (a) Age? Prognosis? Odds of surviving without a ventilator vs. odds of surviving even with a ventilator? Other comorbidities?

Some families of children with microcephaly, particularly in high-income countries, have expressed dismay that the condition is viewed by the media and general public as a catastrophe that warrants prevention at all costs. One parent noted that it is a "little bit hurtful" to know that people look at her children and say "Let's never let this happen to anyone ever again" [10].

16. **Self-reflection: Is this relevant? To the patient in this case, to you as a physician, or to public policy?**
    (a) Creative problem-solving: What would you do with this information? As a physician? As a policymaker?

23.3 Answers to Background Questions

1. **What is the history of Zika virus? What happened with Zika virus activity in Brazil in 2015? What is the current status of Zika activity in the United States? In the world?**
   Zika virus is a mosquito-borne virus that was first identified in humans in Uganda in the early 1950s. Sporadic cases of Zika were reported over the ensuing decades, but the first "outbreak" of Zika was reported on the island of Yap in Micronesia in 2007. Zika took the world stage in 2015 when reports of major outbreaks were seen in Brazil. Over the ensuing months, rapid increases in reported cases of microcephaly and Guillain-Barré syndrome were associated with Zika virus infection. In February 2016, the World Health Organization (WHO) declared Zika and its complications to be a "Public Health Emergency of International Concern," recognizing an urgent need for resources to respond to the epidemic. In November 2016, the WHO lifted the emergency status and committed to a long-term response to the disease [11]. Zika virus has been placed on the WHO's "List of Blueprint Priority Diseases"—diseases that are prioritized for research and development of countermeasures because they "pose a public health risk because of their epidemic potential" and because "there are no, or insufficient, countermeasures" available [12].

As the emphasis shifted to long-term management of this public health threat, the WHO developed a "Zika virus country classification" table that categorizes the level of threat in countries, territories, and sub-national areas. Both Categories 1 and 2 include areas where there is ongoing transmission of Zika virus disease [13].

- **Category 1**: Area with new introduction or reintroduction with ongoing transmission, currently 27 areas
- **Category 2**: Area with either (a) evidence of virus circulation before 2015 or (b) ongoing transmission that is no longer in the new or reintroduction phase but where there is no evidence of interruption in transmission, currently 44 areas, including Brazil
- **Category 3**: Area with interrupted transmission *and* with potential for future transmission, currently 15 areas, including the United States
- **Category 4**: Area with established competent vector (i.e., a mosquito capable of transmitting the virus) but no known documented past or current transmission, currently 62 areas

Based on the most recent data published by the Pan American Health Organization, the highest incidence of Zika is in the Latin Caribbean, where the regional cumulative incidence is approximately 350 per 100,000 persons. Within the region, Saint Barthelemy has a reported 10% incidence of Zika [14].

As of July 2018, no local mosquito-borne Zika virus transmission had been reported in the continental United States in 2018, although in previous years, transmission had been reported in Hawaii, Texas, and Florida [15]. Between 2015 and 2018, there were more than 5700 cases of Zika reported in the United States. Of these, 5400 were attributed to travel-related exposure. More than 37,000 cases have been reported in U.S. territories, including Puerto Rico, with the overwhelming majority attributed to local transmission [16].

2. **How is Zika transmitted? What are the symptoms of Zika infection? What percentage of infected individuals are symptomatic?**

   Zika virus is a mosquito-borne *flavivirus* that is primarily transmitted through the bite of an infected *Aedes* species (*Ae. aegypti* and *Ae. Albopictus*) mosquito. Human-to-human transmission can occur vertically (mother-to-child, both intrauterine and intrapartum transmission) and sexually. The extent of sexual transmission remains to be determined. Some experts believe it has been underestimated as a contribution to disease spread. Zika can be passed from infected people before symptoms appear, while symptomatic, and after symptoms end; and, although difficult to document, the virus may also be passed by a person who carries the virus but never develops symptoms. There are reported cases of laboratory- and transfusion-associated transmission as well. While viral particles have been identified in breastmilk, transmission through breastfeeding has not been documented [17].

   The incubation period for Zika (from bite to infection) is 3–12 days. Approximately 80% of people who are infected with Zika are asymptomatic [18]. Most people who do develop symptoms have mild, self-limiting disease with non-specific symptoms, including maculopapular and pruritic rash, fever, joint and muscle pain, headache, and conjunctivitis. Most symptoms resolve within 2 weeks [18].

   The greatest concerns with Zika infection are its associated complications, including severe neurologic sequelae in children of women infected with Zika during pregnancy (see Background Question 3) and in some infected adults. Neurologic complications of Zika include meningitis, meningoencephalitis, and Guillain-Barré syndrome.

3. **What are the known risks of pregnancy complications in women who are infected with Zika? How common are these complications?**

   Zika virus in pregnancy is known to cause a range of birth defects, most prominently microcephaly, a condition in which brain development is impaired during pregnancy. The virus itself crosses the placenta and targets specific cells (neural progenitor cells and glial cells) in the developing brain, causing cell death, reduced nerve cell proliferation, and, consequently, impaired brain development [19]. Congenital Zika syndrome (CZS) is "a unique pattern of birth defects found among fetuses and babies infected with Zika during pregnancy" [20]. The characteristic findings in this syndrome include severe microcephaly with a partially collapsed skull, decreased brain tissue volume, damage to the retina, joint problems, and increased muscle tone (hypertonia). Infants with CZS have been found to have a range of visual, developmental, and cognitive impairments [20]. The full spectrum of Zika's impact on the developing fetus is unknown as researchers and clinicians are starting to observe developmental delays in children who were born without obvious deformity or deficit.

   Zika is a notifiable disease in the United States, meaning that healthcare practitioners are legally obligated to report cases to their local, territorial, or state health departments, which then must report to the CDC. During the Zika epidemic in 2015, in order to better understand the impact of vertical transmission of Zika virus and to protect the health of infants, the CDC developed and implemented a "Pregnancy and Infant Registry." The registry contains data for women in the U.S. and U.S. territories who were known to have been exposed to Zika virus during pregnancy. As of July 2018, the registry reported that almost 2500 women in the United States and 4900 women in U.S. territories had laboratory evidence of possible Zika infection during pregnancy [21].

   Shapiro-Mendoza and colleagues published an analysis of data in the U.S. territory database, reviewing outcomes from more than 2500 completed pregnancies between January 2016 and April 2017 [22]. The authors conclude that "among completed pregnancies with positive nucleic acid tests confirming Zika virus infections identified in the first, second, and third trimesters of pregnancy, the percentages of fetuses or infants with possible Zika-associated birth defects was 8%, 5%, and 4% respectively" [22]. These percentages may vary as more studies about CZS are completed.

4. **Describe how the incidence of microcephaly has changed in Brazil over the past 5 years. What are some hypotheses regarding why the incidence of Congenital Zika Syndrome (CZS) after Zika infection is higher in Brazil than in other countries?**
Prior to 2015, the historical mean occurrence of microcephaly in Brazil was 2 cases per 10,000 live births [23]. In October 2015, the Brazilian Ministry of Health received a report of an unexpected increase in cases of microcephaly from a northeastern region of Brazil. The following month, the Ministry of Health declared a "Public Health Emergency of National Concern" based on concerns regarding the possible association between Zika virus and severe neurologic disorders in both infants and adults. Oliveira and colleagues have described two waves of Zika activity between January 2015 and November 2016, drawing on data from the Brazilian Ministry of Health's surveillance systems [23]. The authors conclude that more than 1.67 million cases of Zika were reported during that time period, including more than 41,000 cases during pregnancy. Of the completed pregnancies, 1950 infection-related cases of microcephaly were confirmed [23]. Furthermore, the authors report a wide range in incidence rates (3.2–49.9 cases per 10,000 live births) over time and geography, with the highest rates seen in the northeast region of Brazil during the first wave of Zika activity [23]. After the second wave, the incidence of microcephaly dropped precipitously. Other assessments have corroborated these findings of marked regional differences in rates of Zika-caused microcephaly [24].

The reasons for the stark differences in rates have not yet been established. The most important factor with respect to *number* of cases is population density, but this is taken into account when comparing rates (number of cases per measure of the population, such as live births). For differences in incidence rates, one key factor is likely the initial attack rate for the disease: The higher the proportion of people infected, the greater the risk of infection among women who are pregnant at the time. Differences in attack rate could be in part due to the environment (higher incidence rates were reported when conditions were more conducive to mosquito breeding). Beyond environmental factors, however, there is speculation that the severity of the initial infection, the patient's viral load, and the presence of coinfection with other viruses, such as the dengue viruses, may affect the risk of microcephaly [25]. Current *in vitro*, animal, and human data are conflicting on the importance of these factors in the risk of infection, disease, maternal-fetal transmission potential, and fetal outcomes of infection.

5. **What are the current CDC recommendations for women who are identified as being exposed to Zika while pregnant?**
The CDC defines "possible Zika exposure" as "travel to, or residence in an area with risk for mosquito-borne Zika virus transmission or sex with a partner who has traveled to or resides in an area with risk for mosquito-borne Zika virus transmission" [26]. It should be noted that possible exposure is not determined based on a history of symptoms, as many infected people are asymptomatic. CDC recommendations were revised in 2017 in light of the declining prevalence of Zika, which decreased the positive predictive value of the test [26–28].

Current recommendations include the following [26]:
- All pregnant patients should be asked about possible Zika exposure before and during the current pregnancy at every prenatal visit.
- Those who have recent possible exposure *and* symptoms of Zika virus disease should be tested with concurrent Zika virus nucleic acid test (NAT) and serologic testing (IgM) as soon as they are identified as such through 12 weeks after symptom onset.
- Asymptomatic patients with ongoing possible exposure (due to geography) should be tested three times during pregnancy (assuming they are negative each time); optimal timing of testing has not been determined. Testing is not recommended for asymptomatic patients without ongoing possible exposure.
- Those with possible exposure and ultrasound findings consistent with Zika (see Background Question 3) should be tested as above in order to determine the etiology of the fetal anomalies; testing of fetal or placental tissue can be considered in some circumstances.

The laboratory testing algorithm is quite complex: real-time reverse transcriptase-polymerase chain reaction (rRT-PCR) confirms Zika, but a negative result does not rule it out. Previous recommendations and considerations included obtaining IgM testing for Zika as part of preconceptional counseling, in order to determine whether the patient had recently been exposed; however, it was subsequently determined that anti-Zika IgM antibodies may persist beyond 12 weeks, making them less useful for determining recent infection. Results suggestive of Zika infection must be confirmed with a plaque reduction neutralization test for Zika and for other flaviviruses (including dengue) [26].

In addition to the above, pregnant women whose partners may have been exposed to Zika or are traveling to Zika-endemic areas are advised to "consistently and correctly" use condoms for the duration of the pregnancy [27].

6. **What are the current CDC recommendations for people who would like to travel to areas in which Zika is endemic?**

The CDC maintains a Web page with information directed at travelers [29]. This includes the 5-page "Zika: A Guide for Travelers," written in lay language and available in both English and Spanish [29, 30].

The CDC advises that travelers to such areas:
- Protect themselves against mosquito bites (e.g., by using U.S. Environmental Protection Agency [EPA]-registered insect repellent, wearing long pants and long-sleeved shirts, staying in spaces with screens or closed windows and doors) while traveling
- Protect others by avoiding spreading Zika to local mosquitoes through continued use of such measures for 3 weeks after returning
- Use condoms while traveling and afterward to avoid infecting partners: women should do so for at least 2 months and men for at least 6 months (or for the duration of the partner's pregnancy, as noted above)

The CDC also recommends that pregnant people avoid travel to Zika-endemic areas unless absolutely necessary and that both men and women who are planning to conceive also avoid travel to these areas or choose to delay conception. Because there are reported cases of transmission of Zika via semen 6 months after the date of presumed infection, women are advised not to try to conceive with a partner who has traveled to a Zika-endemic area for at least that long, while a woman who has traveled herself could safely conceive 2 months after travel [30].

7. **How have religious groups, such as the Catholic Church, responded to the Zika epidemic? Contrast this response with the response to the HIV epidemic in the**

1980s and 1990s. How have these epidemics affected U.S. policy with regard to provision of reproductive health services internationally?

In the wake of the Zika epidemic, Pope Francis commented that the use of artificial contraception (such as condoms) might be morally justified as the "lesser of two evils" [31, 32]. Other Catholic scholars objected, arguing that it was not true that the birth of a child with microcephaly represented a danger that justified the use of hormonal contraception [33]. While not perhaps a major shift in religious teaching (and indeed, 98% of American Catholic women have used artificial contraception in spite of church teachings), it was nonetheless consequential, particularly because the HIV epidemic had never led to such a pronouncement [34, 35].

In fact, as recently as 2010, Pope Benedict XVI had firmly reiterated his position that condom use was not the appropriate solution to address the HIV epidemic, although the Southern African Catholic Bishops Conference had counseled that the use of condoms to prevent the spread of HIV between discordant spouses was acceptable, on the basis that "everyone has a right to defend one's life against mortal danger" [36, 37]. The positions of the Catholic Church are particularly important in the context of Zika because Catholicism has left lasting cultural influences throughout Latin America.

The nature of epidemics such as HIV, which was predominantly spread through unprotected sexual intercourse, and Zika, which exerts its greatest impact on pregnant women and pregnancy outcomes, posed dilemmas for those who hold particular beliefs restricting sexual and reproductive choices. Because such choices have been politicized, U.S. policy with regard to provision of reproductive health services internationally has tended to change from administration to administration. The "Mexico City policy," also called the "global gag rule," is a policy instituted by President Ronald Reagan in 1984 [38]. It restricts foreign nongovernmental organizations (NGOs) from receiving American aid for family planning programming if they use any funds (including those from non-U.S. sources) to "provide abortion services, counselling, or referrals, or advocate for liberalisation of their country's abortion laws" [39]. The original policy contained exceptions for rape, incest, and the health of the mother.

Historically, this policy has been rescinded by Democratic presidents and reinstated by Republican ones. As of January 2017, it was reinstated with an additional restriction: NGOs that provided even counseling that included abortion as an option would no longer be able to receive funds for HIV prevention and treatment, in addition to family planning funds [39]. The policy has significant implications for NGOs' ability to provide needed services. One study of the policy's impact in Ghana found that the availability of contraceptive services for poor and rural communities declined during the gag rule years and that fertility and abortion rates were higher in gag rule years when compared to non-gag rule years [40].

8. **What are some of the strategies proposed to reduce the risk of Zika infection in countries impacted by the disease? Which ones have been implemented? Were they effective?**

Strategies to reduce the risk of Zika infection can be divided into three key groups: vector control, personal prevention, and vaccination [41].

Vector control includes well-established approaches to reducing mosquito populations near human habitats, such as eliminating standing water and disposing of garbage thoroughly, as well as the use of insecticides (within domiciles, as well as fumigation of cargo at ports to avoid spreading larva) [41]. Vector control also includes newer approaches intended to interfere with mosquito breeding. Such methods include the introduction of the bacteria *Wolbachia*, which reduces the risk of mosquito-to-human transmission and is passed from female mosquitoes to their offspring. *Wolbachia* has been shown to be effective in controlling dengue and is currently being studied for Zika transmission [41]. Another method of this type is genetic tailoring of mosquitoes to decrease reproductive fitness; again, this has been successfully used to control dengue but has not yet been shown to reduce rates of Zika infection [41]. Furthermore, some scientists have expressed concern that this latter technique may inadvertently result in spread of such deleterious genes to non-mosquito populations, disrupting ecosystems [41].

Personal prevention includes the strategies recommended by the CDC for individual travelers (see Background Question 6), as well as the use of bed nets (best known and studied for prevention of malaria) and personal protective equipment for healthcare workers [30, 41].

Personal prevention strategies can be highly effective for individuals, but the extent to which they are effective for populations depends on programmatic support, particularly resources and education. That is, people must be aware of these strategies and able to access the resources they require. Strategies such as staying in climate-controlled environments are therefore not very effective for most people living in Zika-endemic areas, who may not have electricity, much less air conditioning. Applying daily repellent is also not scalable for populations [42]. Programs to distribute insecticide-treated bed nets have been successful in malaria prevention; however, no published studies specifically have been shown to impact Zika prevalence, and there is some evidence that bed nets are less effective for mosquito-borne viruses [43, 44].

Vaccination against Zika virus is currently under active investigation. As of this writing, many vaccines were in preclinical development or had entered early phase trials, and phase 1 trials had yielded promising results in healthy volunteers [45–47].

Finally, prevention might also include *screening of the blood supply* to protect patients who require blood transfusions [41].

It should be noted that these strategies are not Zika-specific in theory or, largely, in execution (with the exception of a Zika vaccine) and are employed to address concerns regarding many emerging arboviruses, including Zika, dengue, yellow fever, and chikungunya, as well as West Nile virus and Japanese encephalitis [42, 48].

23.4 Responses to Discussion Questions

> **Box 23.1 Teaching Tip**
> This case is unique in the book in the sense that learners are asked to imagine themselves in the role of a physician practicing in a resource-limited setting (by World Bank classification, Brazil is an upper-middle-income country). The healthcare provider in this vignette may well have trained at American medical schools and hospitals and may be an immigrant or Brazilian-born, but she or he is expressly *not* described as volunteer or researcher on a temporary assignment.

An ultrasound is concerning for microcephaly consistent with Ms. Silva's history of presumed Zika infection.

✓ 1. **How would you break this news to Ms. Silva and Mr. Oliveira?**
Families very often suspect when bad news is coming. They recognize nonverbal cues, such as a technician leaving the room abruptly and returning with the physician or being seated in a private room when they are usually seen on a noisy open ward. Therefore, it is important not to delay sharing the findings. Such delays will only prolong anxiety and distress.

However, it is also important to understand what Ms. Silva and Mr. Oliveira already know and understand. If Ms. Silva has been concerned about this possibility throughout her pregnancy, the approach will be different than if she was never informed that her previous symptoms were indicative of Zika and that the infection might have repercussions for her pregnancy outcome. In the latter case, the couple may need more explanation of Zika and Congenital Zika Syndrome, as well as time to process this new information. Patients' educational and cultural backgrounds will also influence the optimal approach, as will their prior relationship with the clinician giving the news. For example, if patients do not understand the relationship between head circumference and neurodevelopmental outcomes, it may take some time to grasp that connection, and they will need this time before they can ask essential questions that other patients might ask immediately.

Finally, the potential ambiguity inherent in this situation poses its own challenge. The ultrasound finding is very concerning but it is not unequivocal: The clinician cannot predict with certainty the outcome for the unborn child. This uncertainty can make it more difficult for clinicians to strike the right tone, since if they are perceived as being too negative, it can alienate those patients who are inclined to be more hopeful. On the other hand, clinicians need to be honest so that patients can make informed decisions. This "balancing act" was a key theme identified in a meta-synthesis of 40 studies of breaking bad news in oncology, drawn from 12 Western countries [49].

One practical approach to breaking bad news follows the acronym SPIKES. The following version is slightly modified [50]:

- *Setting:*
  - Give news in person and not over the telephone whenever possible. For this reason, if a difficult diagnosis is suspected, clinicians should set a follow-up appointment to discuss the results in advance, avoiding the possibility that a last-minute request for a face-to-face appointment will prolong the patient's anxiety.
  - Find a reasonably comfortable, quiet space (recognizing that, for some healthcare facilities in resource-limited areas, private consultation rooms and "family meeting" rooms are not the norm).
  - Make sure everyone, including the clinician, is sitting down.
  - If patients have been recently examined, they should be given a chance to dress.
  - Introduce clinician(s) by name and role, if not already known to family.
- *Perception*
  - Assess baseline knowledge, and/or summarize what has happened in the patient's care prior to this point, putting the news in context. However, if the patient or family is clearly already aware of the purpose of the meeting, excessive introductory statements may simply feel like the clinician is stalling.
  - Give the news clearly and succinctly in language likely to be understandable to the patient or family. Do not delay giving the news with "small talk."
  - "I am sorry to tell you that…" is an appropriate way to start.
- *Invitation*
  - Invite the patient to take control of the encounter by asking questions.
  - Ask or follow the patient's lead regarding terminology (e.g., "baby" versus "fetus").
- *Knowledge:*
  - Prioritize what essential information must be conveyed in the first conversation.
  - Avoid medical jargon and complex explanations (unless requested).
  - Pause after each piece of new information to give the patient time to absorb it, and ask more questions; be comfortable with silence.
  - Assess for understanding.
  - If there are medical decisions that must be made quickly, these may need to be presented in the first conversation.
- *Empathy:*
  - Avoid statements like, "I understand what you are going through" or "I know what this must be like for you."

Consider using "This is really hard" instead.
- Observe different reactions between various family members. Ensure that the patient is receiving adequate support from family.
- Ask if there are more questions.
- Summary and strategy:
  - Explain any next steps, including further testing or other options.
  - Make sure the patient has appropriate follow-up scheduled and knows how to reach out if new questions or concern arises before then.

✓ 2. **Self-reflection: How would you discuss the potential risks and options in an appropriate way? How do you avoid conveying your personal biases?**

> **Box 23.2 Teaching Tip**
> Encourage learners to recognize that bias exists in both directions: Healthcare practitioners may feel strongly that microcephaly is a devastating outcome and the risk too great, or they may be strongly opposed to abortion regardless of these circumstances. This concept is also explored, though more superficially, in ▶ Chapter 16, through the lens of a devastating genetic disease.

Ms. Silva and Mr. Oliveira are understandably devastated. The baby is a girl—a much hoped for daughter after three sons. Ms. Silva asks what you would do if it were you, or your significant other, who was in this situation.

Abortion law is complicated in Brazil. It is illegal except in cases of rape or where a woman's life is in danger, although the Supreme Court ruled in 2012 that it was permissible in cases of anencephaly. In late 2016, the Supreme Court ruled that abortion should not be illegal in the first trimester [1, 2]. Despite the laws, abortion is common, with a 2010 study indicating that 20% of Brazilian women had obtained an abortion illegally [3].

✓ 3. **What are Ms. Silva's options?**
   In general, there are three primary options for a pregnant woman with known recent Zika virus infection and good access to health care:

1. Obtain a first-trimester surgical or medical abortion. *This is not an option for Ms. Silva because she is already in her second trimester; even if she had sought an abortion earlier, she may not have been able to find a healthcare practitioner willing to perform one safely, or she may have been unable to afford such a procedure.*
2. Continue with the pregnancy with close monitoring for signs of microcephaly (in the United States, the CDC recommends serial ultrasounds every 3–4 weeks), and consider a second-trimester abortion if there is evidence of congenital anomalies. *Again, this is not an option for Ms. Silva at this point in her pregnancy. Serial monitoring may not have been feasible depending upon how far she lives from the clinic and access to a second-trimester abortion in a country without legal first-trimester abortion is likely to be extremely difficult.*
3. Continue with the pregnancy to term with or without increased monitoring.

Because Ms. Silva is already approaching the third trimester, she has few options. A late abortion (after 24 weeks or, sometimes, an estimation of viability) is difficult to obtain, even in the United States, where the courts have repeatedly upheld that late abortions must be permitted to protect the life or health of the pregnant patient. She may need to consider:
- Where she will deliver, if she has limited access to neonatal care in her area
- How she and her partner will care for a child with significant medical needs
- Whether placing the child for adoption is something they wish to consider

✓ 4. **Self-reflection: How does the legality of the procedure affect your response?**
   (a) **Is it ever appropriate to share your personal experiences (for instance, if you had been in a similar position or had a close friend or family member who had)?**

> **Box 23.3 Teaching Tip**
> The challenges of responding to "What would you do, Doctor?" are discussed in detail in ▶ Chapter 19, in the context of research participation. This case focuses on related issues, specifically practicing medicine in the face of laws that one might find unethical and discussing personal experiences with patients.

One of the nursing assistants suggests that it is cruel to tell Ms. Silva about the microcephaly when there is nothing she can do to prevent it now.

✅ 5. **How would you respond?**
First and foremost, patients are entitled to honest information about their health and well-being, regardless of whether healthcare practitioners think the information is actionable. That is a crucial aspect of respect for persons and for autonomy. Some people may think, "I wouldn't want to know if it were me," but they are not in the patient's position and cannot know what they would choose to do.

In rare circumstances, patients may instruct physicians *not* to disclose test results. For example, an elderly patient might ask the physician to communicate with a son or daughter who is the healthcare proxy regarding results and treatment decisions. In other situations, patients choose not to obtain additional testing when a serious medical condition is suspected. For example, some pregnant patients choose not to undergo routine prenatal screening for fetal anomalies or choose not to receive confirmatory testing when screening is suggestive of a problem. To force patients to receive such testing would violate autonomy; the healthcare practitioner's role is simply to explain the risks and benefits of testing.

In this case, it appears that Ms. Silva has chosen to undergo the ultrasound and presumably desires the results. While it is certainly possible, particularly in cultural contexts in which physicians are treated deferentially, that she was unaware she could decline testing, it would be equally paternalistic to withhold the results from her now.

A 2003 survey included more than 4000 obstetrician-gynecologists (30% of those invited to participate) in Brazil and concluded that the closer they were to a patient, the more likely they were to support her decision to have an abortion. Of all respondents, 41% had helped a patient to get an abortion, and 49% had helped a relative. Among those who had personally experienced unplanned pregnancy, 78% of female physicians had obtained abortions for themselves, and 80% of male physicians had helped a partner to get an abortion. The figures were only slightly lower for physicians who self-described as "very religious," with 70% choosing abortion for themselves or their partners in the face of unwanted pregnancy [4].

✅ 6. **Self-reflection: How do you react to these statistics? Do you think studies in the United States would yield similar findings?** [4]

> **Box 23.4 Teaching Tip**
> The aforementioned study was titled "The Closer You Are, the Better You Understand" [4], and it highlights the double standard that people, including healthcare practitioners, sometimes hold for themselves and others.

✅ 7. **Is this information useful in public policy debates? Why or why not?**
Under ideal circumstances, all public policy decisions are based on high-quality evidence. Policy should be grounded in three categories of evidence: (1) evidence of the problem, (2) evidence of an effective solution, and (3) evidence of the feasibility of implementing the proposed solution. In all cases, the quality of the evidence (i.e., of the data that inform the evidence) should be taken into consideration. Are the data generalizable, accurate, reproducible, free from significant bias, and so forth?

Assuming that the aforementioned survey results are valid (note that the survey response rate and attendant possibility of selection bias is one major limitation), these data indicate that the current policy, banning abortions, is not effective at preventing people from seeking an abortion. For many people, this is evidence enough to stimulate and justify a public policy debate.

Another question that should be asked in any public policy debate is: What are the alternatives? In this case, evidence for the success of the alternative, legalizing abortion, has been established on the global stage.

The survey data cited earlier notably do not address some of the core issues, particularly: Why are people seeking abortions? In Brazil and many other countries, relevant information would include data regarding access to effective methods of contraception, the status of women in the country, and the extent of poverty in the country. Worldwide, according to the World Bank and the United Nations, more than 80% of women have reported contraceptive use; however, more affluent and better educated women are disproportionately likely to have access to contraception on multiple levels [51, 52]. Such women are not only better able to afford contraception, they are also more likely to have the knowledge required to consider it and choose the method best for their circumstances. In addition, they are more likely to be empowered within their communities and relationships to refuse sexual activity and to choose when to bear children.

Although policy should be grounded in evidence and impact individuals equitably, in reality it is common for policies to be spurred by elected officials or highly influential individuals who are personally impacted by an issue. Physicians who are often highly regarded in their communities may use their voices to advocate for or against laws that impact them. Sometimes such advocacy can be a driver of positive change; however, advocates and policy-makers should be attentive to the risk of neglecting problems that lack a well-connected "champion."

In the aforementioned survey, it is clear that physicians are more likely to be able to access an officially banned procedure, but it is not clear whether they are concerned about what happens to those who do not have access through such channels. The pervasive problem of unsafe abortions has been called a "preventable pandemic" [53]. Between 19 and 20 million women, almost exclusively in developing countries, undergo unsafe abortions, and an estimated 70,000 women die from these procedures. In addition to the mortality associated with unsafe abortions, millions more experience complications, including hemorrhage, infection, and infertility [53].

Some countries, most notably Sweden, have imposed legal limits on conscientious objection by physicians and other healthcare practitioners, arguing that many have a monopoly on services, particularly in smaller communities, and that they should therefore offer all legal treatments and procedures within their scope of practice [5, 6].

8. **Do you agree or disagree with this argument? Why or why not?**
Schuklenk and Smalling have written an erudite analysis of this issue [54]. They note that:

> It is not unusual for students in any given bioethics class to offer something like the following defence of conscientious objection rights: 'Remember the Nazi experiments and the abuse of prisoners there and then? It is good that conscientious objection rights exist to protect good doctors refusing to participate in such crimes.'

They then argue that most cases of conscientious objection are not analogous to Nazi Germany, and in fact, "the typical conscientious objector does not object to unreasonable, controversial professional services—involving torture, for instance—but to the provision of professional services that are both uncontroversially legal and that patients are entitled to receive" [54]. A key part of the argument rests on the simple fact that healthcare professionals have voluntarily joined their profession—a profession that includes a scope of practice recognized by both other professionals and by society. In this view, a physician is certainly free to object to contraception or abortion and could easily choose to pursue a specialty that does not involve provision of these services. Similarly, a physician might object to the withdrawal of artificial nutrition and hydration from a dying patient and should probably not choose a career in geriatrics, palliative care, or critical care medicine.

(a) **Is there an ethical difference between a pharmacist who refuses to dispense birth control and an obstetrician-gynecologist who refuses to perform abortions? Why or why not?**

Some would argue yes—that the pharmacist is acting on orders from a physician, providing something to a patient who then administers it herself, while the OB/GYN is directly performing an action that results in the abortion. Others would disagree and argue that everyone in the chain is a moral agent and should be entitled to act on personal conscience.

From a more practical standpoint, retail pharmacists and even some hospital pharmacists rarely have access to patients' full histories and contextual information. The retail pharmacist, unless asking, is unlikely to know whether hormonal contraception is being used to manage a related medical problem (such as menorrhagia, dysfunctional uterine bleeding, etc.) or for contraception, which would seem to complicate their ability to construct an informed conscientious objection in many circumstances.

(b) **Does it matter if the professional with the objection is practicing in a small community where patients do not have alternatives?**

It does indeed matter. When a provider has a monopoly on services in a community, the consequences of conscientious objection can be much greater. In those situations, the burden on the clinician should correspondingly be higher [54]. That is, if the clinician's conscientious objection prevents a patient from receiving the desired, necessary medical care, then the harm to the patient should outweigh the potential harm of violating the clinician's conscience.

Ms. Silva's pregnancy is far too advanced for her to obtain an abortion in-country, even if she were able to afford it, and she does not have the means to go out of country. Three months later, she gives birth to a baby girl, Ana Lucia, who appears healthy but has a head circumference well below the 3rd percentile. At her follow-up visit with you, Ms. Silva asks, "But she could still be normal, right?"

9. **What would you say now?**

As in the initial diagnostic conversation (discussed previously in the context of "breaking bad news"), the clinician needs to balance honesty and realism with hope and compassion. If the clinician has access to statistics regarding the outcomes for children in Ana Lucia's situation, those may be helpful to share while taking into account that many people have difficulty thinking in terms of probabilities. (There is some evidence that patients who exhibit numeracy—the ability to use numerical information—are better at this [55].) At the same time, the clinician should make sure that Ms. Silva is treated with the empathy and joy extended to other new mothers: with congratulations on the birth and compliments regarding her baby.

10. **What are some ways in which Zika disproportionately affects poor women and their families?**

Zika can disproportionately impact poor women and their families in many ways. They may be at risk for greater exposure to the virus and they may be at greater risk for being disproportionately burdened by the sequelae of Zika infection.

With respect to increased risk for exposure to the virus, the vector for the Zika virus is an aggressive human biter that lives near and in houses and breeds in small containers of water (e.g., trash, old tires). People who live in poverty may have living environments that are more conducive to mosquito breeding, including poor water and sanitation conditions. In addition, they may have fewer resources to protect themselves from mosquitoes, for example, less access to repellents, air conditioning, and screens on their windows.

With respect to being disproportionately burdened by the sequelae of Zika, poor women may have less access to birth control or to abortion. Furthermore, if they do have a child CZS, they may have less access to the healthcare and educational services that their child needs to optimize their health status.

(a) **What are some of the challenges the family faces now?**
Ana Lucia very likely has CZS, as described in Background Question 3. Such children have an extremely high likelihood of having significant developmental delay including delays in cognitive, behavioral, communication, and motor development. In a recent article, Anne Wheeler, PhD, concludes:

> Activities of daily living will be compromised, and most children with CZS will require lifelong care. Many will not obtain even the most basic self-help skills; however, many activities of daily living, such as self-feeding, dressing, and toileting, may be obtainable by some with early intervention and ongoing support. Given what we know of other neurologic impairments, individual and environmental variables (such as maternal education, formal and informal support systems, access to quality early-intervention services, and maternal and child nutrition) will have a mediating or moderating effect on infant outcomes [56].

Ms. Silva and Mr. Oliveira will face the challenge of determining who will take care of Ana Lucia. Are they both working? Can they continue to do so? Do they have means to transport her to her medical appointments? In addition to having a child with special needs, the couple will face the challenge of still providing care and attention to their other three children. These factors are likely going to significantly add emotional, physical, and financial stress for the family.

As Ana Lucia grows, she has trouble feeding and eventually begins having seizures. The family lives several hours from the nearest clinic. Once a month, Ms. Silva makes the 8-hour journey (each way) by bus so that Ana Lucia can receive physical therapy. She performs the therapies at home between visits and has quit her job to do so. During those travel days, while Mr. Oliveira is working, their oldest son, Theo, aged 8, looks after his siblings, who are 3 and 4 years old.

Ms. Silva also participates in a support group called Maes de Anjos Unidos (United Mothers of Angels). They advocate for the Brazilian government to provide services in local hospitals, to compensate families for the care they must provide at home, and to offer psychological support for families under stress.

11. **Are your obligations to advocate for this family different if you are the only healthcare practitioner for 20,000 people, rather than a private physician with a patient load of 2000?**
In general, physicians and other healthcare practitioners do have obligations to advocate for patients and families, particularly those who have limited resources and can benefit from additional support in accessing other resources. However, some healthcare practitioners in low- and middle-income countries, as well as those practicing in underserved areas of high-income countries, may already have extraordinary demands placed upon them, such as high patient volumes. A utilitarian perspective on such scenarios would argue that the practitioners should maximize the well-being of the greatest number of individuals. In the case of the lone practitioner caring for 20,000 people, he needs to meet basic medical demands for as many as possible, likely leaving little time for individual patient advocacy. In the case of the private physician, she likely has much more time per patient and can be expected to devote some of it to advocacy.

In both cases, advocacy for establishment of programs to provide more services locally and support families in other ways might ultimately lead to better outcomes for everyone, but healthcare practitioners may have to choose how best to allocate their time. For example, it may be practically, ethically, and even psychologically challenging for the only healthcare practitioner in a community to leave patients to go to the capital and lobby lawmakers for change, knowing that those patients will have limited care in the practitioner's absence.

12. **What obligations does the government have to families of children like Ana Lucia?**
One of the greatest challenges facing Brazil at the height of the Zika epidemic was its financial situation. In December 2015, just 1 month after announcing a public health emergency, the Brazilian government announced a government spending freeze. While one can easily argue that the government has the obligation to provide comprehensive services for children with CZS or any child with developmental delay, it is unlikely that the government will be able to meet such obligations. In the United States, the Individuals with Disability Education Act's (IDEA) Early Intervention program provides funding for supportive services—such as physical, speech, and occupational therapy—for children with a wide spectrum of developmental delays. (▶ Chapter 6 provides more information on early intervention services.)

In May of 2017, Brazil announced the end of the official public health emergency due to the Zika epidemic, although new infections continue to occur (at lower rates) in Brazil, as well as in the United States and many other countries. Many questions remain unanswered, particularly why CZS in Brazil at much higher rates than in other countries with outbreaks.

The end of the emergency is attributed to a reduction in the number of susceptible hosts and to the success of mosquito eradication efforts, including fumigation of around 20 million homes, as well as the release of mosquitoes genetically modified to produce sterile offspring and the deliberate infection of mosquitoes.

13. **What are some of the potential ethical concerns with the interventions used?**
Two major areas of concern included:
  – The potential negative impact of fumigation on the health of those exposed, as well as the long-term environmental impact
  – The potential infringement on autonomy and personal liberty by spraying homes over the occupants' objections

In the United States in the 1940s, DDT was used extensively and was responsible for eradication of malaria. However, it is now known that DDT exposure causes infertility, genitourinary anomalies, breast cancer, diabetes, and cognitive impairments. Its current use in Africa and Asia involves spraying inside homes, leading to higher levels in individuals, although less accumulation in the environment.

Recent research in Upstate New York showed a correlation between aerial mosquito spraying with pyrethroids (intended to reduce the spread of West Nile virus and Eastern equine encephalitis) and an increased prevalence of autism spectrum disorders and pervasive developmental disorders [57]. While this research does not conclusively show that spraying is causative, it does raise a significant question regarding the risks and benefits of pesticide use—questions that also arose with regard to the use of DDT. These concerns are particularly potent if the disease prevented is relatively rare, but the sequelae of pesticides affect large numbers of people for decades after exposure.

(a) **In what other ways are environmental concerns also ethical concerns?**
There are also arguments that since wealthier countries are primarily responsible for global warming (e.g., the United States is one of the largest carbon producers in the world—second overall and seventh per capita), these countries have ethical obligations to address and attenuate its environmental impact. This is

relevant to public health concerns, like the Zika epidemic, because global warming plays a significant role in emerging infections and epidemics, as well as in increasing prevalence of other diseases, including asthma and cancers such as endemic Burkitt lymphoma [58, 59].

(b) **The government was criticized for giving fumigators permission to forcibly enter any building, including private homes. Were they justified in doing so?**
As a general rule, public health authorities in most countries, including the United States, have broad authority to protect the public's health. Any such intervention that compromises individuals' autonomy or puts them at risk should be carefully considered (although in practice, this may not always occur). Public health authorities should choose an intervention that provides an effective solution to the threat and that is the safest and least intrusive for the individuals affected.

Public health authorities have the right to confiscate personal property (e.g., animal hides contaminated with anthrax), require medical treatment or face imprisonment (e.g., tuberculosis), or even isolate entire communities (e.g., Ebola). Because Zika is a mosquito-borne illness and mosquitoes are not confined to any property, many would argue that the effort to fumigate homes was indeed justifiable. In the United States, people are not given a choice when a local health department opts to conduct spraying to mitigate against the threat of Eastern equine encephalitis, for example.

The more pertinent question may thus be the issue of "forcibly entering" people's homes. Public health authorities have an obligation to communicate effectively, even in a crisis. Properly informing people of the plan in advance, providing them options of what they can do, rather than what they cannot do, can help mitigate criticism and promote collaboration. Creating and following through on positive messaging (e.g., "We will provide welcome centers for you and your family to gather while the property is being treated.") are also helpful.

"Funding by crisis" is a critical way of describing how high-income countries address epidemics. That is, many countries, including the United States, do not invest in adequate public health infrastructure (both domestically and globally) that could prevent public health emergencies through early detection but rather cobble together resources when an emergency arises [7].

The Ebola epidemic in West Africa in 2014 ago presented some very different ethical issues. Among these were the use of a still-experimental therapy called ZMapp. During the outbreak, ZMapp was in extremely limited supply and *still in the experimental phase, with little data available about its safety or its effectiveness* [8, 9].

14. **Given that very few of the thousands of patients with Ebola would have been able to get ZMapp, which criteria for selecting who gets ZMapp are ethically justifiable and which are not? Why?**
First, it should be acknowledged that "When thousands of people are confronted with a life-threatening disease, and no specific therapies or preventive measures exist, it can be ethically acceptable to assume greater risks and offer patients unproven interventions" [60].

Under those circumstances, attention then shifts to *who* should have access to these unproven interventions. In many cases, the first consideration would be potential benefit to participants (the need to prevent as many deaths as possible during the current outbreak) coupled with the needs of the study to yield data that will drive further research, innovation, and eventually widespread benefits during future epidemics [61].

For example, the sickest patients might be poor candidates if they are beyond aid, even if the experimental therapy is effective. Not only would their inclusion prevent others from receiving benefit, it might also obscure the efficacy of the drug, leading the study to reach the wrong conclusion. For some experimental therapies, it is known in advance of the trial (based on mechanism of action, preclinical studies, or off-label use of an intervention) that some patients are more likely to benefit. Perhaps they have a polymorphism affecting drug metabolism that seems to make therapy more efficacious or perhaps certain laboratory markers in mice predicted a poorer response. However, in the case of ZMapp, there were no scientifically established criteria for which patients were most likely to benefit (e.g., age, illness severity, organ function, etc.) [8, 9].

In practice, ZMapp was initially tested almost entirely in healthcare workers. Rid and Emanuel note the pros and cons of this approach:

» Because health professionals put themselves at risk to care for patients and could help more patients once recovered, the principles of reciprocity and helping the largest number of people could justify their prioritisation. However, health-care workers are often well-off and have special ties to the medical establishment. Their priority might therefore be viewed as further privileging of the already well-off, especially by contrast with those who provide care without being trained as health professionals [60].

As suggested previously, criteria that are not ethically justifiable include allowing—explicitly or implicitly—more affluent, educated, and well-connected patients to enroll over poor, less educated, or otherwise socially disadvantaged patients. Another justifiable approach might have been the inclusion of local community leaders in early studies, given their importance to their communities and the potential for promoting collaborations [62].

Box 23.5   Personal Perspective

One common issue in population health research and interventions is understanding what will be acceptable to the community in need. Cheikh Ibrahima Niang is a medical anthropologist from Senegal who studied community attitudes toward Ebola during the 2014 epidemic [62]. He noted:

» I knew this was a very big challenge but on arriving in Kailahun (in Eastern Sierra Leone), I found the situation was worse than I thought. Njala, the most Ebola-affected village, was a ghost town. Ebola had killed more than 40 of its residents, nearly a third of the village. Most of the remaining people had fled. Houses were closed, there were a lot of orphans and there was nothing to eat. No one wanted to bring them food, too scared of this unknown, deadly disease. It was very hard.

He then stated:

» We knew that if we could better understand people's beliefs—about Ebola, about the response, about treatment centres, about safe burials—lives would be saved. We realized that communities needed to be more included in making decisions such as where to locate Ebola treatment centres. And most importantly, we sensed that people needed to be heard. Before you can create effective messages, you have to listen first. And even then, communities will translate medical messages into their own terms. You have to give them the knowledge that gives them power to make their own decisions. So we did a lot of listening, without making any noise.

Eventually, he and his team realized:

» It became clear that resistance was a way people affirmed their position when their dignity felt threatened. Nobody wants to die from Ebola. When villagers said, 'Ebola doesn't exist. Ebola is a poison that Westerns are sending us.' You say 'ok,' and then you learn that they do not like the way they have been treated.

» Once heard and understood, communities felt reassured, violence diminished.

(a) **What ethical or scientific concerns might exist if a similar experimental therapy existed for Zika?**
The key difference in the case of the Zika virus is that the primary physical harm is incurring to the developing fetus. Thus, the target population for study of an intervention (e.g., a vaccine to prevent infection or a therapy to ameliorate the neurotropic effects of the virus) would be pregnant people. Notably, consent for research into these two potential interventions is treated differently: A pregnant person who wanted to participate in a vaccine study could consent to do so if she were the target of the intervention, with no input from the other biological parent, while the same patient would need consent from the other parent to enroll in a study of an intervention targeting the fetus [63].

> **Box 23.6   Teaching Tip**
> If learners do not get to this issue on their own, consider asking directly: "Pregnant women have been historically classified as a 'vulnerable' population in research. Do you think this classification is justified ethically?"

Historically, pregnant people have been classified as a "vulnerable" population in research [64, 65]. This classification has not only sharply limited the practice of evidence-based medicine in pregnancy, but it is also thought to have negatively impacted overall enrollment of women in clinical research, as some studies simply excluded any women who could physiologically become pregnant [66]. Failure to perform high-quality research in a given population incurs harm to that population [67]. Instead of being given the option of participation in research and receiving the subsequent benefits of better medicine developed through research, they are effectively participating in an ongoing, uncontrolled experiment without giving consent. That is, when physicians and other healthcare practitioners treat patients based on tradition, anecdote, or experience because of a lack of data to support best practices, the risks may be equal or greater to the risks of participating in a study.

Furthermore, this classification has been critiqued as demeaning and promoting bias: For all other "vulnerable" populations, the vulnerability lies in the potential research participant's inability to perceive voluntary informed consent, either because of age (children), intellectual impairment, or subjugated status (prisoners, chiefly, but also military recruits, employees of organizations conducting the research, etc.) [68]. These features are not applicable to pregnant people, who are as capable of assessing risks and benefits of research participation as of any other medical intervention. (▶ Chapter 7 explores the ethics of consent in pregnancy in more detail.)

In the case of a Zika vaccine, a feasible option is to test the vaccine in nonpregnant adults first; moving to pregnant adults and then children after safety and preliminary efficacy is established while recognizing that safety and efficacy often vary between populations [69]. Any assessment of the efficacy of a treatment for previously acquired Studies on CZS, however, would by definition need to be done in pregnant adults. Although the first-in-human studies could be done in healthy nonpregnant adults, efficacy would not be relevant.

Several months later, you see Mr. Oliveira outside the pediatric intensive care unit. Ana Lucia has been admitted with influenza and staphylococcal pneumonia. He is distraught that she may need to be intubated and mechanically ventilated, but all the ventilators are currently in use.

✓ 15. **Creative problem-solving:** Like the situation with ZMapp discussed above, this is a realistic example of a situation in which bedside rationing of care may occur. If you were working in this intensive care unit, how would you decide which children get placed on ventilators?

> **Box 23.7 Teaching Tip**
> Consider using prompts (e.g., age, prognosis, odds of surviving without a ventilator vs. odds of surviving even with a ventilator, other comorbidities, cognitive function) and asking learners to argue for or against consideration of each. Alternatively, assign them to work together to develop and defend a written plan for rationing, outside of the small group session.

Some families of children with microcephaly, particularly in high-income countries, have expressed dismay that the condition is viewed by the media and general public as a catastrophe that warrants prevention at all costs. One parent noted that it is a "little bit hurtful" to know that people look at her children and say "Let's never let this happen to anyone ever again" [10].

✓ 16. **Self-reflection:** Is this relevant? To the patient in this case, to you as a physician, or to public policy?
   (a) **Creative problem-solving:** What would you do with this information? As a physician? As a policymaker?

References

1. Diniz D, Gumieri S, Bevilacqua BG, Cook RJ, Dickens BM. Zika virus infection in Brazil and human rights obligations. Int J Gynecol Obstet. 2017;136(1):105–10.
2. Ruibal A. Social movements and constitutional politics in Latin America: reconfiguring alliances, framings and legal opportunities in the judicialisation of abortion rights in Brazil. Contemp Soc Sci. 2015;10(4):375–85.
3. Faúndes A. Unsafe abortion–the current global scenario. Best Pract Res Clin Obstet Gynaecol. 2010;24(4):467–77.
4. Faúndes A, Duarte GA, Neto JA, de Sousa MH. The closer you are, the better you understand: the reaction of Brazilian obstetrician-gynaecologists to unwanted pregnancy. Reprod Health Matters. 2004;12:47–56.
5. Heino A, Gissler M, Apter D, Fiala C. Conscientious objection and induced abortion in Europe. Eur J Contracept Reprod Health Care. 2013;18(4):231–3.
6. Fiala C, Gemzell Danielsson K, Heikinheimo O, Guðmundsson JA, Arthur J. Yes we can! Successful examples of disallowing 'conscientious objection' in reproductive health care. Eur J Contracept Reprod Health Care. 2016;21(3):201–6.
7. Higgs S, Morton EW. Funding by crisis is no cure for global health threats [Internet]. Illinois: American Society of Tropical Medicine and Hygiene; 2016 Apr 11 [cited 2018 Aug 2]. Available from: https://www.astmh.org/blog/april-2016-(1)/funding-by-crisis-is-no-cure-for-global-health-thr.
8. Fauci AS. Ebola—underscoring the global disparities in health care resources. N Engl J Med. 2014;371(12):1084–6.
9. PREVAIL II Writing Group. A randomized, controlled trial of ZMapp for Ebola virus infection. N Engl J Med. 2016;375(15):1448–56.
10. Itkowitz C. What this amazing mom of two girls with microcephaly has to say about the Zika scare. The Washington Post [Internet]. 2016 Feb 3 [cited 2018 Aug 2]. Available from: https://www.washingtonpost.com/news/inspired-life/wp/2016/02/03/what-this-amazing-mom-of-two-girls-with-microcephaley-has-to-say-about-zika-scare/?utm_term=.f76e045372cc.
11. World Health Organization. Zika virus disease [Internet]. Geneva: WHO. [cited 2018 Aug 2]. Available: http://www.who.int/emergencies/diseases/zika/en/.
12. World Health Organization. List of blueprint priority diseases [Internet]. Geneva: WHO. [cited 2018 Aug 2]. Available from: http://www.who.int/blueprint/priority-diseases/en/.
13. World Health Organization. Zika virus classification tables [Internet]. Geneva: WHO. [cited 2018 Aug 2]. Available from: http://www.who.int/emergencies/zika-virus/classification-tables/en/.
14. Pan American Health Organization. Zika cumulative cases [Internet]. Washington, D.C. Regional Office for the Americas of the World Health Organization. [cited 2018 Aug 2]. Available from: https://www.paho.org/hq/index.php?option=com_content&view=article&id=12390&Itemid=42090&lang=en.
15. Centers for Disease Control and Prevention. Potential range in US [Internet]. Atlanta (GA): CDC. [cited 2018 Aug 2]. Available from: https://www.cdc.gov/zika/vector/range.html.
16. Centers for Disease Control and Prevention. Zika cases in the United States [Internet]. Atlanta (GA): CDC. [cited 2018 Aug 2]. Available from: https://www.cdc.gov/zika/reporting/case-counts.html
17. Centers for Disease Control and Prevention. Transmission methods [Internet]. Atlanta (GA): CDC. [cited 2018 Aug 2]. Available from: https://www.cdc.gov/zika/prevention/transmission-methods.html.
18. Plourde AR, Bloch EM. A literature review of Zika virus. Emerg Infect Dis. 2016;22(7):1185–92.

19. Wen Z, Song H, Ming GL. How does Zika virus cause microcephaly? Genes Dev. 2017;31:849–61.
20. Centers for Disease Control and Prevention. Microcephaly and other birth defects [Internet]. Atlanta: CDC. [cited 2018 Aug 2]. Available https://www.cdc.gov/zika/healtheffects/birth_defects.html.
21. Centers for Disease Control and Prevention. Zika and pregnancy: pregnant women with any laboratory evidence of possible Zika virus infection, 2015–2018 [Internet]. Atlanta: CDC. [cited 2018 Aug 2]. Available from: https://www.cdc.gov/pregnancy/zika/data/pregwomen-uscases.html.
22. Shapiro-Mendoza CK, Rice ME, Galang RR, et al. Pregnancy outcomes after maternal Zika virus infection during pregnancy- US Territories, January 1 2016-April 25, 2017. Morb Mortal Wkly Rep. 2017;66:615–21.
23. Oliveira WK, Franca GVA, Carmo EH, Duncan BB, Kuchenbecker RS. Schmidt MI. Infection-related microcephaly after the 2015 and 2016 Zika virus outbreaks in Brazil: a surveillance-based analysis. Lancet. 2017;390:861–70.
24. Jaenisch T, Rosenberger KD, Brito C, Brady O, Brasil P, Marques T. Risk of microcephaly after Zika virus infection in Brazil, 2015 to 2016. Bull World Health Organ. 2017;95:191–8.
25. Rodriguez LC, Paixao ES. Risk of Zika-related microcephaly: stable or variable? Lancet. 2017;390:824–6.
26. Oduyebo T, Polen KD, Walke HT, et al. Update: interim guidance for health care providers caring for pregnant women with possible Zika virus exposure—United States (including US territories), July 2017. MMWR. 2017;66(29):781.
27. Oster AM, Brooks JT, Stryker JE, Kachur RE, Mead P, Pesik NT, Petersen LR. Interim guidelines for prevention of sexual transmission of Zika virus—United States, 2016. MMWR. 2016;65(5):1–2.
28. Chen T-H, Staples JE, Fischer M. Zika. In: CDC yellow book. Atlanta (GA): CDC; updated 2017 May 31 [cited 2018 Aug 2]. Available from: https://wwwnc.cdc.gov/travel/yellowbook/2018/infectious-diseases-related-to-travel/zika.
29. Centers for Disease Control and Prevention. Travelers' health: Zika travel information [Internet]. Atlanta (GA): CDC. [cited 2018 Aug 2]. Available from: https://wwwnc.cdc.gov/travel/page/zika-information.
30. Centers for Disease Control and Prevention. Zika: a traveler's guide. Atlanta (GA): CDC. [cited 2018 Aug 2]. Available from: https://wwwnc.cdc.gov/travel/files/zika-travel-brochure-508.pdf.
31. Lee MJ, Edara RS, Clark PA, Myers AT. Zika virus: can artificial contraception be condoned? Int J Infect Dis. 2016;15(1). Available from: https://philpapers.org/rec/LEEZVC.
32. Imperato PJ. The convergence of a virus, mosquitoes, and human travel in globalizing the Zika epidemic. J Community Health. 2016;41(3):674–9.
33. Atkinson GM. Humanae vitae, rape and the Zika virus: five remarks. Nat Cathol Bioethics Q. 2016;16(2):209–14.
34. Guttmacher Institute. Religion and family planning tables [Internet]. New York: The Institute. [cited 2018 Aug 2]. Available from: https://www.guttmacher.org/religion-and-family-planning-tables.
35. Joseph A. Breaking down what the pope's nod to birth control means for public health. STAT News [Internet]. Boston (MA); 2016 Feb 18 [cited 2018 Aug 2]. Available from: https://www.statnews.com/2016/02/18/pope-francis-zika-contraception/.
36. Coleman RG. Pope Francis and the Zika Virus. Health Care Ethics USA. 2016;24(2):1–6.
37. Olowu D. Responses to the global HIV and AIDS pandemic: a study of the role of faith-based organisations in Lesotho. SAHARA-J: J Soc Asp HIV/AIDS. 2015;12(1):76–86.
38. US policy statement for the international conference on population. Popul Dev Rev. 1984;10(3):574–9.
39. Starrs AM. The Trump global gag rule: an attack on US family planning and global health aid. Lancet. 2017;389(10068):485–6.
40. Jones KM. Contraceptive supply and fertility outcomes: evidence from Ghana. Econ Dev Cult Chang. 2015;64(1):31–69.
41. Rather IA, Kumar S, Bajpai VK, Lim J, Park YH. Prevention and control strategies to counter Zika epidemic. Front Microbiol. 2017;8:305.
42. Wilder-Smith A, Gubler DJ, Weaver SC, Monath TP, Heymann DL, Scott TW. Epidemic arboviral diseases: priorities for research and public health. Lancet Infect Dis. 2017;17(3):e101–6.
43. Bhatt S, Weiss DJ, Cameron E, Bisanzio D, Mappin B, Dalrymple U, et al. The effect of malaria control on Plasmodium falciparum in Africa between 2000 and 2015. Nature. 2015;526(7572):207.
44. Fernandes JN, Moise IK, Maranto GL, Beier JC. Revamping mosquito-borne disease control to tackle future threats. Trends Parasitol. 2018;34:359–68.
45. Richner JM, Diamond MS. Zika virus vaccines: immune response, current status, and future challenges. Curr Opin Immunol. 2018;53:130–6.
46. Gaudinski MR, Houser KV, Morabito KM, Hu Z, Yamshchikov G, Rothwell RS, et al. Safety, tolerability, and immunogenicity of two Zika virus DNA vaccine candidates in healthy adults: randomised, open-label, phase 1 clinical trials. Lancet. 2018;391(10120):552–62.
47. Modjarrad K, Lin L, George SL, Stephenson KE, Eckels KH, De La Barrera RA, et al. Preliminary aggregate safety and immunogenicity results from three trials of a purified inactivated Zika virus vaccine candidate: phase 1, randomised, double-blind, placebo-controlled clinical trials. Lancet. 2018;391(10120):563–71.
48. Weaver SC, Charlier C, Vasilakis N, Lecuit M. Zika, Chikungunya, and other emerging vector-borne viral diseases. Annu Rev Med. 2018;69:395–408.
49. Bousquet G, Orri M, Winterman S, Brugière C, Verneuil L, Revah-Levy A. Breaking bad news in oncology: a metasynthesis. J Clin Oncol. 2015;33(22):2437–43.
50. Greiner AL, Conklin J. Breaking bad news to a pregnant woman with a fetal abnormality on ultrasound. Obstet Gynecol Surv. 2015;70(1):39–44.
51. UNICEF State of the World's Children and Childinfo, United Nations Population Division. World contraceptive use, household surveys: contraceptive prevalence, any methods (% women aged 15–49) [Internet]. Washington, D.C.: World Bank Group. [cited 2018 Aug 2]. Available from: https://data.worldbank.org/indicator/SP.DYN.CONU.ZS?view=chart.

52. Department of Economic and Social Affairs, Population Division, United Nations. Trends in contraceptive use worldwide 2015 [Internet]. New York: UN; 2015 [cited 2018 Aug 2]. Available from: http://www.un.org/en/development/desa/population/publications/pdf/family/trendsContraceptiveUse2015Report.pdf.
53. Grimes DA, Benson J, Singh S, Romero M, Ganatra B, Okonofua FE, Shah IH. Unsafe abortion: the preventable pandemic. Lancet. 2006;368(9550):1908–19.
54. Schuklenk U, Smalling R. Why medical professionals have no moral claim to conscientious objection accommodation in liberal democracies. J Med Ethics. 2017;43(4):234–40.
55. Reyna VF, Nelson WL, Han PK, Dieckmann NF. How numeracy influences risk comprehension and medical decision making. Psychol Bull. 2009;135(6):943.
56. Wheeler A. Development of infants with congenital Zika syndrome: what do we know and what can we expect? Pediatrics. 2018;141(s2):S154–60.
57. Hicks SD, Wang M, Fry K, Doraiswamy V, Wohlford EM. Neurodevelopmental delay diagnosis rates are increased in a region with aerial pesticide application. Front Pediatr. 2017;5:116.
58. Baer H, Singer M. Global warming and the political ecology of health: emerging crises and systemic solutions. London: Routledge; 2016.
59. Jamrozik E, Selgelid MJ. Ethics, climate change and infectious disease. In: Bioethical Insights into Values and Policy. Cham, Switzerland: Springer; 2016. p. 59–75.
60. Rid A, Emanuel EJ. Ethical considerations of experimental interventions in the Ebola outbreak. Lancet. 2014;384(9957):1896–9.
61. Joffe S. Evaluating novel therapies during the Ebola epidemic. JAMA. 2014;312(13):1299–300.
62. Niang CI. Ebola diaries: lessons in listening [Internet]. Geneva: WHO; 2015 [cited 2018 Aug 2]. Available from: http://www.who.int/features/2015/ebola-diaries-niang/en/.
63. Institutional Review Board for Health Sciences Research. Vulnerable subjects – pregnant women [Internet]. Charlottesville (VA): University of Virginia. [cited 2018 Aug 2]. Available from: http://www.virginia.edu/vpr/irb/hsr/vulnerable_pregnancy.html.
64. Mastroianni AC, Henry LM, Robinson D, Bailey T, Faden RR, Little MO, Lyerly AD. Research with pregnant women: new insights on legal decision-making. Hastings Cent Rep. 2017;47(3):38–45.
65. Saenz C, Cheah PY, van der Graaf R, Henry LM, Mastroianni AC. Ethics, regulation, and beyond: the landscape of research with pregnant women. Reprod Health. 2017;14(S3) https://doi.org/10.1186/s12978-017-0421-3.
66. Merton V. The exclusion of pregnant, pregnable, and once-pregnable people (a.k.a. women) from biomedical research. Am J Law Med. 1993;19:369–451.
67. Nijjar SK, D'amico MI, Wimalaweera NA, Cooper NA, Zamora J, Khan KS. Participation in clinical trials improves outcomes in women's health: a systematic review and meta-analysis. BJOG. 2017;124(6):863–71.
68. Hurst SA. Vulnerability in research and health care; describing the elephant in the room? Bioethics. 2008;22(4):191–202.
69. Marston HD, Lurie N, Borio LL, Fauci AS. Considerations for developing a Zika virus vaccine. N Engl J Med. 2016;375(13):1209–12.

Further Reading on This Topic

Barrat J. Spillover: Zika, Ebola and beyond [Internet]. Frontline. Arlington: PBS; 2016. Available from: http://www.pbs.org/spillover-zika-ebola-beyond/home/.

Camus A. The plague. New York: Vintage; 2012.

Deniz D. Zika, the film [Internet]. Paraíba, Brazil; 2016 Apr 7. Available from: https://www.youtube.com/watch?v=j9tqt0jaoG0.

Farmer P. Infections and inequalities: the modern plagues. Berkeley/Los Angeles: University of California Press; 2001.

"I Just Want to Help People and See the World"

Andrea V. Shaw

24.1 Background Questions – 478

24.2 Additional Case Information and Questions for Discussion – 479

24.3 Answers to Background Questions – 480

24.4 Responses to Discussion Questions – 484

References – 491

■ **Glossary**

Medical mission—Typically used to describe faith-based endeavors that involve administration of medical services.

Service-learning—A teaching and learning strategy that integrates meaningful community service with instruction and reflection to enrich the learning experience, teach civic responsibility, and strengthen communities (National Service Learning Clearinghouse).

Short-term experience in global health (STEGH)—An experience offered by organizations based in high-income countries (HICs) to participate in a health-related experience (research, service, and/or education) in a lower-middle-income country (LMIC).

Short-term medical mission (STMM)—A similar term describing experiences offered by organizations based in HICs to participate in a health-related experience (research, service, and/or education) in a LMIC.

Voluntourism—A combination of volunteering and tourism that allows an individual to enjoy an experience in international travel while contributing to a local project.

Learning Objectives

1. Understand the potential benefits and risks to all stakeholders involved in a short-term experience in global health (STEGH).
2. Learn to assess the ethics of potential global health experiences and develop a plan for how to undertake a trip in such a way to address global health inequity appropriately.
3. Appraise the obligations of health care practitioners to address health inequalities locally and globally.

> **Case Background**
>
> *You are in your first year of medical school. A family friend tells you that her faith community is organizing a medical mission this summer for a month in a lower-middle-income country (LMIC) by World Bank criteria. The mission will include two doctors, two nurses, a pharmacist, social worker, and three other professionals who are not in the health field. The team will also include five students, and she wonders if you would be interested. The team will work at a clinic that is affiliated with a local church. The participants must pay for their transportation costs, but the organization will pay all the other expenses.*

24.1 Background Questions

> **Box 24.1   Teaching Tip**
>
> For the first five questions, learners should consider a low-income country (LIC) or lower-middle-income country (LMIC) that they have visited, that they plan to visit, or that otherwise interests them. Alternatively, all learners might decide together on a single country to research, or the educators may choose to assign a specific country or countries.

1. Describe the health care system in the LMIC you have chosen. Identify what types of health services are available in public and private facilities.

2. Identify five key health-related indicators (e.g., infant mortality, maternal mortality, under age 5 child mortality, life expectancy, traffic fatality rate, incidence or prevalence of significant diseases, etc.). Compare those indicators for your chosen country to those of the United States. What do these indicators show about health in your chosen country?

3. Describe an example of health inequalities in your chosen country.

4. What is universal health coverage? What is the basic structure by which citizens of your chosen country pay for health care? Has universal health coverage been achieved in the country?

5. The U.S. Centers for Disease Control and Prevention's (CDC) Travelers' Health website contains information about recommended vaccines, a "Healthy Travel" health packing list, and travel health notices. Identify key recommendations for travelers to your chosen country. Does this country have a travel notice related to health? If so, describe that notice.

6. What are community health workers? Find and describe an example of how community health workers have improved health in a LMIC.

7. What role do faith-based organizations (FBO) play in many LMIC health care systems? What role do they play in short-term experiences in global health (STEGH) and short-term medical missions (STMMs)?

8. Choose a nongovernmental organization that is active in global health (for instance, Médecins Sans Frontières). What are key

principles of that organization? What is their role in the countries in which they work?

24.2 Additional Case Information and Questions for Discussion

This is your first trip outside of the continental United States; you are very excited. You only have 1 month off in between your first and second year of medical school. The mission team plans to spend the first week in the capital city to adjust to the local time and explore the city before going out to the more remote area where the affiliate's church and clinic are located. The team will then spend 2 weeks serving at the clinic. The final week of the trip will involve a chance to explore local wildlife on a short safari.

1. How should you prepare for a trip like this? If you have had peers go on trips like this, how did they prepare?

2. Self-reflection: What would you want to know about yourself before committing to the trip?

You are excited about the trip because you know that you will learn a great deal as this is your first experience traveling out of the country. You are aware of great inequalities in the world that lead to much lower life expectancy in LMIC countries. You feel a strong desire to "help" in any way that you can. As a student, you have few skills and a limited knowledge base, but you feel that giving "something will be better than nothing," so you are committed to the work. You are not certain if your future career in medicine will involve work outside of the United States.

3. What would you like to know about the mission before committing to the trip? Consider the interests of all stakeholders involved.

You recognize that you have not had time to study the language, history, or culture beyond reading a travel guide. You would like to learn more about the country, but you are busy with end-of-year exams. Immediately after your last exam, you are attending a friend's wedding and departing for your trip the next day. You also realize that you do not know very much about the partnership between the two churches.

4. Under these circumstances, should you defer your trip? Why or why not?
   (a) How do you decide if you have committed adequate time to prepare for work abroad?
   (b) How do you decide if this is an appropriate stage in your training for participating in such a trip?
   (c) In answering the above, did you consider the questions in the context of their impact on optimizing your learning from the experience or your capacity to positively impact the health of the community you are visiting? How do you weigh these two perspectives?

5. How would you assess if the project that you are considering is part of a meaningful partnership?
   (a) How can you learn in advance how the community truly perceives the work your group does?

6. What does it mean to describe a project or intervention as "sustainable"? How would you ensure this project is sustainable?

7. Do you think providing medical care should be separated from promoting religion? Why or why not?
   (a) Self-reflection: Does secular medical work also promote a worldview?

When you mention your potential trip to a friend, she excitedly tells you about a third-year medical student who learned to place chest tubes in infants during a disaster relief trip after an earthquake in Central America. Many students and some trainees (e.g., medical residents) go abroad with the expectation that they will be able to provide more "hands-on" care than they are allowed within the U.S. training system. Some fully licensed health care practitioners may provide care outside their usual scope of practice (for instance, a surgeon who typically only operates on adults in their home country may plan to operate on children during the mission trip).

8. Is this ethical? Why or why not?

9. **Creative problem-solving**: Given the rising popularity and growing commercial interest in STEGH, how can organizations that sponsor STEGHs ensure that programs and program participants meet basic ethical standards?

10. If you have this type of opportunity and decide to take it, what are your obligations
    (a) To the community you will be visiting?
    (b) To your colleagues and fellow travelers?
    (c) To your home community and patients?

One of the two doctors going on the trip is a dermatologist who has a private clinic, in which he sees affluent patients who primarily pay out of pocket for his services. This practice has done well and afforded him a very comfortable lifestyle, as well as the financial means to go on trips such as these, which he does every year.

11. What are some of the ethical considerations here? Consider issues such as scope of practice and resource allocation.

12. What is our duty to assist and address health inequity globally? Consider this question as an individual, a group of people, an academic institution, and a nation.

13. **Self-reflection**: Is participating in a short-term global experience something that you previously considered when you think about your future career? If so, how has this case influenced your thoughts about such experiences? If not, are you more or less likely to consider doing so now?

24.3 Answers to Background Questions

> **Box 24.2 Teaching Tip**
> We have provided two sample responses for the first five questions: Kenya and Haiti. If the learners choose one country on which to focus, they are likely to develop a more comprehensive understanding of that particular country's health care system and health outcomes after the case discussion. If they choose different countries, they will emerge with a broader but more superficial understanding of different systems and outcomes. If selecting a single country, it is important to recognize whether any of the learners have personal experience with that country (e.g., born there, have family living there, visit regularly, etc.) and to acknowledge that, while data regarding health systems and outcomes may seem overwhelmingly negative, it does not reflect the richness of lives lived in such countries.

1. **Describe the health care system in the LMIC you have chosen. Identify what types of health services are available in public and private facilities.**
   Kenya: Health care is provided through a combination of public, private for-profit, and private not-for-profit facilities and is financed through taxation, insurance, and direct payment (out-of-pocket) [1]. Health care services are arranged in tiers running from level 1 (dispensary, the lowest level of care) to level 6 (referral hospitals, the highest level of care) [1]. The majority of Kenyans receive health care through the public sector, which provides a range of support: from basic primary care at the most rural health dispensaries, staffed by public health nurses, to more complex care at larger health centers and district and provincial hospitals, to the highest level of tertiary care at the national hospital. Private hospitals and faith-based organizations may supplement care where available.
   Haiti: Haiti, the poorest country in the Western Hemisphere, has struggled to sustain basic infrastructure, including that of the health care system. Haiti's health care system currently includes a combination of public, private, and not-for-profit facilities. According to the Pan American Health Organization, a branch of the World Health Organization (WHO), Haiti's health system includes 10 health departments, 42 health district units, and more than 900 facilities (38% public, 42% private, and 20% mixed) [2]. After a devastating earthquake in 2010, the public sector developed a community health care approach. Under this approach, health centers employ community

workers who are supervised by family health teams, which include physicians and nurses. After a pilot study demonstrated improvement in health outcomes, this approach was expanded throughout the country [2]. The not-for-profit health organization Partners in Health is very active in Haiti and reports being the largest nongovernmental health agency in the country with more than 5700 staff in country [3].

2. **Identify five key health-related indicators (e.g., infant mortality, maternal mortality, under age 5 child mortality, life expectancy, traffic fatality rate, incidence or prevalence of significant diseases, etc.). Compare those indicators for your chosen country to those of the United States. What do these indicators show about health in your chosen country?**
Health-related indicators are measures that are used to describe the health of a population. The data in Table 24.1 show how the health care system is or is not able to address major causes of morbidity and mortality within a population using existing resources [4–6]. For example, measles vaccine coverage rates can inform stakeholders about the effectiveness of the health system's vaccine delivery system. Human immunodeficiency virus/acquired immunodeficiency syndrome (HIV/AIDS) mortality rates reflect both prevalence of disease and access to treatment. Both Kenya and Haiti have low life expectancy and high maternal mortality and HIV-/AIDS-related mortality rates compared to the United States. The United States has higher rates of diabetes and tobacco use.

3. **Describe an example of health inequities in your chosen country.**
The WHO defines health inequities as "systematic differences in the health status of different population groups that have significant social and economic costs both to individuals and societies." [7]
Kenya: Health inequities in Kenya involve disparities in access to services for those who are poor and live rurally, including preventive care (e.g., vaccines), health promotion (e.g., prenatal vitamins), and supportive services (e.g., facility with the capability of performing a cesarean delivery when indicated). For example, people from the northeastern province of Nyanza have a lower life expectancy, higher child mortality rate, and higher rate of HIV compared to people living in the central provinces, closer to the capital. [8]
Haiti: There are extreme health inequities in Haiti for those who are poor and who live in a rural setting. In addition to lacking access to health care, poor Haitians also have limited access to education and to clean water. As in most countries,

Table 24.1 Key health-related indicators comparing Kenya, Haiti, and the United States. For comparability, all of the data are the most current data available from the World Health Organization's country statistical profile as well as the WHO's country profiles on diabetes and tobacco [4–6]

	Kenya	Haiti	United States
Life expectancy in years	61	63	79
Maternal mortality per 100,000 live births	400	157	28
Deaths due to HIV/AIDS per 100,000 population	126	73	2.5
Prevalence of diabetes (%)	4.0	6.9	9.1
Measles immunization coverage (%)	93	64	91
Adult tobacco smoking (%)	Men: 19.7　Women: 0.9	Men: 13.6　Women: 5.0	Men: 22.1　Women: 15.5

HIV human immunodeficiency virus, *AIDS* acquired immunodeficiency syndrome educational achievement is strongly associated with positive health outcomes, as are environmental exposures, such as clean water. (This is also the case in Kenya.) According to the Pan American Health Organization, 88% of the urban population has access to safe drinking water compared to only 49% in rural areas [2]. These differences in environmental health contributed to a devastating cholera epidemic that occurred after the 2010 earthquake in Haiti.

4. **What is universal health coverage? What is the basic structure by which citizens of your chosen country pay for health care? Has universal health been achieved?**
The World Health Organization defines universal health coverage as meaning that "all people and communities can use the promotive, preventive, curative, rehabilitative and palliative health services they need, of sufficient quality to be effective, while also ensuring that the use of these services does not expose the user to financial hardship." [9]
Kenya: Kenya has not achieved universal health coverage. Shortly after independence in 1963, Kenya's stated policy was to provide free health care for all through the public sector; however, in the ensuing decades, user fees dramatically increased. Out-of-pocket expenses have been a significant source of health care financing and have created a barrier to health care for much of Kenya's population [1]. In general, the public sector provides a "national health card" to those who are employed, which then allows people to obtain subsidized services at public health care facilities. Those who have private health insurance or other resources generally seek care at private facilities.
Haiti: Haiti has not achieved universal health coverage. Access to care is significantly impacted by the individuals' ability to pay. As described in Background Question 1, Haiti's health care system is fragmented and financed by a combination of governmental and nongovernmental aid, with a high proportion of donor financing. Haiti's health indicators are disproportionately worse than other countries in the region. In 2017, the World Bank issued a report about the status of health care spending in Haiti and found that government funding had plummeted as a proportion of health expenditure. This sharp decline was attributed to increases in international aid after the 2010 earthquake [10]. The report also identified the systemic underfunding of primary care, despite a stated interest in community health centers (as described in Background Question 1) [10].

5. **The Centers for Disease Control and Prevention's Traveler's Health website contains information about recommended vaccines, a "Healthy Travel" health packing list, and travel health notices. Identify key recommendations for travelers to your chosen country. Does this country have a travel notice related to health? If so, describe that notice. While the CDC posts notices related to health concerns, the U.S. state department is the best resource for travel notices related to security concerns.**
Kenya: The CDC recommends that travelers to Kenya are up-to-date on routine vaccines, as well as typhoid, hepatitis A, and yellow fever. In addition, there are specific recommendations for how travelers can reduce their risk of acquiring malaria, cholera, yellow fever, meningitis, hepatitis B, polio, and rabies.
Haiti: The CDC recommends that travelers to Haiti are up-to-date on routine vaccines, as well as typhoid and hepatitis A. In addition, there are specific recommendations for how travelers can reduce their risk of acquiring malaria, cholera, hepatitis B, and rabies. As of July 2018, an Alert Level 2: Practice Enhanced Precautions notice was in effect in Haiti as a result of major hurricanes in 2017. This alert provides recommendations to minimize the risk of illness and injury associated with the damaged infrastructure caused by the storms.

6. **What are community health workers? Find and describe an example of how community health workers have improved health in a LMIC.**

The American Public Health Association (APHA) defines a community health worker as a "frontline public health worker who is a trusted member of and/or has an unusually close understanding of the community served. This trusting relationship enables the worker to serve as a liaison/link/intermediary between health/social services and the community to facilitate access to services and improve the quality and cultural competence of service delivery." [11]

In Kenya, community health workers receive some training in health by the Ministry of Health in Kenya. They are trained to recognize signs and symptoms that may present at home and those that warrant getting to a health care facility for further support. They are armed with certain early treatments—oral rehydration salts for dehydration, bed nets, etc. They often share information about preventative health. Unfortunately, there is no uniform mechanism of support for these community health workers, and local health facilities support them differently in their roles. Locally, people find them an invaluable resource to assist in triaging and managing of health concerns in the community.

7. **What role do faith-based organizations (FBO) play in many LMIC health care systems? What role do they play in short-term experiences in global health (STEGH) and short-term medical missions (STMM)?**
Historically, faith-based organizations have played a major role in the health care sector. Most of the first hospitals and clinics were founded by religious organizations hundreds of years ago [12]. Across the globe, FBOs have significantly contributed to health improvement in many ways, including by increasing capacity through infrastructure, providing direct care services, engaging in community outreach (particularly for vulnerable populations), and advocating for policy change [12]. The role of FBOs is of particular importance in many LMICs, in which the government is less likely to have resources needed to meet the health care needs of the populations they serve. A recent systematic review and meta-analysis of FBO participation in health care delivery described the broad range of FBO involvement, varying based upon on the country and the type of health care (inpatient, outpatient, and so forth). For example, the study concluded that for inpatient care the range of FBO participation ranged from 0 to 45% [13].

With respect to STEGH and STMM, such programs are sponsored by educational institutions, FBOs, secular nongovernmental organizations (NGOs), and even for-profit organizations. The United States, Canada, the United Kingdom, and Australia are the four leading "senders" of participants in STMMs. [14] Historically, the majority of STMMs and STEGHs were organized by FBOs to address the unmet health care needs of the host country; however, secular organizations play an increasingly prominent role in such missions. Almost 50% of medical schools in the United States have global health programs in education, service, and/or research [15]. Although the presence of STMMs and STEGHs has grown, their measurable impact on improving long-term health outcomes has not been well-studied [16].

8. **Choose a nongovernmental organization that is active in global health (for instance, Médécins Sans Frontières). What are key principles of that organization? What is their role in the countries in which they work?**
Médécins Sans Frontières (MSF) is an independent international medical humanitarian organization that focuses on delivering emergency aid to those affected "by conflict, epidemics, disasters, or exclusion from health care." [17] MSF enters an area of devastation, focusing their attention at stabilization (e.g., performing first aid, triaging sick care, treating malnutrition), but they do not focus their efforts on development of local health resources. MSF states that their "actions are guided by medical ethics and the principles of impartiality, independence and neutrality." [17]

24.4 Responses to Discussion Questions

This is your first trip outside of the continental United States; you are very excited. You only have 1 month off in between your first and second year of medical school. The mission team plans to spend the first week in the capital city to adjust to the local time and explore the city before going out to the more remote area where the affiliate's church and clinic are located. The team will then spend 2 weeks serving at the clinic. The final week of the trip will involve a chance to explore local wildlife on a short safari.

✅ 1. **How should you prepare for a trip like this? If you have had peers go on trips like this, how did they prepare?**
   James Dwyer's essay, "Teaching Global Health Ethics," poses eight questions to students, trainees, and other learners interested in participating in short-term experiences in global health [18]:
   1. Have I studied adequately the language of the people I will be working with?
   2. Have I prepared by studying the history, culture, and social structures of the society I will be working in?
   3. Have I committed adequate time for my work abroad?
   4. Am I going at an appropriate stage in my training?
   5. Is the project that I am considering part of a meaningful partnership that is based on respect?
   6. Are the benefits and burdens of the project fairly distributed?
   7. In addition to providing clinical care, am I working with people to remedy structural injustices?
   8. Do I want to go abroad primarily for the adventure and feeling of altruism, or am I concerned enough about health and justice to also work in my own country to change patterns that impede better global health?

> **Box 24.3 Teaching Tip**
> If learners are quiet, ask them to read and consider each of the eight questions, reflecting on how they might practically obtain the necessary information and how it might affect their decision-making and planning.

In addition to reflecting on Dwyer's eight questions, students and trainees considering such a trip should prepare by finding as much information about the country as possible. It is helpful to learn about current events not only from news reports but also from people with knowledge about the country. It is also very important that they understand the history of the destination country: What impact (positive or negative) has their home country had in the history of this country? They will be entering this country as a foreigner and representative of their home country, so this is crucial information.

Trainees should also devote time to understanding the local health system, available resources, and how volunteers and their organization fit into the context of the local community. They should commit to learning how their time there would be most beneficial to the welfare of the community they are visiting, taking into account the priorities of the local population and the sustainability of the project.

Additionally, they should be aware of potential hazards that may present during the trip. Reviewing sources such as the CDC and U.S. State Department websites (see Background Question 5 above) will be helpful.

✅ 2. **Self-reflection: What would you want to know about yourself before committing to the trip?**

> **Box 24.4 Teaching Tip**
> Encourage learners to consider the last of Dwyer's eight questions, and take time to understand their motivations for such a trip.

"I Just Want to Help People and See the World"

You are excited about the trip because you know that you will learn a great deal as this is your first experience traveling out of the country. You are aware of great inequalities in the world that lead to much lower life expectancy in LMIC countries. You feel a strong desire to "help" in any way that you can. As a student, you have few skills and a limited knowledge base, but you feel that "something will be better than nothing," so you are committed to the work. You are not certain if your future career in medicine will involve work outside of the United States.

3. **What would you like to know about the mission before committing to the trip? Consider the interests of all stakeholders involved.**

   Trainees considering such a mission should assess whether this trip has benefit to all parties involved. It is not ethical to form a unilateral agenda on this trip—that is, volunteers should not impose their own values and priorities on the community they intend to serve.

   Other stakeholders and possible interests include:
   - *The church or other philanthropic group sponsoring the mission.* What role, if any, does promotion of specific values or beliefs play in the mission? (This question applies regardless of whether the sponsoring organization is faith-based.) For missions supported by an FBO, an additional consideration is the importance of connecting with others globally around one's faith. Volunteers may feel that this experience will build stronger ties among their local faith communities as well.
   - *The partner church in the destination country.* How are their interests similar or different from the above?
   - *The community in the destination country.* The local community is likely to be concerned with benefits to the entire community and possibly to local health care systems, not just to individuals. While they may be friendly and open about sharing their culture and heritage, they deserve a balance of respect, support, and opportunity equivalent to what they grant to visitors. If this trip takes away time and energy from already limited local resources (e.g., by pulling multilingual individuals away from other roles to serve as interpreters) or has downstream unintended consequences for patients and communities, there may be an imbalance of risk for the local population.
   - *The individual patients and families who will receive services.* It is easy to assume that they simply want good health care for themselves and their families, but the services rendered must be delivered in a sustainable way to be meaningful and ensure a balance of benefit over harm [19].

Box 24.5 Personal Perspective

Several qualitative studies have been undertaken to better understand how short-term global health work is viewed by recipients and other stakeholders. In one such study, in Guatemala, several themes emerged, including the desire for more respect and better coordination between local and visiting physicians, as well as sharing of resources and equipment, and the inability of short-term work to address root causes of poor health [19]. The following two quotes are illustrative:

» [Foreign] surgical teams only work on the tip of the iceberg when it comes to addressing the medical problems of this country. The problems of Guatemala—corruption, lack of resources, lack of education—all come from poverty. So poverty is the root of the problem, and surgery does not address poverty. – *Surgeon working at a large national hospital*

» Guatemalan patients, especially those with less education, tend to put more faith in a blonde haired, blue eyed, white skinned foreign physician than their own Guatemalan physicians. These foreigners show up with their shiny new equipment and do their free surgeries without ever working with any of [the Guatemalan physicians]. U.S. doctors come to Guatemala and practice medicine when and where they want. Guatemalan doctors may have a hard time even entering the U.S., let alone being able to practice medicine there. U.S. physicians are not superior to Guatemalans. I am perfectly capable of taking care of my own people. – *Surgeon working at a private clinic and national hospital*

Short-term missions may be effective if they are closely aligned with the goals and needs of the community and if they take steps to promote sustainability in health care for the community. For instance, a health care practitioner might undertake a short-term mission with the goal of training health care workers in the local community to perform a particular procedure. Some programs abroad have sustained collaborations with universities and hospitals in more developed countries; although any one practitioner may visit only for a few weeks to a few months per year, the same practitioners may return year after year, and the relationship between institutions might allow for longitudinal goals to be achieved.

You recognize that you have not had time to study the language, history, or culture beyond reading a travel guide. You would like to learn more about the country, but you are busy with end-of-year exams. Immediately after your last exam, you are attending a friend's wedding and departing for your trip the next day. You also realize that you do not know very much about the partnership between the two churches.

✓ 4. **Under these circumstances, should you defer your trip? Why or why not?**
The student should think about all the information he has gathered about the country and weigh the local needs against the time he wants to commit to the trip. One way to think about this is imagine being in the position of the people the student is going to try to assist and to then determine whether a quick in-and-out trip would be respectful. The student should keep in mind that the best scenario is a situation in which the local people gain as much from visiting health care practitioners as the practitioners learn from them. This framework can help the student to decide whether it is better to proceed with the trip or to postpone it until he is prepared to commit more time to it.

Thus, under the described circumstances, the best answer would probably be affirmative. The trip should be deferred. The student does not have time to adequately prepare and is at a stage of training in which the trip might be interesting and perhaps (depending on how much time he actually spends interacting with local residents) culturally informative, but he is unlikely to gain much medical knowledge (with no time to study local disease processes) and unlikely to contribute much in the way of medical skills (as a first-year medical student). Students in this position may serve both their own goals and the community's goals better by waiting until they are adequately prepared for international work.

(a) **How do you decide if you have committed adequate time to prepare for work abroad?**
It depends on many things including your goals of the trip, your expectations of your role in the mission, and your own experiences.

For instance, a physician or surgeon might undertake a short-term mission with the goal of training providers or health care workers in the local community to perform a particular procedure. Some programs abroad have sustained collaborations with universities and hospitals in more developed countries; although any one physician may visit only for a few weeks to a few months per year, the same physicians may return year after year, and the relationship between institutions might allow for longitudinal goals to be achieved.

Meaningful research or even quality improvement projects will always require significant time commitments, although in some cases, students, trainees, and other learners may be able to establish a strong mentoring relationship with a researcher who has long-term involvement with a particular site. That mentor might then assist the trainee in identifying one piece of a particular research project that can be addressed in a shorter period of time. Crucial to this question is a dialogue between the student or trainee and her project mentor. The mentor should have an established relationship with the international site and have a good understanding of the site's needs, resources, and expectations for visiting students and trainees. It is also important that the mentor work with the trainee to anticipate potential problems that might delay a project or prevent its completion during a short trip.

Ethical issues related to the conduct of research in resource-limited settings are discussed in greater detail in ▶ Chapter 23.

(b) **How do you decide if this is an appropriate stage in your training for participating in such a trip?**
This depends on the goals of the trip and is something a good mentor can help a student or trainee to determine. Does the student feel confident that the skills she has acquired at this stage of training are adequate to provide the service she envisions providing? For example, can she teach in a medical school, work in a clinic alongside local health professionals, or engage in a campaign to raise awareness in a local community about a particular health problem?

The more closely the student's goals align with the goals of the trip and the needs of the local community, the more appropriate the plan will be to pursue. Taking part in health care activities in resource-limited communities can understandably result in stress or guilt for students and trainees who are not adequately prepared for those activities [15].

In addition, it is important to consider the implications of the experience in the context of the learner's career trajectory, particularly for trainees who are further along in their careers.

(c) **In answering the above, did you consider the questions in the context of their impact on optimizing your learning from the experience or on your capacity to positively impact the health of the community you are visiting? How do you weigh these two perspectives?**
They should both be given significant weight, but the latter cannot be achieved without the former. Students and trainees do not represent "free health care." In fact, they often place unexpected burdens on the people, hospitals, and communities that host them, as they do not yet have the necessary skills to practice independently. A crucial thought experiment is this: Think of any local community in the United States that suffers similar problems as a community in a LMIC, and then imagine volunteering in that community with the added caveat of language barriers abroad. What would the student be able to contribute? If the student cannot picture himself contributing locally, it is likely that his contribution abroad would be similarly limited.

5. **How would you assess if the project that you are considering is part of a meaningful partnership?**
Regardless of where one works in the global human community, people all have similar desires to be treated with respect. From that belief, health care practitioners and trainees must be willing to address the local needs that *are*, rather than the needs perceived from afar. It is important that all visiting practitioners be open and willing to learn from the host community—they know far more about their needs than visitors do. This dialogue is essential to a successful project for both parties.

Prior to the trip, it can be difficult to obtain objective information if the student does not have contact with people who have previously been involved in the mission and/or with people living in the local community. However, asking critical questions about the origins of the mission, its evolution, and its current goals can help to better understand the partnership.

Additional questions might include:
- Was there a needs assessment done to engage the local community and understand perceived needs?
- Are objectives for the projects jointly written by the local church and the local community? Are there any efforts to gather feedback on the current mission and to assess impact or outcomes in the local community?
- How does the group decide what services to offer, what to bring, and what supplies to carry? Do the local clinic or church have a say? What about the larger community?
- What other health care services are available in the community and does the mission coordinate with them in any way?
- Are the services available uniformly to anyone in the area or only those who are members of the local church? How are potential patients who are not affiliated with the local church, or of a different religion entirely, treated?

If the project does not prioritize the needs of the local population and have a commitment to their needs driving the process forward, it is not a partnership built on equality and respect.

(a) **How can you learn in advance how the community perceives the work your group does?**
It is difficult to locate unbiased information addressing a community's perceptions. First, in order to advance the mission of its program, the group you are traveling with may only post or share positive feedback it has received. Second, people from the local community may not feel comfortable providing constructive feedback if they perceive that doing so could jeopardize future resources. Opportunities to learn of these perceptions from the local community are not likely to arise until the student or trainee is already immersed in the situation. If feasible, talking with someone who had recently been in the community can help you learn about this.

6. **What does it mean to describe a project or intervention as "sustainable"? How would you ensure this project is sustainable?**
A project or intervention may be considered sustainable if the group has (1) invested in something that meets the needs of the local community (as above) and (2) has invested with local partners in a way that will allow the project or intervention to continue in the absence of external resources (e.g., personnel, supplies, money from outside the community or government). For example, delivering 2 weeks of chronic medications (e.g., antihypertensives for high blood pressure) to a population that does not have access to medications or routine services may potentially cause more harm than good, if there is no plan in place to allow them access to long-term care and treatment. Understanding local resources and future plans for project evaluation and development is essential to sustainable efforts. Capacity-building interventions seek to develop local resources, including knowledge and expertise, in order to achieve sustainable improvement in health care services.

7. **Do you think providing medical care should be separated from promoting religion? Why or why not?**
Religion is instrumental in many people's lives, and FBOs have a significant presence in many STEGHs, as described earlier. Students (and any volunteers, in fact) should understand what, if any, expectations there are about promoting religion before committing to the trip and ensure that those expectations are consistent with their goals.

A service that is needed by any human individual or community must never be tied to whether or not they share in the belief system of the provider. Health is a globally recognized human right and need for all, and those who have the capacity to provide health care should feel the urge to give it without discrimination on the basis of religion or any other basis. While a person's faith may prompt a desire to serve through medical care, and this is a laudable goal, it is unethical to serve only those who subscribe to that faith.

Patients and their families, especially ones who are ill and most especially those with limited resources or options for accessing care, are in deeply vulnerable positions. It is important that mission trips separate medical care from proselytization to assure that patients are able to seek care without feeling an obligation to participate in religious activities. Even if the mission tries to clearly separate medical care from proselytization, patients may not perceive that separation and may feel an obligation to participate in religious activities. Furthermore, all health care practitioners—across all settings—should take care not to assume that their way of practicing their beliefs or religion mirrors that of someone else. Even individuals who purportedly ascribe to the same organized religion or philosophy often have distinct ways of applying those beliefs to their daily lives.

(a) **Self-reflection: Does secular medical work also promote a worldview?**

When you mention your potential trip to a friend, she excitedly tells you about a third-year medical student who learned to place chest tubes in infants during a disaster relief trip after an earthquake in Central America. Many students and some trainees (e.g., medical residents) go abroad with the expectation that they will be able to provide more "hands-on" care than they are allowed with the U.S. training system. Some fully licensed health care practitioners may provide care outside their usual scope of practice (for instance, a surgeon who typically only operates on adults in their home country may plan to operate on children during the mission trip).

8. **Is this ethical? Why or why not?**
   The principle of justice and the duty to assist are overarching themes that drive many health professionals to travel from resource-rich settings to resource-poor settings to serve. There is often a feeling that "*some* help is better than no help," which may or may not be true in a given situation, depending on local circumstances. This approach needs to be balanced by considerations of justice and non-maleficence. Health care practitioners should only deliver care that they are qualified to provide competently, no matter where they are in the world.

   In this case, if a medical student was working alongside a qualified health care practitioner and assisted with placement of a chest tube, then this would have been comparable to a medical student's role in the United States and thus ethically permissible. On the other hand, if the student were to perform this procedure without adequate supervision, this would not be acceptable. If a surgeon who typically operates on adults plans to operate on children during a STEGH, this difference in experience should be recognized and approved through a local ministry of health or an appropriate overseeing agency (such as a body responsible for licensing medical practitioners within that country or overseeing humanitarian organization in situations where local governments are nonfunctional).

9. **Creative problem-solving: Given the rising popularity and growing commercial interest in STEGH, how can organizations that sponsor STEGHs ensure that programs and program participants meet basic ethical standards?**

10. **If you have this type of opportunity and decide to take it, what are your obligations**
    (a) **To the community you will be visiting?** Loh and colleagues proposed a framework that involves more effort on the part of visiting groups to ensure that the design, implementation, and evaluation of STEGHs is likely to result in meaningful, locally relevant improvements in the receiving community [20]. Students and inexperienced clinicians have the potential to add stress to the system that they

are attempting to support, if their skills are not sufficiently advanced to meaningfully enhance current practice, if they are not familiar with local disease processes and resources, and/or if their presence strains resources for local health care practitioners who are likely already overburdened (e.g., additional assistance in language support, explanation of available resources). Mutual and reciprocal benefit to the community visited is of utmost importance when considering such a project [21].

(b) **To your colleagues and fellow travelers?**
It is important to have an open conversation with everyone involved, in which all participants in such trips are candid about their needs and expectations. It will then be easier to make decisions that are sensitive to the needs of the others. Students' obligations to colleagues should dovetail with both the mission of the trip as well as their own commitments to learning and service.

(c) **To your home community and patients?**
While there will be many in the student's home community who commend the student for giving time and service to such a project, there are likely to be others who are of the opinion that she should not be going abroad. Some may share the concerns reflected here, while others may perceive health care needs in distant countries as "someone else's problem" and argue that medical students have a duty to care for those in the community and country that trained them. However, it is possible to ethically fulfill both obligations to one's home community and a desire to serve internationally.

Beyond the scope of this chapter, the "brain drain" represents the reverse scenario and is a significant problem: high-income countries, such as the United States, frequently recruit nurses, physicians, and other health care practitioners from LMICs, after they have received health professions training subsidized by their communities and government, "draining" human capital from those countries. Within the United States, foreign-trained health care practitioners often provide care to poor and rural communities that are underserved by U.S.-trained clinicians.

One of the two doctors going on the trip is a dermatologist who has a private clinic, in which he sees affluent patients who primarily pay out of pocket for his services. This practice has done well and afforded him a very comfortable lifestyle, as well as the financial means to go on trips such as these, which he does every year.

✓ 11. **What are some of the ethical considerations here? Consider issues such as scope of practice and resource allocation.**
In situations in which a health care practitioner has a very specific skill set, the following issues should be considered:

— **Beneficence and non-maleficence:** Are the dermatologist's skills transferable to the level of clinical care requested and/or needed in the community served by this mission?

— **Sustainability:** Does this physician return to the same community every year? Has he established meaningful relationships in the community? If so, this affords him increased knowledge and trust in the community over time and likely allows him to do more to support sustainable interventions.

— **Justice and resource allocation:** Does this physician support vulnerable populations of need within his U.S.-based practice, or does he limit his service to the mission trip where he feels there is greater need for his services? Is he contributing broadly to the society that supported his training?

Box 24.6   Teaching Tip
Such concerns apply not only to global health but also to other situations in which health care practitioners may choose to primarily provide services in lucrative areas while volunteering in their free time. Consider asking learners to debate the ethics of such scenarios. How do they balance physicians' autonomy with physicians' responsibilities to society (the "social contract" [22])?

✓ 12. **What is our duty to assist and address health inequity globally? Consider this question as an individual, a group of people, an academic institution, and a nation.**
Physicians and other health care practitioners take an oath to support beneficence, non-maleficence, and justice. Inherent in principles of justice is the duty to serve all those in need, regardless of background or status. This drives practitioners to address health equity in their daily work, as well as our collective missions. While the U.S. system of medicine allows for many clinicians to practice in for-profit models, academic institutions and public systems should continue to advocate and care for vulnerable populations.

✓ 13. **Self-reflection: Is participating in a short-term global health experience something that you previously considered when you think about your future career? If so, how has this case influenced your thoughts about such experiences? If not, are you more or less likely to consider doing so now?**

Acknowledgments The author wishes to thank Jok Jok, PhD, for his review and comments on this chapter.

References

1. Munge K, Briggs AH. The progressivity of health-care financing in Kenya. Health Policy Plan. 2014;29(7): 912–20.
2. Pan American Health Organization. Health in the Americas. Country report: Haiti. Washington, D.C.: Regional Office for the Americas of the World Health Organization. [cited 2018 Jun 19]. Available from: https://www.paho.org/salud-en-las-americas-2017/?page_id=131.
3. Partners in Health. Haiti [Internet]. Boston (MA): PIH®. [cited 2018 Aug 30]. Available from: https://www.pih.org/country/haiti.
4. World Health Organization. Country profiles [Internet]. Geneva: WHO. [cited 2018 Jul 4]. Available from: http://www.who.int/countries.
5. World Health Organization. Diabetes country profiles 2016 [Internet]. Geneva: WHO. [cited 2018 Jul 4]. Available from: http://www.who.int/diabetes/country-profiles/en/.
6. World Health Organization. Tobacco control country profiles [Internet]. Geneva: WHO. [cited 2018 Jul 4]. Available from: http://www.who.int/tobacco/surveillance/policy/country_profile/en/.
7. World Health Organization. 10 facts on health inequities and their causes [Internet]. Geneva: WHO. [cited 2018 Jul 6]. Available from: http://www.who.int/features/factfiles/health_inequities/en/.
8. Heifer International. 10 things: inequality in Kenya [Internet]. Little Rock (AR): Heifer International; 2012 Jul 24 [cited 2018 Jul 4]. Available from: https://www.heifer.org/join-the-conversation/blog/2012/july/10-things-inequality-in-kenya.html.
9. World Health Organization. What is universal health coverage? [Internet]. Geneva: WHO. [cited 2018 Jul 1]. Available from: http://www.who.int/health_financing/universal_coverage_definition/en/.
10. Cavagnero E, Cros M, Dunworth A, Sioblom M. Better spending, better care: a look at Haiti's health financing: summary report. Washington, D.C.: World Bank; 2017 [cited 2018 Jul 6]. Available from: http://www.worldbank.org/en/country/haiti/publication/better-spending-better-care-a-look-at-haitis-health-financing.
11. American Public Health Association. Community health workers [Internet]. Washington, D.C.: APHA. [cited 2017 Jul 8]. Available from: https://www.apha.org/apha-communities/member-sections/community-health-workers.
12. Levin J. Partnerships between the faith-based and medical sectors: implications for preventive medicine and public health. Prev Med Rep. 2016;4: 344–50.
13. Kagawa RC, Anglemyer A, Montagu D. The scale of faith based organization participation in health service delivery in developing countries: systemic review and meta-analysis. PLoS One. 2012;7(11):e48457.
14. Caldron PH, Impens A, Pavlova M, Groot W. A systematic review of social, economic and diplomatic aspects of short-term medical missions. BMC Health Serv Res. 2015;15:380.
15. Crump JA, Sugarman J. Ethical considerations for short-term experiences by trainees in global health. JAMA. 2008;300(12):1456–8.
16. Sykes K. Short-term medical service trips: a systematic review of the evidence. Am J Public Health. 2014;104(7):e38–48.
17. Médecins sans Frontières [Internet]. Geneva: MSF. [cited 2018 Jul 8]. Available from: https://www.msf.org/.
18. Dwyer J. Teaching global health ethics. In: Benatar S, Brock G, editors. Global health ethics. Cambridge, UK: Cambridge University Press; 2011. p. 319–28.

19. Green T, Green H, Scandlyn J, Kestler A. Perceptions of short-term medical volunteer work: a qualitative study in Guatemala. Global Health. 2009;5:4.
20. Loh LC, Cherniak W, Dreifuss BA, Dacso MM, Lin HC, Evert J. Short term global health experiences and local partnership models: a framework. Global Health. 2015;11:50.
21. DeCamp M, Lehmann LS, Jaeel P, Horwitch C, Ethics ACP. Professionalism and Human Rights Committee. Ethical obligations regarding short-term global health clinical experiences: an American College of Physicians position paper. Ann Intern Med. 2018; 168(9):651–7.
22. Cruess SR, Cruess RL. Professionalism: a contract between medicine and society. CMAJ. 2000; 162(5):668–9.

Further Reading on this Topic

Bornstein D. How to change the world: Social entrepreneurs and the power of new ideas. New York: Oxford University Press; 2007.

Farmer P. Infections and inequalities: the modern plagues. Berkeley/Los Angeles: University of California Press; 1999.

Farmer P. Pathologies of power: health, human rights, and the new war on the poor. Berkeley/Los Angeles: University of California Press; 2003.

Fink S. War hospital: a true story of surgery and survival. New York: Perseus Books Group; 2003.

Reid TR. The healing of America: a global quest for better, cheaper, and fairer health care. New York: Penguin Group; 2010.

Evaluation and Assessment

Contents

Chapter 25　Evaluating Cases in Context – 495

Chapter 26　A Practical Framework for Learner Assessment – 501

Evaluating Cases inContext

Lauren J. Germain

25.1 **Overview – 496**
25.1.1 What Is Evaluation? – 496

25.2 **Context Evaluations – 496**
25.2.1 What Is Meant by Context, and How Is Context Related to Evaluation? – 496
25.2.2 What Can Be Learned from Context Evaluation That Influences the Presentation of Cases in This Book? – 497

25.3 **Input Evaluations – 497**
25.3.1 What Is an Input Evaluation, and How Is It Useful? – 497

25.4 **Process Evaluations – 497**
25.4.1 What Is Process Evaluation, and How Is It Useful? – 497
25.4.2 What Are Some Examples of Effective Process Evaluation? – 498

25.5 **Product Evaluations – 499**
25.5.1 What Is Product Evaluation and Why Engage in It? – 499

References – 499

© Springer Nature Switzerland AG 2019
A. E. Caruso Brown et al. (eds.), *Bioethics, Public Health, and the Social Sciences for the Medical Professions*, https://doi.org/10.1007/978-3-030-03544-0_25

25.1 Overview

- **Editors' Note**

This chapter is intended for course directors, administrators, and others who are responsible for curricular design, implementation, and evaluation. It builds upon content presented in ▶ Chapters 1 and 2. We recommend reading those chapters first, as well as reviewing at least one of the cases (▶ Chapters 3, 4, 5, 6, 7, 8, 9, 10, 11, 12, 13, 14, 15, 16, 17, 18, 19, 20, 21, 22, 23, and 24). After completion of this chapter, we suggest reading ▶ Chapter 26.

25.1.1 What Is Evaluation?

A widely cited definition of evaluation is "determining the merit, worth, or value of something, or the product of that process" [1]. Evaluation of a case or case series is important to identify areas of strength and weakness, continuously improve processes, justify resources, and determine impact. Learner, facilitator, and environmental characteristics affect behavior and perception, which consequently affect the overall impact of any case on participants. Consideration of these contextual features is necessary in order to maximize the value of each case to stakeholders and appropriately evaluate the effectiveness of a case or case series. This chapter outlines an application of Stufflebeam and Zhang's CIPP (context/inputs/process/products) evaluation model to the cases presented in this book and provides a map for evaluating the effectiveness of cases in different educational settings (◘ Fig. 25.1) [2]. The CIPP model was selected for its focus on continuous program improvement throughout planning, case implementation, and post-case evaluation processes as well as the ways that the model strategically considers essential context factors.

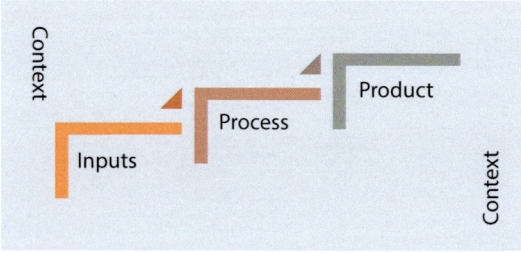

◘ Fig. 25.1 The CIPP model [2]

25.2 Context Evaluations

25.2.1 What Is Meant by Context, and How Is Context Related to Evaluation?

Consideration of the context in which a case is presented is crucial to successful teaching and learning. The context of a case includes learner identities and experiences, facilitator identities and experiences, the local learning environment, community needs and demographics as well as state, regional, and national conditions. Contextual factors shift over time. Simple, effective mechanisms can be put into place to track these shifts and inform evaluation.

The purpose of context evaluations is to "…assess needs, problems, assets and opportunities as well as relevant contextual conditions and dynamics" [2]. In a complex, adaptive system such as an academic medical center, questions such as (1) "What do the learners need?", (2) "What does the healthcare system need?", and (3) "What have the learners and facilitators experienced prior to approaching the case?" yield indispensable information to the implementation of a program. Such information can be used to increase the relevance and value of the cases.

While possibly time-consuming and logistically challenging, meeting with representatives of primary stakeholder groups prior to implementing the cases can be a useful way to collect critical contextual information and lay foundations for ongoing information sharing. For example, the case presented in ▶ Chapter 12 (In the Words of a Refugee) centers on refugee trauma. To optimize the authenticity of the case in context, connecting with representatives from local refugee communities and refugee support organizations may be useful and foster important partnerships. Further, recognizing that learners and facilitators may have experienced (or may currently be experiencing) trauma themselves, connecting with student counseling offices in advance of the case may allow, for example, these offices to proactively set up open counseling hours after the case discussions and for case facilitators to have access to important referral information. When considering primary stakeholder groups, they might include learners, faculty acting as small group facilitators, education administrators, healthcare system administrators, community members, student counseling offices, local organization leaders, and others.

25.2.2 What Can Be Learned from Context Evaluation That Influences the Presentation of Cases in This Book?

Stakeholder engagement in context evaluation can help to elucidate the culture among learners and educators.

Knowing (1) whether the learner culture is more collegial or competitive and (2) what incentivizes a specific group of learners aids in the development of group dynamics and facilitator training. Asking facilitators to reflect on their own ideologies, strengths, challenges, and positions relative to the case content assists in the creation of well-matched facilitator pairings, for those teaching in partnerships. Education administrators may be able to align cases with existing learning experiences and assessments to tie strands of an educational process together and strengthen learner preparation overall. Health system administrators, community members, and local organization leaders may provide valuable insights into the local relevance of the cases and provide critical information to contextualize the cases within one's local environment and increase learner agency.

Awareness of the composition of the learner support network is also important. Building relationships with student counseling offices and other learner support systems is crucial, as learners may be triggered, or negatively affected personally, by the cases (discussed further under the heading "Trigger Warnings and Brave Spaces" in ▶ Chapter 2).

Establishing relationships with stakeholders also allows course directors and others involved in the implementation of these cases to account for substantive context shifts critical to interpreting measures of success of a case or case series. Consider, for example, the case presented in ▶ Chapter 9. Participant engagement in the case and overall educational outcomes may be impacted by context shifts such as a highly publicized debate about vaccines among politicians. Responses to ▶ Chapter 4 may be impacted by perceptions of the local shelter network.

Learners who engage in clinical practice (or observational experiences, such as shadowing of healthcare practitioners) are likely impacted by local patient demographics, the behaviors of those with whom they work (the "hidden curriculum"), and system-level policies and dynamics. Collaborating with stakeholders builds foundations for a healthy, continuous program evaluation that takes context into account.

25.3 Input Evaluations

25.3.1 What Is an Input Evaluation, and How Is It Useful?

An input evaluation considers program feasibility and resource use. This part of the evaluation method examines the plan and what it might take to meet the goal(s) it sets forth. An input evaluation might examine the goals of the case or case series and how likely they are to be met with the proposed plan and within the constraints of the overall educational and healthcare system. Key resources to consider are facilitator availability, time and incentives to teach, faculty development needs, group meeting space(s), curricular time (contact hours available in the learner schedule), and administrative costs. In most institutions, a department or educational curriculum committee will conduct an input evaluation when the presentation of a case or case series is initially proposed.

25.4 Process Evaluations

25.4.1 What Is Process Evaluation, and How Is It Useful?

Process evaluation can be distilled to the term "monitoring." Process evaluation includes collecting and analyzing data that indicate *how* the case or case series is progressing. Evidence collected during process evaluation can be used to understand group dynamics and what is and is not working in each group and case. This process allows leaders—both facilitators and course directors—to make informed, necessary changes in situ. For example, process evaluation may reveal that one group concludes their sessions substantially after other groups. With this datapoint, the course director or facilitator can begin to collect information about why this difference is occurring: Were the learners in the group more engaged than learners in other groups? Were the facilitators struggling to keep time and conversations on track? Was the group spending too much time on topics tangential to the learning objectives?

A particularly effective component of process evaluation is observation, which can be used to answer questions like those in the example above and to collect information about group processes in general. Situating a trained observer in the room during a case discussion to take field notes provides rich descriptive information about a session. This record can also be utilized to triangulate facilitator assessment of learners and more fully understand what happens during any specific case allowing facilitators to quickly make adjustments. Using the example of a group that consistently runs over the allocated time for a case, field notes may provide a record of how the group allocates time and where, specifically, they may be able to adjust conversations in the future so that the group stays on schedule.

A post-case survey, administered to facilitators and learners immediately following a session, is a useful tool when exploring how any individual case functions in context. For example, a post-case survey of facilitators or learners might identify questions that arose in the group regarding a law or policy relevant to the case being discussed. Such information allows course directors to target faculty development to address gaps and other areas of concern. Similarly, such a survey may find that learners were well-prepared for some cases but struggled in others. This information might lead course directors to modify the sequence of cases or add preparatory materials (readings or lectures), or it might lead to the realization that a case discussion was scheduled shortly before a quiz or an exam, forcing students to prioritize coursework.

However, while post-session surveys can be useful, if a series of cases is presented, participants may experience survey fatigue from this method. In these situations, a mid-point assessment (halfway through a case series or course) can be most valuable.

25.4.2 What Are Some Examples of Effective Process Evaluation?

In our institution, employing a process evaluation strategy that includes stakeholder engagement, mid-point surveys, and observations has allowed us to identify challenges, describe what is happening, and respond effectively. The following example illustrates how we have used this process.

Learners in one small group reported, on the mid-point survey and directly to the course director, that two facilitators "didn't get along." A trained observer who was not affiliated with the course but who was familiar to the students was sent to the group to take field notes and was intentionally not made aware of the specific concern. Following the observation, the course leadership team debriefed and read the field notes. As a team, they were able to use the record of events (quotes, behaviors, expressions, etc.) to more fully understand the environment in the session.

The observed session involved the case in ▶ Chapter 21. The two facilitators—by virtue of their different training, experience, frameworks, and roles—approached the case differently; and in the small group discussion, they processed through the similarities and differences in their approaches aloud. One facilitator, a physician, discussed the option of writing a letter to the hypothetical patient's insurance company, advocating for coverage. The other facilitator, who was trained in health humanities, questioned the effect of such a letter, as it could contribute to pathologizing gender dysphoria. Together, the facilitators and learners delved into the ethical and legal foundations of the case. However, the facilitators never stated a conclusive "right answer" to what should be done by the hypothetical resident physician in the case. Several similar conversation patterns occurred in the session.

Given this information, we sought to understand whether the learners in the group perceived conflict when the facilitators discussed differing approaches based on their professional roles and experiences. By engaging learner stakeholders and returning to our context evaluation, we were able to realize that this hypothesis was correct: The learners, in this example, were first-year medical students. Few had previously been team-taught, and they were not aware that the purpose of having multiple facilitators in the sessions was for representation of diverse viewpoints and professional expertise. The team made a more concerted effort to explain facilitator pairing and intent moving forward. Overall, this process evaluation allowed the system to respond to a signal (the midyear survey data), study and describe the issue, and address it effectively.

25.5 Product Evaluations

25.5.1 What Is Product Evaluation and Why Engage in It?

The emphasis of product evaluation is on program outcomes. Product evaluation is primarily guided by the question "Did the program meet its goals?" and focuses on impact. Kirkpatrick's four-level model can be useful for this component of the CIPP process [3]. Kirkpatrick's levels are reaction, learning, behavior, and results [3].

To evaluate *reactions*, one might survey participants and inquire about their responses to the cases or case series. Both facilitator and learner feedback is valuable to this process. Measures of satisfaction and perceptions of utility and efficacy of cases might also be included in such surveys. Reflective writing is another method of collecting data on reactions that also encourages post-session thought. As noted in ▶ Chapter 2, the self-reflection questions embedded in each case can serve as prompts for reflective writing.

To evaluate *learning*, one could add an indirect measure to a survey, asking what participants believe they learned from a session, or a direct measure of learning, such as an exam, objective structured clinical examination (OSCE), case study, or presentation could be used. Direct measures include quizzes, exams, essays, simulations, and other assessments of student learning.

Evaluation of the latter two of Kirkpatrick's levels, *behavior* and *results*, are more challenging and context-specific. Behavioral criteria are measures of learners' authentic choices and actions in their workplaces. In health professions education, this is sometimes referred to as workplace-based assessment. If the goal of a case or case series is to change the way that a group of learners behaves, one might consider collecting data on specific behaviors before or after implementation of a case. Measures might include:

- Number of ethics consultations requested by learners during case studies
- Number of times learners take the socioecological model into account during discussions
- Language used in clinical documentation (e.g., person-first language, patients' preferred gender pronouns)
- Behavioral changes reported in self-reflections, peer assessments, or by clinical coaches or mentors

Similarly, the results level can present a challenge for evaluation because results are dependent on context and, in a complex, adaptive system like an academic medical center, it is difficult to isolate a singular cause of any outcome. That said, meaningful results are often the most sought-after program outcomes, as they serve to illustrate substantial value of an educational intervention such as case-based learning. Results might be measured by changes in patient outcomes, increased system efficiency, increases in satisfaction among a specified group of patients, or other similar measurable outcomes. While imperfect and challenging, striving for any necessary positive impact in the results domain is fundamental to health professions education and, as such, merits conversation from a program development team.

> **Conclusion**
> While the cases in this book have been tested, their effectiveness in unique educational settings and institutions will be impacted by contextual factors that are important to consider. The CIPP evaluation model is a useful framework because it outlines components for ongoing evaluation and process refinement to ensure that the cases achieve maximum impact.

References

1. Scriven M. Evaluation thesaurus. 4th ed. Newbury Park (CA): Sage; 1991. p. 139.
2. Stufflebeam DL, Zhang G. The CIPP evaluation model: how to evaluate for improvement and accountability. New York: Guilford; 2017.
3. Kirkpatrick DL. Evaluation of training. In: Craig RL, editor. Training and development handbook: a guide to human resource development. 2nd ed. New York: McGraw-Hill; 1976. p. 301–19.

A Practical Framework forLearner Assessment

Lauren J. Germain

26.1 Overview – 502

26.2 What Is Assessment? – 502

26.3 Context and Purpose – 502
26.3.1 What Elements of Context Are Relevant to the Implementation of a Learner Assessment System that Maximizes Meaningful Feedback on Participation in Discussion of These Cases? – 502

26.4 Assessment Scales – 503
26.4.1 What Assessment Scale Can Facilitate Meaningful Feedback on Case Participation? – 503
26.4.2 How Might One Incorporate Basic Professionalism or Attendance Concerns into the Assessment System? – 504

26.5 Rater Training and Calibration – 504

26.6 Learner Reflection – 505

26.7 Assessment System Validation Evidence – 505

References – 506

© Springer Nature Switzerland AG 2019
A. E. Caruso Brown et al. (eds.), *Bioethics, Public Health, and the Social Sciences for the Medical Professions*, https://doi.org/10.1007/978-3-030-03544-0_26

26.1 Overview

- **Editors' Note**

This chapter is intended for course directors, small group facilitators, and other educators or administrators responsible for assessing learning and implementing assessment plans. It builds upon content presented in ▶ Chapters 1, 2, and 25. We recommend reading those chapters first; we also suggest reviewing at least one of the cases (▶ Chapters 3, 4, 5, 6, 7, 8, 9, 10, 11, 12, 13, 14, 15, 16, 17, 18, 19, 20, 21, 22, 23, and 24).

26.2 What Is Assessment?

Assessing learners is fundamental for constructing and giving meaningful feedback as well as making determinations about content mastery and/or competency. There are many types of assessments: formative assessments *for* learning and summative assessments *of* learning, direct observation and self-report, and assessments perceived as "objective" and "subjective," among many others. In 2010, medical education leaders met in Ottawa and crafted a consensus statement on the criteria for good assessment. While assessment takes many forms, the document notes that "no single set of criteria for good assessment apply equally well to all situations," which is why this chapter begins with a discussion of context and purpose [1]. Fit for purpose is essential for constructing and implementing assessment systems that produce the feedback necessary to foster learner development and determine whether learners have met the objectives of the case or case series.

26.3 Context and Purpose

26.3.1 What Elements of Context Are Relevant to the Implementation of a Learner Assessment System that Maximizes Meaningful Feedback on Participation in Discussion of These Cases?

Considering the context in which a case is presented is a crucial step in designing an effective learner assessment system. Context elements to consider are (1) whether the case is offered as a single session or part of a series or course, (2) learner positionality, (3) facilitator positionality, and (4) necessary assessment "artifacts."

26.3.1.1 Individual Case or Part of a Series

The cases in this book were designed to effectively galvanize learning and thought when used individually or as part of a case series. The overarching purpose of the learning experience(s) should drive implementation and has implications for assessment. Learner assessment and feedback strategies vary depending on how cases are presented and the purpose of the learning experience(s). By definition, the use of single cases will yield fewer points of observation than a case series, but feedback about whether or not learners have met the learning objectives stated in the case(s) is important in either format.

The learning objectives provide a guide for the assessment of learner performance on any case. The objectives stated at the beginning of each case are most commonly written at the "application," "analysis," "synthesis," or "evaluate" levels of Bloom's taxonomy [2]. Thus, in order to offer useful feedback to learners, facilitators need an assessment framework that acknowledges the behaviors and statements related to application, analysis, synthesis, and evaluation during case discussions. The adapted reporter, interpreter, manager, and educator (RIME) scheme, as described later in this chapter, is a key tool to guide assessment on a single case. Additionally, when implementing a case series, educators can consider a series of formative assessments to drive reflection and growth before a final, summative judgment is rendered about overall performance or whether adding assessments from each small group together is more appropriate for the final grade.

26.3.1.2 Learner Positionality

Prior to implementing a session, it is important for facilitators to think about learner positionality or the views, values, cultures, experiences, and identities of the learners who will be participating. By answering a series of simple questions—including "Who are the learners and what are they looking for in terms of feedback?" and "How does this session differ from the way that the learners typically work together/or learn?"—facilitators will maximize the utility of the assessment.

The "stakes" connected to a session assessment can drive behavior, so educators should think through the "value" of each session. For example, a session assessment that is worth 2% of an overall course grade may be viewed differently by learners than a session or series worth 100% of the overall course grade. Depending on the culture and competing priorities, learners may behave differently due to this value placement. A good process for thinking through these matters is to engage one or more learners as members of the implementation team.

The cases in this book are intended to engage learners with variable levels of experience and expertise from novice to expert. For learners with expert levels of knowledge, though perhaps not knowledge specific to the case, it is important that assessments be completed by facilitators whom they trust as peer experts. When using the framework recommended by this book, it is imperative to remind all learners that the assessments are based on a limited set of contextualized observations and are meant to categorize behaviors and provide feedback, not make judgments about them as professionals.

26.3.1.3 Facilitator Positionality

Lessons from qualitative research processes are directly applicable to the assessment of performance on the cases in the book. The small group facilitators, or other educators acting as raters or graders, are necessarily impacted by their identity, including their expertise and experience, and it is *the facilitators*, not the assessment form, that are the actual instruments of assessment. Thus, transparency with learner about facilitator selection and expertise, as well as information on how they were trained to assess in a reliable manner, is important for developing trust in the feedback they share. During facilitator training, course directors should ask facilitators to reflect on what they value, where their biases may lie, and how their identity might impact their assessment of learners. Together, they should brainstorm strategies to manage bias and determine the appropriate level of self-disclosure that contributes to legitimacy, authenticity, and comfort (further discussed in the section on Rater Training and Calibration later in this chapter).

26.3.1.4 Necessary "Artifacts"

Quite simply, what do all stakeholders need as evidence that the learning objectives were met? As part of a continuing education experience, for example, stakeholders (licensing organizations, employers, etc.) may require a basic form signed by the facilitator acknowledging attendance and/or indirect evidence of impact such as learner self-assessment. Alternatively, as part of a course grade, assessment may require a systematic trail of formal feedback submitted through a learning management system (LMS) or other evaluation system.

26.4 Assessment Scales

26.4.1 What Assessment Scale Can Facilitate Meaningful Feedback onCase Participation?

The cases presented in this book were deliberately constructed for use by stakeholders with different educational and experience levels and for use either individually or as a series. As such, an effective assessment scale requires flexibility, clarity, and conciseness. Pangaro's RIME assessment framework, first defined in 1999 and now used extensively to assess learners in academic medicine, fits all three criteria [3]. As a result of its broad use, many healthcare professionals are already somewhat familiar with the RIME model. RIME is intended to be synthetic, meaning that a single ordinal rating scale encompasses all of the observed behaviors into one summary judgment.

While Pangaro's scale is most often used in the clinical arena of residency and clerkship training in medical education, the developmental framework is appropriate as it gives language to observations of learners as they progress through the "knows" (knowledge), "knows how" (competence), and "shows how" (performance) stages of Miller's pyramid [4]. Our team modified the language of the original scale to include more specific language about small group participation to fit the new context for application.

Definitions of each of the ratings that comprise the modified RIME scale are as follows:
- **Reporter**
  - Takes ownership of reliable, accurate gathering and reporting of relevant information
  - Participates in small group discussions beyond presentations of prep questions at least once per case
  - Is able to answer "what" questions correctly (e.g., can define ethical principles or basic public health concepts)
- **Interpreter**
  - Takes ownership of thinking through problems and has the knowledge, skills, and confidence to offer reasonable suggestions for next steps
  - Consistently participates in small group discussions beyond presentations of prep questions
  - Asks questions to acquire more information when necessary
  - Is able to answer "why" questions
- **Manager**
  - Takes ownership for developing plans to resolve problems (ethical, social, clinical, etc.) with consideration given to multiple stakeholders (patients, families, other healthcare providers, etc.)
  - Integrates knowledge from multiple sources outside the case
  - Makes contributions that are valuable, advance discussion, and do not simply restate points already made
  - Appreciates the nuances of more complex situations (i.e., that there may not always be a right answer)
  - Is able to offer and respond to counterarguments effectively
- **Educator**
  - Fulfills a promise of maintaining expertise in one's self and others
  - Takes ownership of self-correction and self-improvement
  - Demonstrates leadership in teaching and learning
  - Effectively draws on previous knowledge and experience in a way that educates peers
  - Is able to recognize when to defer to a peer's knowledge or experience

26.4.2 How Might One Incorporate Basic Professionalism or Attendance Concerns into the Assessment System?

A recent adaptation of the RIME assessment framework adds a P, for professionalism, to the mnemonic, making it PRIME [5]. Professional behaviors in this definition include "reliability, responsibility, teamwork, respect for patient's values, punctuality, respect for staff and peers, appropriate attire for clinical care, and demeanor and comportment." If it is possible that there may be concerns related to the professional behaviors of learners, educators might choose to add an additional domain to scoring that is available to be earned by all learners or a space to identify when a professionalism concern has arisen.

26.5 Rater Training and Calibration

A shared mental model about the definition and behavioral anchors associated with each rating is necessary if multiple individuals will be facilitating cases and observing learners. Rater training should include a thorough conversation about the purpose of the case(s) or course, the learning objectives, and the rating scale. After discussing the rating scale in depth, course directors can share a series of vignettes (videos, skits, or written descriptions) or a piece of a mock session and ask all of the facilitators to rate the participants in the vignette. The course director should speak frankly about why each facilitator selected each rating and what descriptive feedback they might share with the learner to promote growth and reflection. This process should be continued until raters are consistently in agreement. The rater calibration process should include vignettes, videos, or skits that increase in complexity to pull facilitator attention in multiple directions as is the reality of leading a case. This exercise allows raters to recognize how fatigue, attention, and observation quantity impact their ratings and allow them to address any issues with their co-facilitators. Strategies to manage the pull of attention between facilitation of the case and learner observation include facilitators switching between the roles of facilitator and observer so that each may focus on a primary role or take session breaks at intentional intervals (e.g., every 30 minutes or whenever the facilitators or learners tire).

If the cases are being used in a series with multiple points of assessment and the learners are at a level where their behaviors are likely to change over time, reference points may be useful on your evaluation scales. Examples include the following:
- "Interpreter is the expected level of performance at this time of the year."
- In a course with first-year medical students, "Most students will not reach this level of performance (Educator); most small groups will not have students performing at this level."

These phrases help to calibrate the raters and frame the feedback to the learners.

26.6 Learner Reflection

There is evidence that reflecting on one's performance during the feedback process improves the likelihood that the assessment will impact the learner [6]. Thus, in our institution, we incorporated reflection and self-assessment into the system to maximize impact. Learners were presented with a self-evaluation scale that mirrored the scale that the facilitators used to rate them. The learner self-assessment tool also included a question asking learners how they might adjust their behaviors to reach the next level on the scale.

26.7 Assessment System Validation Evidence

The modified RIME framework presented in this chapter was used to assess the performance of 170 first-year medical students over the course of 1 year. Assessments and written evidence in support of ratings were collected quarterly. As substantial growth in knowledge, reasoning, and team-based learning skills were anticipated, only the final assessment was summative. At the end of the third quarter, students self-assessed and discrepancies between self-assessments and faculty assessments were addressed one-on-one prior to summative judgment.

The data collected over the year, presented in ◘ Fig. 26.1, indicate that the anticipated learner development was detected in the system and that all students progressed beyond reporting by the end of the case series.

Additionally, field notes from a small group observation conducted in the final quarter of the

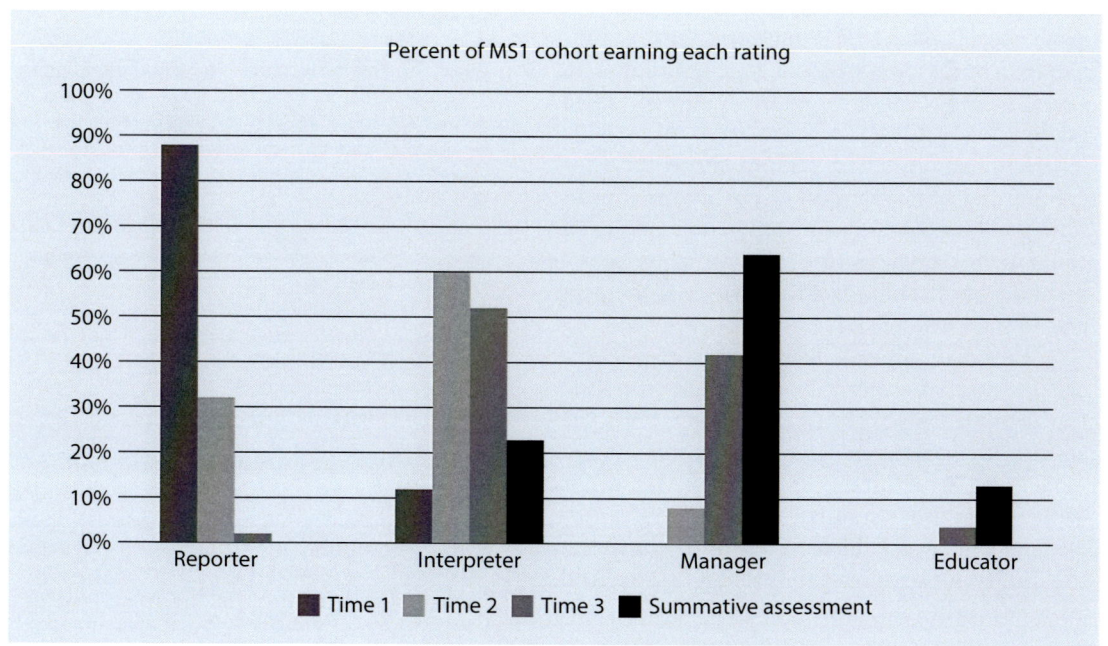

◘ Fig. 26.1 Breakdown of students' longitudinal performance in case-based small group discussions, as assessed by the modified RIME scheme

year provide evidence to substantiate participation quality of the students who earned the top rating.

> **Conclusion**
> The modified RIME scheme is one approach to assessment of learner performance in the cases presented in this book. Consideration of context is crucial to success, and RIME is flexible enough to be used in individual sessions or longitudinal experiences with learners at various levels and within various fields.

■ **Author's and Editors' Note**

The assessment approach presented in this chapter focuses on learner performance during discussions of the cases in this book. However, when implementing cases in a series or as part of a course, it may be appropriate to use additional modes of assessment, the full spectrum of which are beyond the scope of this chapter. While the specified goals of the case series or course drive the choice of assessment method, practical matters such as the number of learners, number of potential assessors, and time for assessment are also important to acknowledge.

For example, written examinations may be suitable for assessing content knowledge and, to a lesser extent, reasoning. Multiple-choice examinations can be administered to any number of learners, and software programs allow for rapid grading of larger cohorts. Essay questions on written examinations often require more time for grading but allow for deeper reflection, expression of knowledge, and application of logic. Script concordance tests require substantial development time but can be rapidly rated once the programming is completed [7–9].

Oral examinations—when the ratio of learners to assessors permits their use—provide learners the opportunity to demonstrate the ability to articulate an argument aloud. Objective structured clinical examinations (OSCE) and other simulated encounters can be designed as spaces for learners to demonstrate the ability to apply the knowledge, skills, and attitudes acquired through case discussions (e.g., an actor might play a patient, similar to one in ► Chapter 18, who is angry that his pain has not been adequately treated, allowing the learner to demonstrate how she would approach the situation) [10].

Longer writing assignments are also more resource-intensive, in terms of faculty time spent grading and/or giving feedback, but can be very useful to learners. Reflective narratives may be spaces for authentic expression of thoughts and what learners are taking away from cases. Research papers or critical analyses of existing literature may showcase critical thinking and understanding of research methodologies.

In summary, depending on program goals and resources, there are many forms of learner assessment that may enhance a course or case series that includes cases from this book. Conversely, the cases can be adapted into assessment tools themselves, as was discussed in ► Chapter 2: Approaches to Using This Book.

References

1. Norcini J, Anderson B, Bollela V, Burch V, Costa MJ, Duvivier R, Galbraith R, Hays R, Kent A, Perrott V, Roberts T. Criteria for good assessment: consensus statement and recommendations from the Ottawa 2010 conference. Med Teach. 2011;33(3):206–14.
2. Bloom BS. Taxonomy of educational objectives, handbook I: the cognitive domain. New York: David McKay Co Inc; 1956.
3. Pangaro L. A new vocabulary and other innovations for improving descriptive training evaluations. Acad Med. 1999;74:1203–7.
4. Miller GE. The assessment of clinical skills/competence/performance. Acad Med. 1999;74(11):1203–7.
5. Holmes AV, Peltier CB, Hanson JL, Lopreiato JO. Writing medical student and resident performance evaluations: beyond performed as expected. Pediatrics. 2014;133(5):766–8.
6. Sargeant JM, Mann KV, van der Vleuten CP, Metsemakers JF. Reflection: a link between receiving and using assessment feedback. Adv Health Sci Educ Theory Pract. 2009;14(3):399–410.
7. Fournier JP, Demeester A, Charlin B. Script concordance tests: guidelines for construction. BMC Med Inform Decis Mak. 2008;8:18.
8. Dory V, Gagnon R, Vanpee D, Charlin B. How to construct and implement script concordance tests: insights from a systematic review. Med Educ. 2012;46(6):552–63.
9. Lubarsky S, Dory V, Duggan P, Gagnon R, Charlin B. Script concordance testing: from theory to practice: AMEE guide no. 75. Med Teach. 2013;35(3):184–93.
10. Harden RM, Lilley P, Patricio M. The definitive guide to the OSCE: the objective structured clinical examination as a performance assessment. Edinburgh: Elsevier; 2016.

Supplementary Information

Glossary – 508

Index – 517

© Springer Nature Switzerland AG 2019
A. E. Caruso Brown et al. (eds.), *Bioethics, Public Health, and the Social Sciences for the Medical Professions*, https://doi.org/10.1007/978-3-030-03544-0

Glossary

Accountable care community Innovative model from the U.S. Centers for Medicare and Medicaid Services that bridges the gap between clinical and community service providers and assesses whether there is an impact on healthcare utilization and cost when health-related social needs are systematically screened for, identified, and addressed.

Accountable care organization (ACO) Group of physicians, hospitals, and other healthcare providers that work together to coordinate services for patients, with the goals of providing appropriate and timely services, reducing duplication of services, and reducing medical errors.

Addiction Defined by the American Society of Addiction Medicine (ASAM) as a "primary, chronic disease of brain reward, motivation, memory and related circuitry"; like other chronic diseases, it involves cycles of relapse and remission and without treatment is progressive, resulting in disability or premature death. Characterized by "inability to consistently abstain, impairment in behavioral control, craving, diminished recognition of significant problems with one's behaviors and interpersonal relationships, and a dysfunctional emotional response."[1]

Adverse childhood experiences Adverse childhood experiences (ACEs) are categorized into three groups: abuse, neglect, and family/household challenges. Each category is further divided into multiple subcategories. As an example, abuse includes emotional, physical, and sexual abuse. Details about ACEs can be found at the U.S. Centers for Disease Control.[2]

Adverse event Unintended injuries or outcomes resulting in harm—including death, disability, or prolonged hospital stays—arising from medical care, including the absence of indicated treatment.

Agent (in epidemiology) Part of the epidemiologic triad; any causative factor (e.g., person, vehicle, pathogen, chemical, object) to which the host must be exposed in order to bring about a particular injurious outcome.

Allostatic load "Wear and tear on the body" that results from repeated or chronic exposure to stress.

Americans with Disabilities Act (ADA) 1990 law that prohibits discrimination against people with disabilities, including in medical services, and requires that disabled persons have access to medical services and medical facilities.

Artificial nutrition and hydration May include any method of administering nutrition or hydration to a patient who cannot take food or fluids independently or with assistance from a caregiver, such as placement of a nasogastric tube for administration of enteral formula or intravenous fluids.

Assessment Testing, measuring, collecting, and combining information and providing feedback.

Bariatric surgery A surgical procedure performed on the stomach or small intestine to induce weight loss among patients who either have a body mass index (BMI) above 40 kg/m^2 or patients who have a BMI between 35 and 40 kg/m^2 with disease pathology related to obesity. These procedures aim to induce long-term weight loss by combining restricted intake with potential malabsorption, thereby improving or resolving comorbidities associated with obesity and improving quality of life.

Battery The legal doctrine that any intervention upon the patient's body without their consent is a violation of the patient's rights.

Bedside rationing Withholding of a medically beneficial service from an individual patient because of that service's cost to someone other than the patient. It is usually at the discretion of the treating physician. Rationing not involving an individual patient and physician is often described as resource allocation.

Beneficence A principle of medical ethics; healthcare practitioners must strive to act in the best interests of their patients.

Best interests standard An approach to surrogate decision-making commonly used in pediatrics; suggests that parents should make decisions for their child that promote that child's best interests, maximizing likely benefits and minimizing likely harms.

Built environment The human-designed spaces in which people live, learn, work, and play, including homes, buildings, streets, public spaces, transportation systems, and overall infrastructure—all of which influence health and wellness.

Chromosomal diseases Occur when the entire chromosome, or large segments of a chromosome, is missing, duplicated, or otherwise altered. Down syndrome is a prominent example of a chromosomal abnormality.

Common Rule The Common Rule refers to federal policy for protection of human subjects in research. It has been

---

1   American Society of Addiction Medicine. Public Policy Statement: Definition of Addiction. ▶ https://www.asam.org/resources/definition-of-addiction.
2   The Center for Disease Control and Prevention. ACEs Definitions. Data and statistics. ▶ https://www.cdc.gov/violenceprevention/acestudy/about.html.

Glossary

in place since 1981 and governs Institutional Review Boards and the like.

Competency Informed by a patient's decision-making capacity, a legal designation, whereby individuals have the right to make their own decisions.

Competitive foods Foods and beverages offered at schools apart from meals served through the federally reimbursed school lunch, breakfast, and after-school snack programs (including food and beverages sold at snack bars, student-run stores, and vending machines).

Confidentiality A physician's ethical duty to respect the privacy of his or her patients, by maintaining their medical information in a secure manner and avoiding release of that information without patient consent.

Conflict of interest A conflict of interest is anything that threatens to come between the physician or investigator (or nurse or other healthcare practitioner) and his/her primary duty to the welfare of the patient or research subject or between the investigator and the integrity of the research.

Confounding variable Any extrinsic factor (sometimes unrecognized) that may influence the effect of a perceived independent variable on a measured dependent variable.

Conscientious objection Refusal to perform a role or responsibility because of moral or other personal beliefs.

Conspiracy Although it varies by jurisdiction, the criminal act of conspiracy is generally defined as an agreement by two or more people to commit a crime, and then one person proceeds with an overt act of criminal activity. Each person is punishable and held to the same standard as the one who may have engaged in the activity, even if they do not know the identities of their co-conspirators. Broadening of the conspiracy statutes at the federal level, as well as the U.S. Supreme Court's decision in *United States vs. Shabani, 513 U.S. 10 (1994)*, has led to a significant increase in the number of incarcerated women, particularly in federal drug crimes, because it allows every participant in a crime to be held liable for the actions of every other participant, regardless of what role they played in the crime.

Decisional capacity The ability to understand and reason about relevant medical information and to reach an informed decision within the context of personal values and goals. Patients who lack capacity cannot make autonomous informed decisions.

Developmental delay Development of a child that falls behind the typical developmental progression.

Diabetes mellitus type 2 Chronic medical condition that affects the way the body metabolizes glucose (sugar) and causes glucose levels to be higher than normal because of the body's resistance to insulin. It is the most common form of diabetes mellitus.

Duty hours Defined by the Accreditation Council for Graduate Medical Education (ACGME) as all clinical and academic activities related to a graduate medical training program (e.g., internship, residency, fellowship), including time spent in direct patient care in any setting, administrative duties related to patient care, arranging for transfer of patient care, physically in the healthcare institution while on call (but not call taken from home), in scheduled academic activities, and in research activities; does not include time spent at home while not engaged in any of the above (including during "home call"); and does not include time spent reading or studying.

Duty to protect Often recognized with "duty to warn"; refers to a clinician's ethical and legal responsibility to protect both their patients and other individuals in the community from predicable or anticipated harm.

Early intervention A federally funded, locally run program that provides evaluations and services for children suspected to have developmental delay.

Ecological study An analysis of potential risk-modifying factors and their associated health outcomes that utilizes population averages (often defined geographically) as individual units of study.

Effect size Quantification of the strength or magnitude of a particular measured effect; often used to describe whether there is a substantive difference between two groups based on the influence of a defined variable.

Emergency exception The rule that in emergencies the patient's informed consent is not required to provide treatment to preserve life and well-being.

Emerging infection An infection that has recently appeared within a population or one with a rapidly increasing incidence or geographic range.

End-of-life care Care that takes place during the final hours, days, or weeks of a patient's life. End-of-life care may include hospice and palliative care.

Environment Part of the epidemiologic triad; extrinsic factors (e.g., societal policies, cultural norms) that may alter exposure and influence the host's susceptibility to a causative agent.

Evaluation Defined by Scriven (1991) as "judging the worth or merit of something or the product of the process."[3]

---

3   Scriven M. Evaluation Thesaurus (4th eds.). Newbury Park, CA: Sage. 1991.

Event-free survival (EFS) Proportion of patients who are still alive and have not experienced a relapse at a defined time point after diagnosis; often contrasted with "overall survival." Time point typically varies based on the disease, although "5-year event-free survival" is commonly reported.

Female genital mutilation The World Health Organization (WHO) defines female genital mutilation (FGM) as comprising "all procedures that involve partial or total removal of the external female genitalia, or other injury to the female genital organs for non-medical reasons."[4] Also called female genital cutting or, historically, female circumcision.

Food desert Neighborhoods or communities whose access to healthy, affordable food sources is limited by distance to food stores or markets, paucity of food stores in a particular area, family income, and/or transportation availability.

Food insecurity Inconsistent access to adequate food due to constrained resources.

Free erythrocyte protoporphyrin (FEP) An enzyme in the heme pathway that becomes elevated in the presence of prolonged high levels of lead in the blood.

Full disclosure In a medical context, this is often meant to include an admission of a mistake, discussion of the error, and a link from the error to harm.

Gender diverse or gender nonconforming A person whose behavior or appearance does not match cultural and societal expectations for what is stereotypically appropriate for their assigned gender. Examples of gender-diverse identities include gender nonbinary (neither male nor female) and gender fluid (more male or more female at different times).

Gender dysphoria (GD) Distress or discomfort that may occur when a person's internal sense of gender does not match the sex assigned at birth, based on anatomy and/or chromosomes.

Gender identity A person's internal sense of being male, female, neither, or both.

Gender incongruence A diagnosis for a transgender person, in other words, a person whose gender identity (internal sense of gender or asserted gender) does not match the gender assigned at birth. This is an International Classification of Diseases-11 (ICD-11) diagnosis, and it replaces the term "transexualism."

Germ cell tumor Type of growth, which can be benign or malignant, that is derived from germ cell tumors. This commonly occurs in the gonads (testes, ovaries) and brain but can occur elsewhere.

Gonadotropin-releasing hormone (GnRH) analogue A class of drugs used for suppression of precocious puberty and to treat hormone-responsive cancers, endometriosis, and uterine fibroids. It is also used for suppression of puberty in adolescents with gender dysphoria.

Harm principle An alternative approach to surrogate decision-making that suggests parents should be allowed to make decisions for a child in accordance with their beliefs and values unless the likely harm from the decision exceeds a certain threshold, often defined as death or irreversible disability.

Health literacy The extent to which patients can access, process, and reason about health information.

High-income country (HIC) A country with a gross national income (GNI) per capita of $12,236 USD or more (World Bank, 2016).

Hospice Hospice care is specifically defined as care for those who are expected to live fewer than 6 months longer.

Host Part of the epidemiologic triad; referring to the injured individual. Hosts have their own intrinsic risk factors that can influence their susceptibility to a causative agent.

Human trafficking The United Nations Office on Drugs and Crime defines human trafficking as "the recruitment, transportation, transfer, harboring or receipt of persons, by means of the threat or use of force or other forms of coercion, of abduction, of fraud, of deception, of the abuse of power or of a position of vulnerability or of the giving or receiving of payments or benefits to achieve the consent of a person having control over another person, for the purpose of exploitation."[5]

Hypoplastic left heart syndrome A rare congenital cardiac defect in which the left side of the heart is underdeveloped or absent.

Implicit (unconscious) bias Attitudes or stereotypes that affect people's understanding, decisions, and actions without conscious knowledge.

Infant mortality Death of an infant before 1 year of age; usually reported as a rate (number of infant deaths per 1000 live births).

---

4 World Health Organization. Female genital mutilation. ► http://www.who.int/news-room/fact-sheets/detail/female-genital-mutilation.

5 United Nations Office on Drugs and Crime. Human trafficking. ► https://www.unodc.org/unodc/en/human-trafficking/what-is-human-trafficking.html.

Glossary

Infectious tuberculosis (TB) disease Refers to a state in which a person has clinical evidence of TB that is considered communicable. Pulmonary and laryngeal TB are communicable (infectious TB disease), whereas TB disease of the brain, spine, and kidneys is generally not considered to be communicable.

Informed consent The doctrine that healthcare practitioners must provide patients with relevant information about their medical condition and the risks, benefits, and burdens of treatment options so that patients can understand and reason about this information to make a decision to consent to or refuse treatment.

Institutional Review Board for the Protection of Human Subjects (IRB) An IRB is a federally mandated institutional committee whose purpose is to protect the safety, confidentiality, and other interests of human subjects in research studies.

Intrauterine fetal demise (IUFD) Sometimes used interchangeably with stillbirth but may be applied as early as 20 weeks gestation.

Jail Refers to a locally operated correctional facility where people who are awaiting trial and sentencing (i.e., they have been charged but not yet tried for a crime) or who have been sentenced and are serving a short sentence (usually for a misdemeanor or low-level offense) are held. Jails also hold inmates awaiting transfer to state or felony prisons.

Just culture A framework for balanced accountability for frontline operators and the organization in which they work that emphasizes learning over blaming and emphasizes system-level changes to reduce bad outcomes.

Known complication A recognized potential harm acknowledged to sometimes results from specific medical care and/or procedures, despite reasonable provider conduct and precautions.

Latent tuberculosis (TB) infection Refers to a state in which a person has been exposed to *Mycobacterium tuberculosis*, as demonstrated by an immune response, but who has no evidence of disease. People with latent tuberculosis (TB) infection cannot transmit TB.

Lead exposure or lead poisoning Any detectable elevation of blood lead level. The U.S. Centers for Disease Control and Prevention (CDC) also sets a reference level of 5 μ(mu)g/dL, above which public health actions should be undertaken.

Legal custody Legal custody grants a parent or guardian the right to make decisions about school, religion, and medical care. The term "medical custody" or "legal medical custody" may be used to imply that a parent has rights and responsibilities for a child's medical care.

Life expectancy Statistical measure of the average length of time a person is expected to live, typically taking into account birth year, current age, and other factors (e.g., gender). Distinguished from measures of longevity and maximum life span. Because life expectancy is an average, it is highly sensitive to infant and childhood mortality (i.e., a life expectancy of 36 years likely does not mean that most people die in their 30s but that a large proportion die in early childhood).

Limitation of intervention Refers to decisions or orders to restrict the types of interventions (both diagnostic and therapeutic) administered to a particular patient. Usually made when certain interventions cannot achieve the patient's goals of care. These commonly include (but are not exclusive to) Do Not Resuscitate (DNR) and Do Not Intubate (DNI) orders.

Low-income country (LIC) A country with a gross national income (GNI) per capita of $1,005 or less (in 2016, World Bank).

Low-income Categorization of working families in the United States who earn less than twice the federal poverty threshold.

Lower-middle-income country (LMIC) A country with a gross national income (GNI) per capita of $1,006–$3,955 or more (in 2016, World Bank).

Maternal mortality Death while pregnant or within specified time frame after delivery or other termination of pregnancy, regardless of duration of pregnancy, from any cause related to or aggravated by the pregnancy or its management, excluding accidental or incidental causes. The World Health Organization (WHO) defines the time frame of 42 days, while the U.S. Centers for Disease Control and Prevention (CDC) uses 1 year. Maternal mortality is usually reported as a rate (number of maternal deaths per 100,000 live births).

Mature minor A person under the age of majority, often an adolescent, who is thought to have the capacity to make a medical decision for herself equivalent to that of an adult. A subjective description that is not legally defined in most states.

Medicaid State-administered program that provides health insurance coverage to eligible low-income adults, children, pregnant women, elderly adults, and disabled persons. Under the Affordable Care Act, states had the option to expand their Medicaid programs to cover persons up to 138% of the federal poverty level.

Medical error A preventable adverse effect of care, which includes acts of omission or commission, contributing to an unintended result, whether or not it is harmful to the patient.

Medical exemption Circumstances where an individual is not required to receive a vaccine that is otherwise required because of an existing medical condition that either places them at higher risk for a vaccine-associated adverse event or would render the specific vaccine ineffective.

Medical futility Medical futility can be defined in many ways but generally is considered when there is no significant likelihood of benefit for a given intervention.

Medical malpractice Can be understood as an act of omission or commission in which a medical profession deviates from an accepted medical standard of care resulting in patient harm. In determining whether an event is medical malpractice, four common criteria are used (the "4 Ds"): Did the patient (1) establish a doctor-patient relationship resulting in a *duty* of care; (2) was the provider *derelict* in providing said care; and (3) did that dereliction *directly* cause (4) physical or emotional *damage* to the patient?

Medical mission Typically used to describe faith-based endeavors that involve administration of medical services.

Medical-legal partnership (MLP) A collaborative partnership between healthcare workers and lawyers aimed at addressing health-harming social and legal needs of patients.

Medicalization The process by which common physical processes and states come to be pathologized within the medical model and become understood as potential or actual illness.

Medicare Federal health insurance program for the elderly, some younger persons with disabilities, and people with end-stage renal disease.

Medication-Assisted Treatment (MAT) Defined by the Substance Abuse and Mental Health Services Administration (SAMHSA) as "the use of FDA-approved medications, in combination with counseling and behavioral therapies, to provide a 'whole-patient' approach to the treatment of substance use disorders."[6]

Microcephaly Condition in which a fetus or infant's head circumference is significantly smaller than expected for age; can be due to abnormal development or cessation of growth; often associated with neurocognitive problems.

Micronutrient deficiency Deficiency of dietary vitamins and/or minerals essential for development, disease prevention, and health (examples include vitamin D and iron deficiencies).

Middle-income country (MIC) A country with a gross national income (GNI) per capita between $1,005 and $12,235 USD.

Migrant Broadly defined as a person living either transiently or permanently in a country outside of his/her own birth country who lacks legal recognition of rights in the host country.

Minority stress Persistently high levels of stress faced by members of stigmatized minority groups.

Mitochondrial disorders Rare disorders caused by mutations in nonchromosomal DNA located within the mitochondria. These disorders can be found to affect any part of the body including the brain and the muscles.

Multifactorial disorders Occur as the result of mutations in multiple genes, frequently coupled with environmental causes. An example of a multifactorial disorder is diabetes.

Multiple sclerosis (MS) Autoimmune disease that is characterized by focal demyelinated plaques within the central nervous system, affecting multiple areas of neurological function, including vision, movement, touch, cognition, and emotion.

Near-miss Any event that could have resulted in patient harm but did not. Near-misses are indistinguishable from adverse events in every way except the outcome.

Negligence Generally understood in a legal context as failure to exercise the care that a reasonably prudent person would exercise in similar circumstances.

Neonatal mortality Death during the first 28 days of life; usually reported as a rate (number of infant deaths per 1000 live births).

Non-governmental organization (NGO) A nonprofit, voluntary citizens' group that does not accept government funds to do their work.

Non-maleficence A principle of medical ethics, often informally referred to as "do no harm"; identifies the physician's responsibility to always weigh risks against benefits and to choose a course of action that ideally limits chance of harm to the patient.

Obese Having a body mass index (BMI) at or greater than the 95th percentile in children or a BMI greater than 30 kg/m² in adults.

Opioid use disorder Defined in the *Diagnostic and Statistical Manual of Mental Disorders (DSM)*, 5th edition, as the repeated occurrence within a 12-month period of at least 2 of 11 problems, including withdrawal, giving up important life events in order to use opioids, and

---

6 Substance Abuse and Mental Health Services Administration. Medication-Assisted Treatment (MAT). ► https://www.samhsa.gov/medication-assisted-treatment.

Glossary

excessive time spent using opioids; a cluster of 6 or more indicates a severe condition.[7]

Opioid-induced hyperalgesia Occurs when persons who are exposed to opioids become more sensitive to stimulation of their pain receptors (nociceptive sensitization), caused by exposure to opioids.

Opioid Class of drugs that includes natural or semisynthetic prescription pain medications (such as codeine, oxycodone, morphine, and methadone), synthetic opioids (such as fentanyl), and illicit drugs (such as heroin).

Overall survival (OS) Proportion of patients still alive at a defined time point after diagnosis; often contrasted with "event-free survival." Overall survival includes patients who have relapsed but are still living. Time point typically varies based on the disease; although "5-year overall survival" is commonly reported, studies involving a disease with a poor prognosis may report "1-year overall survival."

Overweight Having a body mass index (BMI) between the 85th and 95th percentiles in children or a BMI between 25 and 30 kg/m^2 in adults.

Pain Unpleasant sensory and emotional experience, associated with actual or potential tissue damage, or described in terms of such damage.

Palliative care Medical care that focuses on reducing the burden of symptoms and relieving suffering for people who have life-threatening illnesses.

Patient autonomy A principle of medical ethics; refers to the patient's right to be informed and choose among potential courses of medical action.

Patient safety Freedom from accidental injury in a healthcare context.

Perinatal mortality Stillbirth (death of a fetus weighing at least 500 g, or at or beyond 22 weeks completed gestation) or early neonatal death (first 7 days of life); usually reported as a rate (number of such deaths per 1000 births, including both stillbirths and live births).

Personal belief exemption Circumstances where an individual does not receive a vaccine because doing so would be in contradiction with the beliefs or philosophy of the individual and/or the individual's parent or guardian, beyond the scope of faith-based decisions.

Poverty threshold A federal poverty measure, also referred to as the poverty line, used by the U.S. Census Bureau to define and quantify poverty in the United States every year for families of different sizes. Individuals or families are classified as poor if their annual pretax income falls below a certain dollar amount (i.e., poverty threshold). The federal poverty guidelines are a simplified version of the threshold and are used by government programs to determine eligibility.

Preterm labor Regular contractions of the uterus resulting in changes in the cervix that start before 37 weeks of pregnancy. Changes in the cervix include effacement (the cervix thins out) and dilation (the cervix opens so that the fetus can enter the birth canal).

Primary prevention Public health interventions that target an entire population to prevent the development of disease (obesity) or injury.

Prison Refers to a correctional facility where people serve longer sentences once they have been sentenced to a crime. Prisons tend to be reserved for higher-level offenses such as felony convictions, and they are usually operated by the state government or the U.S. Federal Bureau of Prisons. Private prisons are for-profit correctional facilities that are operated by a third party through a contract with a state or federal agency.

Quality-adjusted life year (QALY) A unit of measurement for quality and quantity of life. QALY are given on a scale of 0 to 1. Death being equal to 0 and 1 being equal to a year of life with full, complete, and perfect health. QALY can be given a value between 0 and 1 for years of illness.

Randomized controlled trial A type of research study in which subjects are randomly allocated to groups that receive a specified intervention, or else a control, where subjects receive either a placebo or the currently accepted standard of care.

Rape myth False beliefs about sexual violence, perpetrators, and survivors that often serve to excuse sexual aggression and create hostility toward victims and can bias criminal prosecution. Examples in the United States include that women often lie about rape, that women "ask for it" or increase their chances of being assaulted by how they act or dress or where they go, and that "real victims" behave in predictable ways (such as fighting back physically and seeking medical attention immediately).[8]

Religious exemption Circumstances where an individual does not receive a vaccine that is otherwise required because doing so would violate religious beliefs of the individual and/or the individual's parent or guardian.

---

7   American Psychiatric Association. Diagnostic and Statistical Manual of Mental Disorders, Fifth Edition. Washington, DC: American Psychiatric Association. 2013.

8   Ubuntu! Myths about sexual assault-rape myths. https://iambecauseweare.wordpress.com/myths-about-sexual-assault-rape-myths/.

Re-emerging infection An infection that at one time was within a population, disappeared, and has now returned with a rapidly increasing incidence.

Resettlement agency According to the United Nations, "resettlement agencies help newly arrived refugees settle into local communities and provide a wide range of services that promote self-sufficiency, such as employment and language skill classes. In addition to refugee integration, several of the agencies also provide comprehensive immigration services to assist refugees on their path to becoming permanent residents or U.S. citizens."[9]

Respect for autonomy The principle that patients have the right to make their own decisions and to have those decisions respected.

Right to an open future An alternative approach to surrogate decision-making that suggests parents should not be permitted to make decisions for a child that irrevocably close off certain future options; often applied to decision-making with regard to genetic testing, wherein knowing the results eliminates the child's future choice *not* to know.

RIME scheme Holistic, developmental learner assessment framework created by Lou Pangaro (1999). Each letter of RIME represents a set of learner behaviors and positions: Reporter, Interpreter, Manager, and Educator.[10]

Root-cause analysis A multidisciplinary process of study or analysis that uses a detailed, structured process to examine factors contributing to a specific outcome, such as an adverse event.

Scope of practice Range of procedures, actions, and processes that a healthcare practitioner is permitted to undertake by the terms of the practitioner's professional license or other certification (e.g., specialty board certification) or credentialing (e.g., hospital privileges).

Second victim phenomenon Refers to the psychological harm, trauma, and distress that healthcare practitioners may experience as a result of their involvement in an unanticipated adverse patient event, medical error, and/or patient-related injury.

Secondary prevention Interventions designed to detect existing diseases in early stages in order to halt or reverse disease progress, thereby improving the probability of positive health outcomes. In the context of obesity, secondary prevention focuses on the identification of overweight/obese individuals and weight reduction to prevent disease progression and development of comorbidities over time.

Sentinel event Defined by The Joint Commission as any unanticipated event in a healthcare setting resulting in death or serious physical or psychological injury to a patient or patients, not related to the natural course of the patient's illness.

Short-term experience in global health (STEGH) An experience offered organizations based in HICs to participate in a health-related experience (research, service, and/or education) in a low- and middle-income country (LMIC).

Short-term medical mission (STMM) A similar term describing organizations based in HICs to participate in a health-related experience (research, service, and/or education) in a low- and middle-income country (LMIC).

Sickle cell disease (SCD) Inherited disorder resulting in abnormal hemoglobin molecules that distort the red blood cells into a sickle shape; characterized by hemolytic anemia and intermittent vaso-occlusive events that cause tissue ischemia, leading to acute and chronic pain as well as organ damage.

Single-gene disorders Occur when an alteration occurs in a gene causing one gene to stop working. An example of a single-gene disorder is Huntington disease.

Social determinants of health Conditions in the places where people live, learn, work, and play that affect health risks and outcomes.

Socio-ecological model A model of health that emphasizes that social and environmental factors interact with individual and biological elements to impact health. Includes individual, interpersonal, organizational, community, and policy levels.

Spontaneous vaginal delivery (SVD) Vaginal birth without medical assistance to induce, expedite, or facilitate labor.

Standard of disclosure The legal rule for determining whether healthcare practitioners have met their obligations to disclose information to patients in the informed consent process.

Statistical power The probability of a study to correctly distinguish a measured effect from one of pure chance (i.e., reject the null hypothesis); as statistical power increases, there is a decreasing probability of false-negative results.

Stillbirth Death of a fetus weighing at least 500 g, or at or beyond 22 weeks completed gestation.

Structural racism System in which institutional practices, policies, and cultural norms work to reinforce ways to perpetuate racial group inequity.

Structural violence An umbrella term used by medical social scientists to denote the harm done by social,

---

9 The UN Refugee Agency. US resettlement agencies. ▶ http://www.unhcr.org/en-us/us-resettlement-agencies.html.
10 Pangaro L. A new vocabulary and other innovations for improving descriptive training evaluations. Acad Med. 1999;74:1203–7.

political, and economic structures and institutions; social structures in society that limit some members of society (and thus society itself) from achieving their potential.[11,12]

Substance use disorder Clinically and/or functionally significant impairment, such as health problems, disability, and failure to meet major responsibilities at work, school, or home, due to recurrent use of alcohol and/or drugs.

Swiss cheese model A model of error causation, attributed to James Reason, that suggests that medical errors cause adverse patient outcomes in complex healthcare systems when flaws in the safeguards at multiple care levels align in a way that allows an error to reach a patient.

Systems approach Encourages a focus on the organization (i.e., inputs, processes) rather than the individual when seeking an understanding of categories of outcomes (i.e., adverse events). In a healthcare context, this approach suggests that because most errors reflect predictable human failings in the context of a poorly designed system, focusing on changing the underlying systems is the best way to reduce harm.

Tuberculosis (TB) disease (also referred to as "active TB") Refers to a state in which a person has clinical evidence of TB. While TB disease most often affects the lungs, TB can also affect other parts of the body, such as the brain, spine, and kidneys.

Tertiary prevention Interventions designed to reduce the impact of existing illness with goals of improving functionality, quality of life, and/or life expectancy.

Therapeutic privilege When healthcare practitioners withhold information from patients because it would do more harm than good to tell the patient; an exception to informed consent, disfavored in law and ethics.

Transgender female A person whose sex assigned at birth was male, but who identifies as a female.

Transgender male A person whose sex assigned at birth was female, but who identifies as a male.

Transgender A person whose sex assigned at birth (based on external anatomy and/or chromosomes) does not match their gender identity. It is important to note that transgender individuals may identify as gay, straight, bisexual, or something else. Sexual orientation and gender orientation are separate concepts.

Type 1 error Also referred to as a "false positive"; occurs when a study result suggests a particular effect is present, when it is actually absent (i.e., falsely rejecting the null hypothesis).

Type 2 error Also referred to as a "false negative"; occurs when a study result suggests an absence of a particular effect, when it is actually present (i.e., falsely failing to reject the null hypothesis).

Undervaccinated Circumstances describing children who are receiving vaccine series but who have substantial delays in being current for all recommended doses due to delays for any reason. These are children who are not "up to date" according to the recommended vaccination schedule.

Vaccine hesitancy Refusal of or delay in receiving vaccination despite a recommendation by health authorities.

Vector-borne disease Defined by the World Health Organization (WHO) as "human illnesses caused by parasites, viruses and bacteria that are transmitted by mosquitoes, sandflies, triatomine bugs, blackflies, ticks, tsetse flies, mites, snails and lice"; distinguished from diseases that are transmitted directly from human to human (via respiratory or oral-fecal routes, for instance) and zoonoses, which are transmitted directly from animal to human.

Vertical transmission Infection caused by pathogens that are transmitted from the mother to an embryo, fetus, or neonate during pregnancy or childbirth; human immunodeficiency virus (HIV) is a well-established example of this.

Voluntourism A combination of volunteering and tourism that allows an individual to enjoy an experience in international travel while contributing to a local project.

Vulnerable population In research, these are groups of people perceived to be an increased risk for exploitation, unable to give true voluntary or informed consent, and requiring special protections; they include children, prisoners, and people with cognitive limitations, although there may be other groups for whom the same concerns apply (e.g., economically disadvantaged).

Waiver of informed consent When the patient does not want to know about their diagnosis or treatment options; an exception to informed consent, disfavored in law and ethics.

Weight bias Negative weight-related attitudes, assumptions, and judgments toward people who are overweight, obese, or underweight that result in stigmatization, discrimination, adverse health outcomes, and social inequities.

Weight-for-length Current anthropomorphic standard used to assess nutritional status in the first 2 years of life.

Zone of parental discretion A tool for ethical deliberation by healthcare practitioners and ethicists, intended to help them recognize when a decision for a child rightfully belongs to the parents.

---

11 Galtung J. Violence, Peace, and Peace Research. J Peace Res. 1969;6(3):167–191.
12 Farmer P, Nizeye B, Stulac S, Keshavjee S. Structural Violence and Clinical. PLOS Medicine. 2006;3(10):e449.

Index

A

Abdominal aortic aneurysm (AAA)
- informed consent
  - affirmative legal obligation 41, 57
  - Affordable Care Act 44
  - alcohol/drug use 41, 43, 44, 56
  - creative problem-solving 41, 56
  - cultural background 40, 46, 47
  - decisional (in)capacity 47, 48
  - disability 43, 44
  - disproportionately burdens 42, 58
  - duty to inform 41, 53
  - emergency treatment 50, 51
  - ethical and legal consequences 54
  - ethical model 40, 45
  - experienced provider 53
  - financial conflicts of interest 43, 44
  - health care workers 44
  - health literacy 46, 47
  - HIV epidemic 41, 43, 44, 56
  - level of experience 42, 57
  - patient autonomy 47
  - personal characteristics 41, 43, 44, 56
  - personal religious/ethnic values 47
  - physical/mental impairment 41, 56
  - quality measures 45
  - self-reflection 41, 42, 52, 56
  - skills/equipment 57
  - standard of disclosure 40, 42, 43
  - surgeon's credentials/hospital's surgical program online 53
  - theory and practice of 46
  - therapeutic privilege 50
  - track record 57
  - truth-telling 54
  - waiver 49
  - well-trained qualified physicians 54
- morbidity and mortality rates 55
- physicians *vs.* institutions 56
- risk of death 55
- surgical complication 40, 45
Ability 304
- definition of 297
- healthcare practitioner 298
Abortion laws 455, 461, 464
Accountable care organizations (ACOs) 146, 152, 153
Accountable Health Communities Model 245

Accreditation Council on Graduate Medical Education (ACGME) 390, 402
Acetaminophen 353
Acquired immunodeficiency syndrome (AIDS) 44, 63, 68, 69, 159, 182, 481
Activities of daily living (ADLs) 296, 300, 305, 308, 468
Actuarial fairness 329
Acute lymphoblastic leukemia 188
Acute pain 353
Addiction 350, 354, 358, 359
Advance care planning codes 264
Advanced practice provider (APP) 24
Adverse Childhood Experiences Study (ACES) 232, 238
Adverse event, definition of 338
Advisory Committee of Immunization Practices (ACIP) 172
Advisory Committee on Childhood Lead Poisoning Prevention (ACLPP) 105
Advisory Committee on Immunization Practices 179
Advocacy 10–11
Affordable Care Act (ACA) 40, 44, 146, 153, 155–157, 256, 279, 307, 320
Alcohol use disorder 210, 351, 359
Allergic reaction 339
Alzheimer disease, EOL care
- advance care planning 264
- catholic tradition 258
- children 259
- disclosure/breaking bad news 263
- ethics consultation 254
- faith and culture 254, 265, 266
- healthcare agent 255
- healthcare proxy 253, 258, 260
- Medicare 252, 256
- MOLST/POLST 252, 255, 256
- monetary incentives 264
- orthodox Jewish community 258
- patient's recent statements and past wishes 260–262
- physiologic perspective 254
- religious belief 252, 258
- spouse 259
- surrogate decision maker 255
- training programs for physician 263
- UDDA 252, 257
- for unbefriended older adults 262, 263
- withholding and withdrawing LST 252, 255
Ambulance 219
American Academy of Pediatrics (AAP) 105, 146, 152, 170, 190, 215, 223

American Bar Association (ABA) 257
American College of Obstetricians and Gynecologists (ACOG) 80, 84
American Community Survey 153
American Medical Association (AMA) 80, 84, 257
American Psychiatric Association 416
American Public Health Association (APHA) 482
American Society for Addiction Medicine (ASAM) 355, 356
Americans with Disabilities Act (ADA) 290, 293, 294, 297, 302, 303
Amniocentesis 316, 317, 320, 322, 323
Amyotrophic lateral sclerosis (ALS) 305
Anchoring 7, 341
Angry 360
Anticipatory guidance around diet 440
Anti-Shackling Law 85
Artificial contraception 461
Artificial nutrition and hydration (ANH) 467
- healthcare agent 252, 255
- physiologic perspective 252, 254
- state's law 254, 264
- surrogate decision maker 252, 255
Assessment system validation evidence 505
Assimilated Mennonites 192
Assisted outpatient treatment (AOT) 218
Assisted reproduction 321, 327
Association for Certified Nurse Midwives (ACNM) 131, 132
Association of Women's Health, Obstetric and Neonatal Nurses 89
Asylum seeker 232, 235
Autism 415
Autism spectrum disorders 469
Autonomy 4, 7, 324, 362, 380
Availability bias 9

B

Beck Depression Inventory (BDI) 223
Behavioral Risk Factor Surveillance System (BRFSS) 293
Beneficence 2, 3, 6, 8, 19, 20, 70, 94, 111, 139, 173, 179, 219, 226, 275, 324, 325, 362, 373, 376, 490
Best interests standard (BIS) 186, 189
Bioethics 466
Birth control 455
Bloodless surgery 385
Bloom's taxonomy 502
Bone marrow transplantation 327
Brain death 252, 256, 257

Brain tumor
- best interests standard 186, 189
- biomedical cultural practices 186
- chance of cure 187, 197
- clinical trial 186, 188, 189
- cost 186, 188
- ethical and social issues 186
- event-free survival 193
- family interests 195
- family support 194
- financial burden 187, 198
- harm principle 186, 189
- healthcare practitioner 186, 195
- incidence 186, 188
- inquiry to pursue 193
- mandated reporter 186, 191
- mature minor 186, 190, 191
- medical decisions, without parents' consent 186, 190
- medical neglect 186, 191, 192
- Mennonites 186, 192
- pediatric decision-making 187, 193, 194
- personal medical decisions or advice 188, 199
- radiation therapy 187, 195
- relapse 193
- religious beliefs 188, 200
- right to an open future 186, 189
- risk assessment 187, 196
- salvage therapy 193
- Supreme Court cases 200
- tight to an open future 186

Brain/central nervous system 188
Brave spaces 34
BRCA mutations 329
Breastfeeding 86
Bronze plans 279

C

Capacity-building interventions 488
Cardiopulmonary resuscitation (CPR) 276
Case events 24
Case-based learning (CBL) approach 5, 6, 26
Catholic faith 47
Centers for Disease Control and Prevention (CDC) 65, 149, 293
- Community Guide 444
- growth charts 440
- National Center for Injury Prevention 214
- National Vital Statistics System 212

Centers for Disease Control and Prevention's Traveler's Health Website 482
Centers for Medicaid and Medicare Services (CMS) 40, 44, 146, 147, 152, 153, 157, 241, 245

Central nervous system (CNS) germ cell tumors 186
Certified professional midwives (CPM) 131
Charter on Medical Professionalism 5, 12
Child abuse 192
Child Protective Services (CPS) 182, 191, 199
Child tax credit 152
Childhood obesity
- BMI screening 434
- built and shared environments 435
- calorie-dense and nutrient-poor 446
- community programs 443
- community resources 445
- counseling rates 443
- definition 432
- diabetic intrauterine environment 435
- diagnostics 433
- diet and exercise program 436
- diet/lifestyle 440
- economic and family stability 445
- economic impact 433
- epigenetic modifications 435
- evidence-based (non-surgical) approach 430, 433, 434
- exercise programs 435
- family's diet and lifestyle 431
- family's immigration status 446
- fetal nutrition and environmental exposures 435
- food deserts 430
- growth parameters and BMI 435
- health outcomes 430
- healthcare practitioners 431
- healthy foods 433
- healthy weight management 435
- heritable changes in gene expression 435
- hyperglycemic intrauterine environment 435
- immigration problems 446
- incidence 432
- maternal diabetes 435
- maternal dietary changes 435
- medical, legal, and social resources 447
- metabolic and psychosocial outcomes 433
- MLP 434
- nutrition counseling 443
- in offspring 435
- parental history 435
- patient's distress 441
- physical activity 433
- population level or on health outcomes 445
- poverty 438
- prenatal exposure to glucose and epigenetic modification 430

- prevalence rates 433
- prevention 433
- primary care practice physicians 442
- primary prevention 432, 433, 444
- resources and responsibilities 431
- screening measures 435
- secondary and tertiary prevention 430, 435, 436
- self-reflection 431
- social challenges 443
- social determinants of health 443
- social inequities 433
- socio-ecological model 434
- socioeconomic and interpersonal issues 443
- treatment 433, 446
- vegetable consumption 444
- WIC and SNAP eligibility 432

Children's Health Insurance Program (CHIP) 152
Chilling effect 446
Chorionic villus sampling (CVS) 316, 317, 320, 322
Chromosomal diseases 316
Chronic illnesses 360, 361
Chronic obstructive pulmonary disease 150
Chronic pain 354
Class preparation 24
Clinical Immunization Safety Assessment Project 172
Cognitive "de-biasing" strategies 10
Cognitive bias 127, 134, 135
- anchoring 9
- availability bias 9
- cognitive "de-biasing" strategies 10
- confirmation bias 10
- framing effect 10
- fundamental attribution error 10
- implicit/unconscious 10
- information bias 9
- learners 10
- premature closure 10
- reflective practice 10
- representativeness 10
- screening recommendations 10

Cognitive impairment 382
Cognitive, definition of 293
Columbia-Suicide Severity Rating Scale (C-SSRS) 223
Coma 252, 256
Commission bias 135
Common Rule 390, 393
Communication, Orientation, Mindfulness, Family, Ongoing, Reiterative, and Team (COMFORT) model 277
Community benefit 156
Community design 434
Community health workers 482
Community immunity 176

Community level factors 117
Community level strategies 434
Comparative effectiveness research (CER) 390, 394, 395
Compassionate use 370, 371, 374, 378
Comprehensive Addiction Recovery Act (CARA) 356
Confidentiality 326, 330
Confirmation bias 10, 135
Conflict of interest 375, 376, 379
Confounder 390, 393
Congenital Zika syndrome 463, 468
Congressional Research Service (CRS) 224
Consolidated Omnibus Budget Reconciliation Act (COBRA) 270, 271, 274, 279
Consultation 356
Consumer Product Safety Commission (CPSC) 212
Contact sexual violence 237
Content organization 26
Context elements
- adapted RIME scheme 502
- artifacts 503
- case series 502
- Facilitator Positionality 503
- Learner Positionality 502–503
Context evaluation 496, 497
Context/inputs/process/products (CIPP) model 496, 499
Contextual features 496
Continuous positive airway pressure (CPAP) 252
Contraceptive services 461
Co-pays 146–148, 157, 160
Core concepts 18–21
Cost-benefit analysis (CBA) 280
Cost-effectiveness analysis (CEA) 280
Crack cocaine epidemic 361
Crash Injury Research (CIREN) 215
Creative problem-solving questions 25, 28
Critical congenital heart defects (CCHD) 132
Cultural competence 7
Cultural humility 7
Culture competence 7
Culture of medicine 7
Culture of safety 345
Current Population Survey Annual Social and Economic Supplements 153
Cytomegalovirus (CMV) 63, 68

D

Dactylitis 351, 359
Daily living tasks 306
Data Safety Monitoring Board (DSMB) 397
Dead donor rule 257
Decisional incapacity 47, 48
Decision-making
- vaccination
  - benefits and risks of 167, 173–175
  - circumstances 166, 173
  - communication approaches 166, 170, 171
  - creative problem-solving 168, 182
  - ethical considerations 179
  - ethical dilemma 174
  - family and healthcare practitioner 167, 181
  - healthcare practitioners/medical practices 166, 167, 170, 179
  - HPV 167, 179–181
  - impact rates 166, 169
  - legal dilemma 174
  - market 166, 172
  - misinformation affects, dissemination of 168, 182
  - positive public health impact 168, 181, 182
  - presumptive vaccine recommendations 181, 182
  - prevention 176
  - process 166, 168
  - professional groups 170
  - public health dilemma 175
  - religious exemption and personal belief exemption 167, 177, 178
  - self-reflection 166, 167, 173, 179
  - spectrum of attitudes 166, 169, 170
  - trust 181
  - unvaccinated patients, healthcare practitioners 175
  - vaccine refusal, impact of 167, 176–178
  - VAERS 166, 171, 172
Decision-making and planning 484
Decision-making capacity 383
Decision-making process 6
Declaration of brain death in Orthodox Judaism 252, 257
Deductibles 147, 148, 157
Definitive prenatal genetic testing 319
Dementia 371, 382, 383
Dengue 459
Depression
- age-adjusted suicide rate 208, 211
- antidepressant use 208, 215–217, 219
- assisted outpatient treatment 218
- causes of death 208, 210
- chain of events 221
- C-SSRS 223
- duty to warn laws 214
- ethical implications 210, 226
- firearm ownership 209, 224
- firearm-related injury 208, 214
- gun injury 209, 212, 224
- gun violence 209, 224, 225
- Haddon matrix 221, 222
- homicide 208, 211
- limitation 209, 224
- mental illness 208, 213
- morbidity and mortality 208, 212
- NYSAFE law 214
- outpatient mental health follow-up 225
- patient confidentiality 210, 225
- Patient Health Questionnaire 223
- patient's confidentiality 208, 212
- policy change 210, 226, 227
- policy-level interventions 223
- primary care physicians 217, 218, 223
- principle of beneficence 219
- sequence of events 221
- social-ecological model 222
- stigma 222
- suicide activity 209, 218
- to treat adult patient without consent 208, 212, 213
- unintentional or accidental deaths 208, 211
- USPSTF 222, 223
Design bias 141
Developmental delay 101, 112, 468, 469
Diabetes Prevention Program 435
Diagnosis and healthcare features 18, 22–23
Diagnostic and Statistical Manual (DSM) 416
Dickey Amendment 214
Directly observed therapy (DOT) 62, 64, 65, 73
Director of community services (DCS) 214, 220
Disability 298, 302
- definition of 297, 298
- emotional process 301
- healthcare practitioners 299
- practical accommodations 300, 301
- socio-ecological model 295, 302, 303
- types of 293
- universal design in healthcare setting 294
Disability specialty physicians 295
Disagreement between family members 265
Discussion questions 24, 25
Disease surveillance 103, 171
Disproportionate-share hospital (DSH) payments 157
Division of Children's Services 236
Division of Refugee Assistance 236
Division of Refugee Health 236
Division of Resettlement Services 236
Do not intubate (DNI) 264, 265
Do not resuscitate (DNR) 253, 254, 258, 264, 265
Downstream implications 11

Down syndrome 106, 320
Drug development and approval 370
Drug maker and study investigators 378
Drug seekers 363
Drug-seeking 352, 361
Duchenne muscular dystrophy (DMD) 370, 377, 378, 383, 384
– definition 370
– eteplirsen 379
– genetic defect 380
– physician 380
Duty hours 390
Duty of healthcare 180, 181
Duty to assist 489

E

Early warning scores 337, 345
Earned income tax credit 152
Ebola epidemic 456, 470
Education and culture change 344
Educational background 27
Educational settings 392
Educational tests 392
Ego-dystonic homosexuality 416
Electronic Benefit Transfer (EBT) card 437
Electronic health record (EHR) 252
Electronic medical records (EMR) 239
Emergency admission 220
Emergency Medical Treatment and Labor Act (EMTALA) 70, 155, 156
End-of-life care
– Alzheimer disease
   – advance care planning 264
   – ANH withdrawal (*See* Artificial nutrition and hydration (ANH) withdrawal)
   – brain death 252, 256, 257
   – Catholic tradition 258
   – children 259
   – coma 252, 256
   – disclosure/breaking bad news 263
   – ethics consultation 254, 258, 260, 264, 265
   – faith and culture 254, 265, 266
   – healthcare proxy 253, 254, 258, 260, 263, 264
   – Medicare 252, 256
   – minimally conscious state 252, 256
   – MOLST/POLST 252, 255, 256
   – monetary incentives 264
   – Orthodox Jewish community 258
   – patient's recent statements and past wishes 253, 260–262
   – physiologic perspective 254
   – religious belief 252
   – spouse 259
   – training programs for physician 263
   – UDDA 252, 257
   – for unbefriended older adults 262, 263
   – vegetative state 252, 256
– withholding and withdrawing LST 252, 255
– metastatic breast cancer
   – advance directive 271, 280
   – COBRA 270, 274
   – COMFORT model 277
   – cost-benefit analysis 280
   – cure value 281
   – disclosure/breaking bad news 277
   – DNR status 280
   – good death 283
   – healthcare practitioner responsibility 281
   – high-cost pharmaceutical company 282
   – hospice care 270, 273, 274
   – insurance 279
   – law/policy for drug pricing 282
   – medical futility 270, 274–276
   – motivating family's decision 270, 275
   – NICE 274
   – palliative care 270, 272, 273, 278
   – pricing structure 281
   – private insurance 270, 274, 281
   – public insurance 270, 274
   – SPIKES model 277
   – total healthcare costs 270, 272
   – treatment in UK 279
Environmental hazards 118
Epidemic 62–64, 70, 352
Estelle v. Gamble, 429 U.S. 97 83
Eteplirsen 378–381, 386
– benefits 381
– risks 381
Ethical issues 487
Ethics consultation 254, 258, 260, 263–265
Ethics/social determinants of health 6
Evaluation
– context 496, 497
– definition 496
– input 497
– process 497–499
– product 499
Event phase 221
Event/injury 221
Event-free survival (EFS) 186, 193
Everyday Discrimination Scale 152
Evidence-based (non-surgical) approach 430
Evidence-based practice (EBP) 6, 8
Exclusion prenatal genetic testing 319
Expert knowledge 27
Extensively drug-resistant (XDR) TB 66

F

Faces, Legs, Activity, Cry, Consolability (FLACC) scale 353
Facilitator positionality 503
Facilitators
– challenges 32
– clinical experience 31
– effective teachers 29
– hearing facilitators 31
– interjections 31
– personal experience 31
– preparation 29
– pre-set ground rules 31
– role 30
– scope of knowledge and expertise 31
– self-reflection and lifelong learning 29
Faith-based organizations (FBO) 478, 483
Family and Medical Leave Act (FMLA) 308
Family planning funds 461
Family's home life 442
Fatality Analysis Reporting System (FARS) 215
FBI's Supplementary Homicide Reports (SHR) 212
Federal Civil Rights Act 70
Federal Food, Drug, and Cosmetic Act (FD&C Act) 374
Federal policy for protection of human subjects 390, 393
Federal Poverty Guidelines (FPG) 437
Federal Poverty Level (FPL) 153, 155, 156, 308
Female genital mutilation 232, 234, 237, 242
Firearm policy change 227
Firearm-related research 226
Firearm-related violence 224
Firearms 210
Follow-up questions 24, 25
Food deserts 439
Food insecurity 430, 431, 437–439, 441, 444
Food Stamps, *see* Supplemental Nutrition Assistance Program (SNAP)
Formative/summative assessment 26, 502
Four principles approach 4, 5
Fragile X syndrome 371, 382
Framing effect 10
Free erythrocyte porphyrin (FEP) 101, 104, 108
Free healthcare 487
Full disclosure 336
Fumigators permission 470
Functional technical standards 297
Fundamental attribution error 10
Funding by crisis 456
Futile care 257, 263

G

Gender assignment 413
Gender diverse 410
Gender dysphoria (GD)

Index

- beneficence 417
- best interests standard 417
- case-control study 424
- cohort study 424
- confounding factors 425
- counseling and psychosocial support 422
- definition 410
- derogatory comments and microaggressions 412, 413, 426
- diagnosis 411, 419, 420
- diagnostic code 420
- ethics consultants approach 411, 421
- GnRH agonist therapy 410, 412, 414, 421, 423, 425, 426
- guardian ad litem 420
- harm principle 417
- health disparities 410, 415
- joint decision-making 411, 420
- justice 417
- mature minors 418
- medical treatment 410, 414
- medicare and medicaid 412, 425
- medications 414
- national and international standards of care 411, 421
- non-maleficence 417
- objections overriding 411, 422
- parent/guardian consent 411, 415
- pediatric decision-making 411, 417
- prevalence 410, 413
- psychotherapy 417
- randomized controlled trial 424
- recall bias 425
- respect for autonomy 417
- right to an open future 417
- selection bias 424
- treatment consent 411, 420
- treatment disagreements 412, 423
- values, beliefs, and social norms 411, 418

Gender identity 410, 411, 415, 416, 419
Gender incongruence 410, 413, 419
Gender nonconforming 410
Gender/body divergence 413, 419
Gender-based violence (GBV) 232, 237
Genetic counseling 318, 329
Genetic Information Nondiscrimination Act of 2008 (GINA) 321
Genetic tailoring of mosquitoes 462
Genetic testing 318, 319, 322, 323
Genetic variation 385
Germ cell tumors 186
Global gag rule 461
Global health 480, 484, 485
Global human community 487
Global warming 469
Gonadotropin-releasing hormone (GnRH) analogue therapy 410, 414, 421

Good health 12
Group dynamics and learning 27
Gynecological care 303

H

Haddon matrix 221, 222
Halacha 258
Harm principle (HP) 186, 189
Harvard's Project Implicit 159
Health Care Access Barriers Model (HCAB) 147, 148
Health care systems 433
Health care workers (HCWs) 44, 69
Health department 63, 71
Health diets 441
Health differences 293, 294, 303
Health disparities 12, 293, 294, 303
Health equity 12
Health inequalities 478, 480, 491
- in Haiti 481
- in Kenya 481
Health Insurance Portability and Accountability Act (HIPAA) 62, 67, 80, 85
Health literacy 40, 46, 47, 304
Health professions education 26, 34
Health professions program or concurrent 26
Healthcare practitioner (HCP) 18, 24, 41, 441, 486
Healthcare provider 42
Healthcare systems 358, 359, 362, 480
- ACA 153
- accessing health care 154, 155
- chronic disease prevalence 150
- circumstances 127
- CMS 152, 153
- cognitive biases 127, 134, 135
- creative problem-solving 128
- ethical obligations 127
- financial barrier
  - co-pays and deductibles 147, 148
  - lack of insurance 147
  - purchasing medications 148
  - transportation cost 148
- financial feasibility 157
- future healthcare practitioners 159
- HCAB 147
- healthcare practitioners weigh risks 128, 140
- HLHS 127
- home births
  - ACNM 131
  - ACOG 131, 132
  - birth and infant's death, circumstances of 135, 136
  - creative problem-solving 142
  - healthcare practitioners, types of 130, 131
  - HLHS 134

- hospital *vs.* in-home care 138
- interprofessional communication 142
- laws 137
- legislation mandating pulse oximetry testing 132, 133
- medicalization *vs.* natural process of care 138
- midwife 130, 131, 140
- mothers and families choosing 127, 137, 138
- NICE 132
- outcomes 128, 141, 142
- patient autonomy during pregnancy, fetus protection 138–140
- pregnant patient with decisional capacity 132
- pulse oximetry screening/ultrasound 136, 137
- rates and demographics 128, 129
- risks and benefits 135
- scope of practice 140
- self-reflection 133, 142
- study design 128, 141, 142
- training and level of experience 133
- implicit bias 158, 159
- influence decision-making 134, 135
- initial impression 127, 134
- interprofessional communication 128
- legal non-profit status 156
- life expectancy 149
- LMIC 480
- maternal and infant mortality, rates of 129, 130
- Medicaid eligibility 155, 156
- myths about poverty 154
- non-financial barrier
  - cognitive barrier 148
  - lack of interpretive services 148
  - navigation to 149
  - structural barrier 148
- non-profit and teaching hospitals 156
- poverty 152, 157
- pulse oximetry screening/ultrasound 127
- racism 150–152
- recommendations 128, 141
- self-reflection 126, 128
- unexpected deaths 127, 134
Health-related indicators 478, 481
- Haiti 481
- Kenya 481
- United States 481
"Healthy Corners" program 445
HeLa cells 372
Historic controls 375
Hodgkin lymphoma 188, 191

Index

Home births
- ACNM 131
- ACOG 131, 132
- birth and infant's death, circumstances of 135, 136
- circumstances 127
- cognitive biases 127, 134, 135
- creative problem-solving 128, 142
- ethical obligations 127
- healthcare practitioners weigh risks 128, 140
- healthcare practitioners, types of 130, 131
- healthcare systems, recommendations 128, 141
- HLHS 127, 134
- hospital *vs.* in-home care 138
- influence decision-making 134, 135
- initial impression 127, 134
- interprofessional communication 128, 142
- laws 137
- legislation mandating pulse oximetry testing 132, 133
- maternal and infant mortality, rates of 129, 130
- medicalization vs. natural process of care 138
- midwife 140
- mothers and families choosing 127, 137, 138
- NICE 132
- outcomes 128, 141, 142
- patient autonomy during pregnancy, fetus protection 138–140
- pregnant patient with decisional capacity 132
- pulse oximetry screening/ultrasound 127, 136, 137
- rates and demographics 128, 129
- risks and benefits 135
- scope of practice 140
- self-reflection 126, 128, 133, 142
- study design 128, 141, 142
- training and credentialing options 130, 131
- training and level of experience 133
- unexpected deaths 127, 134

Home care 307
Home visitation services 152
Homicide 208, 210, 211
Hospital care *vs.* in-home 138
Human immunodeficiency virus/acquired immunodeficiency syndrome (HIV/AIDS)
- epidemic 44
- mortality rates 481
- prevention and treatment 461

Human papillomavirus (HPV) 167, 176, 177, 179–181, 303
Human Rights Watch 236
Human subjects research 370
Human trafficking 232, 237
Humanitarian organization 489
Hunger Vital Sign screening tool 438
Huntington disease (HD) 316–318, 326, 329
- caused by 318
- definition of 318
- ethical obligation 330
- family history of 319
- mutation 316, 317, 324, 325
- predictive testing for 321
- right to an open future 329
- symptoms 318

Hypoplastic left heart syndrome (HLHS) 127

I

ICD-10 system 212
Immigrants 232, 235
Immunizations 180
Implicit bias 10, 147, 158, 159
In vitro fertilization (IVF) 316, 319, 321, 326, 328
Individuals with Disability Education Act's (IDEA) Early Intervention program 469
Influence decision-making 134, 135
Information bias 9
Informed component 216
Informed consent 40, 46, 216, 217, 324, 385
Inheritable congenital hypothyroidism 319
Inheritable disease
- pharmacologic management 319
- predisposition testing 319
- pre-implantation genetic diagnosis 319
- prenatal risk assessment 319
- presymptomatic testing 319

In-home care, hospital care vs. 138
Inmate health 80
Input evaluation 497
Institutional review board (IRB)
- approval 390, 392
- human participants 397
- informed consent 391, 397, 398
- scientific inquiry 391, 398, 399

Instrumental activities of daily living (IADLs) 296, 301, 305, 308
Intellectual impairment 383
Internally displaced person 232, 235
International Organization for Migration (IOM) 236
Interpreter services 232, 233, 239–241
Interprofessional perspectives 26
Intimate partner violence (IPV) 237
Intravenous diphenhydramine 359
Investigational device exemption applications (IDEs) 374

Investigational New Drug (IND) 373, 374
Involuntary admission 220
Involuntary commitment 220
Irregular periods (refugee)
- acknowledging distress 239
- clinician challenges 234, 242
- communication 241
- communication in primary language 233
- confidentiality 240, 241
- cultural humility 234, 245
- inquiring traumatic experiences 240
- mental health 234, 243, 244
- nonverbal cues 239
  (*See Also* **Refugee**)
- sexual assault in United States *vs* refugees 244
- traumatic experience inquiry 233

J

Jail deputy 82, 91
Jehovah's Witness faith 47
Johnson v. Kokemoor 44
Joint Commission on Accreditation of Healthcare Organizations (JCAHO) 353
Journal of the American Medical Association 149
Just culture 336
Justice 325, 328, 373, 377, 383, 384
Justice and resource allocation 491

K

Kendra's Law 218
Kennedy Krieger Institute (KKI) 107, 109

L

Land use policies 434
Latent tuberculosis infection (LTBI) 75
Lead exposure 100, 103, 104
Lead poisoning
- adverse health outcomes 100, 104
- advocacy 112, 120
- case control study 109
- CDC guidelines 104, 105
- community level factors 117
- creative problem-solving 103, 119
- determination 112
- developmental delay 101, 106, 112, 113
- distributive justice 120
- early intervention 106
- ethical concerns 101, 109
- findings of the study 100
- Flint study 108
- health concern 111
- health department 116

Index

- healthcare practitioners' ethical and legal obligations 101, 110–112
- Kennedy Krieger study 109
- kind of study 107, 108
- level of prevention 102
- lifelong cognitive impairments 102, 113
- local population, lead levels in 107, 108
- long-term trend 103
- obligations 102, 115
- patient perspective 113
- peer-reviewed literature 119
- policy level factors 117, 118
- pregnant mother and unborn child 115
- primary prevention 100, 105, 114
- public health systems 100, 105–107
- retrospective study 108
- screening for 100, 104–106, 114
- secondary prevention 106, 114
- self-reflection 102, 116
- separating families, risks and benefits of 102, 116
- services 101, 112
- social structures and customs sometimes 118
- socio-ecologic model 102, 116, 117
- structural violence 103, 118, 119
- tertiary prevention 106, 114

Learner assessment
- context elements
  - adapted RIME scheme 502
  - artifacts 503
  - case series 502
  - facilitator positionality 503
  - learner positionality 502–503
- direct observation 502
- formative 502
- objective 502
- self-report 502
- subjective 502
- summative 502

Learner positionality 502–503
Learner reflection 505
Learner satisfaction 27
Learner support network 497
Learner's career trajectory 487
Learning deficits 113
Legal permanent residents (LPRs) 446
Legal system 199
Less-educated parents 358
Leukemia 336
Liaison Committee on Medical Education (LCME) 402
Lifestyle and environmental factors 87
Life-sustaining treatment (LST) 255
Local health system 484
Local law enforcement 219
Local shelter network 497

Localized germinoma 189
Lost-days-of-work barrier 149
Lower-middle-income country (LMIC) 490

M

Major depressive disorder 216
Major Experiences of Discrimination Scale 152
Mandated reporter 186, 191
Maternal diabetes 435
Mature minor 186, 190, 191
McKown v. Lundman (1996) 200
Measles 176
Measles-mumps-rubella (MMR) vaccine 173
Measurement bias 141
Médécins Sans Frontières (MSF) 483
MedEdPortal 6
Medicaid 308, 328, 430
Medical care 467, 479, 488, 489
Medical certification 220
Medical errors 345
- acceptance of responsibility 343
- compensating patients and/or families 339, 340
- definition of 338
- disclosure of 339, 343
- hierarchical structure in medicine 345
- in hospital 339
- and medical malpractice 339
- Medicare 340
- patient- and family-centered care 340
- patient safety 344
- in United States 338

Medical Expenditure Panel Survey (MEPS) 151
Medical futility 270, 274–276
Medicalization process of care *vs.* natural 138
Medical-legal partnership (MLP) 434, 443, 444
Medically unexplained symptoms (MUS) 244
Medical malpractice 339
Medical neglect 191, 192
Medical orders for life sustaining treatment (MOLST) 252, 255, 256, 264
Medical schools 297
Medical skills 486
Medicare 252, 256, 264, 307, 308, 340
Medication-assisted treatment (MAT) 350, 355
Medication non-adherence 362
Medicine practices 5
Medico-legal partnerships 437
Menarche 413

Mennonites 186, 192
Mental Health Parity and Addictions Equity Act 213
Mental hygiene laws (MHLs) 220
Metastatic breast cancer
- advance directive 271, 280
- COBRA 270, 274
- COMFORT model 277
- cost-benefit analysis 280, 281
- cure value 281
- disclosure/breaking bad news 277
- DNR status 280
- good death 283
- healthcare practitioner responsibility 281
- high-cost pharmaceutical company 282
- hospice care 270, 273, 274
- insurance 279
- law/policy for drug pricing 282
- medical futility 270, 275, 276
- motivating family's decision 270, 275
- NICE 274
- palliative care 270, 272, 273, 278
- pricing structure 281
- public and private insurance 270, 274, 281
- SPIKES model 277
- total healthcare cost 270, 272
- treatment in UK 279

Mexico City policy 461
Microcephaly 462, 464, 465
- incidence 454, 459
- signs of 464

Migrant 232, 235
Minimally conscious state 252, 256
Mitochondrial disorders 316
Mobility, definition of 293
Modified RIME framework 505
Mosquito-borne illness 470
Multidrug-resistant (MDR) TB 66
Multifactorial disorders 316
Multiple sclerosis (MS) 290, 291, 300, 303
- characterization 296
- frequency of 296
- prevalence 296
- RRMS 296
- type of 296

Munchausen syndrome 191

N

Naloxone 356
Naloxone Co-Payment Assistance Program (N-CAP) 356
National Alliance on Mental Illness (NAMI) 213
National Commission on Correctional Health Care (NCCHC) 84
National Conference of Commissioners on Uniform State Laws 257

National Death Index (NDI) 151
National Electronic Injury Surveillance System (NEISS) 212
National Health and Nutrition Examination Survey 152
National Health Interview Surveys (NHIS) 150, 151
National Health Service (NHS) 141, 274
National Highway Traffic Safety Administration (NHTSA) 215
National Immunization Program 171
National Institute for Health Care Excellence (NICE) 132, 270, 274, 279
National Institute of Clinical Excellence 274
National Institutes of Health (NIH) 371, 384, 386
National Intimate Partner and Sexual Violence Survey 237
National Longitudinal Study of Youth 151
National Rifle Association (NRA) 214
National School Lunch Program (NSLP) 152, 430, 437
National Survey on Drug Use and Health (NSDUH) 213
National Transgender Discrimination Survey 415
National Vital Statistics System 210
Natural and semi-synthetic opioids 352
Natural process of care *vs.* medicalization 138
Near-miss events 338, 343
Negligence 339
Neuroblastoma 188
Neurological death 257
New Drug Application (NDA) 373
New York State's law 83
Non-adherence to therapy 362
Non-biased information 488
Nonfatal firearm injuries 212
Non-Hodgkin lymphoma 188
Non-maleficence 3, 8, 19, 20, 70, 94, 173, 179, 324, 325, 362, 376, 377, 417, 489–491
Nonsteroidal anti-inflammatory drugs (NSAIDS) 353
Numerical Pain Scale 353
Nuremberg Code 372
Nurse practitioner (NP) 18–24
Nurse practitioners 295
NY SAFE Act 226
NY Secure Ammunition and Firearms Enforcement Act (NYSAFE) 214

O

Obesity
– in adult 430
– in children 430
Objective structured clinical exams (OSCEs) 310

Obligations of physicians 12, 13
Occupational therapists (OT) 296, 300, 301
Occupational therapy 469
Office of Orphan Products Development (OOPD) 375
Office of Refugee Resettlement (ORR) 236
Omission bias 135
Open-disclosure program 344
Opioid use disorder 350, 351, 355, 356, 358, 359, 362
Opioids 350, 361
– chronic pain 354
– Comprehensive Addiction Recovery Act 356
– New York's 911 Good Samaritan Law 356
– pharmaceutical industry's role 354, 355
– physicians and healthcare practitioners 355
– severe acute pain 353
Organ transplantation 363
Organic technical standards 297
Orphan disease 370, 375
Orthodox Jewish community 258
Outbreak 62, 64
Overall survival (OS) 186

P

Pain 350, 353
– acute/chronic 353
– chest pain 360
– complaints 360
– as fifth vital sign 353
– FLACC scales 353
– institution's policy 360
– Numerical Pain Scale 353
– re-education on rating pain 360
Pain Rating Scale 360
Pangaro's RIME assessment framework 503
Pap test 293
Parkinson disease 371, 382
Patient- and family-centered care 6–7, 340
Patient assistance programs 158
Patient autonomy 12, 47
Patient handoffs 390, 393
Patient Health Questionnaire (PHQ) 223
The Patient Protection and Affordable Care Act (ACA) 358
Patients and healthcare practitioners 7
Patients with disability 299, 303
– accessibility 302
– treatment plan 300
Pedagogical approach 26, 29
Pediatric healthcare 361
– *See also* Vaccination
People with disability 290, 301, 304

– effective care for 299
– health differences 293, 294
– health disparities 293, 294
– healthcare practitioners 300
– insurance coverage 306
Person with disability 295, 296
Personal belief exemption 177, 178
Personal perspectives 25
Personal protective equipment (PPE) 69
Pervasive developmental disorders 469
Pew analysis 114
Phenylketonuria (PKU) 318–319
Physical therapists (PT) 296, 300
Physical therapy 469
Physician assessment 310, 311
Physician assistants 18, 295
"Physician Compare" Website 40, 44
Physician orders for life sustaining treatment (POLST) 252, 255, 256
Physician-oriented standard 43
Physicians and mental health care providers 225
Pilot program 245
Pineal germinoma 188, 193
Plain Mennonites 192
Policy level factors 117, 118
Policy-level interventions 223
Post-event phase 221
Poverty 118, 152, 157
Poverty exercise 443
Practicing clinicians 27
Predisposition testing 319
Pre-event phase 221
Pre-implantation genetic diagnosis (PGD) 316, 326, 328
– costs and societal resources 328
– insurance coverage for 320
– IVF 319
Premature closure 10, 341
President's Commission for the Study of Ethical Problems in Medicine and Biomedical and Behavioral Research 257
Presymptomatic testing 319
Preterm labor
– deputy's perspective 82, 92
– different resources and policy 83, 94
– healthcare practitioner's role 94
– HIPAA 80, 85, 86
– humane birth 82, 93
– incarcerated women 80, 84
– inmate health 80, 83
– jail deputy 82, 91
– maternal and infant health outcomes 80, 86
– patient autonomy or privacy 90
– patient care 81, 90
– patient information 88
– patient-practitioner relationship 81, 89
– physicians and healthcare practitioners 81, 89

Index

- principle of equivalence 91
- public safety issues 82, 93
- reproductive justice 82, 92
- right to health 91
- risk factors 80, 81, 87, 88
- safety and patient assessment 81, 91
- second opinion/appeal process 82, 92
- shackling, risk of 80, 84, 85
- sociocultural factors 81, 88
- structural factors 88
- teen pregnancy 80, 86, 87
- zealous representation 83, 94

Preterm labor 22, 80, 81, 87, 88, 91, 93
Prevalence of disability 293
Prevention-focused approach 345
Primacy of patient welfare 12
Primacy of welfare 376
Primary care physicians (PCPs) 217, 218, 223, 232, 240, 243–245, 295, 300
Primary caregivers 309
Prince v. Massachusetts (1944 Child labor laws and Jehovah's Witnesses) 200
Principle of equivalence 91
Principle of justice 489
Private gun ownership 224
Private hospitals and faith-based organizations 480
Problem-based learning (PBL) approach 5, 6
Process evaluation 497–499
Product evaluation 499
Professional behaviors 504
Professional standard 43
Programs for early childhood education 152
Pubertal development 410, 413
Public health emergency 469
Public health study 100, 106, 107
Public health systems 105, 176
Public mass shooting 224
Pulse oximetry testing 132, 133

Q

Quality-adjusted life expectancy (QALE) 151
Quality-adjusted life years (QALY) 113, 151, 280

R

Race 210
Racism 150–152
Randomized controlled trial (RCT) 390
Rape myth 232
Rapid response 337, 345
Rater training and calibration 504, 505
Real-time reverse transcriptase-polymerase chain reaction (rRT-PCR) 460

Refugees
- definition 232, 235
- healthcare practitioner
  - ethics of 245
  - role of 246
  - society's support for 245, 246
- with irregular menses
  - acknowledging distress 239, 240
  - clinician challenges 234, 242
  - communication in primary language 233, 241
  - confidentiality 240, 241
  - cultural humility 234, 245
  - with fewer socioeconomic resources 235
  - inquiring traumatic experiences 240
  - medically unexplained symptoms 234, 244
  - mental health 234, 243, 244
  - nonverbal cues 233, 239
  - optimal care 235
  - sexual assault in United States vs. 244
- medically unexplained symptoms 245
- post-resettlement 232, 237
- resettled refugees 232, 236
- sexual assault in United States vs. 234
- trauma-informed care 232
- traumatic experience 232
- UNHCR global estimates for 235
- in United States 236

Rehabilitation Act 70
Relapsing-remitting multiple sclerosis (RRMS) 296, 297
Religious beliefs 177
Religious exemption 177, 178
Reporting bias 142
Reproductive fitness 462
Reproductive health services 454, 461
Reproductive justice 82, 92
Research participation 382
Resettled refugees 232, 236, 244
Resident work hour restrictions 393, 394
- creative problem-solving 392, 403, 404
- cynical study design 403
- duty hour violations 402
- ethical issues 391, 395
- frequent handoffs 402
- human subjects research protocols 397
- informed consent 391, 396, 397
- intern's autonomy 396
- IRB
  - approval 390, 392
  - human participants 397
  - informed consent 391, 397, 398
  - scientific inquiry 391, 398, 399
- lack of awareness 399

- lack of supervision 391, 399
- legal action 391, 397
- longer duty hours 391, 401
- moral stakeholders 391, 395, 396
- multi-site study 402
- pragmatic study design 403
- research subjects 391, 396
- self-reflection 391, 392, 397, 401, 402
- sleep deprivation and cognitive impairment 391, 400
- study design 392, 402, 404
- supervision of residents 391, 399, 400

Resource-limited communities 487
Respect for persons 373
Respiratory syncytial virus (RSV) 172
Retail pharmacist 467
Right to an open future (ROF) 186, 189
Right to know 324
Rohingya refugees 236
Root-cause analysis 336

S

Salvage therapy 193
Sample size 375
Scale assessment
- educator 504
- interpreter 504
- manager 504
- Pangaro's RIME assessment framework 503
- professional behaviors 504
- reporter 504
- RIME model 503

Scope of practice 322
Second victim phenomenon 336
Section 504 of the Rehabilitation Act of 1973 294
Selection bias 141
Self-directed learning 24, 25, 28, 29
Self-reflection 244, 245
Self-reflection question 24, 28
Sentinel events 338
Serious adverse events (SAEs) 378
Setting, Perception, Invitation, Knowledge, Empathy, Summary and strategy (SPIKES) approach 463, 464
Severe acute respiratory syndromes (SARS) 69
Short-term experiences in global health (STEGH) 478, 479, 483, 489
Short-term medical missions (STMMs) 478, 483
Short-term missions 486
Sickle cell disease (SCD) 327, 350, 352, 360, 361
- best-practice guidelines 360
- complications 360
- on newborn screening 359
- young adults with 361

Single-gene disorders 316
Small group discussion 27
Small group facilitation 29
Social contract 491
Social determinants of health 12, 435, 441, 443
Social justice 12
Social Security Disability (SSDI) 305, 307
Social workers 252, 258, 260, 263, 295
Societal health policies 222
Socio-ecological model 102, 116, 117, 222, 290, 295, 302, 352, 359
- community 295, 303, 364
- individual 295, 302, 364
- interpersonal 295, 302, 364
- organizational 295, 302, 364
- policy 295, 303, 364
Socioeconomic status (SES) 150–152
Somatization, *see* Medically unexplained symptoms (MUS)
Special Supplemental Nutrition Program for Women, Infants, and Children (WIC) 430, 437
Specialty physicians 295
Speech and language pathologists (SLP) 296
Speech therapy 469
Spermarche 413
SPIKES model 277
Spinal muscular atrophy (SMA) 320
Standard prenatal care
- first-trimester screening 320
- positive screen 320
- second-trimester screening 320
Standards of care 325
Stateless person 232, 235
STEGH, *see* Short-term experience in global health
Strategic Advisory Group of Experts on Immunizations (SAGE) 169
Structural competency 11–12
Structural injustices 484
Structural violence 10, 11, 103, 118, 119
Structural/institutionalized racism 11
Student's home community 490
Substance abuse 89
Substance Abuse and Mental Health Services Administration (SAMHSA) 213, 238
Substance abuse disorder 362
Substance use disorder 350, 355, 357–359, 363
Sudden infant death syndrome (SIDS) 130
Summative assessments 502
Supplemental Assistance Program for Women, Infants, and Children (WIC) 152
Supplemental Nutrition Assistance Program (SNAP) 152, 234, 242, 430, 437

Surrogate decision making 252, 255, 259, 260
Survivors of Torture and Preventive Health 236
Sustainability, Global health 490
Swiss cheese model 336
Synthetic opioids 352
Systems approach 336
Systems-based approach 345

Tarasoff exception 213
Tarasoff laws 225
Taste and food quality studies 392
Tay-Sachs disease (TSD) 316, 320, 321, 326
Teaching tips 25
Technical standards 296, 297
Teen pregnancy 80, 86
Temporary Assistance for Needy Families (TANF) 152
Tetraplegia 298
Thelarche 413
Therapeutic misconception 381
Therapeutic privilege 50
Thyroid carcinoma 188
Time requirement 28, 29
Training healthcare workers 486
Transgender 410, 415
Transsexualism 416
Trauma
- in childhood 238
- inquiring traumatic experiences 240
- trauma-informed care 238, 239
Trauma-informed care 232, 238
Treatment for Adolescents with Depression Study (TADS) 217
Trigger warnings 34
Trihalomethanes 107
Tuberculosis
- active disease 62, 65
- CMV infection 63, 68
- directly observed therapy 62, 64, 65, 73
- effective tool 64, 74
- epidemic 62–64, 70
- ethic and legal 63, 70
- ethical tool 63, 64, 71, 74
- examination or treatment 66
- factors 64, 72
- health department 71
- HIPAA 62, 67
- individual's confidentiality 64, 72
- interpersonal 62, 67
- isolation *vs.* quarantine 63, 71, 72
- latent infection 64, 74
- legal action 64, 73
- legal authority 62, 66
- organizational factors 62, 68

- physician responsibilities 63, 64, 70, 75
- policy/procedures 62, 66
- prevention of 64, 75
- public's health 64
- risk factors 62, 65, 66
Tuskegee Syphilis study 394
Type 2 diabetes
- accessing health care 154, 155
- implicit bias 147, 159
- Medicaid 146
- medicaid eligibility 155, 156
- medications 147, 157, 158
- poverty and poor health 157
- prevalence 146, 150
Type II error 375

U.S. Centers for Disease Control and Prevention's (CDC) Traveler's Health Website 478
U.S. Centers for Medicare and Medicaid Services (CMS) 274
U.S. Federal Bureau of Investigation's (FBI) 211
U.S. Food and Drug Administration (FDA) 373–375, 378
U.S. Preventive Services Task Force (USPSTF) 222, 442
U.S. training system 479
UK's National Health Service (NHS) 279
Unconscious bias 10
Under-immunization 169
Uniform Determination of Death Act (UDDA) 252, 257
Uniform Health-Care Decisions Act 276
Unintentional firearm injuries 211
Unintentional injuries 210
United Nations High Commissioner for Refugees (UNHCR) 232, 235, 236
United States Preventive Services Task Force (USPSTF) 105, 118
Universal healthcare coverage 478, 482
Unrealistic optimism 381
Upfront therapy 193
UpToDate 299

V

Vaccination
- circumstances 166, 173
- communication approaches 166, 170, 171
- creative problem-solving 168, 182
- ethical considerations 179
- ethical dilemma 174
- healthcare practitioners or medical practices 166, 167, 170, 179, 181
- HPV 167, 179–181

- impact rates 166, 169
- legal dilemma 174
- market 166, 172
- misinformation affects, dissemination of 168, 182
- positive public health impact 168, 181, 182
- presumptive vaccine recommendations 181, 182
- prevention 176
- process 166, 168
- professional groups 170
- public health dilemma 175
- refusing, benefits and risks of 167, 173, 174
- religious exemption and personal belief exemption 167, 177, 178
- risks 175
- self-reflection 166, 167, 173, 179
- spectrum of attitudes 166, 169, 170
- trust 181
- unvaccinated patients, healthcare practitioners 175
- vaccine refusal, impact of 167, 176–178
- VAERS 166, 171, 172

Vaccine Adverse Event Reporting System (VAERS) 166
Vaccine exemptions 169, 175
Vaccine hesitancy 166, 169, 170
Vaccine refusal 167, 176, 177
Vaccine-preventable disease 169, 173–174, 182
Vegetative state 252, 256
Voluntourism 485, 487, 491
Vulnerable subjects 373

W

Weathering hypothesis 87
Web-based Injury Statistics Query and Reporting System (WISQARS) 211
Weight bias 436
Welfare check 218
WHO growth chart 440
Whole brain-death 258
WIC program nutrition guidelines 444
Willowbrook Hepatitis study 394
Wisconsin v. Yoder (1972) 200
Women, Infants, and Children (WIC) 152, 430, 432, 437, 440, 443, 444, 446
World Health Organization (WHO) 12, 65, 440, 481

Z

Zealous representation 94
Zika virus 69
- balancing act 463
- blood supply screening 462
- CDC recommendations 459, 460
- Centers for Disease Control recommendations 454
- clinical research 472
- comorbidities 456
- complications 458
- DDT exposure 469
- emotional, physical and financial stress 468
- environmental concerns 456
- ethical concerns 456
- experimental therapy 456
- first-trimester surgical/medical abortion 464
- health and well-being 465
- healthcare practitioners 465
- healthcare proxy 465
- history of 454, 457
- incidence 457
- incubation period 458
- laboratory testing algorithm 460
- medical condition 465
- patient advocacy 468
- people's beliefs 471
- personal biases 455
- personal prevention strategies 462
- population health research and interventions 471
- pregnancy complications 454, 458, 459
- prenatal screening for fetal anomalies 465
- preventable pandemic 466
- primary physical harm 472
- psychological support for families 456
- public policy decisions 465, 466
- research participation 465
- risk reduction 454, 461
- rRT-PCR 460
- sporadic cases 457
- symptoms 454
- transmission 454, 457, 458
- vaccination 462
- vectors 461, 467
- voluntary informed consent 472

Zika Virus Country Classification 457
ZMapp 456, 470, 471, 473

PGMO 08/05/2019

DATE DUE

JAN 2 0 2020	
	PRINTED IN U.S.A.